Lecture Notes in Artificial Intelligence 7104

Subseries of Lecture Notes in Computer Science

LNAI Series Editors

Randy Goebel
University of Alberta, Edmonton, Canada
Yuzuru Tanaka
Hokkaido University, Sapporo, Japan
Wolfgang Wahlster
DFKI and Saarland University, Saarbrücken, Germany

LNAI Founding Series Editor

Joerg Siekmann
DFKI and Saarland University, Saarbrücken, Germany

Longbing Cao   Joshua Zhexue Huang
James Bailey   Yun Sing Koh   Jun Luo (Eds.)

New Frontiers in Applied Data Mining

PAKDD 2011 International Workshops
Shenzhen, China, May 24-27, 2011
Revised Selected Papers

 Springer

Series Editors

Randy Goebel, University of Alberta, Edmonton, Canada
Jörg Siekmann, University of Saarland, Saarbrücken, Germany
Wolfgang Wahlster, DFKI and University of Saarland, Saarbrücken, Germany

Volume Editors

Longbing Cao
University of Technology Sydney, NSW, Australia
E-mail: longbing.cao@uts.edu.au

Joshua Zhexue Huang
Jun Luo
Chinese Academy of Sciences, SIAT, Shenzhen, China
E-mail: {zhexue.huang, jun.luo}@siat.ac.cn

James Bailey
The University of Melbourne, VIC, Australia
E-mail: baileyj@unimelb.edu.au

Yun Sing Koh
The University of Auckland, New Zealand
E-mail: ykoh@cs.auckland.ac.nz

ISSN 0302-9743   e-ISSN 1611-3349
ISBN 978-3-642-28319-2   ISBN 978-3-642-28320-8 (eBook)
DOI 10.1007/978-3-642-28320-8
Springer Heidelberg Dordrecht London New York

Library of Congress Control Number: 2012930911

CR Subject Classification (1998): I.2, H.3, H.4, H.2.8, F.1, J.1, I.4

LNCS Sublibrary: SL 7 – Artificial Intelligence

Typesetting: Camera-ready by author, data conversion by Scientific Publishing Services, Chennai, India

Printed on acid-free paper

Springer is part of Springer Science+Business Media (www.springer.com)

Preface

The 15th Pacific-Asia Conference on Knowledge Discovery and Data Mining in Shenzhen, China, hosted five workshops that allowed researchers and participants to discuss emerging techniques and their application domains extending to previously unexplored areas.

The workshops were selected competitively following a call for workshop proposals. Three of them were sequel workshops (BI, QIMIE, DMHM) that were held in the last few years, while two were new (BDM, AI-TCM), exhibiting both the maturity and innovation of the workshops. The following workshops were organized:

- International Workshop on Behavior Informatics (BI 2011) chaired by Longbing Cao, Jaideep Srivastava, Graham Williams, and Hiroshi Motoda. The main goal of this workshop is to provide an international forum for researchers and industry practitioners to share their ideas, original research results, as well as potential challenges and prospects encountered in behavior informatics.
- Quality Issues, Measures of Interestingness and Evaluation of Data Mining Models Workshop (QIMIE 2011) chaired by Stéphane Lallich and Philippe Lenca. The main focus of QIMIE 2011 is on the theory, the techniques and the practices that can ensure the discovered knowledge of a data mining process is of quality.
- Workshop on Biologically Inspired Techniques for Data Mining (BDM 2011) chaired by Shafiq Alam Burki and Gillian Dobbie. The workshop highlights a relatively new but fast-growing area of data mining which is based on optimization techniques from biological behavior of animals, insects, cultures, social behaviors and biological evolution.
- Workshop on Advances and Issues in Traditional Chinese Medicine Clinical Data Mining (AI-TCM 2011) chaired by Josiah Poon, Xuezhong Zhou, and Junbin Gao. This workshop focuses on usage of data mining such that it becomes a complementary tool to assist traditional chinese medicine clinical research.
- Second Workshop on Data Mining for Healthcare Management (DMHM 2011) chaired by Prasanna Desikan, Jaideep Srivastava, and Ee-Peng Lim. This workshop provides a common platform for discussion of challenging issues and potential techniques in this emerging field of data mining for health care management.

This volume contains the proceedings of the five workshops. Also included are papers from the First PAKDD Doctoral Symposium on Data Mining (DSDM'11). We hope that the published papers propel further interest in the growing field of knowledge discovery in databases (KDD).

Setting up workshops such as these takes a lot of effort. We would like to thank the Program Committee Chairs and their Program Committee members for their time and effort spent to guarantee the high quality of the program.

October 2011

James Bailey
University of Melbourne, Australia

Yun Sing Koh
University of Auckland, New Zealand
(PAKDD 2011 Workshop Co-chairs)

International Workshop on Behavior Informatics (BI 2011)
PC Chairs' Message

Due to the behavior implication in normal transactional data, the requirement of deep and quantitative behavior analysis has outstripped the capability of traditional methods and techniques in behavioral sciences. It is more imperative than ever to develop new behavioral analytic technologies that can derive an in-depth understanding of human behaviors beyond the demographic and historical tracking. This leads to the emergence of inter-disciplinary *behavior representation, modeling, mining, analysis and management* (namely, *behavior informatics*). The aim of the Behavior Informatics Workshop is to provide an international forum for researchers and industry practitioners to share their ideas, original research results, as well as potential challenges and prospects encountered in behavior informatics.

Following the call for papers, BI 2011 attracted 40 submissions from 12 different countries, and accepted 16 of them after a double-blind review by at least 3 reviewers for each paper.

The selected papers include the latest work in social network mining, such as user profile modeling, Web user clustering, emotional analysis etc. On behavior mining, papers are related to community detection, group recommendation, and user behavior patterns, etc. Moreover, pattern discovery and clustering are other areas where the authors have utilized and extended behavior informatics.

The Workshop also featured two invited talks: Jeffery Xu Yu on Graph Behaviors: Discovering Patterns When a Graph Evolves, and Jie Tang on Social Prediction: Can We Predict Users' Action and Emotions?

We are very grateful to the General Chair of BI 2011, Philip S. Yu at the University of Illinois at Chicago, who provided continuous support and guidance for the workshop's success. We are particularly thankful to Gang Li at Deakin University, Australia, the Organizing Chair of BI 2011, for the great efforts he made throughout all the phases of organizing BI 2011. Thanks to the authors who made this workshop possible by submitting their work and responding positively to the changes suggested by our reviewers regarding their work. We are also thankful to our Program Committee who dedicated their time and provided us with valuable suggestions and timely reviews. We wish to express our gratitude to the PAKDD 2011 Workshop Chairs and PAKDD 2011 conference organizers who provided us with fantastic support that made this workshop very successful.

Readers are referred to an edited book on *Behavior Computing: Modeling, Analysis, Mining and Decision*, edited by Longbing Cao and Philip S. Yu, published by Springer. Another resource is the Special Issue on Behavior Computing with *Knowledge and Information Systems: An International Journal*, edited by Longbing Cao, Philip S. Yu and Hiroshi Motoda.

Longbing Cao
Jaideep Srivastava
Graham Williams
Hiroshi Motoda

International Workshop on Behavior Informatics (BI 2011)

Organization

General Co-chair

Philip S. Yu   University of Illinois at Chicago, USA

Workshop Chairs

Longbing Cao   University of Technology Sydney, Australia
Jaideep Srivastava   University of Minnesota, USA
Graham Williams   Australian Taxation Office, Australia
Hiroshi Motoda   Osaka University and AFOSR/AOARD, Japan

Organizing Chair

Gang Li   Deakin University, Australia

Program Committee

Sviatoslav Braynov, USA
Guozhu Dong, USA
Gian Luca Foresti, Italy
Xin Geng, P.R. China
Björn Gottfried, Germany
Ken Kaneiwa, Japan
Dionysios Kehagias, Greece
Kazuhiro Kuwabara, Japan
Gang Li, Australia
Ee-Peng Lim, Singapore
Weicheng Lin, Taiwan

Wenjia Niu, P.R. China
Zbigniew Ras, USA
Yasufumi Takama, Japan
David Taniar, Australia
Karl Tuyls, The Netherlands
Jinlong Wang, P.R. China
Lipo Wang, Singapore
Xiaogang Wang, P.R. China
Xiaofeng Wang, P.R. China
Jeffrey Yu, P.R. China
Daniel Zeng, USA

Quality Issues, Measures of Interestingness and Evaluation of Data Mining Models Workshop (QIMIE 2011)

PC Chairs' Message

The Quality Issues, Measures of Interestingness and Evaluation of Data Mining Models Workshop (QIMIE) focusses on the theory, the techniques and the practices that can ensure the discovered knowledge of a data mining process is of quality. Following QIMIE 2009/PAKDD 2009, QIMIE 2011 was organized in association with PAKDD 2011 (Shenzen, China).

QIMIE 2011 would not have been possible without the work of many people and organizations. We wish to express our gratitude to: Telecom Bretagne, the University of Lyon, the PAKDD 2011 Workshop Chairs (James Bailey and Yun Sing Koh), the Chairs and Co-chairs of PAKDD 2011 (Philip S. Yu, Jianping Fan, David Cheung), the PAKDD 2011 Program Committee Co-chairs (Joshua Huang, Longbing Cao, Jaideep Srivastava), and the QIMIE 2011 Program Committee members. Last but not least, we would like to thank all authors of the submitted papers, the Session Chairs and Longbing Cao for his keynote talk.

Each submission was reviewed by at least three members of the Program Committee. The papers presented in these proceedings were selected after a rigorous review process. The QIMIE 2011 program included one keynote speaker and nine regular paper presentations.

Actionable knowledge discovery is considered as one of the great challenges of the next-generation of knowledge discovery in database studies. The keynote talk by Longbing Cao discussed a general evaluation framework to measure the actionability of knowledge discovered, which covers both technical and business performance evaluation. Metrics and strategies for supporting the extraction of actionable knowledge were discussed.

The regular papers were divided into three sessions, namely, *Clustering*, *Data Structure*, and *Patterns and Rules*.

Clustering Session – Ivanescu et al. propose ClasSi, a new ranking correlation coefficient which deals with class label rankings and employs a class distance function to model the similarities between classes. Estivill-Castro introduces a notion of instance easiness to supervised learning and links the validity of a clustering to how its output constitutes an easy instance for supervised learning. An alternative approach for clustering quality evaluation based on the unsupervised measures of Recall, Precision, and F-measure exploiting the descriptors of the data associated with the obtained clusters is proposed by Lamirel et al.

Data Structure Session – Hadzic presents a novel structure-preserving way for representing tree-structured document instances as records in a standard flat data structure to enable the applicability of a wider range of data-mining techniques. To handle fragmentary duplicate pages on the Internet, Fan and Huang fuse some "state-of-the-art" algorithms to reach a better performance.

Patterns and Rules Session – Yang et al. propose an evolutionary method to search for interesting association rules. An efficient model based on the notion of multiple constraints is constructed by Surana et al. The periodic-frequent patterns discovered with this model satisfy a downward closure property. Hence, periodic frequent patterns can be efficiently discovered. Wu et al. propose Ap-epi, an algorithm which discovers minimal occurrences of serial episode and NOE-WinMiner which discovers non-overlapping episodes. To address the problem of assessing the information conveyed by a finite discrete probability distribution, Garriga elaborates a measure of certainty which includes a native value for the uncertainty related to unseen events.

Philippe Lenca
Stéphane Lallich

Quality Issues, Measures of Interestingness and Evaluation of Data Mining Models Workshop (QIMIE 2011)

Organization

Workshop Chairs

Stéphane Lallich   Université Lyon 2, France
Philippe Lenca   Lab-STICC, Telecom Bretagne, France

Program Committee

Hidenao Abe, Japan
Komate Amphawan, Thailand
Jérôme Azé, France
José L. Balezar, Spain
Bruno Crémilleux, France
Sven Crone, UK
Jean Diatta, La Réunion
Thanh-Nghi Do, Vietnam
Joao Gama, Portugual
Ricard Gavaldá, Spain
Salvatore Gréco, Italy
Fabrice Guillet, France
Michael Hahsler, USA
Shanmugasundaram Hariharan, India
Martin Holena, Czech Republic
William Klement, Canada
Stéphane Lallich, France
Ludovic Lebart, France

Philippe Lenca, France
Yannick Le Bras, France
Ming Li, China
Patrick Meyer, France
Amadeo Napoli, France
David Olson, USA
Zbigniew Ras, USA
Jan Rauch, Czech Republic
Gilbert Ritschard, Switzerland
Robert Stahlbock, Germany
Athasit Surarerks, Thailand
Izabela Szczech, Poland
Shusaku Tsumoto, Japan
Kitsana Waiyamai, Thailand
Dianhui Wang, Australia
Gary Weiss, USA
Takahira Yamaguchi, Japan
Min-Ling Zhang, China

Workshop on Biologically Inspired Techniques for Data Mining (BDM 2011)
PC Chairs' Message

For the last few years, biologically inspired data-mining techniques have been intensively used in different data-mining applications such as data clustering, classification, association rule mining, sequential pattern mining, outlier detection, feature selection and bioinformatics. The techniques include neural networks, evolutionary computation, fuzzy systems, genetic algorithms, ant colony optimization, particle swarm optimization, artificial immune system, culture algorithms, social evolution, and artificial bee colony optimization. A huge increase in the number of papers published in the area has been observed in the last decade. Most of these techniques use optimization to speed up the data-mining process and improve the quality of patterns mined from the data.

The aim of the workshop is to highlight the current research related to biologically inspired techniques in different data-mining domains and their implementation in real-life data-mining problems. The workshop provides a platform to researchers from computational intelligence and evolutionary computation and other biologically inspired techniques to get feedback on their work from other data-mining perspectives such as statistical data mining, AI- and machine learning-based data mining.

Following the call for papers BDM 2011 attracted 16 submissions from 8 different countries, with 8 of them accepted after a double-blind review by at least 3 reviewers. The overall acceptance rate for the workshop was 50%.

The selected papers highlight work in PSO-based data clustering and recommender system, PSO-based association rule mining, and genetic algorithm and neural networks-based training schemes for imbalanced classification problems. Another subject covered is the discovery of microRNA precursors in plant genomes based on an SVM detector. Feature selection is another area where the authors proposed a clustering-based framework formed by several unsupervised feature-selection algorithms.

We are thankful to Ajith Abrahm, General Chair of BDM 2011, for his constant motivation and guidance throughout all the phases of organizing BDM 2011. Thanks to the authors who made this workshop possible by submitting their work and responding positively to the changes suggested by our reviewers. We are also thankful to our Program Committee who dedicated their time and provided us with valuable suggestions and timely reviews. We wish to express our gratitude to the Workshop Chairs who were always available to answer our queries and provided us with everything we needed to put this workshop together.

Shafiq Alam
Gillian Dobbie

Workshop on Biologically Inspired Techniques for Data Mining (BDM 2011)
Organization

General Co-chair

Ajith Abraham — Machine Intelligence Research Labs (MIR Labs), Scientific Network for Innovation and Research Excellence (SNIRE), USA

Workshop Chairs

Shafiq Alam Burki — University of Auckland, New Zealand
Gillian Dobbie — University of Auckland, New Zealand

Program Committee

Patricia Riddle, New Zealand
Azzam ul Assar, Pakistan
Mengjie Zhang, New Zealand
Kamran Shafi, Australia
Stephen Chen, USA
Kouroush Neshatian, New Zealand
Will Browne, New Zealand
Ganesh K. Venayagamoorthy, USA
Yanjun Yan, USA
Asifullah Khan, Pakistan
Ke Geng, New Zealand

Ming Li, China
Ismail Khalil, Austria
David Taniar, Australia
Redda Alhaj, Canada
Lean Yu, China
Fatos Xhafa, Spain
Xiao-Zhi Gao, Finland
Emilio Corchado, Spain
Michela Antonelli, Italy
Khalid Saeed, Poland

Workshop on Advances and Issues in Traditional Chinese Medicine Clinical Data Mining (AI-TCM 2011)

PC Chairs' Message

Traditional chinese medicine, also known as TCM, has a long history and is considered as one of the major medical approaches in China. Different from its western counterpart, TCM embraces a holistic rather than a reductionist approach. While medications in western medicine often strive to avoid drug interaction, TCM works oppositely by exploiting the useful interaction among multiple herbs that are used in a formula (prescription). The efficacy of a TCM medication derives from the interaction of these herbs. Furthermore, while western medicine has only recently begun to popularize the prescription of personalized medicine according to a patient's genome, Chinese medical doctors have long been practicing this approach. In TCM practice, every prescription is a clinical trial from the viewpoint of western medical practice. In terms of clinical study, TCM clinical practice is a kind of real-world clinical trial, in which all the practical treatments aim to maximize the effectiveness. Therefore, the clinical data generated in the clinical operations are important to hypothesis generation and clinical guideline development.

Historically in TCM, all this knowledge was accumulated and formulated through experience rather than the support of proper scientific research. But we have witnessed a significant change in recent years. There is currently a strong push in the globalization of TCM and practice. This movement encourages the use of different techniques, such as statistical and data-mining methods, to find the underlying biomedical principles and theories, and to unlock the secrets of this "ancient" conservative domain of TCM.

Massive TCM clinical data are now being collected at an increasing pace. Data mining can help unveil the hidden precious knowledge, such that it becomes a complementary tool to assist TCM clinical research. There are many aspects in TCM, e.g., theory discovery, therapy methodology analysis, symptom detection, prescription analysis, herbal drug design, and intelligent diagnosis. It is anticipated that the sharing of experiences will provide new ideas in research directions and give the stimulus for creative breakthroughs. Furthermore, in recent years, many researchers in computer science and statistics have joined in the TCM data-mining field.

Papers in this workshop explored various issues in the field including the exploration of the regularities in the use of herbs and herb–herb interaction in TCM, which are more TCM-focused. There were also papers on other important topics, namely, the choice of interestingness function and feature representation in TCM, which are more data-mining driven. Hence, the workshop was well-balanced between TCM and data mining topics, and they all led to a good and productive discussion.

Josiah Poon
Xuezhong Zhou
Junbin Gao

Workshop on Advances and Issues in Traditional Chinese Medicine Clinical Data Mining (AI-TCM 2011)

Organization

Workshop Chairs

Josiah Poon   University of Sydney, Australia
Xuezhong Zhou   Beijing Jiaotong University, China
Junbin Gao   Charles Sturt University, Australia

Program Committee

Zhiwei Cao, China

Huajun Chen, China

Jianxin Chen, China

Tao Chen, China

Yi Feng, China

Li-Yun He, China

Caiyan Jia, China

William Jia, Canada

Paul Kwan, Australia

Guozheng Li, China

Shao Li, China

Clement Loy, Australia

Yonghong Peng, UK

Georg Peters, Germany

Simon Poon, Australia

Qun Shao, UK

Daniel Sze, Hong Kong

Nevin Zhang, Hong Kong

Runshun Zhang, China

Second Workshop on Data Mining for Healthcare Management (DMHM 2011)
PC Chairs' Message

Data mining for healthcare management (DMHM) has been instrumental in detecting patterns of diagnosis, decisions and treatments in healthcare. Data mining has aided in several aspects of healthcare management including disease diagnosis, decision-making for treatments, medical fraud prevention and detection, fault detection of medical devices, healthcare quality improvement strategies and privacy. DMHM is an emerging field where researchers from both academia and industry have recognized the potential of its impact on improved healthcare by discovering patterns and trends in large amounts of complex data generated by healthcare transactions.

This workshop was the second in the series of workshops on Data Mining for Healthcare Management held at PAKDD. The emerging interest in this area has led to the success of these workshops. The workshop served as a critical forum for integrating various research challenges in this domain and promoted collaboration among researchers from academia and industry to enhance the state of the art and help define a clear path for future research in this emerging area.

In response to the call for papers, DMHM 2011 received 9 contributions. We would like to thank the authors for their efforts, since it is their submissions that laid the foundation of a strong technical program. Each submission was reviewed by at least three Program Committee members. Four submissions were selected for presentation. The main selection criterion was the quality of the idea. We would like to thank the members of the Program Committee for taking time to provide insightful critique, and thus ensuring the high quality of the workshop.

The PAKDD community responded very enthusiastically to the DMHM 2011 Workshop, and about 15 people attended the workshop, which brought together data-mining researchers and industry practitioners.

The workshop had an invited talk by Vipin Gopal, Director of Clinical Analytics at Humana. This talk gave an overview of the efforts at Humana, a Fortune 100 health benefits company, in developing advanced analytic solutions for positively impacting care management. Predictive modeling solutions that help identify patients who are at high risk for readmissions, and subsequent interventions to reduce the overall readmissions rates, were discussed.

In their paper titled "Usage of Mobile Phones for Personalized Healthcare Solutions", M. Saravanan, S. Shanthi and S. Shalini introduce one mobile-based system known as the Personalized Mobile Health Service System for Individual's Healthcare, which caters to the specific needs of the user without the constraint on mobility.

Wei Gu, Baijie Wang and Xin Wang, in their paper titled "An Integrated Approach to Multi-Criteria-Based HealthCare Facility Location Planning," present an integrated approach to healthcare facility planning. A new health accessibility estimation method is developed in order to capture the current characteristics of preventive healthcare services and the problem is formalized as a multi-criteria facility location model.

In "Medicinal Property Knowledge Extraction from Herbal Documents for Supporting Question Answering Systems," C. Pechsiri, S. Painual and U. Janviriyasopak, aim to automatically extract the medicinal properties of an object, from technical documents as knowledge sources for healthcare problem solving through the question-answering system, especially What-Question, for disease treatment.

Finally, in their paper titled "Robust Learning of Mixture Models and Its Application on Trial Pruning for EEG Signal Analysis," B. Wang and F. Wan present a novel method based on deterministic annealing to circumvent the problem of the sensitivity to atypical observations associated with the maximum likelihood (ML) estimator via a conventional EM algorithm for mixture models.

DMHM 2011 turned out to be a very successful workshop by all measures. The quality of papers was excellent, the discussion was lively, and a number of interesting directions of research were identified. This is a strong endorsement of the level of interest in this rapidly emerging field of inquiry. For more information on the workshop. Please visit the website: http://www-users.cs.umn.edu/ desikan/pakdd2011/accepted.htm

Prasanna Desikan
Ee-Peng Lim
Jaideep Srivastava

Second Workshop on Data Mining for Healthcare Management (DMHM 2011)
Organization

Workshop Chairs

Prasanna Desikan   Boston Scientific, USA
Jaideep Srivastava   University of Minnesota, USA
Ee-Peng Lim   Singapore Management University, USA

Program Committee

Vipin Kumar, USA   Yonghong Tian, China
Joydeep Ghosh, USA   Woong-Kee Loh, Korea
Wei Jiang, Hong Kong   Tieyun Qian, China
V.R.K. Subrahmanya Rao, India   Kalyan Pamarthy, USA
Hui Xiong, USA   Jimmy Liu Jiang, Singapore
Michael Steinbach, USA   Chandan Reddy, USA
Deepa Mahajan, USA   San-Yih Hwang, Taiwan

First PAKDD Doctoral Symposium on Data Mining (DSDM 2011)
Chairs' Message

The First PAKDD Doctoral Symposium on Data Mining (DSDM 2011) was held at Shenzhen, China, May 24, 2011, in conjunction with the 15th Pacific-Asia Conference on Knowledge Discovery and Data Mining (PAKDD 2011). This was the first edition of the doctoral symposium series at annual PAKDD conferences. This symposium aims to provide a forum for PhD students as well as junior researchers with newly received PhD degrees to present their recent work and seek for constructive feedback and advice from senior researchers, and bridge possible research collaborations with other participants.

This year we received nine submissions, where five were from mainland China, one from Hong Kong, one from Taiwan, and two from the USA. The submissions went through a rigorous reviewing process. Most submissions received four reviews. All the reviews were done in an encouraging and advising way with the purpose of providing constructive suggestions to the authors. The DSDM 2011 Chairs examined all the reviews to further guarantee the reliability and integrity of the reviewing process. Finally, five papers were accepted, one of which was granted the Best Paper Award.

To further realize the goal of this doctoral symposium, we held two invited talks from academia and industry, respectively. Moreover, we held a panel session and invited four senior researchers from the data mining community as our panelists. All the participants and panelists discussed a number of issues related to the trends in data-mining research, data mining in multidisciplinary research and other general questions.

DSDM 2011 would not have been a success without the support of many people. We wish to take this opportunity to thank the Program Committee members for their efforts and engagements in providing a rich and rigorous scientific program as well as the suggestions to the authors. We wish to express our gratitude to the PAKDD 2011 Program Committee Chairs Joshua Huang, Longbing Cao and Jaideep Srivastava for their invaluable support and concern. We are also grateful to the invited speakers Jie Tang and Ping Luo for their insightful and inspirational talks, and to the panelists Longbing Cao, Xintao Wu, Jie Tang and Ping Luo for their discussions and constructive advice.

Last but not least, we also want to thank all authors and all symposium participants for their contributions and support. We hope all participants took this opportunity to share and exchange ideas with one another, and to benefit from the discussions at DSDM 2011.

Ming Li
Reynold C.K. Cheng
Mi-Yen Yeh

First PAKDD Doctoral Symposium on Data Mining (DSDM 2011)

Organization

Symposium Chairs

Ming Li   Nanjing University, China
Reynold C.K. Cheng   University of Hong Kong, Hong Kong
Mi-Yen Yeh   Academia Sinica, Taiwan

Program Committee

Deng Cai, China
Chien Chin Chen, Taiwan, ROC
Lei Chen, Hong Kong, China
Jiefeng Cheng, Hong Kong, China
Byron Choi, Hong Kong, China
Chi-Yin Chow, Hong Kong, China
Gao Cong, Singapore
Bi-Ru Dai, Taiwan, ROC
Wei-Shinn Ku, USA
Shou-De Lin, Taiwan, ROC
Fei Tony Liu, Australia

Man-Kwan Shan, Taiwan, ROC
Jie Tang, China
Lei Tang, USA
Ivor Tsang, Singapore
Chi-Yao Tseng, Taiwan, ROC
Raymond Wong, Hong Kong, China
Shan-Hung Wu, Taiwan, ROC
De-Chuan Zhan, China
Min-Ling Zhang, China
Jun Zhu, USA

Table of Contents

International Workshop on Behavior Informatics (BI 2011)

Evaluating the Regularity of Human Behavior from Mobile Phone
Usage Logs.. 3
  Hyoungnyoun Kim and Ji-Hyung Park

Explicit and Implicit User Preferences in Online Dating............... 15
  *Joshua Akehurst, Irena Koprinska, Kalina Yacef, Luiz Pizzato,
  Judy Kay, and Tomasz Rej*

Blogger-Link-Topic Model for Blog Mining 28
  Flora S. Tsai

A Random Indexing Approach for Web User Clustering and Web
Prefetching.. 40
  Miao Wan, Arne Jönsson, Cong Wang, Lixiang Li, and Yixian Yang

Emotional Reactions to Real-World Events in Social Networks......... 53
  Thin Nguyen, Dinh Phung, Brett Adams, and Svetha Venkatesh

Constructing Personal Knowledge Base: Automatic Key-Phrase
Extraction from Multiple-Domain Web Pages....................... 65
  Yin-Fu Huang and Cin-Siang Ciou

Discovering Valuable User Behavior Patterns in Mobile Commerce
Environments 77
  Bai-En Shie, Hui-Fang Hsiao, Philip S. Yu, and Vincent S. Tseng

A Novel Method for Community Detection in Complex Network Using
New Representation for Communities............................. 89
  Yiwen Wang and Min Yao

Link Prediction on Evolving Data Using Tensor Factorization 100
  Stephan Spiegel, Jan Clausen, Sahin Albayrak, and Jérôme Kunegis

Permutation Anonymization: Improving Anatomy for Privacy
Preservation in Data Publication................................ 111
  *Xianmang He, Yanghua Xiao, Yujia Li, Qing Wang,
  Wei Wang, and Baile Shi*

Efficient Mining Top-k Regular-Frequent Itemset Using Compressed
Tidsets   124
    Komate Amphawan, Philippe Lenca, and Athasit Surarerks

A Method of Similarity Measure and Visualization for Long Time
Series Using Binary Patterns   136
    Hailin Li, Chonghui Guo, and Libin Yang

A BIRCH-Based Clustering Method for Large Time Series Databases . . .   148
    Vo Le Quy Nhon and Duong Tuan Anh

Visualizing Cluster Structures and Their Changes over Time by
Two-Step Application of Self-Organizing Maps   160
    Masahiro Ishikawa

Analysis of Cluster Migrations Using Self-Organizing Maps   171
    Denny, Peter Christen, and Graham J. Williams

Quality Issues, Measures of Interestingness and Evaluation of Data Mining Models Workshop (QIMIE 2011)

ClasSi: Measuring Ranking Quality in the Presence of Object Classes
with Similarity Information   185
    Anca Maria Ivanescu, Marc Wichterich, and Thomas Seidl

The Instance Easiness of Supervised Learning for Cluster Validity   197
    Vladimir Estivill-Castro

A New Efficient and Unbiased Approach for Clustering Quality
Evaluation   209
    *Jean-Charles Lamirel, Pascal Cuxac, Raghvendra Mall, and
    Ghada Safi*

A Structure Preserving Flat Data Format Representation for
Tree-Structured Data   221
    Fedja Hadzic

A Fusion of Algorithms in Near Duplicate Document Detection   234
    Jun Fan and Tiejun Huang

Searching Interesting Association Rules Based on Evolutionary
Computation   243
    *Guangfei Yang, Yanzhong Dang, Shingo Mabu,
    Kaoru Shimada, and Kotaro Hirasawa*

An Efficient Approach to Mine Periodic-Frequent Patterns in
Transactional Databases   254
    Akshat Surana, R. Uday Kiran, and P. Krishna Reddy

Algorithms to Discover Complete Frequent Episodes in Sequences   267
    Jianjun Wu, Li Wan, and Zeren Xu

Certainty upon Empirical Distributions   279
    Joan Garriga

Workshop on Biologically Inspired Techniques for Data Mining (BDM 2011)

A Measure Oriented Training Scheme for Imbalanced Classification
Problems   293
    Bo Yuan and Wenhuang Liu

An SVM-Based Approach to Discover MicroRNA Precursors in Plant
Genomes   304
    Yi Wang, Cheqing Jin, Minqi Zhou, and Aoying Zhou

Towards Recommender System Using Particle Swarm Optimization
Based Web Usage Clustering   316
    Shafiq Alam, Gillian Dobbie, and Patricia Riddle

Weighted Association Rule Mining Using Particle Swarm
Optimization   327
    Russel Pears and Yun Sing Koh

An Unsupervised Feature Selection Framework Based on Clustering   339
    Sheng-yi Jiang and Lian-xi Wang

Workshop on Advances and Issues in Traditional Chinese Medicine Clinical Data Mining (AI-TCM 2011)

Discovery of Regularities in the Use of Herbs in Traditional Chinese
Medicine Prescriptions   353
    Nevin L. Zhang, Runsun Zhang, and Tao Chen

COW: A Co-evolving Memetic Wrapper for Herb-Herb Interaction
Analysis in TCM Informatics   361
    Dion Detterer and Paul Kwan

Selecting an Appropriate Interestingness Measure to Evaluate the
Correlation between Syndrome Elements and Symptoms 372
    Lei Zhang, Qi-ming Zhang, Yi-guo Wang, and Dong-lin Yu

The Impact of Feature Representation to the Biclustering of
Symptoms-Herbs in TCM 384
    Simon Poon, Zhe Luo, and Runshun Zhang

Second Workshop on Data Mining for Healthcare Management (DMHM 2011)

Usage of Mobile Phones for Personalized Healthcare Solutions 397
    M. Saravanan, S. Shanthi, and S. Shalini

Robust Learning of Mixture Models and Its Application on Trial
Pruning for EEG Signal Analysis 408
    Boyu Wang, Feng Wan, Peng Un Mak, Pui In Mak, and Mang I Vai

An Integrated Approach to Multi-criteria-Based Health Care Facility
Location Planning 420
    Wei Gu, Baijie Wang, and Xin Wang

Medicinal Property Knowledge Extraction from Herbal Documents for
Supporting Question Answering System 431
    Chaveevan Pechsiri, Sumran Painuall, and Uraiwan Janviriyasopak

First PAKDD Doctoral Symposium on Data Mining (DSDM 2011)

Age Estimation Using Bayesian Process 447
    Yu Zhang

Significant Node Identification in Social Networks 459
    Chi-Yao Tseng and Ming-Syan Chen

Improving Bagging Performance through Multi-algorithm Ensembles ... 471
    Kuo-Wei Hsu and Jaideep Srivastava

Mining Tourist Preferences with Twice-Learning 483
    Chen Zhang and Jie Zhang

Towards Cost-Sensitive Learning for Real-World Applications.......... 494
    Xu-Ying Liu and Zhi-Hua Zhou

Author Index 507

International Workshop on Behavior Informatics (BI 2011)

Evaluating the Regularity of Human Behavior from Mobile Phone Usage Logs

Hyoungnyoun Kim[1,2] and Ji-Hyung Park[1,2]

[1] Department of HCI and Robotics, University of Science and Technology, Korea
[2] Interaction and Robotics Research Center, Korea Institute of Science and Technology, 39-1, Hawolgok-Dong, Seongbuk-Gu, Seoul, Korea
{nyoun,jhpark}@kist.re.kr

Abstract. This paper investigated the relationship between incrementally logged phone logs and self-reported survey data to derive regularity and predictability from mobile phone usage logs. First, we extracted information not from a single value such as location or call logs, but from multivariate contextual logs. Then we considered the changing pattern of the incrementally logged information over time. To evaluate the patterns of human behavior, we applied entropy changes and the duplicated instances ratios from the stream of mobile phone usage logs. By applying the Hidden Markov Models to the patterns, the accumulated log patterns were classified according to the self-reported survey data. This research confirmed that regularity and predictability of human behavior can be evaluated by mobile phone usages.

Keywords: Regularity, predictability, human behavior, mobile phone log, reality mining.

1 Introduction

Prediction of human behavior is a challenging problem in behavior informatics [8] that remains to be solved in various research domains. Predictability has often been related to regularity. Regularity implies consistency of human behavior, which includes events that occur at irregular intervals. On the other hand, predictability can be represented as a degree that expresses a next possible event at a certain time. To predict a next event, the distribution of events that have occurred based on history is necessary. In addition, the causal relation between a previous event and the next event is used for the prediction.

One way to predict human behavior is to model patterns of behavioral information based on the history. Recently, many researchers have focused on the problem of calculating predictability of human behavior [1][2]. They tried to solve this problem using location information. However, it is difficult to analyze human behavior by only one modality. Estimating behavior by only a pattern, such as a location or call-log, is limited. That information cannot reflect other aspects of human behavior. Our research overcomes this limitation by suggesting a method that classifies behavioral patterns from contextually multivariate logs.

L. Cao et al. (Eds.): PAKDD 2011 Workshops, LNAI 7104, pp. 3–14, 2012.

To evaluate the regularity and predictability of human behavior, we make full use of logged contextual information. From the historical logging patterns, both regularity and predictability can be validated through a relative comparison among subjects. In order to construct pattern models for inference, long time logging is required. Moreover, we also have to consider personalized modeling because each subject has his/her own behavioral patterns.

Fortunately, we discovered an interesting feature in *reality mining* data [3], which is the data set that we are trying to deal with. This data set obtains a survey data through self-reporting from subjects. This survey includes questions whether subjects have regular and predictable working schedules. Additionally, the data contains contextual information, such as time, location, and mobile phone activities, that are acquired while using a mobile phone. However, this information is insufficient to express directly the regularity of the working schedule of subjects. We, therefore, attempt to figure other patterns from data acquired through continuous logging. In this paper, we model the unique properties of subjects behavior that are represented in the incrementally logged data, and we evaluate the regularity according to the relationship between phone usage logs and self-reported survey data.

This paper is organized as follows. In Section 2, we briefly survey related work on extracting behavioral patterns from phone usage. After that, the reality mining data is preprocessed and we inspect the characteristics of selected data in Section 3. In Section 4, we propose measures to analyze an incrementally logged contextual data set. Finally, the experimental results on the comparative performance of our classification from the survey data are represented in Section 5.

2   Related Work

Mobile phone use is being used as a device for understanding human behavior with randomness. According to the improvement of sensing interfaces of mobile phones, the performance of recognizing human behavior has increased [4]. Furthermore, several excellent data sets have been archived on a mobile platform by researchers elaborative efforts to discover special properties of human behavior [5][6]. Eagle et al. established a reality mining data set by logging over a hundred peoples mobile usage patterns for nine months [3]. They focused on the relationship among human behaviors and tried to find a common basic behavior pattern across many subjects [6]. Moreover, they presented a visualization method for a social network composited by mobile phone usage [7]. This type of research based on reality mining data is still ongoing in order to discover unique behavior patterns of human beings. Dantu et al. detected changing points of both events and patterns from call records and tried to predict callers behavior [9][10].

On the other hand, Song and Barabasi estimated the predictability of human behavior using phone events and location information [1][2][5]. They used a huge amount of phone usage data with almost 10 million instances. They calculated the entropy-based predictability from those historically logged data. Currently several studies are in progress to construct social networks from mobile phone usage data. To classify subject groups by behavior patterns and to predict their next behavior, these data sets have been widely used [11][12].

Unlike previous studies, this paper aims to evaluate regularity and predictability of behavior based on self-reported survey data of subjects. To this end, we use two discriminative approaches. First, we examine changing patterns of incrementally logged information. The pattern graph varies with the accumulated time. Then the variation of the graph eventually determines the regularity of human behavior. Second, we investigate the complex correlation of multivariate contextual information rather than simple information, such as a single call-log or location information.

3 Data Preprocessing

In order to extract multivariate contextual information from mobile phone usage, we use a reality mining data set [6]. This data includes several contextual items of information, such as cell tower IDs, application usage, and phone status, as well as call logs. It contains continuously acquired data from two months to nine months. Additionally, self-reported survey data about participating subjects is provided. Although the original reality mining data supports both call and location logs, it needs further processing to extract complex contextual information on the subjects. Therefore, we modify the original data to represent behavioral patterns. After preprocessing the data set, we acquire a contextual data set with several attributes as in Table 1. All data is converted to categorical data for easy analysis.

Table 1. Extracted contextual attributes and their contained values

Attributes (number of values)	Values
Day of week (7)	Sunday, Monday, Tuesday, Wednesday, Thursday, Friday, Saturday
Time (8)	Dawn, early morning, morning, noon, afternoon, evening, night, midnight
Interval (8)	10 sec, 1 min, 10 min, 30 min, 1 hour, 3 hours, 6 hours, over 6 hours
Place (6)	Work, home, elsewhere, not defined, no signal, phone off
Application (28)	Camera, menu, phonebook, screensaver, phonebook, etc.
Cell tower (10)	ATTWirelA, T-MobileHome, TMOHome, not defined, etc.
Call (2)	Yes, no

We define event instances as follows in order to create sequential and multivariate instances [13]. First, human behavior composed of asynchronous events, which means changes of a value in an attribute do not influence to the other attributes. For example, the duration of application usage is independent of the location of cell towers. Likewise, there is no guarantee that calls and changes of places occur simultaneously. Second, events occur at irregular intervals. The events are logged only when some changes of attributes are detected. In order to make sequential data from the individual logs, we add information on interval between the logs.

Each attribute has its own sequence of value changes. Hence, determining events in multivariate instances needs to consider correspondence to the changing time of values on other attributes. Until any event appears, an individual attribute reserves the previously extracted value. In a single time line, for instance, location of a subject has changed from L^A to L^B, then L^C as in Fig. 1(a). In the place L^A there was one call, after that the subject received a message in L^B. He/she used two applications of a mobile phone in the place of L^C. Now, the divided events are totally 11, respectively three, three, and five events for the locations. Figure 1(b) shows the changing values of all attributes by the total event instances on one given day.

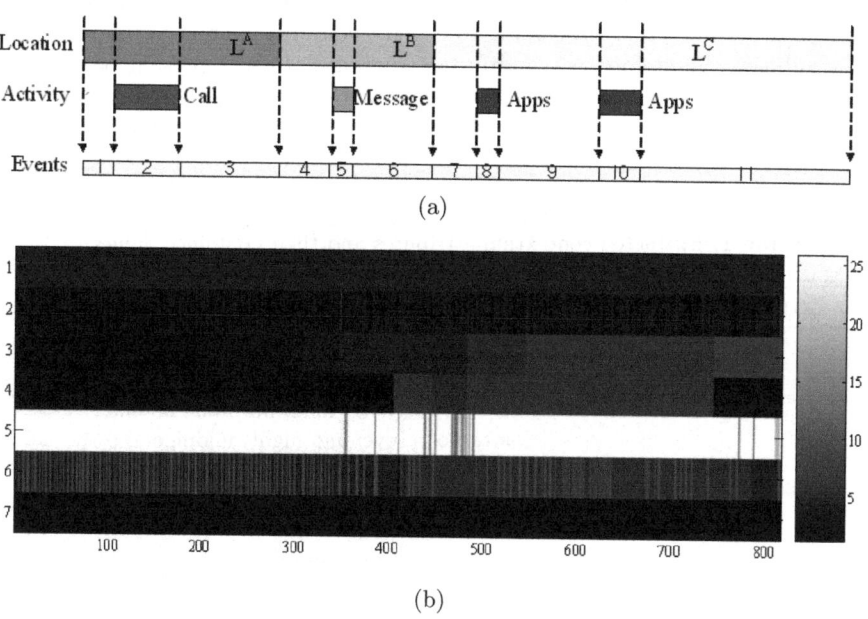

Fig. 1. Event definition in multivariate sequential information. (a) Asynchronous and irregular information is extracted and it composes an event stream. (b) Value changes of seven attributes in a day. The x-axis indicates the indexes of the total events.

Random activities of subjects are reflected in various values of attributes, so that the distribution of values in attributes is not uniform. For a subject, value distributions from all instances are expressed in histograms as in Fig. 2. In case of an attribute related to the day of the week, the distribution is relatively uniform (see Fig. 2(a)). This means that all the events occur regardless of the day of the week. On the other hand, the other attributes have irregular distributions and they show very biased value distribution. Attributes including places, applications, calls, and intervals are especially noticeable. Among them, the distribution of intervals shows a distinct characteristic as shown in Fig. 2(b). The distribution is similar to the scale free graph that Barabasi described [2]. As the interval is longer, the probability of occurrence decreases exponentially.

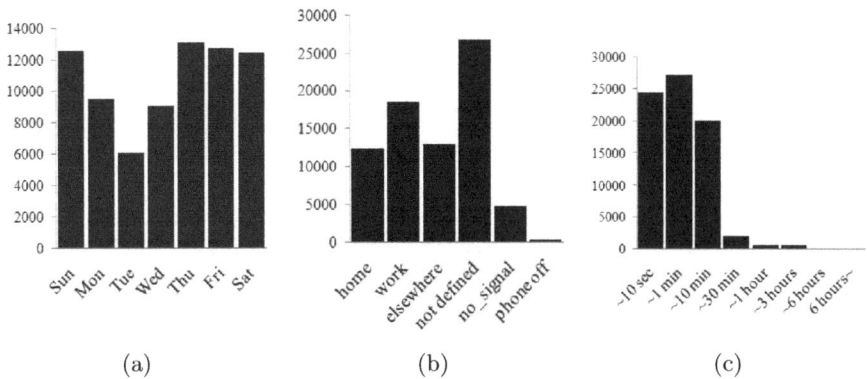

Fig. 2. Value distributions of attributes from all instances are expressed in histograms (a) day of the week (b) place (c) interval between adjacent events

4 Measures

To evaluate the regularity of human behavior from the irregular phone usage data, we additionally investigate structural properties that are latent in the data instances. We separate instances into several types considering properties of the incrementally logged data. All the instances are divided into logging days. Each day has its instances, so that we calculate the distribution of instance types each day. Moreover, we enumerate the accumulated distribution of instances. In addition, we use the accumulated entropy values that have been used for predicting human behavior [1]. By these three measures, we model the mobile phone usage patterns of the subjects.

4.1 Duplication Types

Through the data preprocessing, the incrementally logged data instances are temporally connected. This means that one data instance is composed of two adjacent instances in the time line. In this data structure, we define the relationship between the previously stored data and newly incoming data. If the incoming data is the same as the one of the previously logged data, we define the case as *duplication*. Behavior patterns of subjects can be repeated by the same contextual states, can be changed to other states, and can appear in new contextual states. Considering these properties of behavior patterns, we classify four *duplication types*.

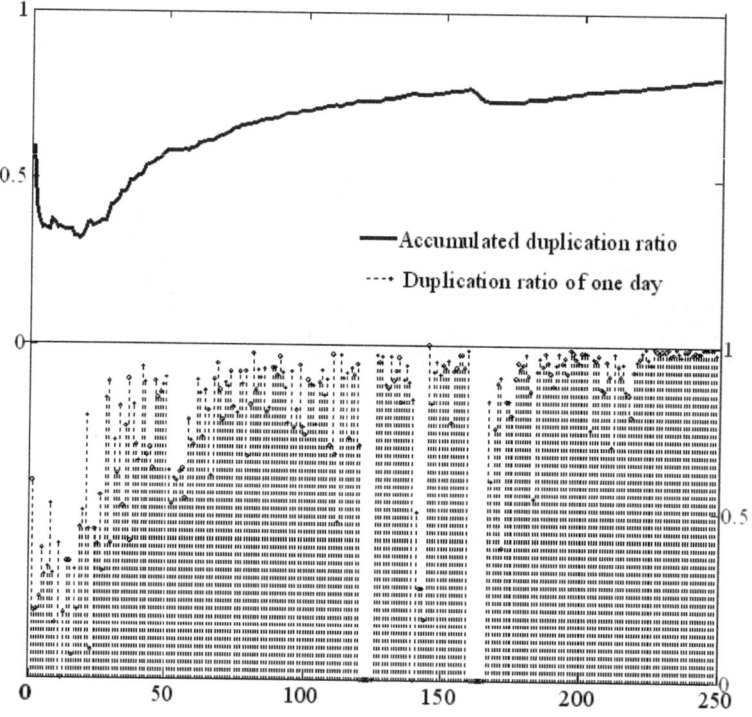

Fig. 3. A graph for changes of the duplication ratio (bottom) and accumulated duplication ratio (top) for a subject. The x-axis indicates the training days and the y-axis indicates the ratio of Type 1 and Type 2 over total instances.

We assumed that every instance in our data set is temporally connected, and all the incoming instances are logged. In this circumstance, *Type 1* data indicates an instance whose all values are matched in the logged data set. In this type, an instance is temporally connected to only one other instance. On the other hand, if an instance in a previous time step is connected to more than two different instances in the next time step, we define the instance as *Type 2*. This occurs when the input pattern is fixed and the output patterns are diverse. In some cases, although the all values of instances have appeared respectively in the previous instances, the combination of values has never been logged before.

Type 3 data includes this case related to an unknown instance. The last case, *Type 4* data, contains unknown values in the incoming instance.

From the definition of types, Type 1 and Type 2 come under the duplicated instances, and Type 3 and Type 4 are classified to instances including new values. Thereupon, we can suppose that if the ratio of Type 1 and Type2 over the total instances is high, then a subjects behavior is determined by the repetitive contextual patterns. That is, we can estimate that the regularity is high.

The accumulated duplication ratio of one subject is represented by a graph as in Fig. 3. The percentage means the proportion of Type 1 and Type 2 over all the instances. The lower part of Fig. 3 shows the duplication ratio on each day, and the upper part represents the change of the accumulated distribution ratio. During the initial logging period, the ratio is low because the size of the accumulated instances is small, and most instances are Type 3 and Type 4 compared to the previously logged instances. However, in a later period, the duplication ratio is almost 100%; accordingly, the accumulated duplication ratio also increases.

From Fig. 3, we discover other characteristics in the behavior pattern of the subject. After the duplication ratio is saturated in 60 days, three holes emerge (120th day, 140th day, and 170th day). These holes imply that the behavior pattern of the subject has changed, so that many new values appear. Consequently, we recognize that the duplication ratio is very low.

4.2   Accumulated Entropy

Predictability for accumulated behavioral data can be simply represented by entropy [14]. In this section, we use accumulated entropy as a measure to estimate the predictability of extracted instances. The entropy of multivariate instances is calculated by Equation 1.

$$H(X_{1:t}) = \sum_{i=1}^{d} -p(x_{1:t}^i)log(p(x_{1:t}^i)) \tag{1}$$

where, $X_{1:t}$ means an instance set logged from the first day to the tth day. d, t, and x_i indicate the size of the dimension of attributes, the accumulated days, and the value of the ith attribute, respectively. Hence, $p(x_{1:t}^i)$ indicates the probabilistic distribution of a value x of the ith attribute from the first day to the tth day. The total entropy H is calculated by summarizing the entropies of all the attributes.

Entropy varies according to data accumulation. The *accumulated entropy* has an interesting pattern compared to the duplication ratio as in Fig. 4. In the initial period, the accumulated entropy increases monotonously. However, as the logging time goes by, it reveals a form opposite to the accumulated duplication ratio. If the instances having new values increase, the entropy also increases, and vice versa. From this pattern, we analyze the repetitive pattern, i.e., the high regularity makes the accumulated entropy low. According to subjects, the

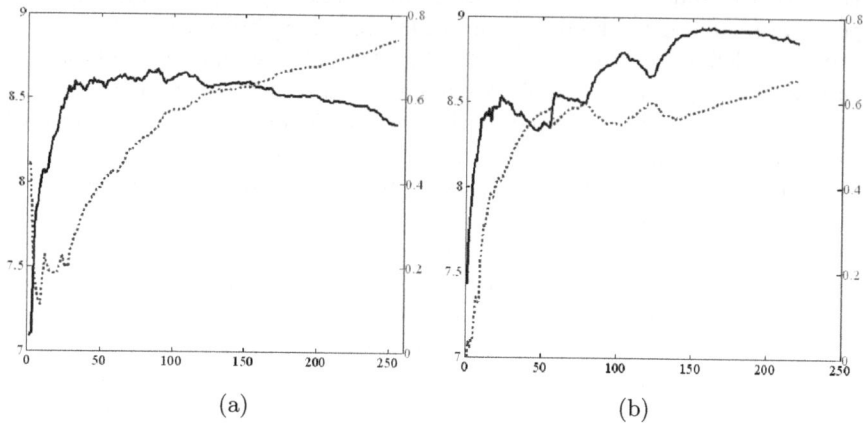

Fig. 4. A graph for changes of accumulated entropy (solid line) and accumulated duplication (dot line) ratio for two different subjects. The x-axis, left y-axis, right y-axis indicate respectively the total training days, entropy, and duplication ratio. (a) subject S8 (b) subject S16.

distribution of the duplication ratio and entropy is represented in various shapes. Figure 4 shows the different graph patterns of the subjects.

5   Evaluation of Survey Data

Determining the regularity of behavioral patterns from instance data is rather difficult. To solve this problem, we require some information. First is the accumulated behavioral data for each subject, and the other is labels for the regularity of the behavior. In this research, consequently, we evaluate whether the self-reported degree of regularity and predictability of subjects can be used for a label to classify the graph patterns of the accumulated contextual logs.

In order to classify behavioral patterns, we use the three measures proposed in Section 4. The three measures are the duplication ratio of one day, the accumulated duplication ratio, and the accumulated entropy. The self-reported survey data represents the regularity and predictability of subjects working schedule to three steps: *very*, *somewhat*, and *not at all*. We assign this degree as a label for grouping subjects. The contextual logs are converted to a sequential pattern by the three measures, and then they are trained using a machine-learning technique. To train sequential changing patterns of behavioral information, we apply the Hidden Markov Models (HMM).

The total number of subjects in the data set was 106. The data of the subjects were inspected through several constraints; the total incremental logging days were over 60 days, all the attributes mentioned in Table 1 were extracted, and the self-reported survey data was not absent. After criteria to extract appropriate subjects were applied, we finally obtained 32 subjects. Then, the subjects were classified by the accumulated logging patterns according to the survey data. In the survey data, the ratio of regularity that is assigned as *very* regular was 18 over 32, and the rest (14 over 32) were assigned as *somewhat* regular. In the case of predictability, the ratio between *very* predictable and *somewhat* predictable was 17:15. No subject assigned as *not at all* regular was satisfied with the above constraints. Figure 5 represents the accumulated duplication ratio for some of subjects, which were separated into two groups of very and somewhat regular. Hence, the two groups according to regularity and predictability were trained respectively using the HMM. For training the HMM, we set parameters as follows. The dimension of data was three, the number of states was seven, and the number of Gaussian mixtures was five.

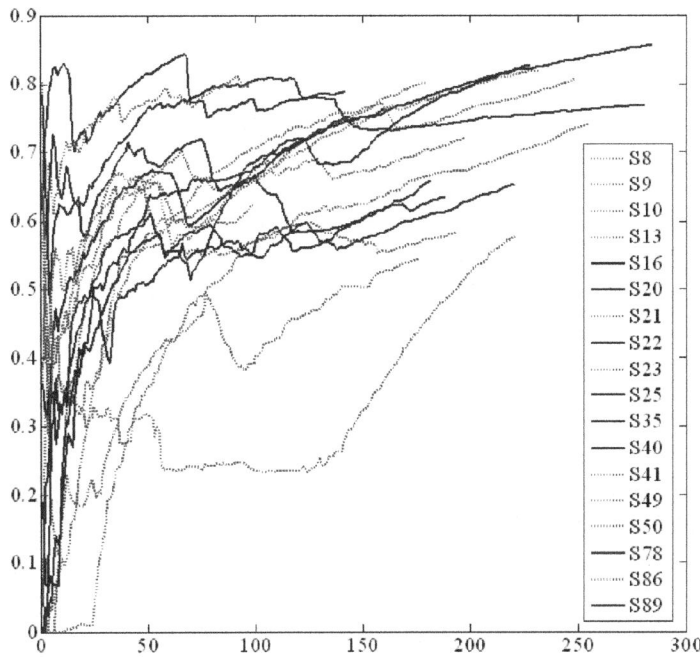

Fig. 5. Accumulated duplication ratio of parts of subjects. The dot lines represent very regular subjects and the solid lines indicate somewhat regular subjects. The x-axis indicates total training days.

After training each HMM model according to subjects groups, we validated the data that was used in training. We anticipated that the individually trained HMM models would well distinguish other subject groups if the assigned labels were appropriate. The validation result was over 80%; the recognition rate for regularity was 81.25%; and the recognition rate for predictability was 81.25%. These rates mean that the precision of the regularity models based on self-reported survey data is high.

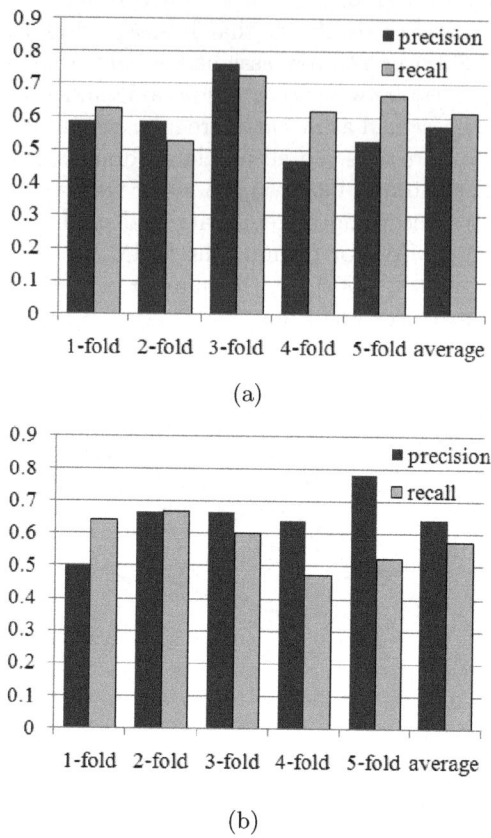

Fig. 6. Classification results of subjects according to (a) predictability and (b) regularity are represented by precision-recall values

Additionally, we classified untrained data by $k \times 2$ cross validation [15]. Here, k was assigned as five, so that total subjects were divided into five subsample sets. We calculated and compared the performance of classification for k-folds by precision and recall values. Figure 6 shows the result of the classification of subjects according to predictability and regularity. For the predictability group, the overall recognition rate was about 60%. Occasionally, the precision-recall was close to 80% but the average precision was 60%. Likewise, the regularity group achieved a similar performance with the predictability group. Through the experiment described in this paper, we could not conclude that the self-reported survey data has a strong connection to our measures for regularity and predictability. However, we explored the possibility that the contextual mobile phone usage pattern can be a measure to evaluate the regularity and predictability of subjects behaviors.

6   Conclusion

In this paper, we investigated the possibility that the regularity of human behavior can be classified through the mobile phone usage with a huge data set. We used incrementally logged multivariate contextual data to analyze the regularity and predictability implied in human behavior. To discover the behavioral patterns, we use both the duplication ratios between logged instances and an incoming instance and entropy for accumulated instances. In addition, we assigned self-reported survey data of subjects as a label for the degree of regularity and predictability. These values constructed a sequential graph with multiple channels and a label. From the experimental results, we determined that the precision of labeling was over 80% and the recognition rate of untrained subjects was about 60%.

Generally, the cost acquiring self-reported survey data is considerably expensive. This research was a trial to obtain an approximate regularity of subjects from mobile phone usage logs. For a more accurate analysis, construction of interfaces that are closely related to human behavior and data sets through the interfaces can be a solution. Furthermore, the degree of regularity would be extended by addition of both historically correlated entropy and intervals of duplicated instances. We anticipate that those various logs will help to determine and predict peoples behavior patterns.

References

1. Song, C., Qu, Z., Blumm, N., Barabasi, A.L.: Limits of predictability in human mobility. Science 327(5968), 1018–1021 (2010)
2. Barabasi, A.L.: Bursts, the hidden pattern behind everything we do. Dutton, New York (2010)
3. Massachusetts Institute of Technology, Reality Mining Project, http://reality.media.mit.edu/
4. Kwok, R.: Personal technology: Phoning in data. Nature 458(7241), 959–961 (2009)
5. Song, C., Qu, Z., Blumm, N., Barabasi, A.L.: Limits of predictability in human mobility - supplementary material (2010)
6. Eagle, N., Pentland, A., Lazer, D.: Inferring social network structure using mobile phone data. In: Proceedings of the National Academy of Sciences, pp. 15274–15278 (2009)
7. Eagle, N., Pentland, A.: Reality mining: sensing complex social systems. Personal and Ubiquitous Computing 10(4), 255–268 (2006)
8. Cao, L.: In-depth Behavior Understanding and Use: the Behavior Informatics Approach. Information Science 180(17), 3067–3085 (2010)
9. Phithakkitnukoon, S., Dantu, R.: Adequacy of data for characterizing caller behavior. In: Proceedings of KDD Inter. Workshop on Social Network Mining and Analysis (2008)
10. Phithakkitnukoon, S., Horanont, T., Di Lorenzo, G., Shibasaki, R., Ratti, C.: Activity-Aware Map: Identifying Human Daily Activity Pattern Using Mobile Phone Data. In: Salah, A.A., Gevers, T., Sebe, N., Vinciarelli, A. (eds.) HBU 2010. LNCS, vol. 6219, pp. 14–25. Springer, Heidelberg (2010)

11. Jensen, B.S., Larsen, J., Hansen, L.K., Larsen, J.E., Jensen, K.: Predictability of Mobile Phone Associations. In: Inter. Workshop on Mining Ubiquitous and Social Environments (2010)
12. Chi, J., Jo, H., Ryu, J.H.: Predicting Interpersonal Relationship based on Mobile Communication Patterns. In: The ACM Conf. on Computer Supported Cooperative Work (2010)
13. Kim, H., Kim, I.J., Kim, H.G., Park, J.H.: Adaptive Modeling of a User's Daily Life with a Wearable Sensor Network. In: Tenth IEEE International Symposium on Multimedia, pp. 527–532 (2008)
14. MacKay, D.J.C.: Information theory, inference, and learning algorithms. Cambridge Univ. Press, Cambridge (2003)
15. Dietterich, T.G.: Approximate statistical tests for comparing supervised classification learning algorithms. Neural Computation 10(7), 1895–1923 (1998)

Explicit and Implicit User Preferences in Online Dating

Joshua Akehurst, Irena Koprinska, Kalina Yacef,
Luiz Pizzato, Judy Kay, and Tomasz Rej

School of Information Technologies, University of Sydney, Australia
firstname.lastname@sydney.edu.au

Abstract. In this paper we study user behavior in online dating, in particular the differences between the implicit and explicit user preferences. The explicit preferences are stated by the user while the implicit preferences are inferred based on the user behavior on the website. We first show that the explicit preferences are not a good predictor of the success of user interactions. We then propose to learn the implicit preferences from both successful and unsuccessful interactions using a probabilistic machine learning method and show that the learned implicit preferences are a very good predictor of the success of user interactions. We also propose an approach that uses the explicit and implicit preferences to rank the candidates in our recommender system. The results show that the implicit ranking method is significantly more accurate than the explicit and that for a small number of recommendations it is comparable to the performance of the best method that is not based on user preferences.

Keywords: Explicit and implicit user preferences, online dating, recommender systems.

1 Introduction

Online dating websites are used by millions of people and their popularity is increasing. To find dating partners users provide information about themselves (*user profile*) and their preferred partner (*user preferences*); an example using predefined attributes is shown in Fig. 1. In this paper we focus on the user preferences, which is an important issue in behavior informatics [8]. We distinguish between *explicit* and *implicit* user preferences. The explicit user preferences are the preferences stated by the user as shown in Fig. 1. The implicit user preferences are inferred from the interactions of the user with other users.

Online dating is a new research area, with only a few published papers in the last year. Kim et al. [1] proposed a rule-based recommender that learns from user profiles and interactions. In another paper of the same group, Cai et al. [2] introduced a collaborative filtering algorithm based on user similarity in taste and attractiveness. McFee and Lanckriet [3] proposed an approach that learns distance metrics optimized for different ranking criteria, with evaluation in online dating. Diaz et al. [4] developed an approach for learning a ranking function that maximises the number of positive interactions between online dating users based on user profiles. In our previous work [5] we defined a histogram-based model of implicit preferences based

L. Cao et al. (Eds.): PAKDD 2011 Workshops, LNAI 7104, pp. 15–27, 2012.

on successful interactions, and showed that it can be used to generate better recommendations than the explicit preferences, when used in a content-based reciprocal recommendation algorithm.

In this paper we re-examine the effectiveness of explicit and implicit user preferences. There are three main differences with our preliminary exploration [5]. Firstly, in this paper we propose a different model of implicit preferences; it is learned from both successful and unsuccessful interactions, as opposed to being inferred from successful interactions only. Secondly, we evaluate the predictive power of the explicit and implicit user preferences in general, not as a part of a specific recommendation system. Thirdly, we propose a new method for using these preferences in our latest recommender, a hybrid content-collaborative system [6]. For a given user, it generates a list of candidates that are likely to have reciprocal interest and then ranks them to produce a shortlist of personalized recommendations so that users are quickly engaged in the search. In [6] the candidate ranking was done using the interaction history of the candidates, whereas in this paper we investigate the use of user preferences.

It is important to note that in the area of preference learning there has been little work on evaluating the relative power of explicit preferences, since this information is not normally available. Our work addresses this shortcoming for the case of online dating, where the explicit preferences are available.

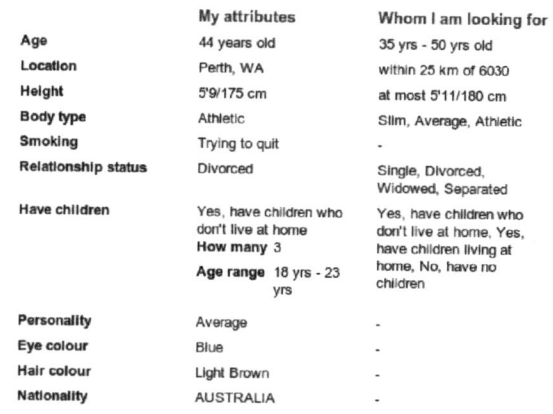

	My attributes	Whom I am looking for
Age	44 years old	35 yrs - 50 yrs old
Location	Perth, WA	within 25 km of 6030
Height	5'9/175 cm	at most 5'11/180 cm
Body type	Athletic	Slim, Average, Athletic
Smoking	Trying to quit	-
Relationship status	Divorced	Single, Divorced, Widowed, Separated
Have children	Yes, have children who don't live at home How many 3 Age range 18 yrs - 23 yrs	Yes, have children who don't live at home, Yes, have children living at home, No, have no children
Personality	Average	-
Eye colour	Blue	-
Hair colour	Light Brown	-
Nationality	AUSTRALIA	-

Fig. 1. User profile and explicit user preferences - example

Our contributions can be summarised as following:

- We propose a new approach for inferring the implicit user preferences. Given a target user U, we use U's previous successful and unsuccessful interactions with other users to build a machine learning classifier that captures U's preferences and is able to predict the success of future interactions between U and new users.
- We investigate the reliability and predictive power of the explicit and implicit user preferences for successful interactions. The results show that the explicit

preferences are not a good predictor of the success of user interaction, while the implicit preferences are a very good predictor.

- We propose an approach for using the user preferences in an existing recommender system, extending our previous work [6]. In particular, we propose to use these preferences for ranking of recommendation candidates. We compare the performance of the explicit and implicit ranking methods with another method that doesn't use user preferences.

Our evaluation was conducted using a large dataset from a major Australian dating website.

This paper is organized as follows. Section 2 describes how the users interact on the website. Section 3 defines the explicit user preferences and introduces our approach for learning the implicit user preferences. Section 4 describes the analysis of the predictive power of the explicit and implicit user preferences. Section 5 explains the proposed approach for using user preferences in a recommender system and discusses the results. Section 6 presents the conclusions and future work.

2   Domain Overview

We are working with a major Australian dating site. The user interaction on this site consists of four steps:

1) Creating a user profile and explicit user preferences – New users login to the web site and provide information about themselves (user profile) and their preferred dating partner (explicit user preferences) using a set of predefined attributes such as the ones shown in Fig. 1.
2) Browsing the user profiles of other users for interesting matches.
3) Mediated interaction – If a user A decides to contact user B, A chooses a message from a predefined list, e.g. *I'd like to get to know you, would you be interested?* We call these messages *Expressions of Interest (EOI)*. B can reply with a predefined message either positively (e.g. *I'd like to know more about you.*), negatively (e.g. *I don't think we are a good match.*) or decide not to reply. When an EOI receives a positive reply, we say that the interest is *reciprocated*.
4) Unmediated interaction – A or B buy tokens from the website to send each other unmediated message. This is the only way to exchange contact details and develop further relationship.

We call an interaction between users A and B a *successful interaction* if: 1) A has sent an EOI to B and B has responded positively to it or if 2) B has sent and EOI to A and A has responded positively to it.

3   User Preferences

3.1   Explicit User Preferences

We define the explicit preferences of a user U as the vector of attribute values specified by U. The attributes and their possible values are predefined by the website.

In our study we used all attributes except *location;* for simplicity we considered only people from Sydney. More specifically, we used 19 attributes: 2 numeric (*age* and *height*) and 17 nominal (*marital status, have children, education level, occupation industry, occupation level, body type, eye color, hair color, smoker, drink, diet, ethnic background, religion, want children, politics, personality and have pets*).

In addition, and again for simplicity, we have removed all interactions between the same sex users and only compared people of opposing genders.

3.2   Implicit User Preferences

The implicit user preferences of a user U are represented by a binary classifier which captures U's likes and dislikes. It is trained on U's previous successful and unsuccessful interactions. The training data consists of all users $U+$ with whom U had successful interactions and all users $U-$ with who U had unsuccessful interactions during a given time period. Each user from $U+$ and $U-$ is one training example; it is represented as a vector of user profile attribute values and labeled as either *Success* (successful interaction with U) or *Failure* (unsuccessful interaction with U). We used the same 19 user profile attributes as the explicit user preferences listed in the previous section. Given a new instance, user U_{new}, the classifier predicts how successful the interaction between U and U_{new} will be by outputting the probability for each class (*Success* or *Failure)* and assigning it to the class with higher probability.

As a classifier we employed NBTree [7] which is a hybrid classifier combining decision tree and Naïve Bayes classifiers. As in decision trees, each node of a NBTree corresponds to a test for the value of a single attribute. Unlike decision trees, the leaves of a NBTree are Naïve Bayes classifiers instead of class labels. We chose NBTree for two reasons. First, given a new instance, it outputs a probability for each class; we needed a probabilistic classifier as we use the probabilities for the ranking of the recommendation candidates, see Section 5. Second, NBTree was shown to be more accurate than both decision trees and Naïve Bayes, while preserving the interpretability of the two classifiers, i.e. providing an easy to understand output which can be presented to the user [7].

4   Are the User Preferences Good Predictors of the Success of User Interactions?

We investigate the predictive power of the explicit and implicit user preferences in predicting the success of an interaction between two users.

4.1   Explicit User Preferences

Data
To evaluate the predictive power of the explicit preferences we consider users who have sent or received at least 1 EOI during a one-month period (March 2010). We further restrict this subset to users who reside in Sydney to simplify the dataset. These

two requirements are satisfied by 8,012 users (called *target users*) who had 115,868 interactions, of which 46,607 (40%) were successful and 69,621 (60%) were unsuccessful. Each target user U has a set of *interacted users* U_{int}, consisting of the users U had interacted with.

Method

We compare the explicit preferences of each target user U with the profile of the users in U_{int} by calculating the number of matching and non-matching attributes.

In the explicit preferences the user is able to specify multiple values for a single nominal attribute and ranges for numeric attributes. For a numeric attribute, U_{int} matches U's preferences if U_{int}'s value falls within U's range or U_{int} has not specified a value. For a nominal attribute, U_{int} matches U's preferences if U_{int}'s value has been included in the set of values specified by U or U_{int} has not specified a value. An attribute is not considered if U has not specified a value for it. The preferences of U_{int} match the profile of U if all attributes match; otherwise, they don't match.

Results

The results are shown in Table 1. They show that 59.40% of all interactions occur between users with non-matching preferences and profiles. A further examination of the successful and unsuccessful interactions shows that:

- In 61.86% of all successful interactions U's explicit preferences did not match U_{int}'s profile.
- In 42.25% of all unsuccessful interactions U's explicit preferences matched the U_{int}'s profile.

Suppose that we use the matching of the user profiles and preferences to try to predict if an interaction between two users will be successful or not successful (if the profile and preferences match -> successful interaction; if the profile and preferences don't match -> unsuccessful interaction). The accuracy will be 49.43% (17,775+39,998 /115,868), and it is lower than the baseline accuracy of always predicting the majority class (ZeroR baseline) which is 59.78%. A closer examination of the misclassifications shows that the proportion of false positives is higher than the proportion of false negatives, although the absolute numbers are very similar.

In summary, the results show that the explicit preferences are not a good predictor of the success of interaction between users. This is consistent with [4] and [5].

Table 1. Explicit preferences - results

	U's explicit preferences and U_{int}'s profile matched	U's explicit preferences and U_{int}'s profile did not match	Total
Successful interactions	17,775 (38.14%)	28,832 (61.86%) (false positives)	46,607 (all successful interactions)
Unsuccessful interactions	29,263 (42.25%) (false negatives)	39,998 (57.75%)	69,261 (all unsuccessful interactions)

4.2   Implicit User Preferences

Data

To evaluate the predictive power of the implicit preferences we consider users who have at least 3 successful and 3 unsuccessful interactions during a one-month period (February 2010). This dataset was chosen so that we could test on the March dataset used in the study of the implicit preferences above. Here too, we restrict this subset to users who reside in Sydney. These two requirements are satisfied by 3,881 users, called *target users*. The training data consists of the interactions of the target users during February; 113,170 interactions in total, 30,215 positive and 72,995 negative. The test data consists of the interactions of the target users during Match; 95,777 interactions in total, 34,958 positive (37%, slightly less than the 40% in the study above) and 60,819 negative (63%, slightly more than the 60% in the study above). Each target user U has a set of *interacted users* U_{int}, consisting of the users U had interacted with.

Method

For each target user U we create a classifier by training on U's successful and unsuccessful interactions from February as described in Section 3.2. We then test the classifier on U's March interactions. This separation ensures that we are not training and testing on the same interactions.

Results

Table 2 summarizes the classification performance of the NBTree classifier on the test data. It obtained an accuracy of 82.29%, considerably higher than the ZeroR baseline of 63.50% and the accuracy of the explicit preferences classifier. In comparison to the explicit preferences, the false positives drop from 61.86% to 30.14%, an important improvement in this domain since a recommendation that leads to rejection can be discouraging; the false negatives drop from 42.25% to 9.97%.

Table 2. Classification performance of NBTree on test set

	Classified as: Successful interactions	Unsuccessful interactions	Total
Successful interactions	24,060 (68.83%)	10,538 (30.14%) (false positives)	34,958 (all successful interactions)
Unsuccessful interactions	6,064 (9.97%) (false negatives)	54,755 (90.03%)	60,819 (all unsuccessful interactions)

In summary, the results show that the implicit preferences are a very good predictor of the success of user interactions, and significantly more accurate than the explicit preferences.

5   Using User Preferences in Recommender Systems

In this section, we propose an approach for using the implicit and explicit user preferences in a recommender system. More specifically, we proposed that they are used to rank the recommendation candidates in our hybrid content-collaborative recommender system [6]. We evaluate the performance of the two methods, compare them with a baseline and also with the currently used ranking method Support which does not use user preferences.

5.1   Hybrid Content-Collaborative Reciprocal Recommender

In [6] we described our hybrid content-collaborative reciprocal recommender for online dating. It uses information from the user profile and user interactions to recommend potential matches for a given user. The content-based part computes similarities between users based on their profiles. The collaborative filtering part uses the interactions of the set of similar users, i.e. who they like/dislike and are liked/disliked by, to produce the recommendation. The recommender is *reciprocal* as it considers the likes and dislikes of both sides of the recommendation and aims to match users so that the paring has a high chance of success.

The process of generating an ordered list of recommendations for a given user U comprises of three key steps, see Fig. 2:

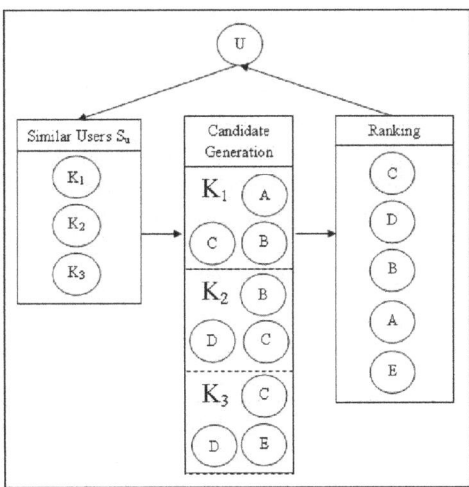

Fig. 2. Recommender process

1. Similar User Generation Based on User Profiles

This step produces a set of K users who have the lowest possible distance to U. For example, in Fig. 1 the set of similar users Su for user U consists of K_1, K_2 and K_3. We use a modified version of the K-Nearest Neighbor algorithm, with seven attributes (*age, height, body type, education level, smoker, have children* and *marital status*) and a distance measure specifically developed for these attributes as described in [6].

2. Candidate Generation Based on User Interactions

This step produces a set of candidate users for recommending to U. For every user in Su, we compute the list of all users with whom they have reciprocal interest with, meaning that these people like U and are also liked by U. For example in Fig. 1 the recommendation candidate set for U is $\{A, B, C, D, E\}$.

3. Candidate Ranking

This step ranks the candidates to provide meaningful recommendations for U. In [6] we use an approach based on the user interactions, in particular the support of Su for the candidates. In this paper we propose two new ranking approaches based on the user preferences: Explicit and Implicit which are based on explicit and implicit user preferences, respectively. We describe the ranking methods in the next section.

5.2   Ranking Methods

5.2.1 Support

This ranking method is based on the interactions between the group of similar users Su and the group of candidates. Users are added to the candidate pool if they have responded positively to at least one Su user or have received a positive reply from at least one Su user. However, some candidates might have received an EOI from more then one Su user and responded to some positively and to others negatively. Thus, some candidates have more successful interactions with Su than others. The Support ranking method computes the support of Su for each candidate. The higher the score, the more reciprocally liked is X by Su. This ranking method is the method used in [6].

For each candidate X we calculate the number of times X has responded positively or has received a positive response from Su, see Table 3. We also calculate the number of times X has responded negatively or has received a negative response from Su. The support score for X is the number of positive minus the number of negative interactions. The higher the score for X, the more reciprocally liked is X by Su. The candidates are sorted in descending order based on their support score.

Table 3. Support ranking - example

Candi-date	# Positive responses of candidate to Su	# Positive responses of Su to candidate	# Negative responses of candidate to Su	# Negative responses of candidate to Su	Support score
A	10	1	4	2	5
B	4	2	4	1	1
C	5	1	1	1	4
D	2	0	6	1	-5

5.2.2 Explicit

This ranking method is based on minimizing the number of non-matching attributes between the candidate profile and the explicit preferences of the target user; the lower the number of non-matches, the higher the candidate ranking. In addition to checking if the candidate satisfies the target user's explicit preferences it also checks the reverse: if the target user satisfies the candidate's explicit preferences. Thus, it minimises the number of reciprocal non-matches.

Table 4. Explicit ranking - example

Candidate	# Matching attributes	# Non-matching attributes	Stage 1: non-match rank	Stage 2 = final ranking: match rank for ties
A	2	0	1	2
B	2	2	2	4
C	4	2	2	3
D	4	0	1	1

We compare each candidate (i.e. its profile) with the explicit preferences of the target user and each target user with the explicit preferences of the candidate. We tally the number of matches and non-matches from both comparisons. The candidates are first sorted in ascending order based on the non-match score (stage 1 ranking). After that candidates with the same non-match score are sorted in descending order based on their match score (stage 2 and final ranking). An example is shown in Table 4.

5.2.3 Implicit

This ranking method uses the classifier generated for each target user U, based upon U's previous interactions. Given a candidate, the classifier gives a probability for the two classes *Success* and Failure (successful and unsuccessful interaction between the candidate and target user, respectively). Candidates are then ranked in descending order based on the probability of class *Success*. Table 5 shows an example.

Table 5. Implicit ranking - example

Candidate	Probability for Class *Success*
A	0.95
B	0.74
C	0.36
D	0.45

5.2.4 Baseline

This ranking method assumes that all candidates have an equal chance of a successful pairing and that any one random selection will give the same chance of success as any other ranking approach. For each candidate pool the candidates are randomly shuffled before being presented to the target user.

5.3   Experimental Evaluation

5.3.1 Data

We used the same data as the data used to learn the implicit user preferences, see Section 4.2. As stated already, it consists of the profile attributes and user interactions of all users who had at least 3 successful and at least 3 unsuccessful interactions during February 2010 and reside in Sydney.

For each run of the experiment, the users who meet the two requirements listed above are considered as part of the test set. Information about a test user's interactions is never included when generating and ranking candidates for that user. This ensures a clean separation between testing and training data.

5.3.2 Evaluation Metrics

For a user U we define the following sets:

- Successful EOI sent by U, *successful_sent*: The set of users who U has sent an EOI where the user has responded positively.
- Unsuccessful EOI sent by U, *unsuccessful_sent*: The set of users who U has sent an EOI where the user has responded negatively.
- Successful EOI received by U, *successful_recv*: The set of users who have sent an EOI to U where U has responded positively.
- Unsuccessful EOI received by U, *unsuccessful_recv*: The set of users who have sent an EOI to U where U has responded negatively.
- All successful EOI for U: *successful=successful_sent+successful_recv*
- All unsuccessful EOI for U: *unsuccessful=unsuccessful_sent+unsuccessful_recv*

For each user in the testing set, a list of N ordered recommendations $N\_recommendations$ is generated. We define the successful and unsuccessful EOI in the set of N recommendations as:

- Successful EOI for U that appear in the set of N recommendations: *successful@N = successful ∩ N_recommendations*.
- Unsuccessful EOI for U that appear in the set of N recommendations: *unsuccessful@N = unsuccessful ∩ N_recommendations*.

Then, the *success rate* at N (i.e. given the N recommendations) is defined as:

$$success\ rate\ @\ N = \frac{\#\,successful\ @\ N}{\#\,successful\ @\ N + \#\,unsuccessful\ @\ N}$$

In other words, given a set of N ordered recommendations, the success rate at N is the number of correct recommendations over the number of interacted recommendations (correct or incorrect).

Each experiment has been run ten times; the reported success rate is the average over the ten runs.

5.4   Results and Discussion

Fig. 3 shows the success rate results for different number of recommendations N (from 10 to 200) and different number of minimum EOI sent by U (5, 10 and 20). Table 6 shows the number of users in the test set for the three different EOI_sent.

The main results can be summarised as follows:

- The three recommenders (using Support, Implicit and Explicit as ranking methods) outperform the baseline for all N and minimum number of EOI.
- The best ranking method is Support, followed by Implicit and Explicit. For a small number of recommendations (N=10-50), Implicit performs similarly to Support. This is encouraging since the sucess rate for a small number of recommendations is very important in practical applications. As N increase the difference between Support and Implicit increases.

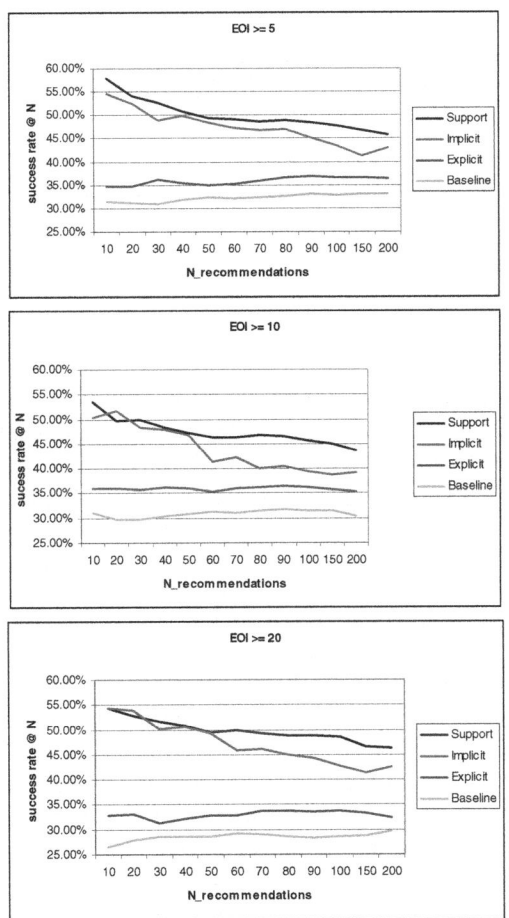

Fig. 3. Success rate for various N and minimum number of EOI sent by U

- Implicit significantly outperforms Explicit for all N and minimum number of EOI. For instance, when the top 10 recommendations are presented (N=10), the success rates are: Implicit=54.59%, Explicit=34.78% for EOI=5; Implicit= 50.45%, Explicit=36.05% for EOI=10; Implicit=54.31%, Explicit=32.95% for EOI=20, i.e. the difference between the two methods is 14.4-21.4%.
- As the number of recommendations N increases from 10 to 200, the success rate for Support and Implicit decreases with 8-12%. This means that the best recommendations are already at the top. Hence, these ranking methods are useful and effective. For Explicit as N increases the success rate doesn't change or even slightly increases in some cases which confirms that the rankig function is less effective, although still better than the baseline.
- A comparison of the three graphs show that as the number of EOI_sent increases from 5 to 20, the success rate trends are very similar.

Table 6. Number of users in the test set for the different number of EOI sent by U

EOI_sent	Number of users
EOI_sent >=5	3,881
EOI_sent >=10	3,055
EOI_sent >=20	1,938

6   Conclusions

In this paper we have reported our study of user preferences in a large dataset from an Australian online dating website.

We first considered the explicit user preferences which consist of the stated characteristics of the preferred dating partner by the user. We showed that the explicit preferences are not a good predictor of the success of user interactions, achieving an accuracy of 49.43%. We found that 61.86% of all successful interactions are with people who do not match the user's explicit preferences and 42.25% of all unsuccessful interactions are with people who match the user's explicit preferences.

We then proposed a novel model of implicit preferences that is learned using a NBTree classifier from both successful and unsuccessful previous user interactions. We showed that it is a very good predictor of the success of user interactions, achieving an accuracy of 89.29%.

We also proposed an approch that uses the explicit and implicit preferences for ranking of candidates in an existing recommender system. The results show that both ranking methods, Explicit and Implicit, outperform the baseline and that Implicit is much more accurate than Explicit for all number of recommendations and minimum number of EOI we considered. For example, when the top 10 recommendations are presented and the minimum number of EOI sent is 5, the success rate of Implicit is 54.59%, the success rate of Explicit is 34.78% and the baseline success rate is 31.35%. In practical terms, the success rate for a small number of recommendations is the most important; for 10-50 recommendations Implicit performs similarly to the best ranking method Support that is not based on user preferences.

Users can benefit from a suitable presentation of their implicit preferences; they can compare the implicit and explicit preferences and adjust the explicit preferences accordingly. We plan to investigate this in our future work.

Acknowledgments. This work is supported by the Smart Services Cooperative Research Centre.

References

1. Kim, Y.S., Mahidadia, A., Compton, P., Cai, X., Bain, M., Krzywicki, A., Wobcke, W.: People Recommendation Based on Aggregated Bidirectional Intentions in Social Network Site. In: Kang, B.-H., Richards, D. (eds.) PKAW 2010. LNCS, vol. 6232, pp. 247–260. Springer, Heidelberg (2010)
2. Cai, X., Bain, M., Krzywicki, A., Wobcke, W., Kim, Y.S., Compton, P., Mahidadia, A.: Collaborative Filtering for People to People Recommendation in Social Networks. In: Li, J. (ed.) AI 2010. LNCS, vol. 6464, pp. 476–485. Springer, Heidelberg (2010)
3. McFee, B., Lanckriet, G.R.G.: Metric Learning to Rank. In: 27th International Conference on Machine Learning, ICML (2010)
4. Diaz, F., Metzler, D., Amer-Yahia, S.: Relevance and Ranking in Online Dating Systems. In: 33rd Int. Conf. on Research and Development in Information Retrieval (SIGIR), pp. 66–73 (2010)
5. Pizzato, L., Rej, T., Chung, T., Koprinska, I., Yacef, K., Kay, J.: Learning User Preferences in Online Dating. In: European Conf. on Machine Learning and Priciples and Practice of Knowledge Discovery in Databases (ECML-PKDD), Preference Learning Workshop (2010)
6. Akehurst, J., Koprinska, I., Yacef, K., Pizzato, L., Kay, J., Rej, T.: A Hybrid Content-Collaborative Reciprocal Recommender for Online Dating. In: International Joint Conference on Artificial Intelligence, IJCAI (in press, 2011)
7. Kohavi, R.: Scaling Up the Accuracy of Naive-Bayes Classifiers: a Decision-Tree Hybrid. In: 2nd International Conference on Knowledge Discovery in Databases, KDD (1996)
8. Cao, L.: In-depth Behavior Understanding and Use: the Behavior Informatics Approach. Information Science 180(17), 3067–3085 (2010)

Blogger-Link-Topic Model for Blog Mining

Flora S. Tsai

Singapore University of Technology and Design,
Singapore 138682
fst1@columbia.edu

Abstract. Blog mining is an important area of behavior informatics because produces effective techniques for analyzing and understanding human behaviors from social media. In this paper, we propose the blogger-link-topic model for blog mining based on the multiple attributes of blog content, bloggers, and links. In addition, we present a unique blog classification framework that computes the normalized document-topic matrix, which is applied our model to retrieve the classification results. After comparing the results for blog classification on real-world blog data, we find that our blogger-link-topic model outperforms the other techniques in terms of overall precision and recall. This demonstrates that additional information contained in blog-specific attributes can help improve blog classification and retrieval results.

Keywords: Blog, blogger-link, classification, blog mining, author-topic, Latent Dirichlet Allocation.

1 Introduction

A blog is an online journal website where entries are made in reverse chronological order. The blogosphere is the collective term encompassing all blogs as a community or social network [19], which is an important area in behavior informatics [2]. Because of the huge volume of existing blog posts (documents), the information in the blogosphere is rather random and chaotic [17].

Previous studies on blog mining [7,13,15,18] use existing text mining techniques without consideration of the additional dimensions present in blogs. Because of this, the techniques are only able to analyze one or two dimensions of the blog data. On the other hand, general dimensionality reduction techniques [16,14] may not work as well in preserving the information present in blogs. In this paper, we propose unsupervised probabilistic models for mining the multiple dimensions present in blogs. The models are used in our novel blog classification framework, which categorizes blogs according to their most likely topic.

The paper is organized as follows. Section 2 describes related models and techniques, proposes new models for blog mining, and introduces a novel blog classification framework. Section 3 presents experimental results on real-world blog data, and Section 4 concludes the paper.

L. Cao et al. (Eds.): PAKDD 2011 Workshops, LNAI 7104, pp. 28–39, 2012.
© Springer-Verlag Berlin Heidelberg 2012

2   Models for Blog Mining

In this section, we propose and apply probabilistic models for analyzing the multiple dimensions present in blog data. The models can easily be extended for different categories of multidimensional data, such as other types of social media.

Latent Dirichlet Allocation (LDA) [1] models text documents as mixtures of latent topics, where topics correspond to key concepts presented in the corpus. An extension of LDA to probabilistic Author-Topic (AT) modeling [11,12] is proposed for blog mining. The AT model generates a distribution over document topics that is a mixture of the distributions associated with the authors [11] . An author is chosen at random for each individual word in the document. This author chooses a topic from his or her multinomial distribution over topics, and then samples a word from the multinomial distribution over words associated with that topic. This process is repeated for all words in the document [12]. Other related work include a joint probabilistic document model (PHITS) [4] which modeled the contents and inter-connectivity of document collections. A mixed-membership model [5] was developed in which PLSA was replaced by LDA as the generative model. The Topic-Link LDA model [8] quantified the effect of topic similarity and community similarity to the formation of a link.

We have extended the AT model for blog links and dates. For the Link-Topic (LT) model, each link is represented by a probability distribution over topics, and each topic represented by a probability distribution over terms for that topic. In the LT model, a document has a distribution over topics that is a mixture of the distributions associated with the links. When generating a document, a link is chosen at random for each individual word in the document. This link chooses a topic from his or her multinomial distribution over topics, and then samples a word from the multinomial distribution over words associated with that topic. This process is repeated for all words in the document.

For the LT model, the probability of generating a blog is given by:

$$\prod_{i=1}^{N_b} \frac{1}{L_b} \sum_{l} \sum_{t=1}^{K} \phi_{w_i t} \theta_{tl} \tag{1}$$

where there are N_b blogs, with each blog b having L_b links. The probability is then integrated over ϕ and θ and their Dirichlet distributions and sampled using Markov Chain Monte Carlo methods.

Likewise, for the Date-Topic (DT) model, each date is represented by a probability distribution over topics, and each topic represented by a probability distribution over terms for that topic, and the representation of dates is performed the same way as the AT and LT model for authors and links. The probability of generating a blog is given by:

$$\prod_{i=1}^{N_b} \frac{1}{D_b} \sum_{d} \sum_{t=1}^{K} \phi_{w_i t} \theta_{td} \tag{2}$$

where there are N_b blogs, with each blog b having D_b dates. The probability is then integrated over ϕ and θ and their Dirichlet distributions and sampled using Markov Chain Monte Carlo methods.

2.1   Blogger-Link-Topic (BLT) Model

Although we have extended the Author-Topic (AT) model for blog links and dates, the existing techniques based on the AT model are not able to simultaneously analyze the multiple attributes of blog documents. In order to solve the problem of analyzing the multiple attributes of blogs, we propose the Blogger-Link-Topic (BLT) model that can solve the problem of finding the most likely bloggers and links for a given set of topics. Figure 1 shows the generative model of the BLT model using plate notation.

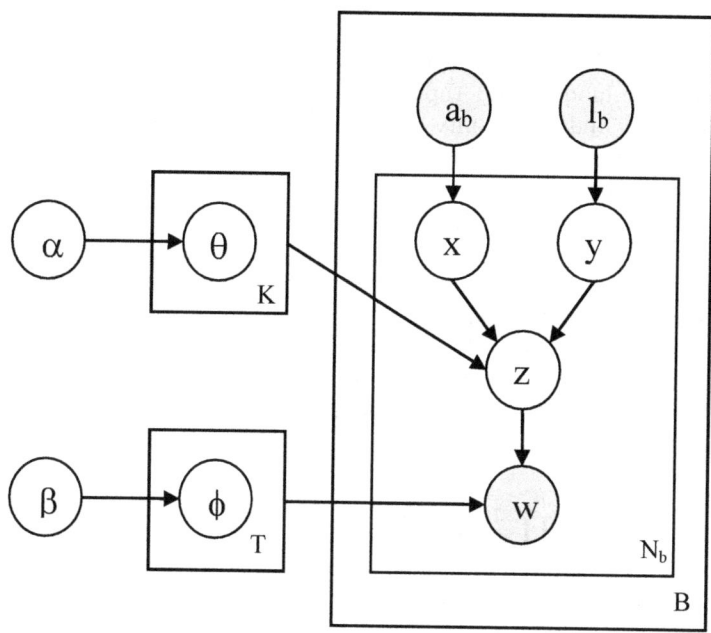

Fig. 1. The graphical model for the blogger-link-topic model using plate notation

In the Blogger-Link-Topic (BLT) model, each blogger and link is represented by a probability distribution over topics, and each topic represented by a probability distribution over terms for that topic. For each word in the blog b, blogger x is chosen uniformly at random from a_b and link y is chosen uniformly at random from l_b. Then, a topic is chosen from a distribution over topics, θ, chosen from a symmetric Dirichlet(α) prior, and the word is generated from the chosen topic. The mixture weights corresponding to the chosen blogger are used to select a topic z, and a word is generated according to the distribution ϕ corresponding to that topic, drawn from a symmetric Dirichlet(β) prior.

We have also extended the BLT model for finding the most likely bloggers and dates for a given set of topics. For the Blogger-Date-Topic (BDT) model, each blogger and date is represented by a probability distribution over topics, and each topic represented by a probability distribution over terms for that topic. Likewise, the model can also be extended to incorporate comments, social network of bloggers, inlinks, and tags.

For the Blogger-Comment-Topic (BCT) model, each blogger and comment is represented by a probability distribution over topics, and each topic represented by a probability distribution over terms for that topic. For the social network of bloggers, the relationships between bloggers can be obtained either directly from the bloggers' list of friends (if available) or indirectly through inferring similar bloggers by the content they publish. Once the network is obtained, then the the Blogger-Network-Topic (BNT) model can be built, where each blogger and social network is represented by a probability distribution over topics, and each topic represented by a probability distribution over terms for that topic. Similarly, incorporating inlinks into the model will need additional processing either through analyzing the outlinks of other blogs that point to the current blog, or by crawling through the link information in the blog network. Then, the Blogger-Inlink-Topic (BIT) model can be built, where each blogger and inlink is represented by a probability distribution over topics, and each topic represented by a probability distribution over terms for that topic. Finally, the Blogger-Tag-Topic (BTT) model can be built from a list of either user-generated tags, subjects, or categories for the blog. In the BTT model, each blogger and tag is represented by a probability distribution over topics, and each topic represented by a probability distribution over terms for that topic.

In order to learn the models for links and dates, we integrated the probability over ϕ and θ and their Dirichlet distributions and sampled them using Markov Chain Monte Carlo methods. Monte Carlo methods use repeated random sampling, which is necessary as our models rely on Bayesian inference to generate the results. Other ways to generate the results include variational methods and expectation propagation.

The BLT model solved the problem of finding the most likely bloggers and links for a given set of topics, a problem that is unique to blog mining. Although it is an extension of the AT model, the model is motivated by the unique structure of blog data and capitalizes on the synergy between bloggers and links. For example, if similar bloggers post similar content, or if similar contents are posted from the same links, the BLT model can model both situations and use this information to categorize the blogs into their correct topics. The BLT model is a better model than LDA and the AT model for the blog mining task because it is able to consider more attributes of blogs than just the blogger or content. The LDA model is limited because it only uses content information, while the AT model only uses blogger and content information, and does not use the additional information provided in links or dates.

In comparison to other models, BLT is different than Link-PLSA-LDA [10] and the mixed-membership model [5] because it does not use PLSA as the generative building block. In the joint probabilistic document model (PHITS) [4] and citation-topic (CT) model [6], the "links" were defined as the relations among documents rather than the actual URL, permalinks, or outlinks. Similarly, in the Topic-Link LDA model [8] defined the formation of a "link" as the effect of topic similarity and community similarity. Thus, our usage of links and the adaptation to blogs are some of the distinguishing characteristics of our model that sets it apart from the ones previously proposed.

The BLT is a general model can excel in blog mining tasks such as classification and topic distillation. The advantage of BLT over simple text classification techniques is that the simple methods will not be able to simultaneously leverage the multiple attributes of blogs. Other blog mining tasks include opinion retrieval, which involved locating blog posts that express an opinion about a given target. For the BLT model to be used in opinion retrieval, the model needs to be first trained on a set of positive, negative, and neutral blogs. Then BLT can be used to detect the most likely blogs which fit the labeled training set. Splog detection is another challenging task that can benefit from the BLT model. To adapt the BLT model to spam posts, trainings needs to be performed on a set of spam and non-spam blogs. After training, the BLT model can be used to leverage the information on bloggers and links to detect the most likely spam and non-spam blogs. Topic distillation, or feed search, searches for a blog feed with a principle, recurring interest in a topic. A subset of this task will be to search for a blog post in a feed related to a topic that the user wishes to subscribe. The BLT model can thus be used to determine the most likely bloggers and links for a given blog post of a given topic.

2.2 Blog Classification Framework

To demonstrate the usefulness of the BLT model, we propose a novel blog classification framework that is able to classify blogs into different topics. In this framework, blog documents are processed using stopword removal, stemming, normalization, resulting in the generation of the term-document matrix. Next, different techniques are implemented to create the document-topic matrix. From the document-topic matrix, the classification into different topics can be performed and evaluated against available relevance judgements.

The heuristics used to develop the blog classification framework can be described as follows:

1. Given the output of the BLT model as a term-topic matrix, a blogger-topic matrix, and a link-topic matrix, the first step is to convert the matrices to a document-topic matrix.
2. Since the document-topic matrix is not normalized, the second step is to normalize the matrices.

3. From the normalized document-topic matrix, the most likely topic for each document (blog) is compared against the ground truth.
4. The corresponding precision and recall can then be calculated.

In order to calculate the document-topic matrices for the various techniques, we propose the following technique. From the results of the Blogger-Link-Topic model, we can obtain a set of document-topic matrices, which can be used to predict the topic given a document. The document-topic matrix (DT_w) based on terms (words) is given by:

$$DT_w = WD' \times WT; \tag{3}$$

where WD is the term-document matrix, and WT is the term-topic matrix. The justification for the equation above is to convert the output of the BLT model WT, which lists the most probably terms for each topic, into a list of the most probable documents per topic, DT_w. Once we have a list of the most likely documents per topic, we can compare them with the ground truth to judge the precision and recall performance.

The document-topic matrix (DT_b) based on bloggers is given by:

$$DT_b = BD' \times BT; \tag{4}$$

where BD is the blogger-document matrix, and BT is the blogger-topic matrix. The justification for the equation above is to convert the output of the BLT model BT, which lists the most probably bloggers for each topic, into a list of the most probable documents per topic, DT_b.

Likewise, the document-topic matrix based on links (DT_l) is given by:

$$DT_l = LD' \times LT; \tag{5}$$

where LD is the link-document matrix, and LT is the link-topic matrix. The justification for the equation above is to convert the output of the BLT model LT, which lists the most probably links for each topic, into a list of the most probable documents per topic, DT_l.

The three document-topic matrices were used to create the normalized document-topic matrix (DT_{norm}) given by:

$$DT_{norm} = \frac{DT_w + DT_b + DT_l}{||DT_w||_{max} + ||DT_b||_{max} + ||DT_l||_{max}} \tag{6}$$

From DT_{norm}, we then predicted the corresponding blog category for each document, and compared the predicted results with the actual categories.

For the author-topic (AT) model, the normalized document-topic matrix DT_{norm} is given by:

$$DT_{norm} = \frac{DT_w + DT_b}{||DT_w||_{max} + ||DT_b||_{max}} \tag{7}$$

and for Latent Dirichlet Allocation (LDA), the normalized document-topic matrix DT_{norm} is just the normalized document-topic matrix (DT_w) based on terms (words):

$$DT_{norm} = \frac{DT_w}{||DT_w||_{max}} \tag{8}$$

The document-topic matrix can be efficiently generated by assuming the sparseness levels of the term-document matrix and the term-topic matrix. Fast sparse matrix multiplication can be performed by moving down both the row and column lists in tandem, searching for elements in the row list that have the same index as elements in the column list. Since the lists are kept in order by index, this can be performed in one scan through the lists. Each iteration through the loop moves forward at least one position in one of the lists, so the loop terminates after at most Z iterations (where Z is number of elements in the row list added to the number of elements in the column list).

Once we obtain the document-topic matrix, the category corresponding to the highest score for each blog document is used as the predicted category, and compared to the actual blog category. Given the normalized document-topic matrix DT and the ground truth topic vector G for each document, the predicted category vector C is calculated by finding the topic corresponding to the maximum value for each row vector $d_i = (w_{i1}, ..., w_{iX})$. Thus, the average precision and average recall can be calculated by averaging the results across all the topics.

3 Experiments and Results

Experiments were conducted on two datasets: BizBlogs07 [3], a data corpus of business blogs, and Blogs06 [9], the dataset used for TREC Blog Tracks from 2006–2008. BizBlogs07 contains 1269 business blog entries from various CEOs's blog sites and business blog sites. There are a total of 86 companies represented in the blog entries, and the blogs were classified into four (mutually exclusive) categories based on the contents or the main description of the blog: Product, Company, Marketing, and Finance [3]. Blogs in the Product category describe specific company products, such as reviews, descriptions, and other product-related news. Blogs in the Company category describe news or other information specific to corporations, organizations, or businesses. The Marketing category deals with marketing, sales, and advertising strategies for companies. Finally, blogs in the Finance category relates to financing, loans, credit information [3].

The Blogs06 collection [9] is a large dataset which is around 24.7 GB after being compressed for distribution. The number of feeds is over 100k blogs while the number of permalink documents is over 3.2 million. For the purpose of our experiments, we extracted a set of topics from the 2007 TREC Blog topic distillation task. The subset of topics were taken from the MySQL database created in [13], which indexed the feeds component. The 6 topics we extracted were 955 (mobile phone), 962 (baseball), 970 (hurricane Katrina), 971 (wine), 978 (music), and 983 (photography). The original number of RSS (Really Simple Syndication) blog items (which correspond to a document) for the 6 topics was 9738; however, after extracting those items with non-empty content, blogger, link, and date information, the total number of items reduced to 2363 documents.

3.1 Blogger-Link-Topic Results

The most likely terms and corresponding bloggers and links from the topics of marketing and finance of the BizBlogs07 collection are listed in Tables 1-2.

Table 1. BLT Topic 3: Marketing

Term	Probability
market	0.03141
compani	0.02313
busi	0.02057
time	0.01027
make	0.00831
Blogger	**Probability**
Chris Mercer	0.12553
FMF	0.12475
Larry Bodine	0.10897
john dodds	0.07022
Manoj Ranaweera	0.04233
Link	**Probability**
merceronvalue.com	0.13012
www.freemoneyfinance.com	0.12931
pm.typepad.com	0.11578
makemarketinghistory.blogspot.com	0.07267
jkontherun.blogs.com	0.05117

As can be seen from the tables, the topics correspond to the four categories of Product, Company, Marketing, and Finance, based on the list of most likely terms. The list of top bloggers correspond to the bloggers that are most likely to post in a particular topic. Usually a blogger with a high probability for a given topic may also post many blogs for that topic. The list of links correspond to the top links for each topic, based on the most likely terms and top bloggers that post to the topic.

3.2 Blog Classification Results

The results of the Blogger-Link-Topic (BLT) model for blog classification is compared with three other techniques: Blogger-Date-Topic (BDT), Author-Topic (AT), and Latent Dirichlet Allocation (LDA). The BizBlogs07 classification results for precision and recall were tabulated in Tables 3 and 4. The Blogs06 classification results are shown in Tables 5 and 6.

Table 2. BLT Topic 4: Finance

Term	Probability
monei	0.01754
year	0.01639
save	0.01576
make	0.01137
time	0.00895
Blogger	**Probability**
FMF	0.49493
Jeffrey Strain	0.06784
Tricia	0.05400
Chris Mercer	0.04487
Michael	0.04177
Link	**Probability**
www.freemoneyfinance.com	0.50812
www.pfadvice.com	0.06948
www.bloggingawaydebt.com	0.05526
merceronvalue.com	0.04588
jkontherun.blogs.com	0.04498

Table 3. Blog Precision Results for BizBlogs07

	Product	Company	Marketing	Finance	Average
BLT	*0.8760*	*0.4415*	*0.8996*	*0.9569*	**0.7935**
BDT	*0.6718*	*0.2151*	*0.4349*	*0.9540*	**0.5690**
AT	*0.8682*	*0.3434*	*0.8922*	*0.9626*	**0.7666**
LDA	*0.8760*	*0.4566*	*0.7026*	*0.8534*	**0.7222**

Table 4. Blog Recall Results for BizBlogs07

	Product	Company	Marketing	Finance	Average
BLT	*0.7829*	*0.6842*	*0.7576*	*0.8397*	**0.7661**
BDT	*0.8301*	*0.6129*	*0.5691*	*0.4510*	**0.6158**
AT	*0.8000*	*0.6741*	*0.7533*	*0.7512*	**0.7447**
LDA	*0.7655*	*0.5789*	*0.5555*	*0.8652*	**0.6913**

Table 5. Blog Precision Results for Blogs06

	Mobile	Baseball	Katrina	Wine	Music	Photo	Average
BLT	*0.8549*	*0.9159*	*0.7083*	*0.8418*	*0.4961*	*0.5512*	**0.7281**
BDT	*0.9604*	*0.8097*	*0.8095*	*0.6461*	*0.5741*	*0.0271*	**0.6378**
AT	*0.8264*	*0.9867*	*0.3125*	*0.8338*	*0.7566*	*0.4995*	**0.6861**
LDA	*0.8484*	*0.9425*	*0.7262*	*0.6997*	*0.5959*	*0.1175*	**0.6550**

The BLT model obtained better overall precision and recall results for both BizBlogs07 and Blogs06 data. For Blogs06, the worst performing category was Photography, due to the general nature of the topic. The Baseball category had the highest overall precision, while the Wine category had the highest overall

Table 6. Blog Recall Results for Blogs06

	Mobile	Baseball	Katrina	Wine	Music	Photo	Average
BLT	0.6419	0.8449	0.6247	0.9874	0.9034	0.3970	**0.7332**
BDT	0.4771	0.7821	0.5540	0.9060	0.9583	0.1250	**0.6337**
AT	0.7054	0.6778	0.5526	0.9094	0.6315	0.6617	**0.6897**
LDA	0.5554	0.4863	0.7011	0.8502	0.8527	0.3071	**0.6255**

recall. The reasons could also be due to the specialized nature of both categories. For both of the datasets, the best to worst performing models were the same, with BLT achieving the best results overall, followed by AT, LDA, and BDT. This confirms our theory that blogger and link information can help to improve results for the blog mining classification task.

3.3   Results on Co-occurrence of BLT and AT Models

As BLT is based on the AT model, the two models will co-occur if all the links are identical. That is, if all the blog data come from the same link, which we define as the root domain. This could conceivably occur if all the blogs were obtained from the same source. In the case of identical links, the output of the BLT model will be identical to the AT model. We confirm this in the following set of experiments for the 2007 TREC Blog data, where all the links were changed to the same name: "www.goobile.com". The output of the first two topics in the BLT model is shown in Tables 7–8. In the tables, the topic numbers correspond to the numbers in the 2007 TREC Blog Track. The output of the AT model is exactly the same as the BLT model, except for the link information which is not included in the AT model. However, the case shown is an exception rather than rule, as it is highly unlikely that all blogs will have the same source. Therefore, the BLT model is still useful to analyze the additional link information not provided in the AT model.

Table 7. BLT Topic 955: Mobile Phone

Term	Probability
mobil	0.02008
phone	0.01816
servic	0.00865
camera	0.00828
imag	0.00790
Blogger	**Probability**
administrator	0.07436
alanreiter	0.06120
dennis	0.05517
admin	0.05174
data	0.05124
Link	**Probability**
www.goobile.com	1.0000

Table 8. BLT Topic 962: Baseball

Term	Probability
year	0.02003
game	0.01229
time	0.01169
plai	0.01069
basebal	0.01066
Blogger	**Probability**
bill	0.19567
scott	0.07846
yardwork	0.04825
dave	0.04246
scottr	0.03421
Link	**Probability**
www.goobile.com	1.0000

4   Conclusion

In this paper, we have proposed a probabilistic blogger-link-topic (BLT) model based on the author-topic model to solve the problem of finding the most likely bloggers and links for a given set of topics. The BLT model is useful for behavior informatics, which can help extract discriminative behavior patterns from high-dimensional blog data. The BLT results for blog classification were compared to other techniques using blogger-date-topic (BDT), author-topic, and Latent Dirichlet Allocation, with BLT obtaining the highest average precision and recall.

The BLT model can easily be extended to other areas of behavior informatics, such as to analyze customer demographic and transactional data, human behavior patterns and impacts on businesses. In addition, the BLT model can help in the analysis of behavior social networks handling convergence and divergence of behavior, and the evolution and emergence of hidden groups and communities.

References

1. Blei, D.M., Ng, A.Y., Jordan, M.I.: Latent dirichlet allocation. J. Mach. Learn. Res. 3, 993–1022 (2003)
2. Cao, L.: In-depth behavior understanding and use: the behavior informatics approach. Information Science 180, 3067–3085 (2010)
3. Chen, Y., Tsai, F.S., Chan, K.L.: Machine learning techniques for business blog search and mining. Expert Syst. Appl. 35(3), 581–590 (2008)
4. Cohn, D., Hofmann, T.: The missing link – a probabilistic model of document content and hypertext connectivity. In: Advances in Neural Information Processing Systems, vol. 13, pp. 430–436 (2001)
5. Erosheva, E., Fienberg, S., Lafferty, J.: Mixed-membership models of scientific publications. Proceedings of the National Academy of Sciences of the United States of America 101(suppl. 1), 5220–5227 (2004)

6. Guo, Z., Zhu, S., Chi, Y., Zhang, Z., Gong, Y.: A latent topic model for linked documents. In: SIGIR 2009: Proceedings of the 32nd International ACM SIGIR Conference on Research and Development in Information Retrieval, pp. 720–721. ACM, New York (2009)

7. Liang, H., Tsai, F.S., Kwee, A.T.: Detecting novel business blogs. In: ICICS 2009: Proceedings of the 7th International Conference on Information, Communications and Signal Processing (2009)

8. Liu, Y., Niculescu-Mizil, A., Gryc, W.: Topic-link lda: joint models of topic and author community. In: ICML 2009: Proceedings of the 26th Annual International Conference on Machine Learning, pp. 665–672. ACM, New York (2009)

9. Macdonald, C., Ounis, I.: The TREC Blogs06 collection: Creating and analysing a blog test collection. Tech. rep., Dept of Computing Science, University of Glasgow (2006)

10. Nallapati, R., Cohen, W.: Link-PLSA-LDA: A new unsupervised model for topics and influence of blogs. In: Proceedings of the International Conference on Weblogs and Social Media (ICWSM). Association for the Advancement of Artificial Intelligence (2008)

11. Rosen-Zvi, M., Griffiths, T., Steyvers, M., Smyth, P.: The author-topic model for authors and documents. In: AUAI 2004: Proceedings of the 20th Conference on Uncertainty in Artificial Intelligence, pp. 487–494. AUAI Press, Arlington (2004)

12. Steyvers, M., Smyth, P., Rosen-Zvi, M., Griffiths, T.: Probabilistic author-topic models for information discovery. In: KDD 2004: Proceedings of the Tenth ACM SIGKDD International Conference on Knowledge Discovery and Data Mining, pp. 306–315. ACM, New York (2004)

13. Tsai, F.S.: A data-centric approach to feed search in blogs. International Journal of Web Engineering and Technology (2012)

14. Tsai, F.S.: Dimensionality reduction techniques for blog visualization. Expert Systems With Applications 38(3), 2766–2773 (2011)

15. Tsai, F.S., Chan, K.L.: Detecting Cyber Security Threats in Weblogs using Probabilistic Models. In: Yang, C.C., Zeng, D., Chau, M., Chang, K., Yang, Q., Cheng, X., Wang, J., Wang, F.-Y., Chen, H. (eds.) PAISI 2007. LNCS, vol. 4430, pp. 46–57. Springer, Heidelberg (2007)

16. Tsai, F.S., Chan, K.L.: Dimensionality reduction techniques for data exploration. In: 2007 6th International Conference on Information, Communications and Signal Processing, ICICS, pp. 1568–1572 (2007)

17. Tsai, F.S., Chan, K.L.: Redundancy and novelty mining in the business blogosphere. The Learning Organization 17(6), 490–499 (2010)

18. Tsai, F.S., Chen, Y., Chan, K.L.: Probabilistic Techniques for Corporate Blog Mining. In: Washio, T., Zhou, Z.-H., Huang, J.Z., Hu, X., Li, J., Xie, C., He, J., Zou, D., Li, K.-C., Freire, M.M. (eds.) PAKDD 2007. LNCS (LNAI), vol. 4819, pp. 35–44. Springer, Heidelberg (2007)

19. Tsai, F.S., Han, W., Xu, J., Chua, H.C.: Design and Development of a Mobile Peer-to-Peer Social Networking Application. Expert Syst. Appl. 36(8), 11077–11087 (2009)

A Random Indexing Approach for Web User Clustering and Web Prefetching

Miao Wan[1], Arne Jönsson[2], Cong Wang[1], Lixiang Li[1], and Yixian Yang[1]

[1] Information Security Center, State Key Laboratory of Networking and Switching Technology, Beijing University of Posts and Telecommunications, P.O. Box 145, Beijing 100876, China
[2] Department of Computer and Information Science, Linköping University, SE-581 83, Linköping, Sweden
wanmiao120@163.com, arnjo@ida.liu.se

Abstract. In this paper we present a novel technique to capture Web users' behaviour based on their interest-oriented actions. In our approach we utilise the vector space model Random Indexing to identify the latent factors or hidden relationships among Web users' navigational behaviour. Random Indexing is an incremental vector space technique that allows for continuous Web usage mining. User requests are modelled by Random Indexing for individual users' navigational pattern clustering and common user profile creation. Clustering Web users' access patterns may capture common user interests and, in turn, build user profiles for advanced Web applications, such as Web caching and prefetching. We present results from the Web user clustering approach through experiments on a real Web log file with promising results. We also apply our data to a prefetching task and compare that with previous approaches. The results show that Random Indexing provides more accurate prefetchings.

Keywords: Web user clustering, User behavior, Random Indexing, Web prefetching.

1 Introduction

Web users may exhibit various types of behaviours associated with their information needs and intended tasks when they are navigating a Web site. These behaviours are traced in Web access log files. Web usage mining [2], which is an important topic in behavior informatics [3] and a branch of Web Mining [1], captures navigational patterns of Web users from log files and has been used in many application areas [8,9,10,11].

Clustering in Web usage mining is used to group together items that have similar characteristics, and user clustering results in groups of users that seem to behave similarly when navigating through a Web site. In recent years, clustering users from Web logs has become an active area of research in Web Mining. Some standard techniques of date mining such as fuzzy clustering algorithms [4], first-order Markov models [5] and the Dempster-Shafer theory [6] have been introduced to model Web users' navigation behaviour and to cluster users based on

L. Cao et al. (Eds.): PAKDD 2011 Workshops, LNAI 7104, pp. 40–52, 2012.

Web access logs. Generally, these techniques capture stand alone user behaviours at the page view level. However, they do not capture the intrinsic characteristics of Web users' activities, nor quantify the underlying and unobservable factors associated with specific navigational patterns. Latent variable models, such as LSA [14], have been widely used to discover the latent relationship from web linkage information and can be utilized to find relevant web pages for improving web search efficiency [7].

Random Indexing [12] is an incremental word space model proposed as an alternative to LSA. Since 2000, it has been studied and empirically validated in a number of experiments and usages in distributional similarity problems [12,13]. However, few of the Random Indexing approaches have been employed into the field of Web mining, especially for the discovery of Web user access patterns.

In this paper we propose a Web user clustering approach to prefetch Web pages for grouped users based on Random Indexing (RI). Specially, segments split by "/" in the URLs are used as the units of analysis in our study. To demonstrate the usability of RI for user cluster detection, we also apply our methodology to a real prefetch task by predicting future requests of clustered users according to the common pages they accessed. Our clustering and prefetching approaches based on Random Indexing are compared to a popular Web user clustering method named FCMdd [4], and a new proposed clustering algorithm called CAS-C [9]. The experimental results show that the RI-based Web user clustering technique present more compact and well-separated clusters than FCMdd and CAS-C, and get higher prefetching accuracy as well.

2   Random Indexing (RI)

Random Indexing is an incremental word space model based on Kanerva's work on sparse distributed representations [12,15]. The basic idea of Random Indexing is to accumulate context vectors based on the occurrence of words in contexts. This technique can be used with any type of linguistic context, is inherently incremental, and does not require a separate dimension reduction phase. The Random Indexing technique can be described as a two-step operation:

Step 1. A unique d-dimensional *index vector* is assigned and randomly generated to each context (e.g. each document or each word). These index vectors are sparse, high-dimensional, and ternary, which means that their dimensionality (d) is on the order of hundreds, and that they consist of a small number(ϵ) of randomly distributed +1s and -1s, with the rest of the elements of the vectors set to 0. In our work each element is allocated with the following probability [13]:

$$\begin{cases} +1 & with \ probability \ \frac{\epsilon/2}{d} \\ 0 & with \ probability \ \frac{d-\epsilon}{d} \\ -1 & with \ probability \ \frac{\epsilon/2}{d} \end{cases}$$

Step 2. *Context vectors* are produced by scanning through the text. As scanning the text, each time a word occurs in a context, that context's d-dimensional index vector is added to the context vector for the word. Words are thus represented by d-dimensional context vectors that are the sum of the index vectors of all the contexts in which the word appears.

The Random Indexing technique produces context vectors by noting co-occurring events within a context window that defines a region of context around each word, and the number of adjacent words in a context window is called the context window size, l. For example, the term t_n in a '2+2' sized context window, c_m, is represented by:

$$c_m = [(w_{n-2})(w_{n-1})t_n(w_{n+1})(w_{n+2})].$$

Here $l = 2$, and the context vector of t_n in c_m would be updated with:

$$C_m = R(w_{n-2}) + R(w_{n-1}) + R(w_{n+1}) + R(w_{n+2}),$$

where $R(x)$ is the random index vector of x. This process is repeated every time we observe t_n in our data, adding the corresponding information to its existing context vector C. If the context c_m is encountered again, no new index vector will be generated. Instead the existing index vector for c_m is added to C to produce a new context vector for t_n.

3   Random Indexing Based Web User Clustering

The procedure of Web user clustering based on Random Indexing is illustrated in Figure 1 and will be outlined in more detail in this section.

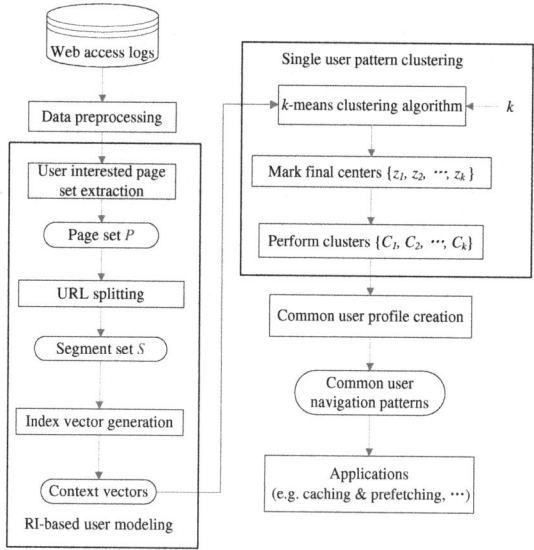

Fig. 1. Working flow of Web user clustering approach based on Random Indexing

3.1   Data Preprocessing

The first part of Web user cluster detection, called preprocessing, is usually complex and demanding. Generally, it comprises three domain dependent tasks: data cleaning, user identification, and session identification.

Data Cleaning. Depending on application and task, Web access logs may need to be cleaned from entry request pages. For the purpose of user clustering, all data tracked in Web logs that are useless, such as graphical page content (e.g. jpg and gif files), which are not content pages or documents, need to be removed.

User Identification. Identifying different users is an important issue of data preprocessing. There are several ways to distinguish individual visitors in Web log data which are collected from three main sources: Web servers, proxy servers and Web clients. In our experiments we use user identification from client trace logs because users in these logs are easily traced via different user IDs.

Session Identification. After individual users are identified, the next step is to divide each user's click stream into different segments, which are called sessions. Most session identification approaches identify user sessions by a maximum time-out. If the time between page requests exceeds a certain limit of access time, we assume a user is starting a new session. Many commercial products use 30 minutes as a default timeout [2]. We will also use 30 minutes in our investigations.

3.2   User Modelling Based on Random Indexing

After all the Web logs are preprocessed, the log data should be further analysed in order to find common user features and create a proper user model for user clustering.

Navigation set of Individual Users. Pages requested by a user only a very small period, such as one session, and not visited anymore, represent temporary user interest and are filtered out in this step. Pages with very low hit rates in the log files that only reflect the personal interest of individual users are also removed based on the pre-set hit number. This leaves us with a user interest page set $P = \{URL_1, URL_2, \ldots, URL_m\}$ composed of the remaining m requested URLs. Each element in P is successfully visited more than the pre-set number of times. Based on P we create a navigation set for individual users, $U = \{U_1, U_2, \ldots, U_n\}$, which contains pages requested by each user.

Random Indexing for Each User. The form of a webpage's URL can contain some useful information. According to the hierarchical structure of most Web sites, the URLs at different levels can be reflected in the sequence of segments split by "/". For example, `http://cs-www.bu.edu/faculty/gacs/courses/cs410/Home.html` may represent that it is the homepage of a course named "cs410" and this course is provided by someone called "gacs" who is a faculty of the department of computer science. Based on this assumption, we can split all the URLs in the user interest page set, P, by "/" and create a set of segments, S, which contains all the segments that have occurred in P. For each segment, a d-dimensional index vector s_i ($i = 1, 2, \ldots, m$, where m is the total number of segments) is generated. Then, as scanning the navigation set U, for each segment appearing in one user, update its zero-initialised context vector u_j ($j = 1, 2, \ldots, n$, where n is the total number of users) by adding the random index vectors of the segments in the context window where the size of the context window is pre-set. Finally, a set of individual users' navigation patterns, which forms a $n \times d$ matrix $A = \{u_1, u_2, \ldots, u_n\}^T$, is created with each row as the context vector u_j of each single user.

3.3 Single User Pattern Clustering

After random indexing of user's transaction data, the single user patterns in matrix A will be clustered by the k-means clustering algorithm. The k-means clustering algorithm [16] is a simple, conventional, clustering technique which aims to partition n observations into k clusters in which each observation belongs to the cluster with the nearest mean. It is a partition-based clustering approach and has been widely applied for decades.

4 Clustering Validity Measures

In order to evaluate the performance of the proposed clustering algorithm and compare it to other techniques we use two cluster validity measures. Since the cluster number should be pre-set for the clustering algorithms, in this paper we first use a relative criterion named SD to estimate the number of clusters for the clustering algorithms before we evaluate their performances. Furthermore, as the access log is an un-marked data set, we choose another validity measure, called CS, to evaluate the performance of different clustering algorithms without any prior knowledge of the data set.

- The SD index combines the average scattering for clusters and the total separation between clusters. For each k input, the $SD(k)$ is computed as

$$SD(k) = Dis(k_{max}) \cdot Scat(k) + Dis(k), \tag{1}$$

where k_{max} is the maximum number of input clusters and influences slightly on the value of SD [17].

$Scat$ is the average scattering within one cluster and is defined as:

$$Scat(k) = \frac{1}{k} \sum_{i=1}^{k} \frac{\|\sigma(C_i)\|}{\|\sigma(X)\|}, \tag{2}$$

where $\sigma(S)$ represents the variance of a data set S. X is the entire data set.

Dis is the total scattering (separation) between clusters and is given by the following equation:

$$Dis(k) = \frac{D_{max}}{D_{min}} \sum_{i=1}^{k} \left(\sum_{j=1}^{k} \|z_i - z_j\| \right)^{-1}, \tag{3}$$

where $D_{max} = max(\|z_i - z_j\|)$ and $D_{min} = min(\|z_i - z_j\|)$ $(\forall i, j \in 1, 2, 3, \ldots, k)$ are the maximum and minimum distances between cluster centers.

Experiments show that the number of clusters, k, which minimizes the SD index can be considered as an optimal value for the number of clusters present in the data set [17].

– The CS index computes the ratio of *Compactness* and *Separation*.

A common measure of *Compactness* is the intra-cluster variance within a cluster, named *Comp*:

$$Comp = \frac{1}{k} \sum_{i=1}^{k} \|\sigma(C_i)\|. \tag{4}$$

Separation is calculated by the average of distances between the centers of different clusters:

$$Sep = \frac{1}{k} \sum \|z_i - z_j\|^2, \quad i = 1, 2, \ldots, k-1, \quad j = i+1, \ldots, k. \tag{5}$$

It is clear that if the dataset contains compact and well-separated clusters, the distance between the clusters is expected to be large and the diameter of the clusters is expected to be small. Thus, cluster results can be compared by taking the ratio between *Comp* and *Sep* :

$$CS = \frac{Comp}{Sep}. \tag{6}$$

Based on the definition of CS, we can conclude that a small value of CS indicate compact and well-separated clusters.

5 Experiments

In this section, we present an experiment using the RI-based Web user clustering approach, and give a detailed investigation of results using the method. We use MatLab for our experiments.

The results produced by our Web user clustering algorithm can be used in various ways. In this section we will illustrate how it can be used for prefetching and caching, which means that URL objects can be fetched and loaded into the Web server cache before users request them.

5.1 Preprocessing of Data Source

The data source for the Web user clustering algorithm is the Web site access log of the Computer Science department at Boston University [18] which is available at The Internet Traffic Archive [19]. It contains a total of 1,143,839 requests for data transfer, representing a population of 762 different users.

We use the part of the logs during the period of January and February 1995. According to the item of 'user id' in the log data, we selected 100 users in the step of user identification. After access log preprocessing, we get 1005 sessions from these 100 users. The User IDs are renumbered, and each one of them have been assigned an identification number between 1 and 100.

5.2   Parameter Setting Investigations

In this subsection we present results from our investigations on the impacts of some key parameters and assign initial values for them.

Cluster Number. First we need to find the proper k values for each clustering algorithm in our work.

We have conducted k-means clustering experiments by measuring SD values for various values of k. The experiment is performed 50 times with 9 different values of k (from 2 to 10) and 6 different dimensions (from $d = 100$ to 600). The results of the k-investigations are given in Figure 2.

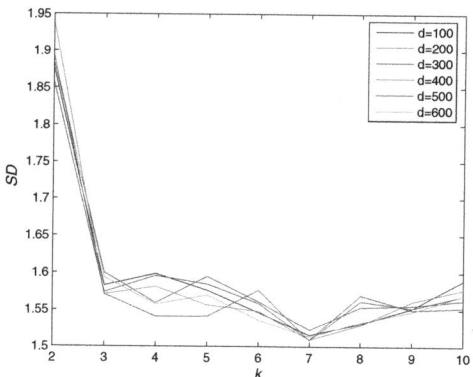

Fig. 2. Comparisons of SD measures for k-means clustering in RI-based tasks

As seen in Figure 2, the SD index is marginally influenced by the dimension of the index vectors. The distribution of SD values for different values of d is similar for most user pattern matrixes and the minimum SD is found at the same k for all dimension settings. Thus, we set $k = 7$ as the optimal cluster number for the k-means clustering algorithm.

We perform similar experiments for the FCMdd and CAS-C algorithms as depicted in Figures 3 and 4. We use $k = 8$ for FCMdd and $k = 7$ for CAS-C.

Dimensionality. In order to evaluate the effects of increasing the dimensionality to the performance of Random Indexing in our work, we computed the values of CS with d ranging from 50 to 600. The performance measurers are reported using average values over 30 different runs. The results are depicted in Figure 5.

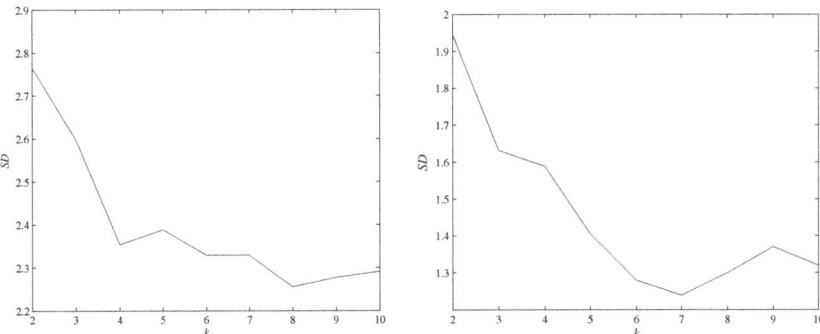

Fig. 3. Comparison of SD for FCMdd in Web user clustering

Fig. 4. Comparison of SD for CAS-C in Web user clustering

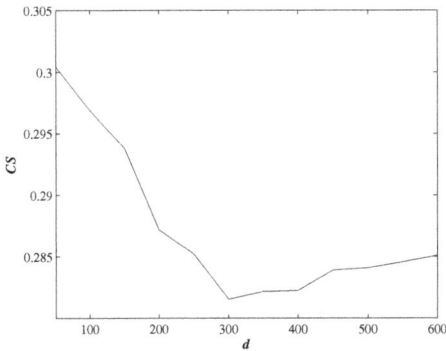

Fig. 5. The influence of various d values to RI-based Web user clustering

From Figure 5 we can see that for $d = 300$ we have the smallest CS. Thus, $d = 300$ is chosen as the dimension of the index vectors used by Random Indexing.

Other Parameters. Two more parameters need values: the number of +1s and -1s in the index vector, ϵ, and the context window size, l. We will use $\epsilon = 10$ as proposed by [20] and $l = 1$ as the URLs are rather short.

To summarise, we will use initial values of parameters in our RI-based experiments as presented in Table 1.

Table 1. Parameter selection in the RI-based approach

k	d	l	ϵ
7	300	1	10

5.3   Common User Profile Creation

After the log data are processed by Random Indexing, the single user navigation pattern matrix A will be clustered by the k-means algorithm.

The RI-based Web user clustering is compared to that generated using FCMdd and CAS-C. Table 2 presents the values of CS for different clustering techniques.

Table 2. Values of CS for different clustering approaches

Methods	k	$Comp$	Sep	CS
$FCMdd$	8	2.2401	5.8465	0.3832
$CAS - C$	7	2.2380	7.1574	0.3127
RI	7	2.1978	7.8054	0.2816

As shown in Table 2, the RI-based clustering algorithm gets the smallest $Comp$ with the largest Sep, and of course, the best CS value. We can conclude that the RI-based approach performs better than FCMdd and CAS-C for clustering Web users.

Based on the clustering results, we build a common user profile as seen in Table 3.

5.4   Prefetching for User Groups

Web caching and Web prefetching are two important techniques used to reduce the noticeable response time perceived by users [21]. For an effective prefetching scheme, there should be an efficient method to predict users' requests and proper prefetching and caching strategies. Various techniques, including Web Mining approaches [22,23], have been utilised for improving the accuracy of predicting user access patterns from Web access logs, making the prefetching of Web objects more efficient. Most of these techniques are, however, limited to predicting requests for a single user only. Predicting groups of users interest have caught little attention in the area of prefetching.

Prefetching Rule. According to the common user profiles created by the RI-based technique, FSMdd and CAS-C, we set up prefetching experiments to prefetch URL requests for users in each cluster. The prefetch rule is defined as follows:

For each cluster, let $P = \{p_1, p_2, \ldots, p_m\}$ be a set of Web pages in the Web server. In this paper, the prefetch rule is defined as an implication of the form $\{p_1, p_2, \ldots, p_i\} \xrightarrow{c} \{q_1, q_2, \ldots, q_j\}$, where $P_1 = \{p_1, p_2, \ldots, p_i\}$ is the page set that users requested in January and February, $P_2 = \{q_1, q_2, \ldots, q_j\}$ is the page set to be prefetched in March, $P_2 \subseteq P_1 \subseteq P$, and c is the portion (or ratio) of users who have requested P_2 in January and February. We use $c = 0.5$ [9] for our prefetch task, which means that more than or equal to 50% of the users pages in one cluster that have been requested in January and February will be prefetched for March.

Table 3. Common user profile created by Web user clustering algorithm using the RI-based approach. The CN column represents the cluster number.

CN	Members	Common user requests
1	4, 19, 33, 40, 67, 90	cs-www.bu.edu/, cs-www.bu.edu/courses/Home.html, cs-www.bu.edu/faculty/heddaya/CS103/HW/1.html, cs-www.bu.edu/faculty/heddaya/CS103/HW/2.html, cs-www.bu.edu/faculty/heddaya/CS103/Home.html, cs-www.bu.edu:80/, sundance.cso.uiuc.edu/Publications/Other/Zen/zen-1.0_toc.html, www.ncsa.uiuc.edu/SDG/Software/Mosaic/StartingPoints/ NetworkStartingPoints.html, www.ncsa.uiuc.edu/demoweb/url-primer.htm
2	6, 61, 71, 83	cs-www.bu.edu/, cs-www.bu.edu/courses/Home.html, cs-www.bu.edu/staff/Home.html, cs-www.bu.edu/staff/TA/biddle/www/biddle.html, cs-www.bu.edu/staff/TA/dmc/www/dmc.html, cs-www.bu.edu/staff/TA/joyceng/home.html, cs-www.bu.edu/staff/people.html, cs-www.bu.edu:80/
3	7, 8, 11, 21, 26, 30, 66, 89	cs-www.bu.edu/, cs-www.bu.edu/courses/Home.html, cs-www.bu.edu/pointers/Home.html, cs-www.bu.edu/students/grads/Home.html, cs-www.bu.edu/students/grads/oira/Home.html, cs-www.bu.edu/students/grads/oira/cs112/hmwrk1.html, cs-www.bu.edu/students/grads/oira/cs112/hmwrk2.html, cs-www.bu.edu/students/grads/oira/cs112/hmwrkgd.html, cs-www.bu.edu/students/grads/oira/cs112/node1.html, cs-www.bu.edu:80/, cs-www.bu.edu:80/students/grads/oira/cs112/
4	1, 9, 12, 17, 24, 25, 31, 34, 35, 42, 50, 59, 72, 76, 78, 81, 82, 84, 97, 99, 100	cs-www.bu.edu/, cs-www.bu.edu/courses/Home.html, cs-www.bu.edu:80/
5	10, 44, 56, 74, 87, 93	cs-www.bu.edu/, cs-www.bu.edu/courses/Home.html, cs-www.bu.edu/faculty/Home.html, cs-www.bu.edu/students/grads/Home.html, cs-www.bu.edu/students/grads/tahir/CS111/, cs-www.bu.edu:80/
6	2, 3, 5, 14, 16, 20, 22, 23, 27, 28, 29, 32, 36, 37, 38, 39, 41, 43, 45, 46, 47, 48, 49, 51, 52, 53, 54, 55, 57, 58, 60, 62, 63, 64, 68, 69, 70, 73, 75, 77, 79, 80, 85, 86, 91, 92, 94, 95, 96, 98	cs-www.bu.edu/, cs-www.bu.edu/courses/Home.html, cs-www.bu.edu/students/grads/tahir/CS111/
7	13, 15, 18, 65, 88	cs-www.bu.edu/, cs-www.bu.edu/faculty/Home.html, cs-www.bu.edu/faculty/crovella/Home.html, cs-www.bu.edu/faculty/crovella/courses/cs210/, cs-www.bu.edu/faculty/crovella/courses/cs210/hwk1/hwk1.html, cs-www.bu.edu/faculty/crovella/courses/cs210/reading.html, cs-www.bu.edu/pointers/Home.html , cs-www.bu.edu:80/, cs-www.bu.edu:80/faculty/crovella/courses/, cs-www.bu.edu:80/faculty/crovella/courses/cs210/

Results and Analysis. Four parameters are used to investigate the performance of our prefetching task: (1) *hits* which indicate the number of URLs that are requested from the prefetched URLs, (2) *precision* which is the ratio of hits to the number of URLs that are prefetched, (3) *recall* which is the ratio of hits to the number of URLs that are requested, and (4) $F_{0.5}$ which considers both the precision and the recall to test the accuracy. Since our prefetch strategy only predicts common URLs within one user cluster, we can't make sure that all requests from a single user are prefetched. Therefore precision is valued higher than recall for prefetching. As a result, we choose $F_{0.5}$ to measure the prefetching accuracy which weights precision twice as much as recall.

Table 4 gives the overall experimental comparison of prefetching for FCMdd, CAS-C and Random Indexing.

Table 4. Overall result comparisons for FCMdd, CAS-C and RI-based prefetchings

Algorithms	Number of cluster detected	Overall precision	Overall recall	$F_{0.5}$
FCMdd	8	0.7039	0.4434	0.6299
CAS-C	7	0.7062	0.4168	0.6201
RI	7	0.7540	0.5311	0.6956

Comparing the top two lines to the last row of Table 4, we can see that the results in the proposed prefetching task achieve a total average precision of 75.40% and a total recall of 53.11%, which are all higher than 70.62% of CAS-C and 44.34% using FCMdd. The $F_{0.5}$ value from RI, 0.6956, is therefore larger than 0.6201 of CAS-C and 0.6299 of FCMdd. We can thus conclude that prefetching based on Random Indexing provides a user request prediction that is better than using FCMdd or CAS-C.

To summarize, Random Indexing can improve the quality of user request prediction and perform better results than FCMdd and CAS-C.

6   Conclusions

This paper focuses on discovering latent factors of user browsing behaviours based on Random Indexing and detecting clusters of Web users according to their activity patterns acquired from access logs. Experiments are conducted to investigate the performance of Random Indexing in Web user clustering tasks. The experimental results show that the proposed RI-based Web user clustering approach could be used to detect more compact and well-separated user groups than previous approaches. Based on common profiles of detected clusters, prediction and prefetching user requests can be done with encouraging results.

Acknowledgments. This work is supported by the National Basic Research Program of China (973 Program) (Grant No. 2007CB311203), the National Natural Science Foundation of China (Grant No. 60805043), the Natural Science Foundation of Beijing, China (Grant No. 4092029), Huo Ying-Dong Education Foundation of China (Grant No. 121062), Specialized Research Fund for the Doctoral Program of Higher Education (No. 20100005110002) and Santa Anna IT Research Institute.

References

1. Etzioni, O.: The world-wide Web: quagmire or gold mine? Communications of the ACM 39(11), 65–68 (1996)
2. Cooley, R., Mobasher, B., Srivastava, J.: Data preparation for mining world wide web browsing patterns. J. Knowl. Inf. Syst. 1(1), 5–32 (1999)
3. Cao, L.: In-depth Behavior Understanding and Use: the Behavior Informatics Approach. Information Science 180(17), 3067–3085 (2010)
4. Krishnapuram, R., Joshi, A., Nasraoui, O., Yi, L.: Low-complexity fuzzy relational clustering algorithms for web mining. IEEE Transaction of Fuzzy System 4(9), 596–607 (2003)
5. Cadez, I., Heckerman, D., Meek, C., Smyth, P., Whire, S.: Visualization of Navigation Patterns on a Website Using Model Based Clustering. Technical Report MSR-TR-00-18, Microsoft Research (March 2002)
6. Xie, Y., Phoha, V.V.: Web User Clustering from Access Log Using Belief Function. In: Proceedings of K-CAP 2001, pp. 202–208 (2001)
7. Hou, J., Zhang, Y.: Effectively Finding Relevant Web Pages from Linkage Information. IEEE Trans. Knowl. Data Eng. 15(4), 940–951 (2003)
8. Paik, H.Y., Benatallah, B., Hamadi, R.: Dynamic restructuring of e-catalog communities based on user interaction patterns. World Wide Web 5(4), 325–366 (2002)
9. Wan, M., Li, L., Xiao, J., Yang, Y., Wang, C., Guo, X.: CAS based clustering algorithm for Web users. Nonlinear Dynamics 61(3), 347–361 (2010)
10. Berendt, B.: Using site semantics to analyze, visualize, and support navigation. Data Mining and Knowledge Discovery 6(1), 37–59 (2002)
11. Ansari, S., Kohavi, R., Mason, L., Zheng, Z.: Integrating e-commerce and data mining: Architecture and challenges. In: Proceedings of ICDM 2001, pp. 27–34 (2001)
12. Kanerva, P., Kristofersson, J., Holst, A.: Random Indexing of text samples for Latent Semantic Analysis. In: Proceedings of the 22nd Annual Conference of the Cognitive Science Society, p. 1036 (2000)
13. Sahlgren, M., Karlgren, J.: Automatic bilingual lexicon acquisition using Random Indexing of parallel corpora. Journal of Natural Language Engineering, Special Issue on Parallel Texts 6 (2005)
14. Landauer, T., Dumais, S.: A solution to Plato problem: the Latent Semantic Analysis theory for acquisition, induction and representation of knowledge. Psychological Review 104(2), 211–240 (1997)
15. Kanerva, P.: Sparse distributed memory. The MIT Press, Cambridge (1988)
16. MacQueen, J.: Some Methods for Classification and Analysis of Multivariate Observations. In: Proceedings of the 5th Berkeley Symposium on Mathematical Statistics and Probability, pp. 281–297 (1967)
17. Halkidi, M., Vazirgiannis, M., Batistakis, Y.: Quality Scheme Assessment in the Clustering Process. In: Zighed, D.A., Komorowski, J., Żytkow, J.M. (eds.) PKDD 2000. LNCS (LNAI), vol. 1910, pp. 265–276. Springer, Heidelberg (2000)
18. Cunha, C.A., Bestavros, A., Crovella, M.E.: Characteristics of WWW Client Traces, Boston University Department of Computer Science, Technical Report TR-95-010 (April 1995)
19. The Internet Traffic Archive. http://ita.ee.lbl.gov/index.html
20. Gorman, J., Curran, J.R.: Random indexing using statistical weight functions. In: Proceedings of EMNLP 2006, pp. 457–464 (2006)

21. Teng, W., Chang, C., Chen, M.: Integrating Web Caching and Web Prefetching in Client-Side Proxies. IEEE Trans. Parallel Distr. Syst. 16(5), 444–455 (2005)
22. Nanopoulos, A., Katsaros, D., Manolopoulos, Y.: Effective Prediction of Web-User Accesses: A Data Mining Approach. In: Proceeding of Workshop WEBKDD (2001)
23. Wu, Y., Chen, A.: Prediction of web page accesses by proxy server log. World Wide Web 5, 67–88 (2002)

Emotional Reactions to Real-World Events
in Social Networks

Thin Nguyen*, Dinh Phung, Brett Adams, and Svetha Venkatesh

Curtin University
thin.nguyen@postgrad.curtin.edu.au,
{d.phung,b.adams,s.venkatesh}@curtin.edu.au

Abstract. A convergence of emotions among people in social networks is potentially resulted by the occurrence of an unprecedented event in real world. E.g., a majority of bloggers would react *angrily* at the September 11 terrorist attacks. Based on this observation, we introduce a *sentiment index*, computed from the *current mood* tags in a collection of blog posts utilizing an affective lexicon, potentially revealing subtle events discussed in the blogosphere. We then develop a method for extracting events based on this index and its distribution. Our second contribution is establishment of a new bursty structure in text streams termed a *sentiment burst*. We employ a stochastic model to detect bursty periods of moods and the events associated. Our results on a dataset of more than 12 million mood-tagged blog posts over a 4-year period have shown that our sentiment-based bursty events are indeed meaningful, in several ways.

Keywords: Emotional reaction, sentiment index, sentiment burst, bursty event.

1 Introduction

Social media is a new type of media where readers, along with their conventional roles as information consumers, can be publishers, editors, or commentators. The blogosphere, a popular representative of the new decentralized and collaborative media model, is where people can participate, express opinions, mediate their own content, and interact with other users. The user-generated content in social media tends to be more subjective than other written genres. This opens new opportunities for studying novel pattern recognition approaches based on sentiment information such as those initially reported in [4,7].

Research in opinion mining and sentiment analysis, which commonly deals with the subjective and opinionated part of data, has recently attracted a great deal of attention [10]. Sentiment information in social media has been used to explain real world settings, such as mapping the proportion of *anxious* posts in Livejournal with the trend of the S&P 500 index in stock market [6], or predicting the 2009 German election based on political sentiment contained in Twitter [13].

Due to the growing scale of the blogosphere, detecting bursty events in the media is an emerging need and important topic in behavior informatics [3]. Existing work

---

* Corresponding author.

L. Cao et al. (Eds.): PAKDD 2011 Workshops, LNAI 7104, pp. 53–64, 2012.

has looked at simple burst detection of a term in a collection of documents, including the popular state-machine developed by Kleinberg [9] (KLB). Kleinberg observes that certain topics in his emails may be more easily characterized by a sudden increase in message sending, rather than by textual features of the messages themselves, and he calls these high intensity periods of sending *bursts*. A term is in bursty time when it grows in intensity for a period.

In this paper, we build a *sentiment index* based on the moods tagged in Livejournal posts, which can potentially be used as important sentimental signals that lead up to detection of important events. The distribution of this index over certain cyclic periods also helps learning periodic events discussed in the online diaries or bloggers' habits. We assume that bursty irregularities in occurrences of certain sentiment patterns in the blogosphere provide strong hints about significant events in the sphere. Thus, we utilize the bursty detection algorithm KLB [9] to detect bursty moods in a large-scale collection of blog posts and the corresponding events over a period of four years.

Our contribution in this paper is twofold. First, we construct a novel scheme of sentiment indices for extracting events in the blogosphere. Second, we introduce the concept of *sentiment burst* and employ the stochastic model KLB for detecting this structure in text streams. This provides the basis for detecting associated bursty events. To our knowledge, we are the first to consider this form of sentiment-based burst detection.

The rest of this paper is organized as follows. In Section 2, we propose a sentiment index as a signal for event detection in blogosphere. Section 3 presents the results of bursty pattern discovery of moods and related facts and figures, and is followed by concluding remarks.

2   Sentiment Index and Event Indicators

Here, we describe a method which uses the emotion information in the '*current mood*' tag assigned to a blog post to discover significant events mentioned in the blogosphere. For the rest of the paper, we will use data extracted from Livejournal[1] for our analysis, but the method is applicable to similar datasets. Our dataset includes more than 12 million blog posts tagged with mood labels from a set of 132 moods predefined by Livejournal. These posts were created between 1 May 2001 – 31 December 2004.

2.1   Sentiment Index for Event Detection

Livejournal allows users to tag a new post with their *current mood*. Its set of 132 predefined moods covers a wide spectrum of emotion. Examples are *cheerful* or *grateful* to reflect happiness, *discontent* or *uncomfortable* for sadness, and so on. Figure 1 contains a tag-cloud of moods in our dataset. A sudden increase in tagging of certain moods could be due to a real-world event. E.g., a furry of *angry* posts might occurs following 9/11 event. To detect the trend of mood patterns used in a period, which is termed *sentiment index*, we summarize the emotion quantity of the moods in the period.

One widely accepted emotion measure used by psychologists in text analysis is *valence*, which indicates the level of *happiness* a word conveys. To measure the valence

---

[1] http://www.livejournal.com

Fig. 1. A cloud visualization of moods tagged in the blog dataset

value a mood label conveys, we adapt the affective norms for English words (ANEW) [2], which contains 1034 English words with assigned values for *valence* and *arousal*. These two dimensions are also used in the circumplex model of affect [11,12], where emotion states are conceptualized as combinations of these two factors. We then use the valence values as building blocks to compute the sentiment index for extraction of events. Those moods which are not in the ANEW lexicon are assigned the valences of their siblings or parents from Livejournal's shallow mood taxonomy[2].

Formally, the sentiment index is defined as follows. Denote by $\mathcal{B} = \{\mathcal{B}_1, \ldots, \mathcal{B}_T\}$ the collection of the blog dataset, where $\mathcal{B}_t$ denotes the set of blog posts arriving at day t, and T is the total days of the corpus. Denote by $\mathcal{M} = \{sad, happy,...\}$ the set of moods predefined by Livejournal. Each blogpost $b \in \mathcal{B}$ in the corpus is labeled with a mood $l_b \in \mathcal{M}$. Denote by v_m the valence value of the mood $m \in \mathcal{M}$. We then formulate the sentiment index I_t at the day t^{th} as

$$I_t = \frac{\sum v_{l_b} | b \in \mathcal{B}_t}{|\mathcal{B}_t|}$$

The average sentiment index $\overline{I}$ is computed as

$$\overline{I} = \frac{\sum_{t=1}^{T} I_t}{T}$$

We define $\delta_t = I_t - \overline{I}$, the sentiment deviation from the sentiment mean for day t^{th}. A plot of this value for the corpus period is shown in Figure 2a.

We conjecture that extreme deviations δ_t can potentially lead to the discovery of important events. Thus, for each year we examine the time when the sentiment index reaches its maximum or minimum. We extract the time when the extremes occur as signalling events to be derived. To get insights into the content of these events, we retrieve all the blog posts during the corresponding periods to form a collection of

---

[2] http://www.livejournal.com/moodlist.bml

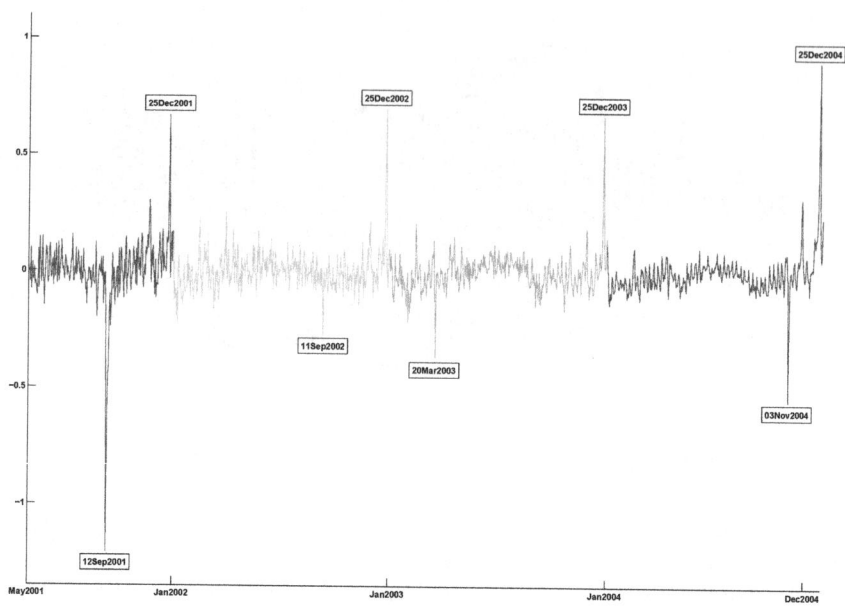

(a) Sentiment deviation.

Year	Events extracted by our proposed sentiment index	Entities extracted by our proposed sentiment index	CNN reports on top stories
2001	class america york center planes buildings war second pentagon trade hit watching school crashed towers tower plane world tv building	America NYC U.S. US Washington American New York City WTC DC Pennsylvania Pearl Harbor Manhattan Pentagon Pittsburgh United States CNN World Trade Center Bush New York	The terrorist attacks (*ranked first*)
2002	country target died remember america lives lost september watching war family bless 11th world families attacks american tv plane	USA Middle East Afghanistan Pennsylvania United States WTC Solution Bush New York Pentagon Iraq America U.S. American US	9/11 anniversary (*ranked tenth*)
2003	support saddam american world president iraq nation of america free bush peace country war iraq troops	IRAQ AMERICAN SADDAM BUSH US AMERICA	War in Iraq (*ranked first*)
2004	election vote years bush ohio country kerry win votes president voting voted	Kerry US Iraq American United States John Kerry Bush America	2004 election (*ranked first*)

(b) The events detected for the days of lowest sentiment.

Fig. 2. Sentiment deviation and the events detected in the lowest sentiment days

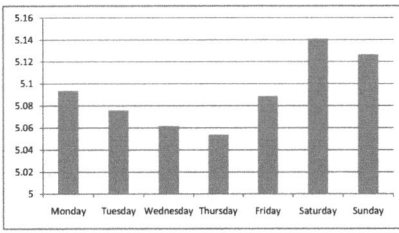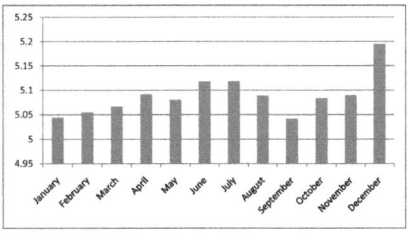

Fig. 3. Weekly and annual patterns of sentiment indices

documents, upon which topic extraction or NLP techniques can be applied to describe its content.

We use Latent Dirichlet Allocation (LDA) [1] to infer latent topics. Since exact inference is intractable for LDA, we use Gibbs sampling, proposed in [8], for learning blogpost–topic distribution. To extract entities corresponding to *person*, *location*, and *organization* in the text, we use the Stanford Named Entity Recognizer (Stanford NER) [5].

Our results show that the highest valence sentiment at the annual granularity reoccurs on the 25th December – Christmas Day. The top probability topics learnt by LDA from the set of related posts in these days also dominantly mention Christmas. These results are quite intuitive, as one would expect many bloggers would be '*happy*' during this period of the year.

In contrast, we find that all the lowest sentiment days are in the time of sobering events. Both topics and entities (Figure 2b) returned from running LDA and Stanford NER on the blog posts on these days mention the top stories reported by CNN[3]. Notably, all the extracted events are ranked first in the CNN lists, except the event detected in 2002 which is ranked tenth.

To examine periodic events or bloggers' behaviors, we aggregate distributions of the valence values by weekdays and by months. As shown in Figure 3, the sentiment index is lowest on *Thursday* and highest on *weekends*. When looking at monthly patterns, *December* is top in the index in a month of many celebrated days, such as Christmas and New Year, while the hole in *September* is caused by 9/11.

2.2 Event Indicators

In Section 2.1, we have considered sentiment in the blogosphere as a whole, aggregating over all moods, to detect sentiment-based events at a global level. However, each individual mood might have its own implications for detecting events. For example, a big drift in the mood '*shocked*' might indicate a catastrophic event, unlike a drift in the mood '*curious*'. Also, some moods might have cyclic emerging patterns over time, e.g. '*thankful*' on Thanksgiving Day or '*loved*' on Valentine's Day, while others may

---

[3] http://edition.cnn.com/SPECIALS/2001/yir/, http://edition.cnn.com/SPECIALS/2002/yir/, http://edition.cnn.com/SPECIALS/2003/yir/, http://edition.cnn.com/SPECIALS/2004/yir/

20000
18000
16000
14000
12000
10000
8000
6000
4000
2000
0
Monday
Tuesday
Wednesday
Thursday
Friday
Saturday
Sunday
DRUNK
WORKING
STRESSED
BUSY
RUSHED

0.3
0.25
0.2
0.15
0.1
0.05
0
Jan-02
Mar-02
May-02
Jul-02
Sep-02
Nov-02
Jan-03
Mar-03
May-03
Jul-03
Sep-03
Nov-03
Jan-04
Mar-04
May-04
Jul-04
Sep-04
Nov-04
COLD
HOT
SICK

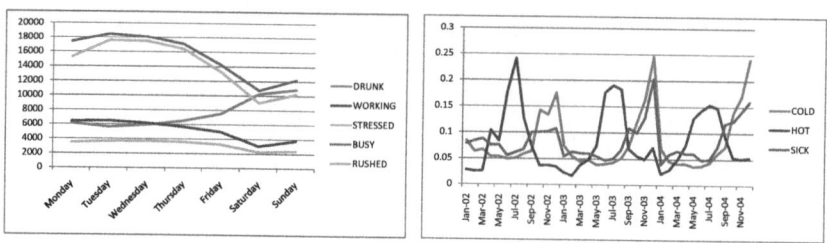

Fig. 4. The set of the lowest entropy moods. Left: The numbers of being tagged by weekdays; Right: The monthly proportion of being tagged.

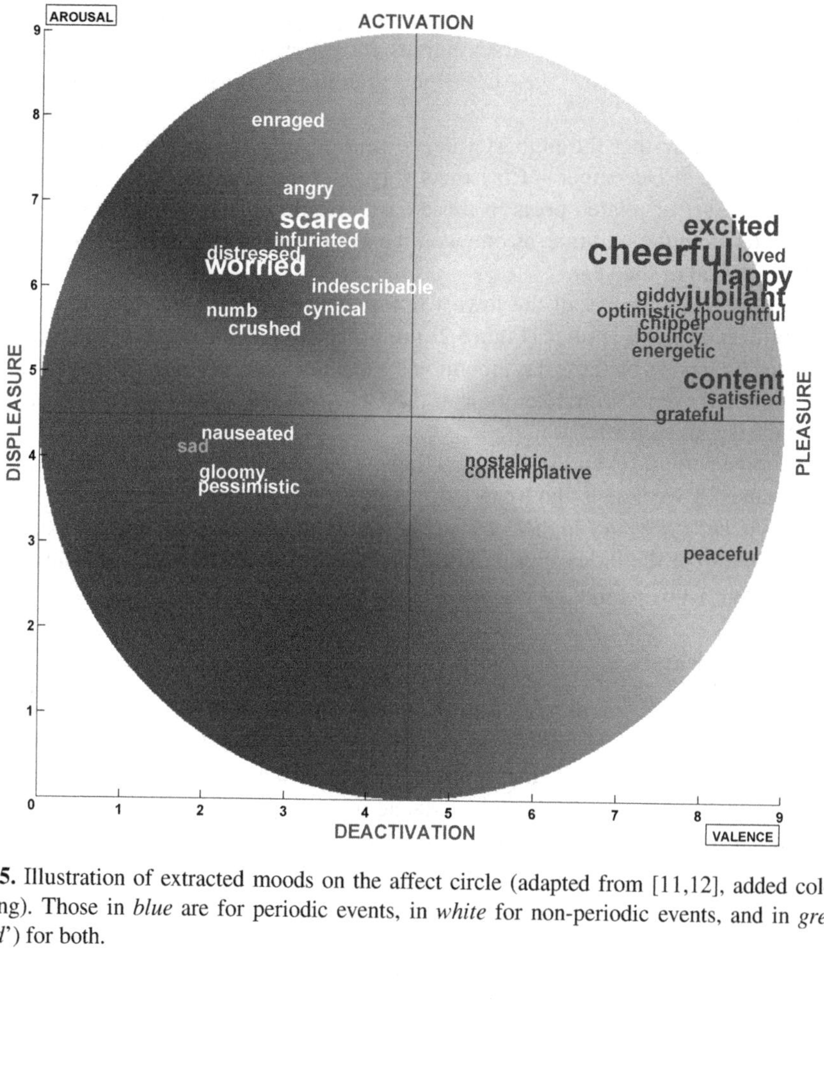

AROUSAL
ACTIVATION
DEACTIVATION
DISPLEASURE
PLEASURE
VALENCE
enraged
angry
scared
infuriated
distressed
worried
indescribable
numb
cynical
crushed
excited
cheerful
loved
happy
giddy
jubilant
optimistic
thoughtful
chipper
bouncy
energetic
content
satisfied
grateful
nauseated
sad
gloomy
pessimistic
nostalgic
contemplative
peaceful

Fig. 5. Illustration of extracted moods on the affect circle (adapted from [11,12], added color-coding). Those in *blue* are for periodic events, in *white* for non-periodic events, and in *green* ('*sad*') for both.

Table 1. The set of moods found to reach their peaks during the time of periodic events

Day	Mood	Event
15 February	loved	Valentine's Day
12 September	sad	9/11 event
24 December	excited	Christmas
25 December	cheerful, chipper, giddy, grateful, happy, jubilant, peaceful	Christmas
26 December	content, satisfied	Christmas
31 December	bouncy, contemplative, energetic, nostalgic, optimistic, thoughtful	New Year

Table 2. The set of moods found to reach their peaks during the time of aperiodic events

Day	Mood	Event
11 Sep 2001	pessimistic, scared	The 9/11 attacks
12 Sep 2001	angry, crushed, distressed, enraged, indescribable, infuriated, nauseated, numb, sad, worried	The 9/11 attacks
11 Sep 2002	cynical, gloomy, indescribable, sad	9/11 anniversary
20 Mar 2003	scared, worried	Iraq war
03 Nov 2004	angry, crushed, cynical, distressed, enraged, gloomy, infuriated, nauseated, numb, pessimistic, scared, worried	2004 US election

be increasingly tagged when non-periodic events happen, e.g., '*shocked*' for the 9/11 event. Motivated by this, we wish to detect which moods can potentially be used as good indicators for periodic and aperiodic events, in a data-driven manner.

Indicative Moods for Periodic Events or Habits. We conjecture that a periodic event or bloggers' habits can cause a cyclic rise of tagging certain moods. Thus, those moods having low entropy – computed on the distribution of times they are tagged by a cycle, e.g. weekdays or months – potentially indicate periodic event patterns.

On the weekday case, we denote by $x^m = \{x_1^m, \ldots, x_7^m\}$ the numbers of blog posts tagged with the mood m on {*Monday,...,Sunday*} over the period. This vector is normalized so that $\sum_{i=1}^{7} x_i^m = 1$. Then, the entropy for this proportion is computed by

$$\mathcal{H}(m) = -\sum_{i=1}^{7} x_i^m \log_2(x_i^m)$$

The data used in this analysis ranges from 01 May 2001 to 27 December 2004 (191 weeks). The five lowest entropy moods detected are shown in Figure 4. '*Busy*' and '*stressed*' are quite similar in tagging distribution, peaking on Tuesday and gradually decreasing for the following days ('*Tuesday at 11:45 is most stressful time of the week*'[4]). '*Working*' and '*rushed*' are mostly used at the beginning of the week. '*Drunk*'

---

[4] http://www.telegraph.co.uk/news/newstopics/howaboutthat/
5113653/Tuesday-at-1145-is-most-stressful-time-of-the-week-
survey-suggests.html

is far different with the others since it is at the holes during early of the week, reaching the peaks at weekends.

In the same way, we detect for which months a given mood is increasingly tagged and infer periodical events with respect to the months. The data for this experiment ranges from 01 January 2002 to 31 December 2004 (36 months). As shown in Figure 4, while quite similar to '*sick*' tagging distribution ('*Cold noses reduce ability to fight virus attacks*'[5]), '*cold*' alternates with '*hot*' across a year.

We also hypothesize that annual events can lead to a sudden increase in tagging certain moods. For example, as shown in Table 1, '*loved*' is found in top tagging on Valentine's Day, or '*optimistic*' for New Year. Mostly these moods have valence larger than 7, implying *happiness*, located on the right of the affect circle [11,12], as shown in Figure 5. '*Sad*' is also found in its top tagging repeatedly on the 12th September of the year, but for an irregular event – the 9/11 attacks.

Indicative Moods for Non-periodic Events. In Section 2.1, we observe that all of the non-periodic important events happen in the days of lowest sentiment indices. This might be explained by a sudden tagging of low valence moods in the days. In this section, we wish to detect moods which potentially indicate non-periodic events.

As shown in Table 2, fourteen moods are found at their maximum when non-periodic events, which are detected in Section 2.1, happen. Two of them, '*scared*' and '*worried*', match three events (they both miss the event in 2002 – 9/11 anniversary). The others match at least two events.

Different to most moods indicative of periodic events, these moods have extremely low *valence*, below 3.21, and many of them are high in *arousal*. These moods are situated on the left of the affect circle, implying *sadness* (Figure 5).

3   Mood-Based Burst Detection and Bursty Event Extraction

In Section 2, we have detected the start time of an event. In this section, we use the algorithm (KLB) [9] to discover bursty intervals of moods, and thus infer bursty periods of related events.

3.1   Bursty Event Detection

The essential idea of KLB is to model bursting as a generative process using a finite state-machine. Denote by s the state of the automaton, in the simple two-state model, one state is responsible for generating blog posts during non-burst period ($s = 0$) and another state is when the burst occurs ($s = 1$). Assume that there are n time points in our corpus, r_t is the number of blog posts tagged with a given mood at time t out of a total of d_t. The emission probability of the pair $\{r_t, d_t\}$ is modeled as a Binomial distribution together with a state-transition cost. A state sequence $\mathbf{q} = (q_1, ..., q_n)$ of minimum total cost can be computed by dynamic programming. We refer readers to [9] for details of the algorithm.

---

[5] http://edition.cnn.com/2005/HEALTH/11/14/cold.chill/index.html

Fig. 6. A cloud of bursty moods

Table 3. The set of moods found bursty in the 9/11 event and the related topics found in the blog posts tagged with the bursty moods in the bursty periods

Mood	Bursty period	Top topic
P*ssed off	11Sep – 13Sep	world trade died center war country terrorists lost united states lives american die innocent attack government buildings live middle pentagon
Shocked	11Sep – 15Sep	world country families family lost lives america shock tragedy trade innocent pray children events prayers die victims horrible scared ones
Angry	11Sep – 16Sep	world plane planes center trade crashed towers pentagon tower twin scared tuesday low canton hour wtc hijacked tv building hit
Sympathetic	11Sep – 19Sep	world lost pray families lives goes peace country family wish terrible died stop victims stand helping terrorism horrible loss tragedy
Worried	11Sep – 21Sep	war world big america city end lost country lives president pray nation ones bush scared heard died scary planes fact
Enraged	11Sep – 25Sep	lost heart family nation center wtc pentagon planes buildings war prayers act lives thoughts trade lucky reason race crashed hell

A mood is considered bursty over a period if in the interval the corresponding state sequence is in high state. From the bursty intervals of given moods, we study if any bursty events are associated. We retrieve the blog posts tagged with the bursty moods in the bursty time to learn the events correlated. We use LDA for learning topics discussed in the content, as in Section 2.

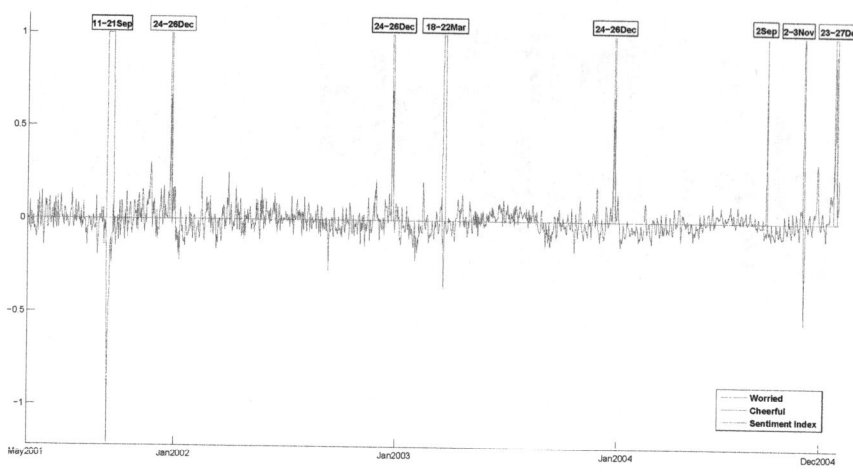

Fig. 7. Examples of moods found bursty throughout the time of the peaks and the holes of the sentiment index

3.2 Experimental Results

We apply a two-state KLB automaton on the corpus to detect which moods are bursty and their bursty periods. 76 moods are detected bursty, resulting in 298 bursty intervals. '*Drunk*', '*cold*', and '*hot*', as shown in Figure 6, have many bursty times but they are not found bursty in the time of the events detected in Section 2.

Other moods are found bursty in the time of the abnormal events. For example, six moods are found bursty on the 9/11 attacks and all top topics returned by LDA on the posts tagged with these moods during their bursty time mention the event (Table 3). Consistent with the finding in Section 2, a majority of these moods are low in *valence* and high in *arousal*.

Those moods found to have peaks during the periodic events in Section 2 are also detected bursty through these days. For example, '*cheerful*', which has peaks on the 25th December in four years of the corpus, is also found bursty four times during Christmas (Figure 7). The top topics in the blog posts tagged with this mood in these bursty time are closely related to Christmas.

The correlation between the extreme points of the sentiment index and the bursty time of some moods (examples shown in Figure 7) complements the process of detecting events. While the former can help find the start time of events, the latter can help determine the events' intervals.

Based on the proposed method, we implemented a system for querying events associated with bursts during specific time-periods using moods as keywords. Figure 8 displays screen-shots from the Web user interface (Web UI). A user can enter a set of moods (8a) and the software subsequently returns bursts associated with those moods along time-line for the specified duration. For example, a query of '*angry, sad*' and duration '11th *Sept 2001* to 31st *Dec 2004*' results in 3 and 8 bursty events corresponding to the moods '*angry*' and '*sad*' respectively. Figure 8b provides a detailed result

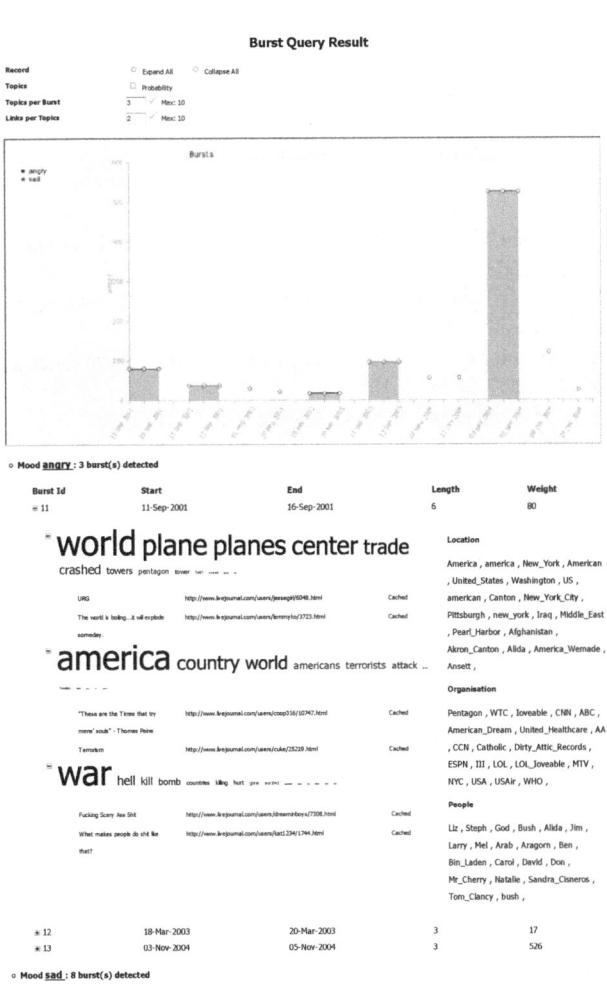

(a) The input Web UI.

(b) The output Web UI.

Fig. 8. A system for querying bursty moods and related events

including topics and entities for the event '*9/11 attacks*' and the time-line for two other events associated with the mood '*angry*', including '*Iraq war*' and '*US election*'.

4   Conclusion

We have investigated the novel problem of detecting sentiment-based bursty events in the blogosphere using mood tags. We proposed an approach to compute a sentiment index which can be used to extract subtle events in the stream of blogposts. We also found some moods themselves can be used as strong indicators of real-world events. While the sentiment indices can help find the start time of events, the bursty periods of moods returned by a state-machine can help to infer the bursty intervals of related events. The results have shown that the emotion information tagged can be exploited to detect meaningful events in blogs. We have also implemented a prototype Web interface for querying sentiment-based bursty events. Our proposed approach is generic and can be applicable to other streaming data.

References

1. Blei, D., Ng, A., Jordan, M.: Latent Dirichlet allocation. Journal of Machine Learning Research 3, 993–1022 (2003)
2. Bradley, M., Lang, P.: Affective norms for English words (ANEW): Instruction manual and affective ratings. University of Florida (1999)
3. Cao, L.: In-depth behavior understanding and use: the behavior informatics approach. Information Science 180, 3067–3085 (2010)
4. Coontz, R.: Blogs: Happiness barometers? Science 325, 5941 (2009)
5. Finkel, J., Grenager, T., Manning, C.: Incorporating non-local information into information extraction systems by Gibbs sampling. In: Proc. of the Association for Computational Linguistics Conference, p. 370 (2005)
6. Gilbert, E., Karahalios, K.: Widespread worry and the stock market. In: Procs. of the Int. AAAI Conf. on Weblogs and Social Media, ICWSM (2010)
7. Giles, J.: Blogs and tweets could predict the future. The New Scientist 206(2765), 20–21 (2010)
8. Griffiths, T., Steyvers, M.: Finding scientific topics. Proceedings of the National Academy of Sciences 101(90001), 5228–5235 (2004)
9. Kleinberg, J.: Bursty and hierarchical structure in streams. Data Mining and Knowledge Discovery 7(4), 373–397 (2003)
10. Pang, B., Lee, L.: Opinion mining and sentiment analysis. Foundations and Trends in Information Retrieval 2(1-2), 1–135 (2008)
11. Russell, J.: A circumplex model of affect. Journal of personality and social psychology 39(6), 1161–1178 (1980)
12. Russell, J.: Emotion, core affect, and psychological construction. Cognition & Emotion 23(7), 1259–1283 (2009)
13. Tumasjan, A., Sprenger, T., Sandner, P., Welpe, I.: Predicting elections with Twitter: What 140 characters reveal about political sentiment. In: Procs. of the Int. AAAI Conference on Weblogs and Social Media, ICWSM (2010)

Constructing Personal Knowledge Base: Automatic Key-Phrase Extraction from Multiple-Domain Web Pages

Yin-Fu Huang and Cin-Siang Ciou

National Yunlin University of Science and Technology,
123 University Road, Section 3, Touliu, Yunlin, Taiwan 640
huangyf@yuntech.edu.tw

Abstract. In the paper, we proposed a general framework that could automatically extract key-phrases from a collection of web pages concerning a specific topic with the help of The Free Dictionary and then construct a personal knowledge base. Both the base and visual feature in a web page are used to calculate the weight of each candidate phrase. The system extracts top p% key-phrases for each web page based on these two features and then generates a term set using union operators. Next, the system builds the relationships between terms in the term set by referencing The Free Dictionary, and then generates a list of terms sorted by weights. With the top q terms specified by users, a semantic graph can be constructed to present the part of a personal knowledge base, which shows the relationships between terms from the same domain. Finally, the experimental results show that the key-phrases generated by the proposed extractor are with good quality and acceptable for humans.

Keywords: key-phrase extraction, semantic graph, learning mechanism, term correlation, POS.

1 Introduction

With the rapid development and growth of digital information, everyone can create and share electronic documents easily on the Internet. In general, a searching website is organized into different domains such as music, comics, cars, etc. Each domain includes many articles concerning a specific topic. However, it is still not efficient for users to retrieve information related to their preferences, which is an important topic in behavior informatics [12]. If each domain is represented with some key-phrases, it would be much convenient for users to understand the corresponding domain and then determine what they want. Although key-phrases are useful in information retrieval, it is not practical to manually mark them in a collection of web pages because it is a time consuming and tremendous work. Therefore, developing a key-phrase extraction system that can automatically extract key-phrases from a collection of web pages is necessary.

L. Cao et al. (Eds.): PAKDD 2011 Workshops, LNAI 7104, pp. 65–76, 2012.

Till now, several key-phrase extraction systems [1-9] have been proposed based on different techniques: supervised algorithms [3, 4, 6, 8, 9] or unsupervised algorithms [2, 5]. A supervised algorithm is to classify a candidate phrase into key-phrase or non-key-phrase after the training, by using a corpus of documents with corresponding author-assigned key-phrases. On the other hand, an unsupervised algorithm usually assigns a numeric score to candidate phrases by considering various features.

Besides, most previous work focused on key-phrase extraction from a document. In this paper, we propose a novel approach which extracts key-phrases from a domain. A domain consists of a collection of web pages concerning a specific topic, from which the proposed approach can generate a list of key-phrases. Moreover, it could work on various domains. First, the system extracts top p% key-phrases for each web page, then collects all the key-phrases from all web pages of the same domain, and finally generates a term set using union operators. Next, the system builds the relationships between terms in the term set by referencing The Free Dictionary, and then generates a list of terms sorted by weights. Finally, with the top q terms specified by users, a semantic graph can be constructed and shows the relationships between terms from the same domain.

The remainder of the paper is organized as follows. A novel approach for key-phrase extraction is proposed in Section 2. Section 3 discusses the experimental results. Finally, we make conclusions in Section 4.

2   System Framework

In this section, we propose the personal knowledge base construction framework as shown in Fig. 1, which consists of the following 8 components: *preprocessor, candidate phrase extractor, feature calculation, refinement, correlation matrix generator, term ranking, semantic graph constructor, learning mechanism*. Each component is explained in detail in the following subsections.

2.1   Preprocessor

The preprocessor parses web pages and records each term appearing in particular tags, e.g. <title>, <b>, <h1> to <h6>, <i>, <a>, and <meta keywords>. The recorded information would be used as a reference for weighting terms and determining how important candidate phrases are. If a candidate phrase appears in particular tags, it will be given more weight than those not appearing. After parsing, the preprocessor transforms web pages into plain texts; in other words, it removes HTML tags, scripts, etc., and only text blocks are preserved. Then, we employ a state-of-the-art part-of-speech (i.e., POS) tagger [10] to annotate words in the plain texts automatically with part-of-speech tags. Therefore, the preprocessor records each word and its corresponding POS tag. This tagging step is important and necessary because we use linguistic information to help candidate phrase selection.

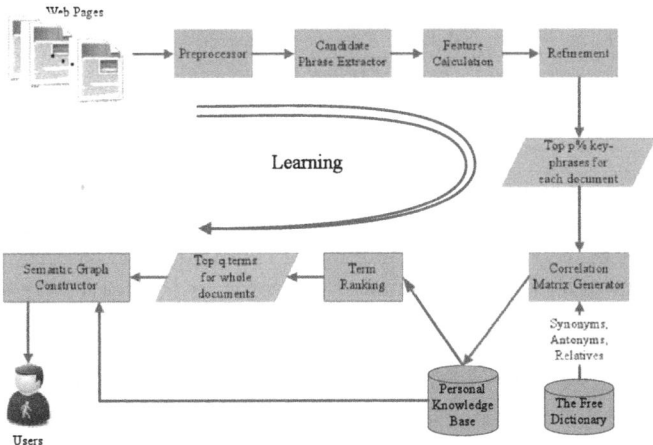

Fig. 1. System framework

2.2 Candidate Phrase Extractor

A phrase makes more sense than a word in the semantic interpretation, so that key-phrases are more suitable than keywords for the document summary. In general, a phrase is a contiguous sequence of words without intervening punctuations; thus, punctuations can be viewed as phrase boundaries used in detecting initial phrases. However, here we still keep apostrophes and hyphens considered as the equivalents to alphabets if they appear between alphabets, e.g. "batter's on-base percentage". Therefore, the candidate phrase extractor finds out initial phrases according to the above-mentioned rule, and then the extra three rules as shown below are applied to filter candidate phrases from initial phrases.

(1) Candidate phrases cannot begin or end with a stop word.
(2) Candidate phrases are limited to a certain maximum length.
(3) Candidate phrases must be noun phrases.

2.3 Feature Calculation

Two features used to calculate the weight of each candidate phrase are proposed. One is the *base feature* measuring the importance of a phrase in a web page based on three attributes (i.e., the frequency of a phrase, the length of a phrase, and the position of a phrase in a web page). Another is the *visual feature* measuring the significance of a phrase in a web page based on its visual appearance.

Base Feature. The base feature weight of a phrase P in a web document D could be calculated as follows:

$$W_{base}(P, D) = \left(0.5 + 0.5\frac{freq(P, D)}{max_{1 \leq i \leq ED}freq(P_i, D)}\right) \times log_2\left(\frac{E_D}{G_D + 1} \times \left(1 + \frac{|P|}{10}\right)\right)$$
$$\times \left(0.5 + 0.5\frac{max_{1 \leq i \leq ED}post(P_i, D) - post(P, D)}{max_{1 \leq i \leq ED}post(P_i, D)}\right)$$

where freq(P, D) is the number of times that P occurs in D, $max_{1 \leq i \leq ED}$ freq(P$_i$, D) is the maximum frequency of all phrases in D, ED is the number of all candidate phrases in D, GD is the number of candidate phrases with the length more than one in D, |P| is the number of words in P, post(P, D) is the first position where P occurs in D, and $max_{1 \leq i \leq ED}$ post(Pi, D)is the maximum position value of all phrases in D.

The *base feature* consists of three components. The first component is related to the frequency of a phrase, which is a weight in the range [0.5, 1]. The second component is related to the length of a phrase. The last component is related to the position of a phrase. The rationale is that a candidate phrase occurring earlier is more important than that occurring later in a web page and its weight is in the range [0.5, 1].

Visual Feature. Here, we consider two kinds of texts: special texts and anchor texts. Special texts include heading, bold, and italicized text, which are usually used to highlight or emphasize certain important information on web pages. Anchor texts are hyperlinks via which more relevant descriptive information would be provided.

The *visual feature* weight of a phrase P in a web document D could be calculated as follows:

$$W_{visual}(P,D)=\begin{cases} \frac{1}{5}\left(\frac{\sum_{i=1}^{E_D} W_{base}(P_i,D)}{E_D}\right) \times freq_s(P,D), & if\ P\ is\ a\ special\ text. \\ \frac{1}{10}\left(\frac{\sum_{i=1}^{E_D} W_{base}(P_i,D)}{E_D}\right) \times freq_a(P,D) & if\ P\ is\ an\ anchor\ text. \\ \left(\frac{freq_s(P,D)}{5} + \frac{freq_a(P,D)}{10}\right)\times\frac{\sum_{i=1}^{E_D} W_{base}(P_i,D)}{E_D} & if\ P\ are\ both. \end{cases}$$

where$freq_s(P, D)$ is the number of times that P appears as a special text, $freq_a(P, D)$ is the number of times that P appears as an anchor text, $\frac{\sum_{i=1}^{E_D} W_{base}(P_i,D)}{E_D}$ is the average base feature weight of all candidate phrases.

The visual feature weight of a phrase P can be considered as the extra one in calculating the final weight of P. A candidate phrase has the visual feature weight only if it appears in particular tags. In general, a base feature is more important than a visual feature, so that the visual feature weight of a candidate phrase cannot be higher than the base feature weight of its own. That is why we take the average base feature weight of all candidate phrases and divide it by 5 or 10.

Weight Calculation. The system gives each candidate phrase a different weight based on the base feature, the visual feature, and whether it appears in title or meta-keyword tags. Here, two features are combined to provide a final weight, and it is normalized by the maximum of all candidate phrases so that the value is in the range $[0, 1]$.

The final weight of a phrase P in a web document D could be calculated as follows:

$$W_{final}(P, D) = \begin{cases} 1, & if\, a\, phrase\, is\, in\, title\, or\, meta-keyword\, tags. \\ W_{base} + W_{visual}, & if\, a\, phrase\, is\, a\, visual\, text. \\ W_{base}, & otherwise. \end{cases}$$

Finally, the system would generate a list of candidate phrases sorted by weights.

2.4 Refinement

For a list of candidate phrases sorted by weights, the system would eliminate sub-phrases from the list if they do not have higher weights than their super-phrases. In other words, a sub-phrase is kept in the list if it has higher weight than its super-phrase; otherwise, it would be removed from the list. After the refinement, the system allows users to specify the top p% key-phrases for the return.

2.5 Correlation Matrix Generator

For each web page of the same domain, the system extracts the top p% key-phrases. Thus, if the size of a key-phrase list is large, relatively more key-phrases are extracted. Then, we collect all the key-phrases from all web pages of the same domain, and generate a term set using union operators. Next, the correlation matrix generator would build the relationships between terms in the term set by referencing The Free Dictionary [11].

Relative Matrix. There are ten sub-dictionaries in The Free Dictionary, but only two of them, called *Dictionary/thesaurus* and *Wikipedia Encyclopedia*, are used as references. The synonyms and antonyms of terms are collected from *Dictionary/thesaurus*, and the relatives from *Dictionary/thesaurus* and *Wikipedia Encyclopedia*. The correlation matrix generator sends each term in the term set to The Free Dictionary and gets back its synonyms, antonyms, and relatives. These results could be built in a *Relative Matrix*, as shown in Fig. 2.

Key Terms	Synonyms	Antonyms	Relatives
KT_1	$T_2, T_{15}, T_{16} \ldots$	$\ldots$	$T_3, T_4, T_6, T_8, T_{10}, T_{11}$
KT_2	$T_1, T_{14}, T_{15} \ldots$	$\ldots$	$T_4, T_5, T_7, T_9, T_{10}, T_{11}, T_{12}, T_{13}$
...	$\ldots$	$\ldots$	$\ldots$

Fig. 2. *Relative Matrix* where KT and T represent the term, and KT_i is the same as T_i

Correlation Calculation. The *Relative Matrix* generated in the last step is used for correlation calculations. The correlation between terms could be calculated as follows:

$$Correlation_{RM}(KT_x, KT_y) = \beta \times Closely\_Related\_Relation$$
$$+(1-\beta) \times Relative\_Relation$$
$$where\, 0.5 \leq \beta \leq 1$$

For each pair of terms, they have an entry value for *Closely Related Relation* and another entry value for *Relative Relation*. The former is to calculate synonyms and antonyms relationships between a pair of terms, whereas the latter is to calculate relativity relationships between them. Here, *Closely Related Relation* is considered as more important than *Relative Relation*, since the former has stronger relationships than the latter. They can be assigned values according to the following rules.

(1) Synonym relationships exist between KT_x and KT_y
    Case 1: If KT_x is the synonym of KT_y, or vice versa,
        then $Correlation_{RM}(KT_x, KT_y) = 1$
        e.g., T_2 is the synonym of KT_1, and KT_2 is the same as T_2.
    Case 2: otherwise, If KT_x and KT_y have the same synonym,
        then $Closely\_Related\_Relation = 1$
        e.g., KT_1 and KT_2 have the same synonym T_{15}.

Key Terms	Synonyms	...
KT_1	T_2, T_{15}, T_{16}	...
KT_2	T_1, T_{14}, T_{15}	...

(2) Antonym relationships exist between KT_x and KT_y
    Case 1: If KT_x is the antonym of KT_y, or vice versa,
        then $Correlation_{RM}(KT_x, KT_y) = 0.5$
        e.g., T_3 is the synonym of KT_4, and KT_3 is the same as T_3.

Key Terms	...	Antonyms	...
KT_3	...	...	...
KT_4	...	$T_3,...$	...

Case 2: otherwise, If KT_x and KT_y have the same antonym,
        then $Closely\_Related\_Relation = 1$
        e.g., KT_4 and KT_5 have the same antonym T_{15}.

Key Terms	...	Antonyms	...
KT_4	...	$T_{15},...$	...
KT_5	...	$...,T_{15}$	...

(3) Relative relationships exist between KT_x and KT_y

Case 1: If KT_x is the relative of KT_y, or vice versa,

then *Relative_ Relation* $= 1$

e.g., T_6 is the relative of KT_5, and KT_6 is the same as T_6.

Key Terms	...	Relatives
KT_5	...	$...,T_6,...$
KT_6	...	...

Case 2: otherwise,

$$Relative\_Relation = \frac{|R_{Tx} \cap R_{Ty}|}{|R_{Tx} \cup R_{Ty}|}$$

where R_{Tx} and R_{Ty} are the sets of relatives for KT_x and KT_y, respectively. e.g., *Relative_ Relation* for KT_6 and KT_7 as shown below is equal to 3/9.

Key Terms	...	Relatives
T_6	...	$T_3, T_4, T_8, T_{10}, T_{11}$
KT_7	...	$T_4, T_5, T_9, T_{10}, T_{11}, T_{12}, T_{13}$

After correlation calculation for each pair of terms, the system generates a correlation matrix. This matrix is a symmetric matrix, and all elements on the main diagonal are equal to 1. It represents the relationships between all terms and is stored in a personal knowledge base.

2.6 Term Ranking

In the correlation matrix, each term has correlation values with others. The value determines how close the relationships of a pair of terms are. Here, we also use a threshold γ to judge whether a pair of terms are close enough. If a term has strong relationships with many terms, the term is considered very important in the term set. Thus, we use the following formula to weight each term in the term set.

$$weight(T_k) = \sum_{j=1,j \neq k}^{n} t_{kj}, \; such\,that\,t_{kj} \geq \gamma$$

where t_{kj} is the correlation value between terms T_k and T_j.

After generating a list of terms sorted by weights, the system allows users to specify the top q terms for the return.

2.7 Semantic Graph Constructor

With the top q terms, a semantic graph can be constructed to present the part of a personal knowledge base, which shows the relationships between terms for the web

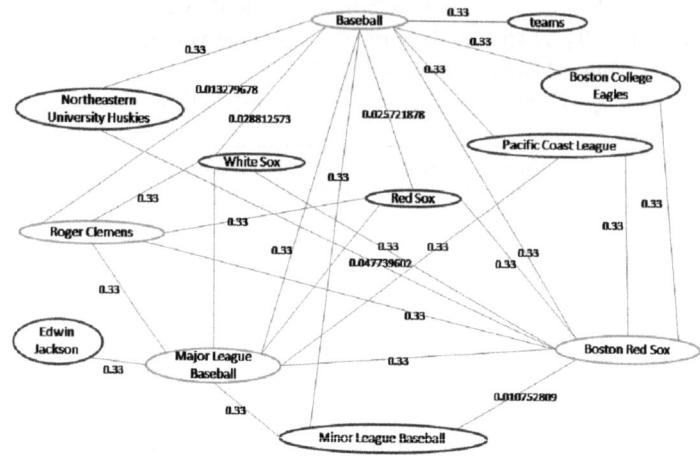

Fig. 3. Semantic graph

pages from the same domain, as illustrated in Fig. 3. The edge values in the semantic graph indicate the strength of the relationships between terms. By the way, users can click any nodes (or terms) in the graph to show the contents of terms with the help of Google search engine or Wikipedia.

2.8   Learning Mechanism

With the constructed semantic graph, new web pages could be exploited by a learning mechanism. Here, some existing information during the learning need not be constructed from scratch. For example, the correlation matrix generator only handles the new terms from new web pages. As new web pages increase, the personal knowledge base would become larger through learning.

3   Evaluation and Experiments

This section describes how to evaluate the key-phrase extraction algorithm using web pages. Five evaluators are invited to judge the key-phrases generated by our extractor; i.e., assigning a numeric score to each key-phrase. The key-phrase extraction system was implemented in Java, and the experiments were conducted on an Intel Core 2 Duo E7200 2533MHz CPU with 2G main memory in Window XP professional.

3.1   Datasets and Measures

In Experiment 1, through Google and Yahoo search engines, each evaluator selects 10 web pages which he/her is interested in from each of 10 specific domains.

These domains are diverse, including baseball, music, computer networking, military, dancing, gambling, car, comics & animation, investing, and religion & spirituality. Each evaluator would judge those key-phrases (or personal knowledge base) extracted from the web pages which he/she selects for himself/herself.

In Experiment 2, our method is compared with KEA [8] on the key-phrase extraction for each document. Also, 10 web pages are selected from each of the same 10 specific domains. Based on these 100 web pages, each evaluator judges the key-phrases generated by our method and KEA.

In [6], Turney defined an acceptable percentage measure for rating each key-phrase as "good" or "bad". In [13], Zhang et al. also used the acceptable percentage measure for evaluation, but the rating in the measure is a 1-to-5 scale (1: not related, 2: poorly related, 3: fairly related, 4: well related, and 5: strongly related). Besides, they also used a quality value measure. Here, these two measures are adopted to evaluate the quality of the key-phrases generated by our system as follows. The acceptable percentage measure:

$$ m_1 = \frac{n_3 + n_4 + n_5}{\sum_{i=1}^{5} n_i} $$

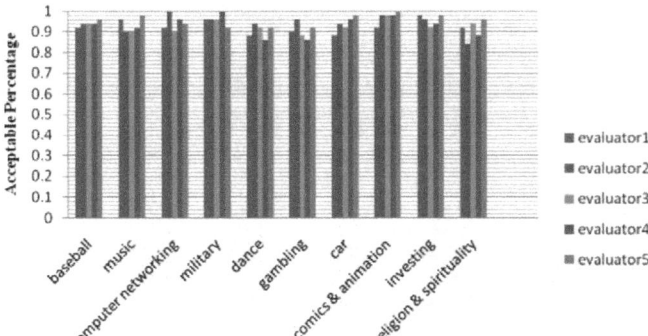

Fig. 4. Acceptable percentages of all domains

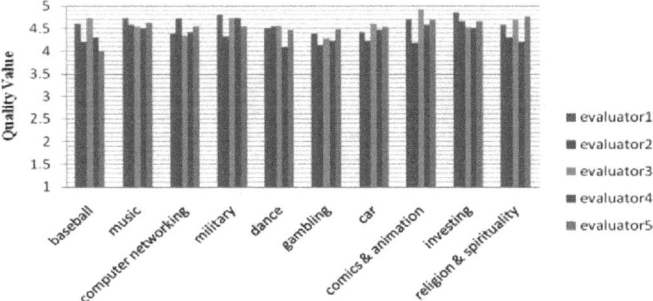

Fig. 5. Quality values of all domains

The quality value measure:

$$m_2 = \frac{\sum_{i=1}^{5} n_i \times i}{\sum_{i=1}^{5} n_i}$$

where n1, n2, n3, n4, and n5 are the number of key-phrases with a score of 1, 2, 3, 4, and 5, respectively.

3.2   Experiment 1: Evaluating Personal Knowledge Base

In the experiment, in general, the system extracts the top 0.1% key-phrases for each web page of a specific domain. If the number of the top 0.1% key-phrases is more than 25, it only extracts the top 25 key-phrases for this web page. However, if the number of the top 0.1% key-phrases is less than 10, it extracts the top 10 key-phrases. Therefore, the number of the extracted key-phrases for each web page is in the range [10, 25]. Besides, the threshold β is set to 0.67 because Closely Related Relation is more important than Relative Relation. Another threshold γ is set to 0.01 so that the system can judge whether a pair of terms are close enough. Finally, for each domain, the system extracts top 50 key-phrases for the evaluators to assign numeric scores to them, based on the relatedness to the most essential topic.

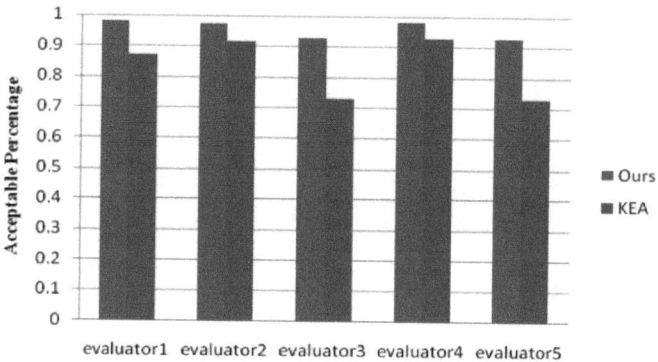

Fig. 6. Average acceptable percentage of all web pages for each evaluator

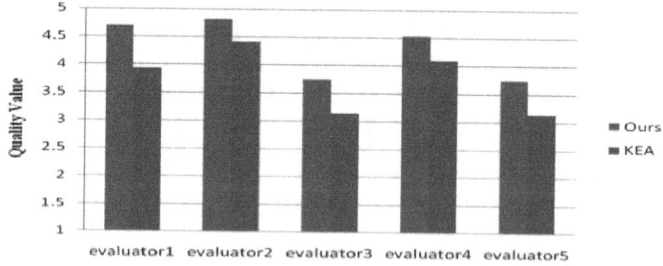

Fig. 7. Average quality value of all web pages for each evaluator

As shown in Fig. 4 and Fig. 5, we found that the acceptable percentages of each domain are more than 85% and the quality values of each domain are more than 4. This means that our extractor works well on various domains.

3.3 Experiment 2: Comparing with KEA

In the experiment, the system extracts the top 0.1% key-phrases for each web page. If the number of the top 0.1% key-phrases is more than 100, it only extracts the top 100 key-phrases for this web page. Besides, the threshold β and γ values are the same as those in Experiment 1. Finally, after performing correlation calculation, the system extracts top 20 key-phrases to compare with the top 20 key-phrases extracted by KEA. As shown in Fig. 6 and Fig. 7, the results indicate that the key-phrases extracted by our method are more acceptable for humans than those done by KEA, and our quality is also better than KEA. For the key-phrase extraction in our system, we not only give weight to the terms appearing in particular HTML tags, but also build the correlation matrix between terms. That is why our system has better performances than KEA.

4 Conclusions

In this paper, we proposed a novel approach that uses The Free Dictionary to help key-phrase extraction from a collection of related web pages and then constructs a personal knowledge base. The system first extracts top p% key-phrases for each web page using two features: base feature and visual feature. Then, we collect all the key-phrases from all web pages of the same domain, and generate a term set using union operators. Next, the system builds the relationships between terms in the term set by referencing The Free Dictionary, and then generates a list of terms sorted by weights. With the top q terms specified by users, a semantic graph can be constructed to present the part of a personal knowledge base, which shows the relationships between terms from the same domain. By the way, users can click any nodes (or terms) in the graph to show the contents of terms with the help of Google search engine or Wikipedia. Finally, the experimental results indicate that the key-phrases generated from the web pages that users prefer in our extractor are with good quality and acceptable for humans, and the extractor also works well on various domains.

Acknowledgments. This work was supported by National Science Council of R.O.C under Grant NSC99-2220-E-224-001.

References

1. D'Avanzo, E., Magnini, B.: A Keyphrase-based Approach to Summarization: the LAKE System at DUC-2005. In: Document Understanding Workshop (2005)
2. El-Beltagy, S.R., Rafea, A.: KP-Miner: a Keyphrase Extraction System for English and Arabic Documents. Information Systems 34(1), 132–144 (2009)

3. HaCohen-Kerner, Y.: Automatic Extraction of Keywords from Abstracts. In: Palade, V., Howlett, R.J., Jain, L. (eds.) KES 2003. LNCS, vol. 2773, pp. 843–849. Springer, Heidelberg (2003)
4. HaCohen-Kerner, Y., Gross, Z., Masa, A.: Automatic Extraction and Learning of Keyphrases from Scientific Articles. In: Gelbukh, A. (ed.) CICLing 2005. LNCS, vol. 3406, pp. 657–669. Springer, Heidelberg (2005)
5. Kumar, N., Srinathan, K.: Automatic Keyphrase Extraction from Scientific Documents Using N-gram Filtration Technique. In: 8th ACM Symposium on Document Engineering, pp. 199–208 (2008)
6. Turney, P.D.: Learning Algorithms for Keyphrase Extraction. Information Retrieval 2(4), 303–336 (2000)
7. Turney, P.D.: Coherent Keyphrase Extraction via Web Mining. In: 20th International Joint Conference on Artificial Intelligence, pp. 434–439 (2003)
8. Witten, I.H., Paynter, G.W., Frank, E., et al.: KEA: Practical Automatic Keyphrase Extraction. In: 4th ACM Conference on Digital Libraries, pp. 254–255 (1999)
9. Zhang, K., Xu, H., Tang, J., Li, J.: eyword Extraction Using Support Vector Machine. In: Yu, J.X., Kitsuregawa, M., Leong, H.-V. (eds.) WAIM 2006. LNCS, vol. 4016, pp. 85–96. Springer, Heidelberg (2006)
10. Schmid, H.: Probabilistic Part-of-speech Tagging Using Decision Trees. In: International Conference on New Methods in Language Processing, pp. 44–49 (1994)
11. The Free Dictionary, http://www.thefreedictionary.com/
12. Cao, L.: In-depth Behavior Understanding and Use: the Behavior Informatics Approach. Information Science 180(17), 3067–3085 (2010)
13. Zhang, Y., Milios, E., Zincir-Heywood, N.: Narrative Text Classification for Automatic Key Phrase Extraction in Web Document Corpora. In: 7th Annual ACM International Workshop on Web Information and Data Management, pp. 51–58 (2005)

Discovering Valuable User Behavior Patterns in Mobile Commerce Environments

Bai-En Shie[1], Hui-Fang Hsiao[1], Philip S. Yu[2], and Vincent S. Tseng[1]

[1] Department of Computer Science and Information Engineering,
National Cheng Kung University, Taiwan, ROC
[2] Department of Computer Science, University of Illinois at Chicago,
Chicago, Illinois, USA
{brianshie,karolter1130}@gmail.com, psyu@cs.uic.edu,
tsengsm@mail.ncku.edu.tw

Abstract. Mining user behavior patterns in mobile environments is an emerging topic in data mining fields with wide applications. By integrating moving paths with purchasing transactions, one can find the sequential purchasing patterns with the moving paths, which are called *mobile sequential patterns* of the mobile users. Mobile sequential patterns can be applied not only for planning mobile commerce environments but also analyzing and managing online shopping websites. However, unit profits and purchased numbers of the items are not considered in traditional framework of mobile sequential pattern mining. Thus, the patterns with high utility (i.e., profit here) cannot be found. In view of this, we aim at integrating mobile data mining with utility mining for finding high utility mobile sequential patterns in this study. A novel algorithm called $UMSP_L$ (*high Utility Mobile Sequential Pattern mining by a Level-wised method*) is proposed to efficiently find high utility mobile sequential patterns. The experimental results show that the proposed algorithm has excellent performance under various system conditions.

Keywords: Utility mining, mobility pattern mining, mobile environments, high utility mobile sequential pattern.

1 Introduction

Data mining refers to the process of discovering potentially useful information from large databases. In behavior informatics [17], previous studies have discovered many kinds of user behavior patterns for different applications, such as cross marketing in business domains [1, 2], websites design and management [4, 15], and mobile environments planning [6, 9, 10, 11, 12, 16]. Among these issues, mining user behavior patterns in mobile environments plays an emerging role in the last decade since mobile devices and wireless applications have become one of the most popular communication media in the world. With a series of users' moving logs recorded by the mobile devices with GPS services, we can acquire the moving paths of mobile users. Combining moving logs and payment records, mobile transaction sequences that are the sequences

L. Cao et al. (Eds.): PAKDD 2011 Workshops, LNAI 7104, pp. 77–88, 2012.

of moving paths with purchased transactions can be generated. There exists useful information recorded in mobile transaction sequences. Yun et al. [16] first proposed a framework combining moving paths and sequential patterns to find mobile sequential patterns, i.e., the sequential patterns with their moving paths, in mobile transaction sequences. For example, a mobile sequential pattern <{<A; clothes><C; lipsticks>}; ABC> means the customers often move through the path <ABC> and bought clothes and lipsticks in A and C, respectively. If the shopkeepers acquire this pattern, they can prepare promotions of lipsticks when they meet the customers who had bought clothes in A to raise the customers' desires to purchase.

However, mobile sequential patterns cannot reflect the actual profit of items in the databases. Valuable user behavior patterns, such as the patterns with purchasing diamond rings, may not be discovered since their frequencies are not enough. Utility mining [3, 5, 7, 8, 13, 14] is proposed for solving this problem in the traditional transaction databases. Given pre-defined utility values, which may be the importance, interestingness or profits of the items, utility mining is to find the high utility patterns, which are the patterns with high utility values, from the databases. From this viewpoint, we can realize that pushing utility mining into the framework of mobility pattern mining is an essential topic. If decision makers know which patterns are more valuable, they can choose more proper reactions based on the useful information. For example, assume there exists two user behavior patterns: UBP_1 = <{<A; clothes> <H; lipsticks>}; ABCDH> and UBP_2 = <{<A; clothes> <H; diamond rings>}; AEFGH>. The two patterns show that although these customers moved from A to H, they may buy different items since they passed through different paths. In general, we can know the profits of diamond rings are much higher than lipsticks. If the shopkeepers in H know both the two patterns, they can prepare some promotions or activities for UBP_2 to attract the customers who buy clothes in A and go through the path AEFGH, such as to raise their desires to purchase diamond rings in H.

Although there exist a number of prominent research works about mobility pattern mining and utility mining, there is no work done on combination of the two topics. In view of this, we attempt to integrate mobile sequential pattern mining with utility mining for finding high utility mobile sequential patterns in this paper. By the combination of high utility patterns and moving paths, real profit of a mobile sequential pattern can be found. High utility mobile sequential patterns are more crucial in many domains, such as mobile commerce environments, metropolitan planning and online shopping websites which sell a wide selection of merchandise in different web pages. A novel algorithm, named $UMSP_L$ (high Utility Mobile Sequential Pattern mining by a Level-wised method), is proposed to efficiently find high utility mobile sequential patterns. The experimental results show that the proposed algorithm $UMSP_L$ not only outperforms the compared algorithm but also bears good scalability. We expect that the useful patterns can bring novel and insightful information in mobile commerce environments. To the best of our knowledge, this is the first work which integrates the two topics to find the high utility mobile sequential patterns with actual acquired utilities.

The remaining of this paper is organized as follows. Related work is briefly reviewed in Section 2. Problem definitions of this work are given in Section 3.

In Section 4, the proposed algorithm, named UMSP$_L$, is addressed in detail. Experimental evaluation is shown in Section 5. The conclusions and future work are given in Section 6.

2   Related Work

Extensive studies [1, 2] have been proposed for finding frequent patterns, including association rules [1] and sequential patterns [2]. Mining user behavior patterns in the mobile environments [6, 9, 10, 11, 12, 16] is an emerging topic in this field. Tseng et al. [12] first proposed SMAP-Mine algorithm to find customers' mobile access patterns. However, in different time periods, popular services may be totally different. Thus, Lee et al. [6] proposed T-Map algorithm to find temporal mobile access patterns in different time intervals. On the other hand, Yun et al. [16] proposed a framework which combined moving paths and sequential patterns to find mobile sequential patterns. By the above researches, although we can know the patterns in different time intervals, defining time intervals is not an easy work yet. In view of this, Tseng et al. [11] proposed TMSP-Mine algorithm to automatically find proper time intervals of mobile sequential patterns based on the genetic algorithm. On the other hand, the character of customers is also important in the mobile environment. Different groups of customers may bring different patterns. Thus, Lu et al. [9] proposed a framework to find the cluster-based mobile sequential patterns. The customers whose moving paths and transactions are similar will be clustered into the same cluster. The found patterns may be closer to real customer behaviors by this method.

However, the profits of items are not considered in the above researches. In frequent pattern mining fields, a new topic raised for conquering this problem, that is, utility mining [3, 5, 7, 8, 13, 14]. In utility mining, each item may have different profit. UMining algorithm [14] proposed by Yao et al. applies an estimation method to prune the search space. However it cannot capture the complete set of high utility itemsets since some high utility patterns may be pruned during the mining process. Among these researches, Liu et al. [8] proposed Two-Phase algorithm which uses the transaction-weighted downward closure property to maintain the downward closure property in utility mining. Although Two-Phase algorithm can reduce the search space of utility mining, it still generates too many candidates. Thus, Li et al. [7] proposed an isolated items discarding strategy to reduce the number of candidates by pruning isolated items during the level-wise searches.

There are also other frameworks for mining high utility itemsets such as tree-based framework [3, 13]. Ahmed et al. [3] proposed a structure named IHUP-Tree which maintains essential information about utility mining. It avoids scanning database multiple times and generating candidates in the mining process. Although IHUP-Tree achieves a better performance than Two-Phase, it still produces too many HTWUIs. Therefore, Tseng et al. proposed an algorithm, named UP-Growth [13], which applies several strategies during the mining processes. By the proposed strategies, the estimated utilities are effectively decreased in the proposed tree structure named UP-Tree in the mining processes and the number of HTWUIs is further reduced. Therefore, the performance of utility mining can be improved.

3   Preliminaries and Definitions

In this section, we first define basic notations for mining high utility mobile sequential patterns in mobile environments in detail, and then address the problem statement of this work.

Let $L = \{l_1, l_2, ..., l_p\}$ be a set of *locations* in the mobile commerce environment and $I = \{i_1, i_2, ..., i_g\}$ be a set of *items* sold in the locations. An *itemset* is denoted as $\{i_1, i_2, ..., i_k\}$, where each item $i_v \in I$, $1 \leq v \leq k$ and $1 \leq k \leq g$. Given a *mobile transaction sequence database* D, a *mobile transaction sequence* $S = <T_1, T_2, ..., T_n>$ is a set of transactions ordered by time, where a *transaction* $T_j = (l_j; \{[i_{j1}, q_{j1}], [i_{j2}, q_{j2}], ..., [i_{jh}, q_{jh}]\})$ represents that a user made T_j in l_j, where $1 \leq j \leq n$. In T_j, the *purchased quantity* of item i_{jp} is q_{jp}, where $1 \leq p \leq h$. A *path* is denoted as $l_1 \, l_2 ... \, l_r$, where $l_j \subseteq L$ and $1 \leq j \leq r$.

Definition. (Utility of a loc-item in a mobile transaction sequence) A *loc-item*, denoted as $<l_{loc}; i_{je}>$, stands for the item i_{je} is happened in the location l_{loc}, where $l_{loc} \in L$ and $i_{je} \subseteq I$. The *utility* of a loc-item $<l_{loc}; i_{je}>$ in a mobile transaction sequence S_j is denoted as $u(<l_{loc}; i_{je}>, S_j)$, that is defined as $q_{je} \times w(i_{je})$, where $w(i_{je})$ is the *unit profit* of item i_{je}, which is recorded in a *utility table*.

Take the mobile transaction sequence database D in Table 1 and the utility table in Table 2 for example, $u(<A; i_1>, S_1) = 3 \times 1 = 3$.

Definition. (Utility of a loc-itemset in a mobile transaction sequence database) A *loc-itemset*, denoted as $<l_{loc}; \{i_1, i_2, ..., i_g\}>$, stands for the itemset $\{i_1, i_2, ..., i_g\}$ is happened in l_{loc}, where $l_{loc} \in L$ and $\{i_1, i_2, ..., i_g\} \subseteq I$. The utility of a loc-itemset $<l_{loc}; \{i_{j1}, i_{j2}, ..., i_{jg}\}>$ in a mobile transaction sequence S_j is denoted as $u(<l_{loc}; \{i_{j1}, i_{j2}, ..., i_{jg}\}>, S_j)$ and defined as $\sum_{k=1}^{g} u(<l_{loc}; i_{jk}>, S_j)$. The utility of a loc-itemset $Y = <l_{loc}; \{i_{j1}, i_{j2}, ..., i_{jg}\}>$ in a mobile transaction sequence database D is denoted as $u(Y)$ or $u(<l_{loc}; \{i_{j1}, i_{j2}, ..., i_{jg}\}>)$ and defined as $\sum_{(Y \subseteq S_j) \wedge (S_j \in D)} u(Y, S_j)$.

Table 1. An example of mobile transaction sequence database D

SID	Mobile Transaction Sequences	SU
S_1	(H; {[i_8, 1]}), (A; {[i_1, 3]}), (B; null), (C; {[i_2, 2], [i_3, 5]}), (D; {[i_4, 2]}), (E; null), (F; {[i_5, 1]})	69
S_2	(A; {[i_1, 4]}), (B; null), (C; {[i_3, 10]}), (D; null), (E; null), (F; {[i_5, 1]}), (G; {[i_7, 4]})	60
S_3	(A; {[i_1, 1]}), (w; null), (C; {[i_3, 10]}), (D; {[i_4, 3]}), (E; null), (F; {[i_5, 2]})	100
S_4	(A; {[i_1, 3]}), (B; null), (C; {[i_2, 2], [i_3, 5]}), (D; null), (E; {[i_6, 10]}), (F; {[i_5, 1]}), (G; {[i_7, 2]})	60

Table 2. An example of utility table

Item	i_1	i_2	i_3	i_4	i_5	i_6	i_7	i_8
Utility	1	5	3	11	18	1	2	1

For example in Table 1, the utility of the loc-itemset $<C; \{i_2, i_3\}>$ in S_1 is calculated as $u(<C; \{i_2, i_3\}>, S_1) = u(<C; i_2>, S_1) + u(<C; i_3>, S_1) = 2\times5 + 5\times3 = 25$. The utility of $<C; \{i_2, i_3\}>$ in the database D is calculated as $u(<C;\{i_2, i_3\}>) = u(<C; \{i_2, i_3\}>, S_1) + u(<C; \{i_2, i_3\}>, S_4) = 25 + 25 = 50$.

Definition. (Utility of a loc-pattern in a mobile transaction sequence database) A *loc-pattern X*, which is denoted as $<l_1; \{i_{11}, i_{12}, ..., i_{1g}\}><l_2; \{i_{21}, i_{22}, ..., i_{2g}\}> ... <l_m; \{i_{m1}, i_{m2}, ..., i_{mg}\}>$, is a list of loc-itemsets. The utility of a loc-pattern X in S_j is denoted as $u(X, S_j)$ and defined as $\sum_{\forall Y \in X} u(Y, S_j)$. The utility of a loc-pattern X in D is denoted as $u(X)$ and defined as $\sum_{(X \subseteq S_j) \wedge (S_j \in D)} u(X, S_j)$.

For example in Table 1, the utility of the loc-pattern $<A; i_1><C; \{i_2, i_3\}>$ in S_1 is calculated as $u(<A; i_1><C; \{i_2, i_3\}>, S_1) = u(<A; i_1>, S_1) + u(<C; \{i_2, i_3\}>, S_1) = 3 + 25 = 28$. The utility of $<A; i_1><C; \{i_2, i_3\}>$ in D is calculated as $u(<A; i_1><C; \{i_2, i_3\}>) = u(<A; i_1><C; \{i_2, i_3\}>, S_1) + u(<A; i_1><C; \{i_2, i_3\}>, S_4) = 28 + 28 = 56$.

Definition. (Support and utility of a moving pattern) A *moving pattern* is composed of a loc-pattern and a path. It is recorded by the form as $<\{<l_1; \{i_{11}, i_{12}, ..., i_{1g}\}><l_2; \{i_{21}, i_{22}, ..., i_{2g}\}> ... <l_m; \{i_{m1}, i_{m2}, ..., i_{mg}\}>\}; l_1 l_2 ... l_m>$. The *support* of a moving pattern P, denoted as $sup(P)$, is defined as the number of mobile transaction sequences which contains P in D. The utility of a moving pattern P, denoted as $u(P)$, is defined as the summation of utilities of the loc-patterns of P in the mobile transaction sequences which contain the path of P in D.

Definition. (High utility mobile sequential pattern) Given a *minimum support threshold* δ and a *minimum utility threshold* ε, a moving pattern P is called a *mobile sequential pattern* if $sup(P) \geq \delta$. Further, P is called a *high utility mobile sequential pattern*, which is abbreviated as *UMSP*, if $sup(P) \geq \delta$ and $u(P) \geq \varepsilon$. Moreover, the length of a pattern is the number of loc-itemsets in this pattern. A pattern with length k is denoted as k-pattern.

For example in Table 1, the support of the moving pattern $<\{<A; i_1><C; \{i_2, i_3\}>\}$; ABC>, which is denoted as $sup(<\{<A; i_1><C; \{i_2, i_3\}>\}$; ABC$>)$, is 2. The utility of $<\{<A; i_1><C; \{i_2, i_3\}>\}$; ABC> in D is calculated as $u(<\{<A; i_1><C; \{i_2, i_3\}>\}$; ABC$>)$ $= u(<A; i_1><C; \{i_2, i_3\}>, S_2) + u(<A; i_1><C; \{i_2, i_3\}>, S_3) = 56$. If minimum support threshold δ is set to 2 and minimum utility threshold ε is set to 50, the moving pattern $<\{<A; i_1><C; \{i_2, i_3\}>\}$; ABC> is a 2-high utility mobile sequential pattern (2-UMSP).

After addressing the problem definition of mining high utility mobile sequential patterns in mobile environments, we introduce the *sequence weighted utilization* and *sequence weighted downward closure* property in mobile transaction sequence databases, which are extended from [8].

Definition. (Sequence utility of a mobile transaction sequence) The *sequence utility* of mobile transaction sequence S_j, which is the sum of the utilities of all items in S_j, is defined as $SU(S_j) = \sum_{<l_{loc}; i_{je}> \subseteq S_j} u(<l_{loc}; i_{je}>, S_j)$.

Definition. (Sequence weighted utilization of patterns) The *sequence weighted utilization*, abbreviated as SWU, of a loc-itemset Y is defined as $SWU(Y) = \sum_{(Y \subseteq S_j) \wedge (S_j \in D)} SU(S_j)$. Moreover, the sequence weighted utilization of a loc-pattern X is defined as $SWU(X) = \sum_{(X \subseteq S_j) \wedge (S_j \in D)} SU(S_j)$. In addition, the sequence weighted utilization of a moving pattern P is defined as the summation of SWUs of the loc-patterns of P in the mobile transaction sequences which contain the path of P in D.

Definition. (High sequence weighted utilization pattern) A pattern Z is called a *high sequence weighted utilization pattern*, abbreviated as WUP, if $sup(Z) \geq \delta$ and $SWU(Z) \geq \varepsilon$. Similarly, in the following paragraphs, *high sequence weighted utilization loc-itemset*, *high sequence weighted utilization loc-pattern* and *high sequence weighted utilization mobile sequential pattern* are abbreviated as WULI, WULP and WUMSP, respectively.

For example in Table 1, the sequence utility of the mobile transaction sequence S_6 is computed as $SU(S_3) = u(<A; i_1>, S_3) + u(<C; i_3>, S_3) + u(<D; i_4>, S_3) + u(<F; i_5>, S_3) = 1 \times 1 + 10 \times 3 + 3 \times 11 + 2 \times 18 = 100$. The sequence weighted utilization of the loc-itemset $<D; i_4>$ is computed as $SWU(<D; i_4>) = SU(S_1) + SU(S_3) = 69 + 100 = 169$. The sequence weighted utilization of the loc-pattern $<A; i_1><C; \{i_2, i_3\}>$ is computed as $SWU(<A; i_1><C; \{i_2, i_3\}>) = SU(S_1) + SU(S_4) = 69 + 60 = 129$. The sequence weighted utilization of the moving pattern $<\{<A; i_1><C; \{i_2, i_3\}>\}; ABC>$ is computed as $SWU(<\{<A; i_1><C; \{i_2, i_3\}>\}; ABC>) = SWU(<A; i_1><C; \{i_2, i_3\}>, S_1) + SWU(<A; i_1><C; \{i_2, i_3\}>, S_4) = 69 + 60 = 129$. If $\delta = 2$ and $\varepsilon = 50$, the loc-itemset $<D; i_4>$ is a 1-WULI, the loc-pattern $<A; i_1><C; \{i_2, i_3\}>$ is a 2-WULP, and the moving pattern $<\{<A; i_1><C; \{i_2, i_3\}>\}; ABC>$ is a 2-WUMSP.

Theorem. (Sequence weighted downward closure property, SWDC). For any pattern P, if P is not a WUP, any superset of P is not a WUP.

Proof: By the definition about SWU, for a pattern P, $SWU(P)$ is larger than or equal to $SWU(P')$, where P' is a superset of P. If $SWU(P)$ is less than ε, $SWU(P')$ is also less than ε. Similarly, by the definition about support, $sup(P)$ is larger than or equal to $sup(P')$. If $sup(P)$ is less than δ, $sup(P')$ is also less than δ. By the above two conditions, if $SWU(P) < \varepsilon$ or $sup(P) < \delta$, $SWU(P')$ or $sup(P')$ must be less than ε or δ, respectively. ∎

Problem Statement. Given a mobile transaction sequence database D, a pre-defined utility table, a user-specified minimum utility threshold ε and a user-specified minimum support threshold δ, the problem of mining high utility mobile sequential patterns from D is to discover all high utility mobile sequential patterns whose supports and utilities are larger than or equal to the two thresholds ε and δ, respectively.

4   Proposed Method: UMSP$_L$

The workflow of the proposed method is shown in Figure 1. The proposed algorithm *UMSP$_L$ (high Utility Mobile Sequential Pattern mining by a Level-wised method)* consists of four steps. The first three steps are to find WUMSPs by the SWDC property. In step 1, the original database is scanned several times to find all WULIs and the WULIs are mapped to specific identities in a mapping table. Note that the mapped WULIs are 1-WULPs. Then in step 2, the original database is transformed into a trimmed database by mapping the WULIs to their new identities in the mapping table.

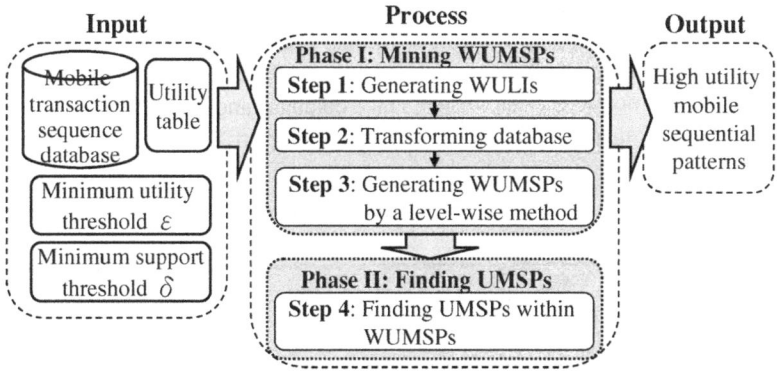

Fig. 1. Framework of the proposed algorithm UMSP$_L$

Subroutine: Generate WUMSP (Step 3 of algorithm UMSP$_L$)
Input: All 1-WULPs, a trimmed database D$_T$, a minimum support threshold δ and a minimum utility threshold ε
Output: WUMSPs

1. Join the 1-WULPs to form candidate 2-WULPs and then store them into 2-candidate trees.
2. Perform an additional scan of D$_T$ to check the supports, SWUs and paths of all candidate 2-WULPs, and then generate 2-WULPs. Assume there is a candidate 2-WULP X, if sup(X) ≥ δ and SWU(X) ≥ ε, X is a 2-WULP. Generate 2-WUMSPs by joining the 2-WULPs with their corresponding paths in the 2-candidate trees.
3. Generate candidate 3-WULPs by combining the 2-WULPs of two 2-WUMSPs if the path of one 2-WUMSP contains another. Store the candidate 3-WULPs into 3-candidate trees.
4. Perform the same process as 2 to find 3-WUMSPs.
5. Generate candidate k-WULPs, where k > 3, by combining the (k-1)-WULPs of the two (k-1)-WUMSPs whose paths are equal to each other. Store the generated candidate k-WULPs into k-candidate trees.
6. Recursively perform the processes 4 and 5 until no candidate WULP is generated.

Fig. 2. The framework of the proposed algorithm UMSP$_L$

After this step, the loc-items which are impossible to be the elements of high utility mobile sequential patterns are removed from the database. Subsequently, the trimmed database is utilized to find the WUMSPs by the proposed level-wised method in the third step. This step is the key step to the mining performance and its procedure is shown in Figure 2. Finally in the fourth step, the WUMSPs are checked to find high utility mobile sequential patterns by an additional scan of the original database.

Next, we use an example to describe the process of the proposed algorithm $UMSP_L$ in detail. Take the mobile transaction sequence database and the utility table in Table 1 and Table 2 for example. Assume the minimum support threshold δ is set as 2 and

Table 3. An example of mapping table

WULIs	$A;i_1$	$C;i_2$	$C;i_3$	$D;i_4$	$F;i_5$	$G;i_7$	$C;\{i_2,i_3\}$
After Mapping (1-WULPs)	$A;t_1$	$C;t_2$	$C;t_3$	$D;t_4$	$F;t_5$	$G;t_6$	$C;t_7$

Table 4. An example of trimmed database D_T

SID	Sequences of Loc-itemsets	Path
S_1'	$<A; t_1; 1>, <C; \{t_2; t_3; t_7\}; 3>, <D; t_4; 4>, <F; t_5; 6>$	ABCDEF
S_2'	$<A; t_1; 1>, <C; t_3; 3>, <F; t_5; 6>, <G; t_6; 7>$	ABCDEFG
S_3'	$<A; t_1; 1>, <C; t_3; 3>, <D; t_4; 4>, <F; t_5; 6>$	AWCDEF
S_4'	$<A; t_1; 1>, <C; \{t_2; t_3; t_7\}; 3>, <F; t_5; 5>, <G; t_6; 7>$	ABCDEFG

the minimum utility threshold ε is set as 100. In the first step, WULIs whose support and SWU are larger than or equal to δ and ε are generated by the processes similar to [8]. In this example, WULIs $<A;i_1>, <C;i_2>, <C;i_3>, <D;i_4>, <F;i_5>, <G;i_7>$ and $<C;\{i_2, i_3\}>$ are generated. (For simplification, the braces of 1-items are omitted.) Then these WULIs are mapped into a mapping table as shown in Table 3.

In the second step, by using the mapping table, the original database D is mapped into the trimmed database D_T as shown in Table 4. In D_T, the original mobile transaction sequences are parsed into the sequences of loc-itemsets and paths. Note that if there is no item in the start or end location of a path, the location will be trimmed. In other words, the paths in D_T must start and end with loc-itemsets. In Table 4, the last number in a loc-itemset stands for the position of the loc-itemset in the path. Take S_1' for example, $<F; t_5; 6>$ means that t_5 is happened in F, and F is in the sixth position of the path.

In step 3, the candidate 2-WULPs are generated by joining the 1-WULPs in the mapping table. At the same time, they are stored into k-candidate trees (k is the length of the patterns) which are shown in Figure 3. Each k-candidate tree stores the candidate k-WULPs whose last loc-itemsets are the same. When inserting a candidate WULP into a candidate tree, the last loc-itemsets of the WULP will be recorded in the root node of the candidate tree, and the other loc-itemsets are then stored in the tree by their original order sequentially. After constructing 2-candidate trees, an additional scan of D_T is performed to check the path support and SWU of each candidate 2-WULP and to form the paths in moving patterns. In this example, three 2-WULPs in Figure 3 (a), i.e., <A;t₁><G;t₆>, <C;t₃><G;t₆> and <F;t₅><G;t₆>, are generated, and then three 2-WUMSPs, i.e., {<A;t₁><G;t₆>; ABCDEFG}, {<C;t₃><G;t₆>; CDEFG} and {<F;t₅><G;t₆>; FG}, are generated.

After generating 2-WUMSPs, candidate 3-WULPs are generated. For two 2-WUMSPs X and Y, they can generate a 3-WULP if the path of X contains that of Y, and vise versa. For example, since the path ABCDEFG contains CDEFG, the candidate 3-WULP <{<A;t₁><C;t₃><G;t₆>}> is generated by the two 2-WUMSPs <{<A;t₁><G;t₆>}; ABCDEFG> and <{<C;t₃><G;t₆>}; CDEFG> and it is inserted into the 3-candidate tree in Figure 3 (b). After constructing the trees, an additional database scan of D_T is performed to generate 3-WUMSPs. In this case, three patterns,

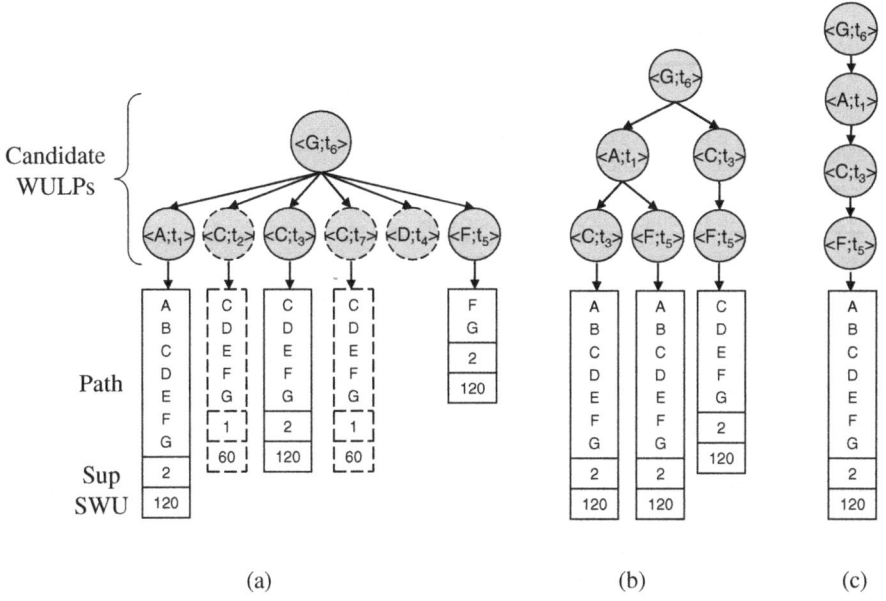

(a)   (b)   (c)

Fig. 3. (a) 2-candidate tree of <G:t6>   (b) 3-candidate tree of <G:t6> (c) 4-candidate tree of <G:t6>

i.e., {<A;t1><C;t₃><G;t₆>; ABCDEFG}, {<A;t₁><F;t₅><G;t₆>; ABCDEFG} and {<C;t₃><F;t₅><G;t₆>; CDEFG}, are generated. Finally, candidate k-WULPs (where k≥4) are generated by combining two (k-1)-WUMSPs whose paths are equal to each other. For example, since the paths of the two 3-WUMSPs <{<A;t₁><C;t₃><G;t₆>}; ABCDEFG> and <{<A;t₁><F;t₅><G;t₆>}; ABCDEFG> are the same, a new candidate 4-WULP <{<A;t₁><C;t₃><F;t₅><G;t₆>}> is generated and inserted into the 4-candidate tree. The processes will be recursively executed until no candidate moving pattern is generated.

After all WUMSP are generated, an additional scan of database will be performed to check real high utility mobile sequential patterns in step 4. The WUMSPs whose utilities are larger than or equal to the minimum utility threshold are regarded as high utility mobile sequential patterns and be outputted.

5   Experimental Evaluations

In this section, the performance of the proposed algorithm $UMSP_L$ is evaluated. The experimental environment is a PC with a 2.4 GHz Processor, 2 GB memory and Microsoft Windows Server 2003 operating system. The algorithms are implemented in Java. The default settings of the parameters are listed in Table 5. In the following experiments, the proposed algorithm is compared with the algorithm MSP which is extended from [16]. The processes of MSP are as follows: First, the mobile sequential patterns whose supports are larger than or equal to the minimum support threshold are generated by the algorithm TJ_{PF} in [16]. Then, an additional check of utility value by

Table 5. Parameter settings in experiments

Parameter Descriptions	Default
D: Number of mobile transaction sequences in the dataset	50k
P: Average length of mobile transaction sequences	20
T: Average number of items per transaction	2
N: Size of mesh network	8
w: Unit profit of each item	1~1000
q: Number of purchased items in transactions	1~5

scanning the original database once is performed for finding high utility mobile sequential patterns.

The first part of the experiments is the performance under varied minimum support thresholds. The experiments are performed on the dataset D50k.P20.T2.N8 and the minimum utility threshold is set as 1%. The results are shown in Figure 4 (a). It can be seen that the execution time of $UMSP_L$ outperforms MSP, especially when the minimum support threshold is low. This is because $UMSP_L$ used two thresholds at the same time such that more candidates are pruned in the mining process.

The second part of the experiments is the performance under varied minimum utility thresholds. The experiments are also performed on the dataset D50k.P20.T2.N8 and the minimum support threshold is set as 0.5%. The results are shown in Figure 4 (b). In this figure, when the minimum utility threshold is higher, the performance of $UMSP_L$ is better. On the other hand, when the minimum utility threshold is lower, the execution time of $UMSP_L$ is closer to that of MSP. This is because when the minimum utility threshold is lower, fewer candidates could be pruned in the mining process. When the minimum utility threshold is below 0.4%, almost no candidate is pruned. Thus, the performance of the two algorithms is almost the same.

The third part of the experiments is the performance under varied number of mobile transaction sequences. The experiments are performed on the dataset Dxk.P20.T2.N8. The minimum support threshold is set as 0.5% and the minimum utility threshold is set as 1%. The results are shown in Figure 4 (c). In this figure, we can observe that both algorithms show to keep good scalability, that is, when the number of mobile transaction sequences is larger, the execution time of the two algorithms is increased linearly. However, the proposed algorithm $UMSP_L$ still outperforms the compared algorithm.

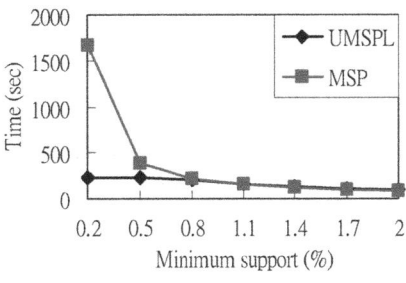
(a) Varying minimum support thresholds

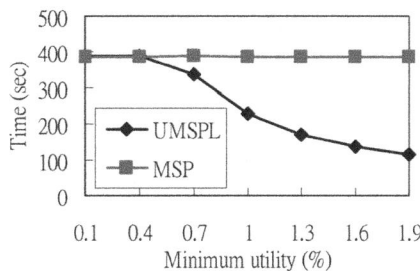
(b) Varying minimum utility thresholds

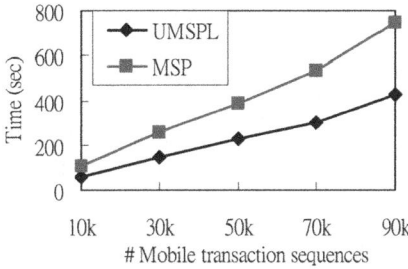
(c) Varying number of transaction sequences

Fig. 4. The performance of the two algorithms under different conditions

6   Conclusions

In this paper, we proposed a novel algorithm named $UMSP_L$ for finding high utility mobile sequential patterns. This paper addressed a new research issue which combines mobility pattern mining and utility mining. The experimental results showed that the proposed $UMSP_L$ algorithm not only outperformed the compared algorithm MSP in different conditions but also delivered good scalability. For future work, we will develop further algorithms based on other frameworks for this problem and perform detailed experiments under various conditions.

Acknowledgment. This research was supported by National Science Council, Taiwan, R.O.C. under grant no. NSC99-2631-H-006-002 and NSC99-2218-E-006-001.

References

1. Agrawal, R., Srikant, R.: Fast Algorithms for Mining Association Rules. In: Int'l Conf. on Very Large Data Bases, pp. 487–499 (1994)
2. Agrawal, R., Srikant, R.: Mining Sequential Patterns. In: IEEE Int'l Conference on Data Mining, pp. 3–14 (1995)

3. Ahmed, C.F., Tanbeer, S.K., Jeong, B.-S., Lee, Y.-K.: Efficient Tree Structures for High Utility Pattern Mining in Incremental Databases. IEEE Transaction on Knowledge and Data Engineering 21(12), 1708–1721 (2009)
4. Chen, M.-S., Park, J.-S., Yu, P.S.: Efficient data mining for path traversal patterns. IEEE Transactions on Knowledge and Data Engineering 10(2), 209–221 (1998)
5. Chu, C.J., Tseng, V.S., Liang, T.: Mining Temporal Rare Utility Itemsets in Large Databases Using Relative Utility Thresholds. Int'l Journal of Innovative Computing Information and Control 4(11), 2775–2792 (2008)
6. Lee, S.C., Paik, J., Ok, J., Song, I., Kim, U.M.: Efficient mining of user behaviors by temporal mobile access patterns. Int'l Journal of Computer Science Security 7(2), 285–291 (2007)
7. Li, Y.-C., Yeh, J.-S., Chang, C.-C.: Isolated items discarding strategy for discovering high utility itemsets. Data & Knowledge Engineering 64(1), 198–217 (2008)
8. Liu, Y., Liao, W.-K., Choudhary, A.: A fast high utility itemsets mining algorithm. In: Utility-Based Data Mining (2005)
9. Lu, E.H.-C., Tseng, V.S.: Mining cluster-based mobile sequential patterns in location-based service environments. In: IEEE Int'l Conf. on Mobile Data Management (2009)
10. Lu, E.H.-C., Huang, C.-W., Tseng, V.S.: Continuous Fastest Path Planning in Road Networks by Mining Real-Time Traffic Event Information. In: Int'l Symposium on Intelligent Informatics (2009)
11. Tseng, V.S., Lu, E.H.-C., Huang, C.-H.: Mining temporal mobile sequential patterns in location-based service environments. In: IEEE Int'l Conf. on Parallel and Distributed Systems (2007)
12. Tseng, V.S., Lin, W.C.: Mining sequential mobile access patterns efficiently in mobile web systems. In: Int'l Conf. on Advanced Information Networking and Applications, pp. 867–871 (2005)
13. Tseng, V.S., Wu, C.-W., Shie, B.-E., Yu, P.S.: UP-Growth: An Efficient Algorithm for High Utility Itemsets Mining. In: ACM SIGKDD Conf. on Knowledge Discovery and Data Mining, pp. 253–262 (2010)
14. Yao, H., Hamilton, H.J.: Mining itemset utilities from transaction databases. Data & Knowledge Engineering 59, 603–626 (2006)
15. Yun, C.-H., Chen, M.-S.: Using pattern-join and purchase-combination for mining web transaction patterns in an electronic commerce environment. In: IEEE Annu. Int. Computer Software and Application Conference, pp. 99–104 (2000)
16. Yun, C.-H., Chen, M.-S.: Mining Mobile Sequential Patterns in a Mobile Commerce Environment. IEEE Transactions on Systems, Man, and Cybernetics-Part C: Applications and Reviews 37(2) (2007)
17. Cao, L.: In-depth Behavior Understanding and Use: the Behavior Informatics Approach. Information Science 180(17), 3067–3085 (2010)

A Novel Method for Community Detection in Complex Network Using New Representation for Communities

Wang Yiwen and Yao Min

College of Computer Science, Zhejiang University, Hangzhou 310027, China

Abstract. During the recent years, community detection in complex network has become a hot research topic in various research fields including mathematics, physics and biology. Identifying communities in complex networks can help us to understand and exploit the networks more clearly and efficiently. In this paper, we investigate the topological structure of complex networks and propose a novel method for community detection in complex network, which owns several outstanding properties, such as efficiency, robustness, broad applicability and semantic. The method is based on partitioning vertex and degree entropy, which are both proposed in this paper. Partitioning vertex is a novel efficient representation for communities and degree entropy is a new measure for the results of community detection. We apply our method to several large-scale data-sets which are up to millions of edges, and the experimental results show that our method has good performance and can find the community structure hidden in complex networks.

Keywords: community detection, complex network, adjacency matrix.

1 Introduction

A network is a mathematical representation of a real-world complex system and is defined by a collection of vertices(nodes) and edges(links) between pairs of nodes. The modern science of networks has brought significant advances to our understanding of complex systems, which is an interdisciplinary endeavor, with methods and applications drawn from across the natural, social, and information sciences. There has been a surge of interest within the physics community in the properties of networks of many kinds, including the Internet, the world wide web, citation networks, transportation networks, software call graphs, email networks, food webs, and social and biochemical networks. Most of these networks are generally sparse in global yet dense in local. The view that networks are essentially random was challenged in 1999 when it was discovered that the distribution of number of links per node of many real networks is different from what is expected in random networks.

One of the most relevant features of graphs representing real systems is community structure. Communities, also called clusters or modules, are groups of

L. Cao et al. (Eds.): PAKDD 2011 Workshops, LNAI 7104, pp. 89–99, 2012.

vertices which probably share common properties or play similar roles within the graph. Such communities, can be considered as fairly independent compartments of a graph, playing a similar role like, e. g., the tissues or the organs in the human body. Detecting communities is of great importance in sociology, biology and computer science, in particular in behavior informatics [4], disciplines where systems are often represented as graphs. Identifying meaningful community structure in social networks is inherently a hard problem. Extremely large network size or sparse networks compound the difficulty of the task, despite the huge effort of a large interdisciplinary community of scientists working on it over the past few years.

Detecting communities is of great importance in sociology, biology and computer science, disciplines where systems are often represented as graphs[7]. One important application of community detection is for web search. While web is constantly growing, web search is becoming more and more a complex and confusing task for the web user. The vital question is which the right information for a specific user is and how this information could be efficiently delivered, saving the web user from consecutive submitted queries and time-consuming navigation through numerous web results[8]. Most existing web search engines return a list of results based on the query without paying any attention to the underlying users interests. Web community detection is one of the important ways to enhance retrieval quality of web search engine. How to design one highly effective algorithm to partition network community with few domain knowledge is the key to network community detection.

2   Related Work

There have been many various approaches and algorithms to analyze the community structure in complex networks, which use methods and principles of physics, artificial intelligence, graph theory and so on. Most of the algorithms use adjacency matrix to present the complex network. Complex network theory uses the tools of graph theory and statistical mechanics to deal with the structure of relationships in complex systems. A network is defined as a graph $G = (N, E)$ where N is a set of nodes connected by a set of edges E. We will refer to the number of nodes as n and the number of edges as m. The network can also be defined in terms of the *adjacency matrix* $G = A$ where the elements of A are

$$A_{ij} = \begin{cases} 1 \ \textit{if node } i \textit{ and node } j \textit{ are connected,} \\ 0 \ \textit{otherwise.} \end{cases} \tag{1}$$

As discussed at length in two recent review articles [7,9] and references therein, the classes of techniques available to detect communities are both numerous and diverse; they include hierarchical clustering methods such as single linkage clustering, centrality-based methods, local methods, optimization of quality functions such as modularity and similar quantities, spectral partitioning, likelihood-based methods, and more.

The Kernighan-Lin algorithm [1] is one of the earliest methods proposed and is still frequently used, often in combination with other techniques. The authors were motivated by the problem of partitioning electronic circuits onto boards: the nodes contained in different boards need to be linked to each other with the least number of connections. The procedure is an optimization of a benefit function Q, which represents the difference between the number of edges inside the modules and the number of edges lying between them. But this algorithm need to know the size of two network communities, otherwise, this flaw enables it very difficult to actual application in network analysis.

In general, it is uncommon to know the number of clusters in which the graph is split, or other indications about the membership of the vertices. In such cases, hierarchical clustering algorithms [2] is proposed, that reveal the multilevel structure of the graph. Hierarchical clustering is very common in social network analysis, biology, engineering, marketing.

Spectral clustering is able to separate data points that could not be resolved by applying directly k-means clustering. The first contribution on spectral clustering was a paper by Donath and Hoffmann [3], who used the eigenvectors of the adjacency matrix for graph partitions. The Laplacian matrix is by far the most used matrix in spectral clustering.

Above algorithms, divisive method and agglomerative method should be introduced. GN algorithm [5], which is one kind of divisive methods, through removing the edge one by one which weight is highest(this weight is the number of short-path passing through each node in the network) until entire network is divided into more and more small part. But GN algorithm has no function to measure the quality of one community partition.

The first problem in graph clustering is to look for a quantitative definition of community. No definition is universally accepted. As a matter of fact, the definition often depends on the specic system at hand and/or application one has in mind [16]. The most popular quality function for community detection is the modularity of Newman and Girvan [6]. It is based on the idea that a random graph is not expected to have a cluster structure, so the possible existence of clusters is revealed by the comparison between the actual density of edges in a subgraph and the density one would expect to have in the subgraph if the vertices of the graph were attached regardless of community structure. This expected edge density depends on the chosen null model,i. e. a copy of the original graph keeping some of its structural properties but without community structure. Modularity can then be written as follows

$$Q = \frac{1}{2m} \sum_{ij} (A_{ij} - P_{ij}) \delta(C_i, C_j) \tag{2}$$

where the sum runs over all pairs of vertices, A is the adjacency matrix, m the total number of edges of the graph, and P_{ij} represents the expected number of edges between vertices i and j in the null model. The δ-function yields one if vertices i and j are in the same community ($C_i = C_j$), zero otherwise.

Modularity is by far the most used and best known quality function, otherwise a number of Modularity-based methods have been proposed in recent years. But this measure essentially compares the number of links inside a given module with the expected values for a randomized graph of the same size and same degree sequences.

The current algorithms are successful approaches in community detection. However there are some drawbacks of current algorithms.

1. Most of these algorithms have time complexities that make them unsuitable for very large networks.

2. Some algorithms have data structures like matrices, which are hard to implement and use in very large networks.

3. Some algorithms also need some priori knowledge about the community structure like number of communities, where it is impossible to know these values in real-life networks.

3   Proposed Method

In this section, we introduce a novel method for community detection in complex network, which owns several outstanding properties, such as efficiency, robustness, broad applicability and semantic. The method proposed is based on partitioning vertex and degree entropy. Partitioning vertex is a new efficient representation for communities and degree entropy is a new measure for the results of community detection.

3.1   Partitioning Vertex

There are a number of recent studies focused on community detection in complex network, but all of them use a naive representation for communities in which each community is assigned to precisely a unique number. For example, a network consists 9 vertices is divided into 3 communities. The result of community detection is $\{1(2), 2(2), 3(1), 4(1), 5(1), 6(2), 7(3), 8(3), 9(3)\}$. So community 1 contains vertices $\{3, 4, 5\}$, community 2 contains vertices $\{1, 2, 6\}$, community 3 contains vertices $\{7, 8, 9\}$. We must record each vertex is assigned to which community when using naive representation and there two critical defects. One is that it is supposed to lose effectiveness and occupy too much space when the number of vertices comes to millions or even larger. The other defect is that naive representation does no good to understand the structure of network better.

For the two reasons above, we proposed a novel representation for communities: partitioning vertex. If we choose n vertices in the network as partitioning vertices, such as $v_1, v_2, ..., v_n$, there are totaly 2^n corresponding communities. Each community can be represented as a n-bit binary string, and each bit stands for that every corresponding partitioning vertex is connected with the community or not. For example, as shown in Figure. 1, the network has 12 vertices and two partitioning vertices is selected, vertex 9 and vertex 10. The whole network is divided into 4 communities, which are community 00(not connected with vertex 9 and vertex 10) with vertices$\{11, 12\}$, community 01(connected with vertex 9, not connected with vertex 10) with vertices{1, 2, 3, 0}, community 10(connected with vertex 10) with vertices {6, 7, 8, 9}, community 11(connected with vertex 9 and vertex 10) with vertices{4, 5}.

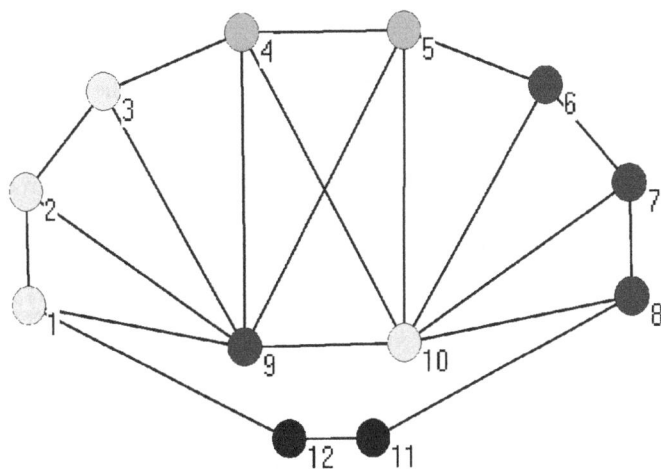

Fig. 1. Example for Partitioning Vertex

There are two outstanding advantages when using partitioning vertex to present communities. One is that you do not have to record each vertex is assigned to which community, only the selected partitioning vertices are needed. So the community detection can be changed to find out partitioning vertices, which is much easier than finding out every vertex is assigned to which community. The other advantage is that every community itself can give information about why the vertices are clustered. The vertices in the same community are all connected with some partitioning vertex or not. In reality, the vertices of complex network have their own specific meanings, and the edges between vertices are also rich in semantic information. When using semantic information in the network and partitioning vertices, we can tell what the semantic information of communities, which help us understand the network better.

It is obvious that a community is no restricted when using a naive representation. But what about the partitioning vertex way? The only difference is that the structurally equivalent vertices(have the same relationships to all other vertices) are in the same community. It is suitable in community detection, for the structurally equivalent vertices probably share common properties or play similar roles in the network.

3.2   Degree Entropy

If we adopt partitioning vertex to present communities, the community detection is changed to find out suitable partitioning vertices. And the most important thing is how to measure the quality of community detection when using partitioning vertex. We proposed degree entropy here to cover it, which is a quantitative function as shown below.

$$DE(v_i) = (degree(v_i)/m) * log_2(1/(degree(v_i)/m))$$ (3)

Here, v_i stands for a vertex in a network, and $degree(v_i)$ stands for the degree of vertex(the number of vertices that are connected with the vertex), m stands for the number of edges in the network. When we say vertex A is connected with vertex B, we get the information that the edge connected vertex A and vertex B is one of the edges connected with vertex B. And the degree entropy is designed to quantitatively determine information. When the degree entropy of vertex B is larger, we get more information. In order to quantitatively determine the information of communities when using partitioning vertices, we definite the degree entropy of community as shown below.

$$DE(c_i) = \sum_{i}^{n}(DE(pv_i))$$ (4)

The degree entropy of community is a sum of degree entropy of every partitioning vertices. Here, c_i stands for a community in a network, n stands for the number of partitioning vertices, $DE(pv_i)$ stands for the degree entropy of partitioning vertex i. We use the degree entropy for community to measure the quality of community detection. So the purpose of community detection is maximizing the degree entropy, as larger degree entropy implies more information.

3.3   The Method

The degree distribution was one of the most popular issues of concern in complex network. More importantly, the scientists find that the degree distribution of complex network obey the power law or just called scale-free property. Scale-free property means that there some vertices owning quite high degree, which is suitable for using partitioning vertices to present communities. So we adopt partitioning vertices in the community detection in complex network, and the community detection is changed to select partitioning vertices.

As the degree entropy of community is introduced to measure the quality of community detection, we must select suitable partitioning vertices to maximize the degree entropy of community. The number of partitioning vertices must be fixed first, such as n, for lager degree entropy of community can be obtained just only using one more partitioning vertex. So which n vertices in a network should be selected to obtain the maximum degree entropy of community? Since the degree entropy of community is a sum of degree entropy of every partitioning vertex, we only need to select n vertices with largest degree entropy. The degree entropy changes with the value of $(degree(v_i)/m)$ is shown in Figure. 2. And the largest degree entropy is obtained when the value of $(degree(v_i)/m)$ is $1/e$, where e is the base of natural logarithm and approximately equal to 2.718281828.

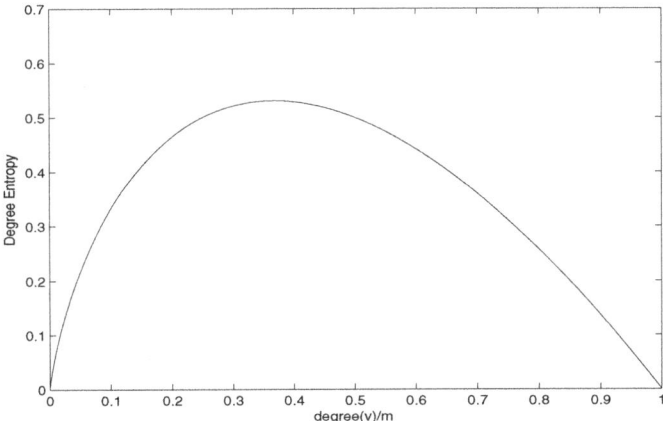

Fig. 2. Degree Entropy changes with the value of degree(v)/m, m stands for the number of edges in the network, degree(v) stands for the degree of vertex

We can select n vertices with largest degree entropy as partitioning vertices according to Figure. 2. In most cases, the value of $(degree(v_i)/m)$ is smaller than $1/e$, where degree entropy is a monotonically increasing function of $(degree(v_i)/m)$. Since m is a constant stands for the number of edges in a network, degree entropy a monotonically increasing function of $degree(v_i)$ too. So we can just select the top degree n vertices as partitioning vertices.

The main steps of our proposed method can be described as follows:

Step1: Set the only parameter n, the number of partitioning vertices. n partitioning vertices can divide the network into most 2^n communities. More communities can be obtained when using larger n. The n is set according to meet the need of user.

Step2: Calculate the degree of every vertex in the network. Our method owns broad applicability for the only information needed is degree.

Step3: Choose the top degree n vertices as partitioning vertices.

4 Experiments

In this section we apply our method to various of networksand all the graphs are drawn with a free software Pajek [10]. It is difficult to compare our method with other methods since there is no universally accepted measure for the results of community detection. The method is tested on the Zachary Karate Club [11], Java Compile-Time Dependency [12], High Energy Particle Physics (HEP) literature [13] and Stanford web graph [14] networks. In each case we find that our method reliably detects the structures.

4.1 Zachary Karate Club

Zachary Karate Club is one of the classic studies in social network analysis. Over the course of two years in the early 1970s, Wayne Zachary observed social interactions between the members of a karate club at an American university. He built network of connections with 34 vertices and 78 edges among members of the club based on their social interactions. By chance, a dispute arose during the course of his study between the clubs administrator and the karate teacher. As a result, the club splits into two smaller communities with the administrator and the teacher being as the central persons accordingly. Figure. 3 shows the detected two communities by our method which are mostly matched with the result of Zacharys study(only 3 vertices are different). The number of partitioning vertices is set to 1 as Zachary Karate Club is known to divided into 2 communities. The selected partitioning vertex is 34, which exactly stands for the club's administrator.

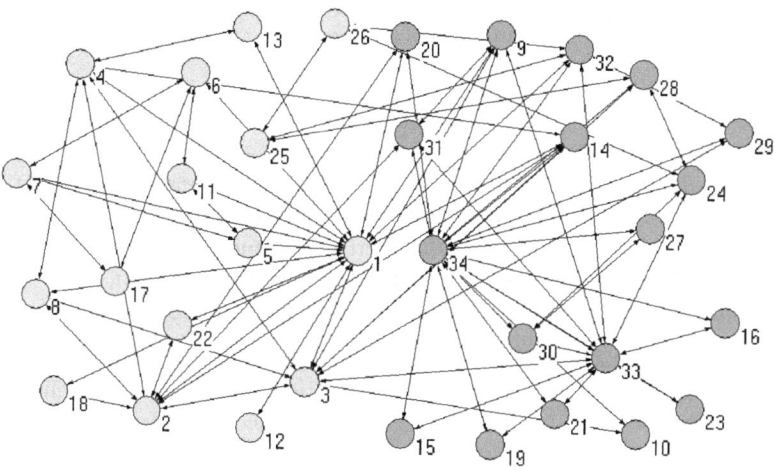

Fig. 3. Community Detection Results of Zachary Karate Club

4.2 Java Compile-Time Dependency

The Java compile-time dependency graph consists of nodes representing Java classes and directed edges representing compile-time dependencies between two classes. For example, the class java.util.Calendar depends (among others) on the class java.util.Hashtable. The data provided contains the dependencies for all classes under the java.* packages for JDK 1.4.2. Dependencies between a provided class and one outside the java.* realm are not shown. Figure. 4 shows the original Java compile-time dependency graph and Figure. 5 shows the detected four communities by our method. There are 1538 vertices and 8032 arcs in the Java compile-time dependency network. The number of partitioning vertices is set to 2 and the whole network is divided into four communities.

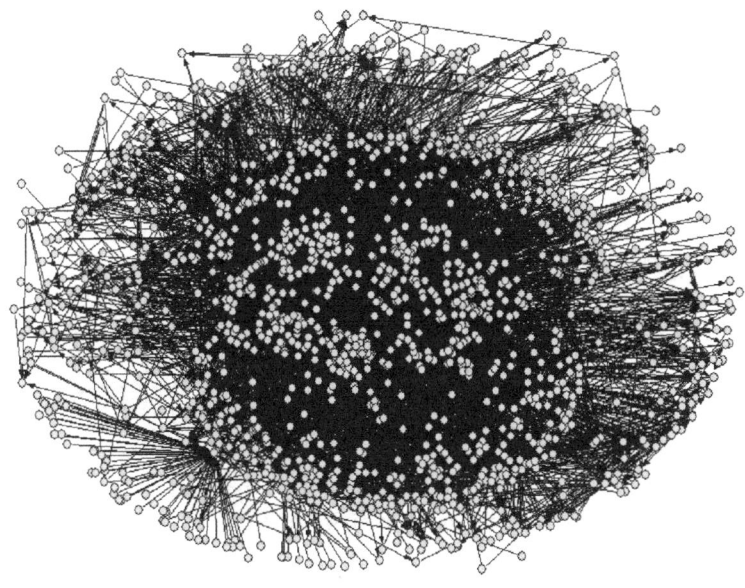

Fig. 4. Java Compile-Time Dependency

Two selected partitioning vertices are vertex 5 (java.lang.String) and vertex 20 (java.lang.Object). The semantic information for different communities is shown as the following Table. 1. The most important property of our method is semantic. The communities detected by our method own their semantic information, which helps people to understand the structure of complex network much better.

Table 1. Semantic Information for Experiment of Fig. 5

Community	Color	Semantic Information
00	Yellow	not depend on
01	Green	depend on java.lang.String
10	Red	depend on java.lang.Object
11	Blue	depend on both java.lang.String and java.lang.Object

4.3 HEP Literature and Stanford Web Graph

An efficient community detection method must can deal with large-scale networks. In this section, our method is tested on HEP literature network and Stanford web graph.The HEP literature network is a citation data from KDD Cup 2003, a knowledge discovery and data mining competition held in conjunction with the Ninth Annual ACM SIGKDD Conference. The Stanford Linear Accelerator Center SPIRES-HEP database has been comprehensively cataloguing the HEP literature online since 1974, and indexes more than 500,000 high-energy physics related articles including their full citation tree. It is a directed

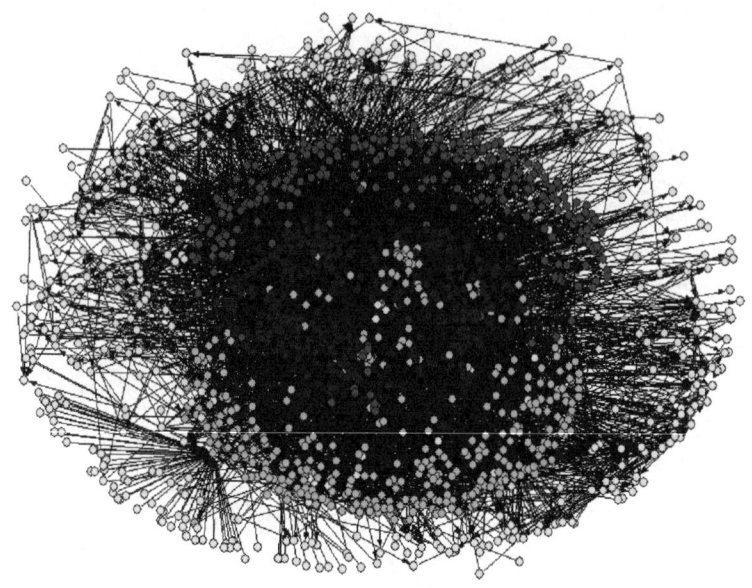

Fig. 5. Community Detection Results of Java Compile-Time Dependency

network with 27240 vertices and 342437 arcs. The units names are the arXiv
IDs of papers; the relation is X cites Y. Stanford web graph was obtained from
a September 2002 crawl (281903 pages, 2382912 links). The matrix rows represent the inlinks of a page, and the columns represent the outlinks. Suitable
visualizations of the two graphs are difficult for they own too many vertices and
edges. We choose fast Modurlity algorithm [15] which proposed by Newman to
compare cost time with our method. The results is shown in Table. 2.

Table 2. Cost Time (seconds)

Network	Number of Vertices	Number of Edges	Our Method	Fast Modurlity
Java Dependency	1539	8032	0.091	4.2
HEP literature	27240	342437	2.7	failed
Stanford web graph	281903	2382912	3.8	failed

5 Conclusion

In this paper, we have mainly proposed an novel method for community detection
in complex network. Identifying communities in complex networks can help us
to understand and exploit the networks more clearly and efficiently. The method
is based on partitioning vertex and degree entropy. Partitioning vertex is a novel
efficient representation for communities and degree entropy is a new measure
for the results of community detection. The method owns several outstanding

properties, such as efficiency, robustness, broad applicability and semantic. We apply our method to several large-scale data-sets which are up to millions of edges, and the experimental results show that our method has good performance and can find the community structure hidden in complex networks.

For the future work, we will continue our research by focusing on the applying our method to online social networks, such as Twitter, Facebook and Myspace. We will also search for more refined theoretical models to describe the structure of complex network.

References

1. Kernighan, B.W., Lin, S.: An efficient heuristic procedure for partitioning graphs. Bell Syst. Tech. J. 49, 291–307 (1970)
2. Hastie, T., Tibshirani, R., Friedman, J.H.: The Elements of Statistical Learning. Springer, Berlin (2001)
3. Donath, W., Hoffman, A.: Lower bounds for the partitioning of graphs. IBM J. Res. Dev. 17(5), 420–425 (1973)
4. Cao, L.: In-depth Behavior Understanding and Use: the Behavior Informatics Approach. Information Science 180(17), 3067–3085 (2010)
5. Girvan, M., Newman, M.E.J.: Community structure in social and biological networks. Proc. Natl Acad. 99, 7821–7826 (2002)
6. Girvan, M., Newman, M.E.J.: Community Structure in Social and Biological Networks. PNAS 99, 7821–7826
7. Fortunato, S.: Community detection in graphs. ArXiv:0906.0612v2 physics.soc-ph (January 25, 2010)
8. Garofalakis, J., Giannakoudi, T., Vopi, A.: Personalized Web Search by Constructing Semantic Clusters of User Profiles. In: Lovrek, I., Howlett, R.J., Jain, L.C. (eds.) KES 2008, Part II. LNCS (LNAI), vol. 5178, pp. 238–247. Springer, Heidelberg (2008)
9. Porter, M.A., Onnela, J.-P., Mucha, P.J.: Communities in networks. Notices of the American Mathematical Society 56, 1082–1097, 1164-1166 (2009)
10. http://vlado.fmf.uni-lj.si/pub/networks/pajek/default.htm
11. Zachary, W.: An information flow model for conflict and fission in small groups. Journal of Anthropological Research 33, 452–473 (1977)
12. http://gd2006.org/contest/details.php
13. http://vlado.fmf.uni-lj.si/pub/networks/data/hep-th/hep-th.htm
14. Kamvar, S.D., Haveliwala, T.H., Manning, C.D., Golub, G.H.: Exploiting the Block Structure of the Web for Computing PageRank. Preprint (March 2003)
15. Newman, M.E.J.: Fast algorithm for detecting community structure in networks. Physical Review E 69, 066133 (2004)
16. Fortunato, S.: Community detection in graphs. Physics Reports 486, 75–174 (2010)

Link Prediction on Evolving Data
Using Tensor Factorization

Stephan Spiegel[1], Jan Clausen[1], Sahin Albayrak[1], and Jérôme Kunegis[2]

[1] DAI-Labor, Technical University Berlin
Ernst-Reuter-Platz 7, 10587 Berlin, Germany
{stephan.spiegel,jan.clausen,sahin.albayrak}@dai-labor.de
http://www.dai-labor.de
[2] University of Koblenz-Landau
Universitätsstraße 1, 56072 Koblenz, Germany
kunegis@uni-koblenz.de
http://west.uni-koblenz.de/

Abstract. Within the last few years a lot of research has been done on large social and information networks. One of the principal challenges concerning complex networks is link prediction. Most link prediction algorithms are based on the underlying network structure in terms of traditional graph theory. In order to design efficient algorithms for large scale networks, researchers increasingly adapt methods from advanced matrix and tensor computations.

This paper proposes a novel approach of link prediction for complex networks by means of multi-way tensors. In addition to structural data we furthermore consider temporal evolution of a network. Our approach applies the canonical *Parafac* decomposition to reduce tensor dimensionality and to retrieve latent trends.

For the development and evaluation of our proposed link prediction algorithm we employed various popular datasets of online social networks like *Facebook* and *Wikipedia*. Our results show significant improvements for evolutionary networks in terms of prediction accuracy measured through mean average precision.

Keywords: Link Prediction Algorithm, Temporal Network Analysis, Evolving Data, Multi-way Array, Tensor Factorization.

1 Introduction

The growing interest in large-scale networks is originated from the increasing number of online social platforms and information networks. Many studies have scrutinized static graph properties of single snapshots, which lacks of information about network evolution [12,15]. Especially for the challenge of link prediction it is necessary to observe trends within time, which are determined by the addition and deletion of nodes. Most network evolution models are based on the average node degree or the growing function of the network size [13,14,15,16,17,19].

L. Cao et al. (Eds.): PAKDD 2011 Workshops, LNAI 7104, pp. 100–110, 2012.

In our approach we capture temporal trends within a multi-way array, also known as tensor, where time is represented as a separate dimension.

Multi-way data analysis is the extension of two-way data analysis to higher-order data, and is mostly used to extract hidden structures and capture underlying correlations between variables in a multi-way array [4]. Instead of only considering pairwise relations between variables (e.g. {user, item}, {user,time} and {item,time}), multi-way data analysis attempts to discover latent factors among multiple objects [5,23]. The most popular multi-way models in literature are *Tucker* and *Parafac*, which both factorize higher-order data into low-dimensional components [4].

This study employs the *Parafac* model to estimate the probability of an edge to be added within an observed network. The examined data is mainly derived from online social and information networks, and is commonly represented as $i \times j \times k$ multi-way array; where i and j represent the number of users or items respectively, and k denotes the number of points in time capturing the date of interaction. Our main purpose is to predict whether and how much a user is likely to rate an item and to which probability a user is going to interact with someone else. However, there are many other behavior-related applications for link prediction, such as predicting the web pages a web surfer may visit on a given day, the places that a traveler may fly to in a given month, or the patterns of computer network traffic [2,1,8,22]. Behavior-oriented link prediction is an important task in behavior informatics [6].

The rest of the paper is organized as follows. Section 2 gives a general introduction to tensor factorization, whereas Section 3 describes our own link prediction algorithm. The experimental results are presented in Section 4. Finally, Section 5 concludes this paper.

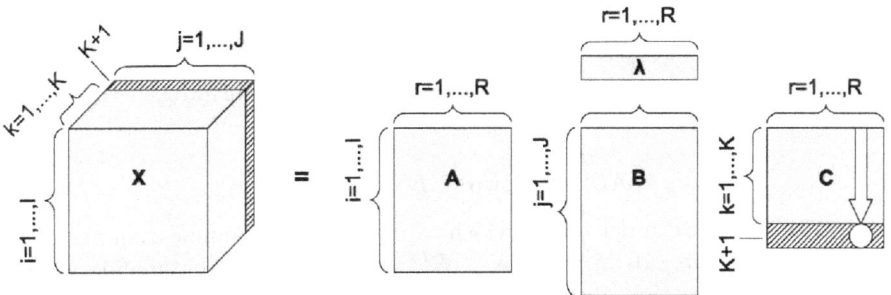

Fig. 1. CP Decomposition of a 3rd-order Tensor:
$$X \in \mathbb{R}^{I \times J \times (K+1)} \Rightarrow \lambda \in \mathbb{R}^{R}; A \in \mathbb{R}^{I \times R}; B \in \mathbb{R}^{J \times R}; C \in \mathbb{R}^{(K+1) \times R}$$

2   Tensor Decomposition

An Nth-order tensor is the product of N vector spaces, each of which has its own coordinate system [10]. Usually first-order and second-order tensors are referred to as vectors and matrices respectively. This paper mainly investigates third-order tensors, also known as multi-way arrays, which can be imagined as a three-dimensional cube like illustrated in Figure 1.

Tensor decompositions first emerged in psychometrics and chemometrics, but recently were adopted to other domains, including data mining and graph analysis. They aim to reveal underlying linear structures and are often employed to reduce noise or data dimensionality [4,10]. Factorization techniques like the CP (CanDecomp/ParaFac) and Tucker model can be considered higher-order generalization of the matrix singular value decomposition (SVD) and principal component analysis (PCA) [2,10]. However, the CP model is more advantageous in terms of interpretability, uniqueness of solution and determination of parameters [4]. Following Kolda [10] the CP mode-3-model can be expressed as either a sum of rank-one tensors (each of them an outer product of vectors a_r, b_r, c_r and a weight λ_r) or factor matrices (refer to Figure 1):

$$ X \approx \sum_{r=1}^{R} \lambda_r (a_r \circ b_r \circ c_r) \equiv [\lambda; A, B, C] \tag{1} $$

with X representing our original tensor and R specifying the number of rank-one components. For the derivation of our proposed link prediction model we use the equivalent factor representation (1), where all columns of matrices A, B, and C are normalized to length one with the weights absorbed into the vector $\lambda \in \mathbb{R}^R$. Each of the factor matrices refer to a combination of the vectors from the rank-one components, i.e.

$$ A = [a_1\ a_2\ \cdots\ a_R] \equiv \begin{bmatrix} a_{1,1} & a_{1,2} & \cdots & a_{1,R} \\ a_{2,1} & a_{2,2} & \cdots & a_{2,R} \\ \vdots & \vdots & \ddots & \vdots \\ a_{I,1} & a_{I,2} & \cdots & a_{I,R} \end{bmatrix} \tag{2} $$

and likewise for B and C. In terms of frontal slices the introduced three-way model (1) can also be written as [10]:

$$ X_k \approx AD^{(k)}B^T \quad \text{with} \quad D^{(k)} = \text{diag}(\lambda_: c_{k,:}) \tag{3} $$

If we consider a third-order tensor which captures intercommunication patterns of i users over k periods of time ($X \in \mathbb{R}^{I \times I \times K}$), the k-th frontal slide would comprise all information about user interactions of date k [3]. Our own link prediction approach estimates future user correlations ($K + 1$) by analysing the temporal trend kept in the observed time slices.

Note that, according to the *Spectral Evolution Model* [11], Equation (3) can also be considered as a joint decomposition of the respective time slice matrices. In the course of time the spectrum captured in matrix D will change, whereas the eigenvectors of matrices A and B stay the same [11].

3   Link Prediction

As might be expected, the term link prediction is characterized as prognosis of future edges within complex networks, where edges may express relations between individuals or abstract objects respectively [17,21]. In general, link prediction techniques are grouped into neighbor-based and link-based algorithms, which either consider adjacent nodes (e.g. *Common Neighbors & Preferential Attachment*) or paths between nodes (e.g. *Katz Algorithm*) to estimate the probability of future edges [17]. Recently there emerged some novel approaches that make use of psychological concepts (e.g. *Theory of Balance & Theory of Status*) to predict links [13,14]. Another group of semantic algorithms utilize attributes of nodes to make prognostications [16,19]. However, many popular link prediction techniques merely scrutinize static graph properties of single snapshots, giving a lack of information about network evolution [12,15].

Our own link prediction approach differs from common techniques in that we consider multiple network snapshots with temporal trends captured in the time factor matrix C, given by the previously introduced CP tensor decomposition (refer to Equation 1 and Figure 1). Furthermore, we employ exponential smoothing to extrapolate future points in time based on the latent temporal trends retrieved by the tensor factorization. The combination of canonical *Parafac* decomposition with exponential smoothing was inspired by *E. Acar* and *T. G. Kolda* [2], who suggest the *Hold-Winters* method to analyse trends and seasonality of temporal data captured in tensors.

In statistics, exponential smoothing is a technique that can be applied to time series data, either to produce smoothed data for presentation, or to make forecasts. The time series data themselves are a sequence of observations. The raw data sequence is often represented by x_t, and the output of the exponential smoothing algorithm is commonly written as s_t which may be regarded as our best estimate of what the next value of x will be. When the sequence of observations begins at time $t = 0$, the simplest form of exponential smoothing is given by the expressions:

$$s_1 = x_0 \tag{4}$$
$$s_{t+1} = \alpha x_t + (1 - \alpha)s_t$$

where α is the smoothing factor $(0 < \alpha < 1)$, which assigns exponentially decreasing weights over time. For further examinations we balance the influence of most recent terms in time series and older data on the extrapolated points in time $(\alpha = 0.5)$. By direct substitution of the defining Equation (4) for simple exponential smoothing back into itself we find that:

$$s_{t+1} = \alpha x_t + (1 - \alpha)s_t \tag{5}$$

$$= \alpha x_t + \alpha(1 - \alpha)x_{t-1} + (1 - \alpha)^2 s_{t-2}$$
$$= \alpha x_t + \alpha(1 - \alpha)x_{t-1} + \alpha(1 - \alpha)^2 x_{t-2} + \cdots + (1 - \alpha)^t x_0$$

In our link prediction algorithm we employ the exponential smoothing technique to estimate the $K + 1$-th frontal slice of tensor X, containing information about future correlations between the variables held in the first and second mode (e.g. users or items). Infered from Equation (3), computation of the $K + 1$-th frontal slice requires to predict vector $c_{K+1,:}$, which can be derived from the temporal trend captured in the columns of factor matrix C (as shown in Figure 1 & 2). The exponential smoothing for each entry of vector $c_{K+1,:}$ can be formulated as:

$$c_{K+1,r} = \alpha c_{K,r} + \alpha(1 - \alpha)c_{K-1,r} + \alpha(1 - \alpha)^2 c_{K-2,r}$$
$$+ \alpha(1 - \alpha)^3 c_{K-3,r} + \cdots + (1 - \alpha)^K c_{1,r} \quad (6)$$

$$(r = 1, ..., R \quad \text{and} \quad 0 < \alpha < 1)$$

Given the extrapolated vector c_{K+1}, we are not only able to estimate the $K + 1$-th frontal slice (refer to Equation 3), but also single links $x_{i,j,K+1}$ that are likely to be established in the future:

$$x_{i,j,K+1} \approx \sum_{r=1}^{R} \lambda_r(a_{i,r} \cdot b_{j,r} \cdot c_{K+1,r}) \quad (7)$$

However, we need to keep in mind that the CP model is an approximation of our original data [3]. The precision of the estimated links depends on various factors like data composition and sparsity, the range of values as well as the number of rank-one components, which are all discussed in the following section.

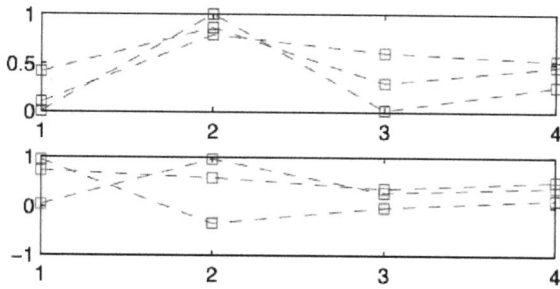

Fig. 2. Temporal Trends in Time Factor Matrix of Wiki-Edit-Ar and Facebook-Links dataset, where the 4th point is extrapolated by Exponential Smoothing

4 Evaluation

In order to evaluate the precision of our link prediction algorithm we need to split the observed data into a training set, which is used as input for the tensor analysis, and a test set, which is compared against the estimated results. The split is either based on a predefined percentage or can be done according a

Table 1. Split of Sample Dataset

	UserID	ItemID	TimeStamp	Tag
	20	1747	200610	1
Training Set	21	5524	200712	1
	25	5067	200705	1
	31	1091	200611	1
75% Observed	⋮	⋮	⋮	⋮
	20	7438	200803	1
Test Set E^+	21	5525	200807	1
X of 25% Obs.	⋮	⋮	⋮	⋮
	31	6388	200812	0
Test Set E^-	43	5411	200804	0
X of 25% Obs.	⋮	⋮	⋮	⋮

determined point in time. As a matter of course the overall precision depends on the proportion of training and test data. In our case three-quarter of the data are used for training and one-quarter for testing ($\frac{75}{25}$%).

Note that usually just a definite number of random entries (E) contained in the test set are used for evaluation. Due to the fact that our link prediction algorithm also estimates missing edges, the random selection needs to incorporate the same amount of both zero and non-zero entries ($E = E^- \cap E^+$, with $\|E^-\| = \|E^+\|$). Table 1 illustrates the split of a sample dataset, which comprises user-item tags and their corresponding time stamps.

Assuming our sample data of Table 1, the training set is used to construct a three-way array capturing tags on items given by users at a specific point in time ($X \in \mathbb{R}^{I \times J \times K}$). Based on the previously introduced mathematical model we are now able to make predictions for tags situated in the future ($[K + 1]$-th frontal slice). In order to calculate the precision of our link prediction algorithm, we can compare the estimated values with those contained within our test set (for both known and missing edges).

As might be expected, the accuracy of link prediction also vary according to the precision measure chosen. Due to its robustness *Mean Average Precision (MAP)* is a frequently used measure in the domain of information retrieval (IR) and machine learning (ML). Contrary to most single-value metrics which are based on an unordered list of values, *MAP* emphasizes ranking relevant values higher. The reording of an observed user vector ($X_{i,:,K+1}$ for $i = 1, ..., I$) according the estimates is illustrated in Table 2.

Given a reordered user vector we are able to calculate its average precision by summing up the precision values for each cut-off point in the sequence of ranked results. Average precision can be formulated as:

$$AP = \frac{1}{N} \sum_{r=1}^{N} P(r) \tag{8}$$

Table 2. Reordered Sample User Vector

Index	1	2	3	4		2	4	1	3
Observed	0	1	0	1	$\Rightarrow$	1	1	0	0
Estimated	.5	.9	.3	.8		.9	.8	.5	.3
		unordered					reordered		

where r is the rank, N is the number of cut-off points (alias non-zero entries) and $P(r)$ precision[1] at a given cut-off rank. The mean of all average precision values for each user vector equals our final measure called *MAP*.

All examined datasets and their according properties are listed in Table 3. Besides different information networks like *Wikipedia* and *Bibsonomy*, we furthermore scrutinized rating data of *MovieLens* and *Epinions*, social network data of *Facebook* as well as *Enron* communication data and other graph like data about *Internet-Growth*. In general, we can classify these datasets into bipartite graphs, which capture correlations between two different sets of elements (e.g. users & items), and unipartite graphs, which give information about interrelation of elements from a single set. Related to tensors, sets of elements are regarded as seperate modes, or in other words dimensions. The categorie of the scrutinized datasets (bipartite or rather unipartite graph) can be infered from the structure column given in Table 3.

Table 3. Properties of Examined Datasets

DATASET	STRUCTURE	#ENTRIES	#MODE1	#MODE2	#SLICES
wiki-edit-viwiki	$user \times page \times time$	2262679	13766	303867	~ 4
wiki-edit-skwiki	$user \times page \times time$	2526392	7229	215638	~ 6
wiki-edit-glwiki	$user \times page \times time$	1315066	2850	91594	~ 7
wiki-edit-elwiki	$user \times page \times time$	1569075	8049	97149	~ 8
wiki-edit-arwiki	$user \times page \times time$	4000735	25692	510033	~ 4
movielens-10m-ut	$user \times tags \times time$	95580	2795	12553	~ 4
movielens-10m-ui	$user \times item \times time$	95580	3097	6367	~ 8
movielens-10m-ti	$tags \times item \times time$	95580	12775	6190	~ 4
internet-growth	$page \times page \times time$	104824	20689	20689	~ 3
facebook-wosn-wall	$user \times user \times time$	876993	30839	30839	~ 14
facebook-wosn-links	$user \times user \times time$	1545686	57356	57356	~ 13
epinions	$user \times user \times time$	19793847	91596	91596	~ 5
enron	$user \times user \times time$	1149884	64145	64145	~ 9
bibsonomy-2ut	$user \times tags \times time$	2555080	4804	167963	~ 8
bibsonomy-2ui	$user \times item \times time$	2555080	1345	335335	~ 4
bibsonomy-2ti	$tags \times item \times time$	2555080	155264	571768	~ 2

As mentioned earlier, this study mainly concentrates on three-dimensional datasets, with time being modeled in one of the modes. In order to make reliable link predictions for the future, we need to reduce the sparsity of our three-way

---

[1] Fraction of the elements retrieved that are relevant to the user's information need.

array by squeezing the time dimension. Assuming that the precision relies upon connectedness of the examined graph, we limit the number of time slices in our tensor model by the following equation:

$$\#slices = \frac{\#entries}{2 * max(\#mode1, \#mode2)} \tag{9}$$

Typically a graph is connected if each node has a degree of at least two [9], implying that each fiber [10] of the first and second mode should have at minimum two non-zero entries (satisfied by Equation 9). Note that the number of slices correlates with the maximum number of possible rank-one components in our CP model (represented by variable R in Equation 1).

In addition to time compression (9) we furthermore ensure connectedness of a graph via reduction to its core component [12,15]. By cutting off all periphery nodes with weak connections to the core, we are able to further improve the precision of our link prediction approach. This is due to the fact that we cannot infer something meaningful about a periphery node without an existing path to the core component of the observed network.

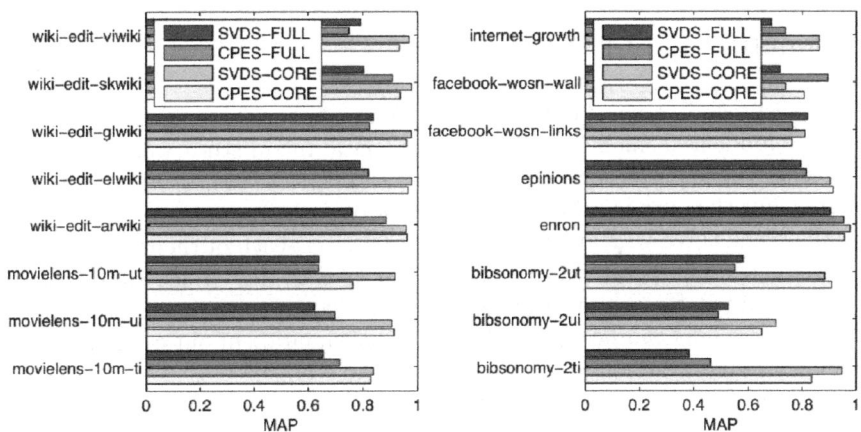

Fig. 3. Precision of Link Prediction

The mean average precision of our proposed link prediction algorithm *(CPES alias Canonical Parafac Decomposition with Exponential Smoothing)* for each examined dataset is illustrated in Figure 3. In order to compare the performance of our approach with an appropriate baseline algorithm, we furthermore scrutinized a matrix factorization method using *Singular Value Decomposition (SVD)*, which has been proven to be an effective technique in many data applications [2,7,18,20]. For the evaluation of the matrix *SVD* technique it is necessary to collapse the time dimension of our original three-dimensional tensor, resulting in a flat two-dimensional structure (represented by M) that accumulates all entries of the first and second mode.

SVD factorizes our two-way array $M_{i \times j}$ into three factor matrices containing the left-singular vectors (U), the singular values (S) and right-singular vectors (V) respectively [2,20]. In case of a reduction to dimension k, the product (M_k) of the resulting matrices represent the best rank-k approximation of our original two-dimensional data structure, picking the k-largest singular values (refer to Equation 10). Similar to the CP model we can estimate missing values of our original matrix (or rather unknown links in the observed complex network) by multiplication of the single factors:

$$M_k = U_{i \times k} \cdot S_{k \times k} \cdot V_{k \times j}^T \tag{10}$$

Our comparision in Figure 3 reveals that on an average CPES performes better than SVD for **full** datasets. In case of a reduction to the **core** component, CPES achieves almost as good results as SVD. Furthermore we can observe that for both link prediction algorithms the *core* variant is superior. This is due to the fact that *full* networks are more sparse and contain numerous periphery nodes that are not connected with the core. All in all, our proposed multi-way model can compete with existing link prediction algorithms based on matrix factorization techniques, such as matrix SVD [2,7,18,20].

5   Conclusion and Future Work

This paper presents a novel link prediction approach based on tensor decomposition and exponential smoothing of time series data. In our study we mainly focus on three-dimensional tensors, which capture evolution of social and information networks. For the purpose of analysing temporal trends in evolving networks, we decompose the three-way model into factor matrices; each of them giving information about one seperate mode. Trends captured in the "time" factor matrix are used as input for exponential smoothing to extrapolate future points. The extended CP model can be regarded as an approximation of our original tensor, including estimates of future links (refer to Figure 1).

Our proposed CPES link prediction algorithm was tested on various datasets like *Facebook* and *Wikipedia*. As showed in the evalution part, our approach achieves satisfying results for most of the examined network data, and beyond that reaches precision scores which are comparable to standard matrix-based prediction algorithms. Furthermore we could observe that CPES performes better for preliminary reduced network data, or in other words dense tensors.

The main contribution of this paper is our introduced multi-way model, which incorporates tensor decomposion as well as exponential smoothing techniques to estimate future links in complex networks. As far as we know there does not exist any comparable approach that extends the CP model in a way that it can be used to predict time slices which capture future correlations between nodes of an evolutionary network.

In future work we want to scrutinize the performance of our CPES algorithm for evolutionary data with more than three dimensions. For instance, acceleration data collected by a mobile phone sensor typically comprise four dimensions, containing gravity information for each of the three axes in an euclidian space as well as a timestamp for each sample. In case of data fusion we could imagine even more than four variables being held in a tensor. We believe that our link prediction algorithm is reliable as long as the data is sufficient, implying that a higher number of dimensions result in more sparse tensors.

References

1. Acar, E., Çamtepe, S.A., Yener, B.: Collective Sampling and Analysis of High Order Tensors for Chatroom Communications. In: Mehrotra, S., Zeng, D.D., Chen, H., Thuraisingham, B., Wang, F.-Y. (eds.) ISI 2006. LNCS, vol. 3975, pp. 213–224. Springer, Heidelberg (2006); Ref:16
2. Acar, E., Dunlavy, D.M., Kolda, T.G.: Link prediction on evolving data using matrix and tensor factorizations. In: ICDMW 2009: Proceedings of the 2009 IEEE International Conference on Data Mining Workshops, pp. 262–269. IEEE Computer Society, Washington, DC, USA (2009); Ref:13
3. Acar, E., Dunlavy, D.M., Kolda, T.G., Mørup, M.: Scalable tensor factorizations with missing data. In: SDM 2010: Proceedings of the 2010 SIAM International Conference on Data Mining, pp. 701–712. SIAM (2010); Ref:15
4. Acar, E., Yener, B.: Unsupervised multiway data analysis: A literature survey. IEEE Trans. on Knowl. and Data Eng. 21(1), 6–20 (2009); Ref:11
5. Bader, B.W., Harshman, R.A., Kolda, T.G.: Temporal analysis of social networks using three-way dedicom. Sandia Report (June 2006); Ref:12
6. Cao, L.: In-depth behavior understanding and use: the behavior informatics approach. Information Science 180, 3067–3085 (2010)
7. Dumais, S.T., Furnas, G.W., Landauer, T.K., Deerwester, S., Harshman, R.: Using latent semantic analysis to improve access to textual information. In: CHI 1988: Proceedings of the SIGCHI Conference on Human Factors in Computing Systems, pp. 281–285. ACM, New York (1988)
8. Georgii, E., Tsuda, K., Schölkopf, B.: Multi-way set enumeration in real-valued tensors. In: DMMT 2009: Proceedings of the 2nd Workshop on Data Mining using Matrices and Tensors, pp. 1–10. ACM, New York (2009); Ref:02
9. Janson, S., Knuth, D.E., Luczak, T., Pittel, B.: The birth of the giant component. Random Struct. Algorithms 4(3), 233–359 (1993)
10. Kolda, T.G., Bader, B.W.: Tensor decompositions and applications. SIAM Review (June 2008); Ref:14
11. Kunegis, J., Fay, D., Bauckhage, C.: Network growth and the spectral evolution model. In: Proc. Int. Conf. on Information and Knowledge Management (2010)
12. Leskovec, J.: Networks, communities and kronecker products. In: CNIKM 2009: Proceeding of the 1st ACM International Workshop on Complex Networks Meet Information & Knowledge Management, pp. 1–2. ACM, New York (2009); Ref:04
13. Leskovec, J., Huttenlocher, D., Kleinberg, J.: Predicting positive and negative links in online social networks. In: WWW 2010: Proceedings of the 19th International Conference on World Wide Web, pp. 641–650. ACM, New York (2010)

14. Leskovec, J., Huttenlocher, D., Kleinberg, J.: Signed networks in social media. In: CHI 2010: Conference on Human Factors in Computing Systems (2010)
15. Leskovec, J., Kleinberg, J., Faloutsos, C.: Graphs over time: densification laws, shrinking diameters and possible explanations. In: KDD 2005: Proceedings of the Eleventh ACM SIGKDD International Conference on Knowledge Discovery in Data Mining, pp. 177–187. ACM, New York (2005); Ref:07
16. Leskovec, J., Lang, K.J., Dasgupta, A., Mahoney, M.W.: Statistical properties of community structure in large social and information networks. In: WWW 2008: Proceeding of the 17th International Conference on World Wide Web, pp. 695–704. ACM, New York (2008)
17. Liben-Nowell, D., Kleinberg, J.: The link prediction problem for social networks. In: CIKM 2003: Proceedings of the Twelfth International Conference on Information and Knowledge Management, pp. 556–559. ACM, New York (2003)
18. Liben-Nowell, D., Kleinberg, J.: The link prediction problem for social networks. In: CIKM 2003: Proceedings of the Twelfth International Conference on Information and Knowledge Management, pp. 556–559. ACM, New York (2003)
19. Ma, N., Lim, E.-P., Nguyen, V.-A., Sun, A., Liu, H.: Trust relationship prediction using online product review data. In: CNIKM 2009: Proceeding of the 1st ACM International Workshop on Complex Networks Meet Information & Knowledge Management, pp. 47–54. ACM, New York (2009)
20. Spiegel, S., Kunegis, J., Li, F.: Hydra: a hybrid recommender system [cross-linked rating and content information]. In: CNIKM 2009: Proceeding of the 1st ACM International Workshop on Complex Networks Meet Information & Knowledge Management, pp. 75–80. ACM, New York (2009)
21. Strogatz, S.H.: Exploring complex networks. Nature 410, 268–276 (2001)
22. Sun, J., Tao, D., Faloutsos, C.: Beyond streams and graphs: dynamic tensor analysis. In: KDD 2006: Proceedings of the 12th ACM SIGKDD International Conference on Knowledge Discovery and Data Mining, pp. 374–383. ACM, New York (2006); Ref:09
23. Symeonidis, P., Nanopoulos, A., Manolopoulos, Y.: Tag recommendations based on tensor dimensionality reduction. In: RecSys 2008: Proceedings of the 2008 ACM Conference on Recommender Systems, pp. 43–50. ACM, New York (2008); Ref:10

Permutation Anonymization: Improving Anatomy for Privacy Preservation in Data Publication

Xianmang He, Yanghua Xiao*, Yujia Li, Qing Wang, Wei Wang, and Baile Shi

School of Computer Science, Fudan University
{071021057,shawyh,071021056,wangqing,wangwei1,bshi}@fudan.edu.cn

Abstract. Anatomy is a popular technique for privacy preserving in data publication. However, anatomy is fragile under background knowledge attack and can only be applied into limited applications. To overcome these drawbacks, we develop an improved version of anatomy: permutation anonymization, a new anonymization technique that is more effective than anatomy in privacy protection, and meanwhile is able to retain significantly more information in the microdata. We present the detail of the technique and build the underlying theory of the technique. Extensive experiments on real data are conducted, showing that our technique allows highly effective data analysis, while offering strong privacy guarantees.

Keywords: privacy preservation, generalization, anatomy, algorithm.

1 Introduction

The information age has witnessed a tremendous growth of personal data that can be collected and analyzed. Organizations may need to release private data for the purposes of facilitating data analysis and research. For example, medical records of patients may be released by a hospital to aid the medical study. Assume that a hospital wants to publish records of Table 1, which is called as microdata (*T*). Since attribute *Disease* is sensitive, we need to ensure that no adversary can accurately infer the disease of any patient from the published data. For this purpose, any unique identifier of patients, such as *Name* should be anonymized or excluded from the published data. However, it is still possible for the privacy leakage if adversaries have certain background knowledge about patients. For example, if an adversary knows that Bob is of age 65 and Sex M, s/he can infer that Bob's disease is Emphysema since Age together with Sex uniquely identify each patient in Table 1. The attribute set that uniquely identify each record in a table is usually referred to as a quasi-identifier(QI for short) [1, 2] of the table.

---

* Correspondence Author. This work was supported in part by the National Natural Science Foundation of China (No.61003001 and No.61033010) and Specialized Research Fund for the Doctoral Program of Higher Education (No.20100071120032).

L. Cao et al. (Eds.): PAKDD 2011 Workshops, LNAI 7104, pp. 111–123, 2012.

Table 1. Microdata

Name	Age	Sex	Disease
Bob	65	M	Emphysema
Alex	50	M	Cancer
Jane	70	F	Flu
Lily	55	F	Gastritic
Andy	90	F	Dyspepsia
Mary	45	M	Flu
Linda	50	F	Pneumonia
Lucy	40	F	Gastritic
Sarah	10	M	Bronchitis

Table 2. A 4-diversity Table

GID	Age	Sex	Disease
1	[50-90]	F/M	Emphysema
1	[50-90]	F/M	Cancer
1	[50-90]	F/M	Flu
1	[50-90]	F/M	Gastritic
1	[50-90]	F/M	Dyspepsia
2	[10-50]	F/M	Flu
2	[10-50]	F/M	Pneumonia
2	[10-50]	F/M	Gastritic
2	[10-50]	F/M	Bronchitis

Table 3. Anatomy

Age	Sex	Group-ID
65	M	1
50	M	1
70	F	1
55	F	1
90	F	1
45	M	2
50	F	2
40	F	2
10	M	2

QIT

Group-ID	Disease
1	Emphysema
1	Flu
1	Cancer
1	Dyspepsia
1	Gastritis
2	Flu
2	Bronchitis
2	Gastritic
2	Pneumonia

ST

Table 4. Permutation Anonymization

Age	Sex	Group-ID
50	F	1
90	M	1
70	F	1
65	F	1
55	M	1
50	M	2
10	F	2
45	M	2
40	F	2

PQT

Group-ID	Disease
1	Emphysema
1	Flu
1	Cancer
1	Dyspepsia
1	Gastritis
2	Flu
2	Bronchitis
2	Gastritic
2	Pneumonia

PST

One way to overcome above threat is anatomy [3]. In a typical anatomy-based solution, we first need to divide tuples into subsets (each subset is referred to as a QI-group). For example, tuples in Table 1 can be partitioned into two subsets {*Bob, Alex, Jane, Lily, Andy*} (with Group-ID 1) and {*Mary, Linda, Lucy, Sarah*} (with Group-ID 2), as indicated by the group ID(GID) in Table 2. Then, we release the projection of microdata on quasi-identifiers as a quasi-identifier table(QIT), and projection on sensitive attribute as a sensitive table(ST). The group ID is also added to QIT and ST. In this way, QI-attributes and sensitive attributes are distributed into two separate tables. For example, by anatomy, Table 1 is separated into two tables as shown in Table 3.

Now, given the published table of anatomy, an adversary with background knowledge of Bob can only infer that Bob belongs to group 1 from QIT and has only probability 1/5 to infer Bob's actual disease from ST. On the other hand, anatomy captures the exact QI-values, which retains a larger amount of data characteristics. Consequently, *anatomy is believed to be a anonymization solution of high data quality compared to generalization-based solutions that lead to heavy information loss* [3,4]. We illustrate this in Example 1.

Example 1. Suppose, we need to estimate the following query:
SELECT COUNT(∗) FROM T WHERE AGE ∈ [40, 70] AND SEX='F' AND DISEASE='Flu'.

If T is the original microdata, i.e. Table 1, we get the accurate result, which is 1. However, if we evaluate the query from the result produced by a generalization based approach, large gross error will occur. Table 2 is a typical result by generalizing Table 1. In Table 2, each QI-attribute in the QI-group is replaced by a less specific form. For example, age 65 is replaced by $[50, 90]$, sex M is replaced by F/M. When estimate the above query from Table 2, without additional knowledge, the researcher generally assumes uniform data distribution in the generalization Table 2. Consequently, we obtain an approximate answer $\frac{1}{4} + \frac{1}{8} = \frac{3}{8}$, which is much smaller than the accurate result. However, when we evaluate the same query from Table 3, we have the result result $\frac{2}{5} + \frac{2}{4} = 0.9$, which is quite close to the accurate result.

1.1 Motivations

However, anatomy is still vulnerable to certain kind of background knowledge attack. For example, suppose an adversary knows that Bob is a 65-year-old male whose record is definitely involved in the microdata, the adversary can only find out that Bob is one of the first five records. With a random guess, the adversary's estimate of the probability that Bob has Emphysema is $\frac{1}{5}$. However, if adversaries have acquired the background knowledge about the correlations between Emphysema and the non-sensitive attributes Age and Sex, e.g., *'the prevalence of emphysema was appreciably higher for the 65 and older age group than the 10-64 age group for each race-sex group'* and *'the prevalence was higher in males than females and in whites than blacks'*, the adversary can infer that Bob has Emphysema rather than other diseases with high confidence.

In reality, numerous background knowledge, such as well-known facts, demographic information, public record and information about specific individuals etc., can be available to adversaries. Furthermore, in general, it is quite difficult for the data publisher to know exactly the background knowledge that will be used by an adversary. As a result of these facts, background knowledge attack arises to be one of great challenges for anatomy-based data anonymization solutions.

Besides the frangibility under background knowledge attack, limited applications of anatomy also motivate us to improve it. Note that in an anatomy solution, all QI-values are precisely disclosed, which is not permitted in those applications where presence attack is a critical concern (In a presence attack, an adversary with QI-values of an individual wants to find out whether this individual exists in the microdata). In some other real applications such as location-based services (LBS) [5, 6], where all QI-attributes themselves are sensitive, the anatomy is not applicable any more.

Thus, we may wonder *whether we can improve the present anatomy techniques so that the final solution is more safe under background knowledge attack, and simultaneously can be applied into more applications such as those concerning presence attack or LBS*. To address this issue, we will propose *permutation anonymization* (PA) as an improved version of present anatomy solution. In following texts, we will first formalize the major problem addressed in this paper and the main solution: PA in Section 2. In Section 3, we present an anonymization algorithm to implement PA. In Section 4, related works are reviewed.

In Section 5, we experimentally evaluates the effectiveness of our technique. Finally, Section 6 concludes the paper.

2   Preliminaries

In this section, we will first give the basic notations that will be used in the following texts. Then, we give the formal definition about permutation anonymization and present theoretic properties about this technique. We close this section by the definition of the major problem that will be addressed in this paper.

2.1   Basic Notations

Given a microdata table $T(A_1, A_2, \cdots, A_n)$ that contains the private information of a set of individuals, and has n attributes $A_1, \cdots, A_n$, and a sensitive attribute (SA) A_s. A_s is categorical and every attribute $A_i (1 \leq i \leq n)$ can be either numerical or categorical. All attributes have finite and positive domains. For each tuple $t \in T, t.A_i (1 \leq i \leq n)$ denotes its value on A_i, and $t.A_s$ represents its SA value.

A *quasi-identifier* $QI = \{A_1, A_2, \cdots, A_d\} \subseteq \{A_1, A_2, \cdots, A_n\}$ is a minimal set of attributes, which can be joined with external information in order to reveal the personal identity of individual records. A *QI-group* of T is a subset of the tuples in T. A *partition* P *of* T is a set of disjoint QI-groups $QI_j (1 \leq j \leq m)$ whose union equals T (Namely, $T = \bigcup_j^m QI_j$). An *anonymization principle*(AP) is a constraint on a SA attribute. A partition P satisfies an AP if the SA attribute of every QI-group in P satisfies the constraint posed by the AP. Most notable principles include k-anonymity [1, 2], l-diversity [7], etc.

A permutation on a set V, is an one-to-one mapping from V to itself. If $|V| = N$, there are overall $N!$ permutations on V. Let α be a permutation on a set V of tuples, we use $\alpha(t_i)$ to denote the image of t_i under α. We use S_V to denote the set of all permutations on V. If V is a QI-group of microdata T, for example, $V = QI_i$, we can independently uniformly select a permutation from S_{QI_i} at random.

2.2   Permutation Anonymization

Now we are ready to give the formal definition about Permutation Anonymization, which is given in the following definition.

Definition 1 (Permutation Anonymization(PA)). *Let T be a table consisting of QI-attributes $A_i (1 \leq i \leq d)$ and sensitive attribute A_s. Given a partition P with m QI-groups on T, permutation anonymization is a procedure with (T, P) as input, which produces a quasi-identifier table PQT and a sensitive table PST satisfying following conditions:*

(1)PQT is a table with schema $(A_1, A_2, \cdots, A_d, \text{Group-ID})$ such that for each tuple $t \in QI_i$, PQT has a tuple of the form: $\left(\alpha_{i_1}(t).A_1, \alpha_{i_2}(t).A_2, \right.$

$\cdots, \alpha_{i_d}(t).A_d, i)$, where $\{\alpha_{i_j} : (1 \leq j \leq d)\}$ is independently uniformly selected from S_{QI_i} at random.

(2)PST is a table with schema $(Group\text{-}ID, A_s)$ such that for each tuple $t \in QI_i$, PST has a record of the form: $(i, \alpha_{i_s}(t).A_s)$, where α_{i_s} is uniformly selected from S_{QI_i} at random.

Example 2. Given Table 1 and one of its partition suggested in Table 2, one valid result of PQT and PST produced by permutation anonymization is shown in Table 4. Suppose we need to evaluate the same query in Example 1, from Table 4, we have $\frac{4}{5} \times \frac{3}{5} + \frac{3}{4} \times \frac{2}{4} = 0.855$, which is quite close to the result of the anatomy approach.

Similar to anatomy, PA also produce two tables. The main difference between anatomy and permutation is that: *anatomy directly releases all the QI-values without extra treatment, while PA releases attribute values after random permutation.* In this sense, anatomy can be considered as an improvement of anatomy. Random permutation of PA accounts for the a variety of advantages of PA compared over anatomy. One of them is the stronger privacy preservation of PA. As shown in Theorem 1 (due to space limitation, all proofs are omitted in this paper), an adversary has small probability to infer the sensitive value of a victim. Compared to naive anatomy under the same partition on microdata (where $Pr\{t.A_S = v\} = \frac{c_j(v)}{|QI_j.A_s|}$), the probability of privacy leakage of PA is significantly small. Besides above advantages, PA also provides good enough data utility, which in most cases is close to data quality of the anatomy approach. We illustrate this in Example 2.

Theorem 1. *Let T be a table with QI-attributes $A_i(1 \leq i \leq d)$ and sensitive attribute A_s, let P be a partition with m QI-groups on T. From an adversary's perspective, for any tuple $t \in QI_j$, $Pr\{t.A_S = v\} = \frac{c_j(v)}{|QI_j.A_s| \cdot \prod_{i=1}^{d} |QI_j.A_i|}$, where $c_j(v)$ is the number of tuples in QI_j with A_s as v, and $|QI_j.A_i|$ is the number of distinct values of QI_j on attribute A_i.*

Moreover, PA can be applied into applications such as LBS [5,6]. In these applications, all attributes are sensitive. Permutation anonymization allows privacy preservation for such applications by permutating all the attributes. PA also exhibits stronger privacy preservation for presence attack, which is shown in Lemma 1 and illustrated in Example 3.

Lemma 1. *From an adversary's perspective, the probability (δ_t) to find out whether an individual t exists in the QI-group QI_j by the presence attack is at most $\delta_t = \frac{\prod_i^d |n_{i_j}|}{|QI_j|^d}$, where n_{i_j} denotes the number of the value $t.A_i$ on attribute A_i in QI_j.*

Example 3. We explain Theorem 1 and Lemma 1 using Table 4. Suppose the victim is $Bob = \langle 65, M \rangle$ whose age and sex are available to adversaries, then from adversary's perspective, the probability that Bob contacted Emphysema is $\frac{10}{5 \times 2 \times 5} = \frac{1}{5}$, and the probability that Bob exists in QI_1 of Table 4 is $\frac{1}{5} \times \frac{2}{5} = \frac{2}{25}$.

2.3   Preserving Correlation

In this section, we discuss the data correlation between QI-attributes and sensitive attribute. The following theorem 2 establishes the lower bound of the RCE (see the Equation 3) achievable by the approach PA.

The combination of attributes define a d-Dimension space DS. Every tuple in the table can be mapped to a point in DS. We model $t \in T$ as an approximate density function (pdf) $\eta_t(x) : DS \longrightarrow [0, 1]$:

$$\eta_t(x) = \begin{cases} 1, \text{if}(x = t) \\ 0, \text{otherwise} \end{cases} \tag{1}$$

Now, we discuss published PA-tables. Assume QI_j as the QI-group containing the tuple t (in the underlying l-diverse partition). Let $v_1, v_2, \cdots, v_\lambda$ be all the distinct A_s values in the QI-group. Denote $c(v_h)(1 \leq h \leq \lambda)$ as the count value in the PST corresponding to v_h. The reconstructed pdf $\eta_t^{PA}(x)$ of t is

$$\eta_t^{PA}(x) = \begin{cases} c(v_1) \times \delta_t, \text{if } x = (t.A_1, t.A_2, \cdots t.A_d, v_1) \\ \cdots\cdots \\ c(v_\lambda) \times \delta_t, \text{if } x = (t.A_1, t.A_2, \cdots t.A_d, v_\lambda) \\ 0, \text{otherwise} \end{cases} \tag{2}$$

In the Equation 2, the δ_t is the probability of presence attack of tuple t. Given an approximate pdf η^{PA} (Equation 2), we quantify its error from the actual η_t(Equation 1), the re-construction error (RCE)(see [3]):

$$RCE = \sum_{\forall t \in T} Err(t) = \sum_{\forall t \in T} \int_{x \in DS} (\eta_t^{PA}(x) - \eta_t(x))^2 dx. \tag{3}$$

Theorem 2. *RCE (Equation 3) is at least $\sum_{t \in T}(1 + l \times \delta_t^2 - 2\delta_t)$, for any pair of PQT and PST, where $|T|$ is the cardinality of the microdata T, δ_t is the probability of presence attack of tuple t.*

2.4   Problem Definition

Using PA, we can implement different security models, one of them is l-diversity (given in Definition 2), which is widely used in previous researches about privacy preservation and will be one of major objectives of this paper. Another aspect of privacy preservation is data utility. In general, high data quality or less information loss is expected. In this paper, we use normalized certainty penalty (Definition 3) to measure the information loss. Now, we are ready to give the formal definition about the problem that will be addressed in this paper.

Definition 2 (l-diversity [7]). *A generalized table T^* is l-diversity if each QI-group $QI_j \in T^*$ satisfies the following condition: let v be the most frequent A_s value in QI_j, and $c_j(v)$ be the number of tuples $t \in QI_j$, then $\frac{c_j(v)}{|QI_j|} \leq \frac{1}{l}$.*

Definition 3 (Normalized Certainty Penalty(NCP) [8]). *Suppose a table T is anonymized to T^*. In the domain of each attribute in T, suppose there exists a global order on all possible values in the domain. If a tuple t in T^* has range $[x_i, y_i]$ on attribute $A_i (1 \leq i \leq d)$, then the normalized certainty penalty in t on A_i is $NCP_{A_i}(t) = \frac{|y_i - x_i|}{|A_i|}$, where $|A_i|$ is the domain of the attribute A_i. For tuple t, the normalized certainty penalty in t is $NCP(t) = \sum_i^d NCP_{A_i}(t)$. The normalized certainty penalty in T is $\sum_{t \in T^*} NCP(t)$.*

In general, a table T together with a partitioning P on T implicitly implies a generalization T^*, which is obtained by replacing each numeric value of $t.A_i$ by a range $[Min_i, Max_i]$ of $QI_i.A_i$ (Min_i, Max_i are the minimal and maximal value of $QI_i.A_i$), replacing each categorial value $t.A_j$ by $QI_i.A_j$. In following texts, to simplify description, we use 'NCP of a table T and its partitioning P' to denote the NCP of their implicit generalization T^*. In some contexts with confusion, table T is also omitted.

Example 4. We calculate NCP for Table 2. Note that the domain of $\langle Age, Sex \rangle$ are $\langle [10-90], \{F, M\} \rangle$. The NCP of Table 2 is $\frac{90-50}{90-10} \times 5 + \frac{50-10}{90-10} \times 4 + \frac{2}{2} \times 9 = 13\frac{1}{2}$.

Definition 4 (Problem Definition). *Given a table T and an integer l, we aim to generate PST and PQT for T by PA so that PST is l-diversity and NCP of PQT is minimized.*

3 Generalization Algorithm

In this section, we will propose an algorithm to implement PA. By problem definition, we can see that the key to solve the problem is to find an appropriate partition of T so that l-diversity can be achieved and information loss can be minimized. We will first present the detail of the partitioning step, which produces QI-groups $G_1, G_2, \cdots, G_n$ satisfying l-diversity. After this step, we use a populating step is to implement permutation anonymization essentially.

3.1 The Partitioning Step

In this subsection, we will present a simple yet effective partitioning algorithm, which runs linearly and produce a partitioning satisfying l-diversity. The detailed procedure is presented in Figure 1.

The principle l-diversity demands that: the number of the most frequent A_s value in each QI-group G_i can't exceed $\frac{|G_i|}{l}$. Motivated by this, we arrange the tuples in T to a list ordered by its A_s values, then distribute the tuples in L into $G_i (1 \leq i \leq g)$ a round-robin fashion. The resulting partitioning is guaranteed to be l-diversity, which is stated in Theorem 3. (If table T with sensitive attribute A_s satisfies $\max\{c(v) : v \in T.A_s\} > \frac{|T|}{l}$, then there exists no partition that is l-diversity.)

Input: Microdata T, parameter l
Output: QI-groups G_j that satisfy l-diversity;
Method:
1. If $\max\{c(v) : v \in T.A_s\} \geq \frac{|T|}{l}$, Return;
2. Hash the tuples in T into groups $Q_1, Q_2, \cdots, Q_\lambda$ by their A_s values;
3. Insert these groups $Q_1, Q_2, \cdots, Q_\lambda$ into a list L in order;
4. Let $g = \frac{|T|}{l}$, set QI-groups $G_1 = G_2 = \cdots = G_g = \emptyset$;
5. Assign tuple $t_i \in L$ $(1 \leq i \leq |L|)$ to G_j, where $j = (i \bmod g) + 1$

Fig. 1. A naive partitioning algorithm

Theorem 3. *If table T with sensitive attribute A_s satisfies $\max\{c(v) : v \in T.A_s\} \leq \frac{|T|}{l}$ (where $c(v)$ is the number of tuples in T with sensitive value v), the partition produced by our partitioning algorithm fulfills l-diversity.*

Theorem 4. *The complexity of the naive partitioning algorithm is $O(|T|)$, where $|T|$ denotes the cardinality of microdata T.*

The above partitioning algorithm takes no account of information loss. To reduce information loss, we will first preprocess the microdata following the idea: *distribute tuples sharing the same or quite similar QI-attributes into the same sub-tables.* Then for each sub-table T_i, we call the naive partitioning algorithm to partition T_i into QI-groups.

The detailed preprocessing procedure is presented in Figure 2. Initially, S contains T itself (line 1); then, each $G \in S$ is divided into two generalizable subsets G_1 and G_2 such that $G_1 \cup G_2 = G$, $G_1 \cap G_2 = \emptyset$ (line 5-7). Then for each new subset, we check whether $G_1(G_2)$ satisfies l-diversity (line 8). If both are generalizable, we remove G from S, and add G_1, G_2 to S; otherwise G is retained in S. The attempts to partition G are tried k times and tuples of G are randomly shuffled for each time (line 3-4). Our experimental results show that most of G can be partitioned into two sub-tables by up to $k = 5$ tries. The algorithm stops when no sub-tables in S can be further partitioned.

In the above procedure, the way that we partition G into two subsets G_1 and G_2 is influential on the information loss of the resulting solution. For this purpose, we artificially construct two tuples $t_1, t_2 \in G$ with each attribute taking the maximal/minimal value of the corresponding domains, and then insert them G_1 and G_2 separately (line 6). After this step, for each tuple $w \in G$ we compute $\Delta_1 = NCP(G_1 \cup w) - NCP(G_1)$ and $\Delta_2 = NCP(G_2 \cup w) - NCP(G_2)$, and add tuple w to the group that leads to lower penalty (line 7). After successfully partitioning G, remove the artificial tuples from G_1 and G_2 (line 8).

Theorem 5. *The average complexity of the partitioning step is $O(|T|log|T|)$.*

3.2   The Populating Step

For each QI-group $QI_j (1 \leq j \leq m)$ generated by the previous partitioning step, we independently uniformly generate $d + 1$ (d is the number of QI-attributes)

Input: A microdata T, **integers** k **and** l
Output: A set S **consisting of sub-tables of** T;
Method:
/* the parameter k is number of rounds to partition G*/
1. $S = \{T\}$;
2. While($\exists G \in S$ that has not been partitioned)
3.   For $i = 1$ to k
4.   Randomly shuffle the tuples of G;
5.   Set $G_1 = G_2 = \emptyset$;
6.   Add tuple t_1 (t_2) of extremely maximal (minimal) value to G_1 (G_2);
7.   For any tuple w
              compute Δ_1 and Δ_2.
              If($\Delta_1 < \Delta_2$) then Add w to G_1, else add w to G_2;
8.   If both G_1 and G_2 satisfy l-diversity
              remove G from S, and add $G_1 - \{t_1\}, G_2 - \{t_2\}$ to S, **break**;
9.Return S;

Fig. 2. The preprocessing algorithm

permutations on QI_j, i.e. $\alpha_1, ... \alpha_m, \alpha_s$. Then, for each tuple $t \in QI_j$, insert a tuple $\langle \alpha_1(t).A_1, \cdots, \alpha_d(t).A_d, j \rangle$ into PQT, and insert $\langle j, \alpha_s(t).A_S \rangle$ into PST. The above procedure is repeated until all QI-groups have been processed. Clearly, this step run linearly. Hence, the overall anonymization algorithm runs in $O(|T|log(|T|))$.

4   Discussions and Related Work

Limited by space, we only present the related work about the anonymization techniques. Previous anonymization techniques including data perturbation [9], condensation [10], clustering [11], Swapping [12]. These techniques are employed to hide the exact values of the data. However, it may not be suitable if one wants to make inferences with 100% confidence. Authors of paper [10] proposed the condensation method, which releases only selected statistics about each QI-group. In essence, authors of paper [13] provide another version of Anatomy [3]. Data swapping [12] produces an alternative table by interchanging the values (of the same attribute) among the tuples in T, however, it is not designed with linking attacks in mind. Consequently, data swapping can't promise the prevention of such attacks. In addition, [4] proposes an interesting anonymization technique, ANGEL, to enhance the utility for privacy preserving publication.

5   Experiments

In this section, we experimentally evaluate the effectiveness (data quality) and efficiency (computation cost) of the proposed technique. For these purposes, we utilize a real data set CENSUS containing personal information of 500k American adults. The data set has 9 discrete attributes as summarized in Table 5.

Table 5. Summary of attributes

Attribute	Number of distinct values	Types
Age	78	Numerical
Gender	2	Categorical
Education	17	Numerical
Marital	6	Categorical
Race	9	Categorical
Work-class	10	Categorical
Country	83	Numerical
Occupation	50	Sensitive
Salary-class	50	Sensitive

Table 6. Parameters and tested values

Parameter	Values
l	4,6,8,10
cardinality n	100k, 200k, 300k, 400k, 500k
number of QI-attributes d	3, 4, 5, 6, 7
query dimensionality w	$4, 5, ..., d+1$
expected selectivity s	0.05, 0.10, ..., 0.4

In order to examine the influence of dimensionality and sensitive value distribution, we create five tables from CENSUS, denoted as SAL-3,··· SAL-7, respectively. Specially, SAL-d ($3 \leq d \leq 7$), treats the first d attributes in Table 6 as the QI-attributes, and Salary-class as the sensitive attribute A_s. For example, SAL-3 is 4D, containing QI-attributes Age, Gender, and Education.

In the following experiments, we compare permutation against anatomy from two aspects: (i) utility of the published tables for data analysis, and (ii) cost of computing these tables. For anatomy, we employ the algorithm proposed in Ref [3]. We use the executable code that was downloaded from the author's homepage. Table 6 summarizes the parameters of our experiments.

All experiments are conducted on a PC with 1.9 GHz AMD CPU and 1 gigabytes memory. All the algorithms are implemented with Microsoft Visual C++ 2008.

5.1 Accuracy

Anonymized data is often used for analysis and data mining. We measure the data utility by answering aggregate queries on published data, since they are the basic operations for numerous mining tasks (e.g., decision tree learning, association rule mining, etc.). Specifically, we use queries of the following form:

SELECT COUNT (*) FROM SAL-d WHERE $A_1 \in b_1$ AND $A_2 \in b_2$ AND ··· AND $A_w = b_w$.

Here, w is a parameter called the query dimensionality. $A_1, ..., A_{w-1}$ are $w-1$ arbitrary distinct QI-attributes in SAL, but A_w is always Salary-class, $b_i (1 \leq i \leq w-1)$ is a random interval in the domain of A_i. The generation of $b_1, \cdots, b_{w-1}$ is governed by another parameter termed volume s, which is a real number in $[0, 1]$, and determines the length (in the number of integers) of $b_i (1 \leq i \leq w)$ as $\lfloor |A_i| \cdot s^{1/(w+1)} \rfloor$. Apparently, the query result becomes larger given a higher s. We derive the estimated answer of a query using the approach explained in [13]. The accuracy of an estimate is gauged as its relative error. Namely, let act and est be the actual and estimated results respectively, the relative error equals $|act - est|/act$. The average error rate is computed in answering a workload which contains 1000 queries.

Now, we explore the influence of l on data utility. Towards this, we vary l from 4 to 10. The result is shown in figure 4. The error increases with the growth of l. This is expected, since a larger l demands stricter privacy

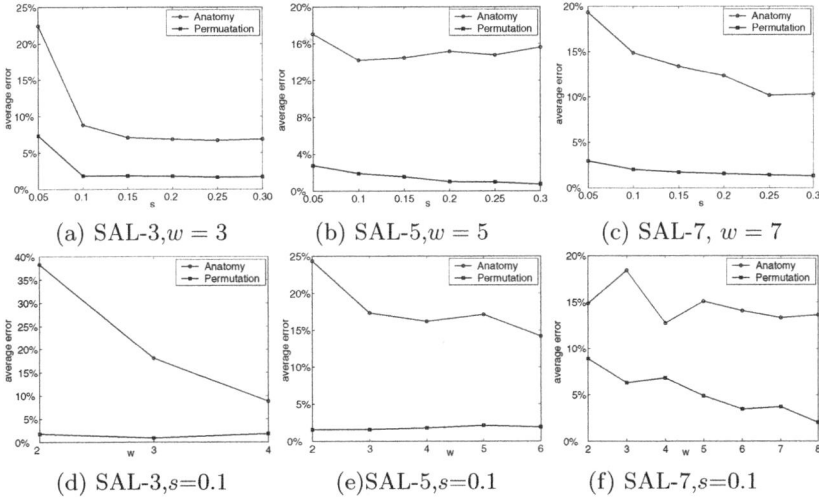

Fig. 3. Average error accuracy vs. parameters s and w

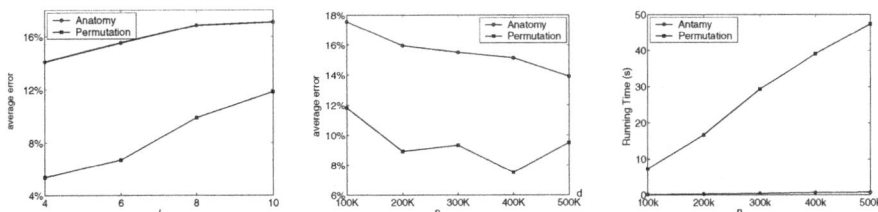

Fig. 4. Average error accu- **Fig. 5.** Average error accu- **Fig. 6.** Running time vs. n,
racy vs. l, $s = 0.1$ racy vs. n, $l=10$ $l=10$

preservation, which reduces data utility. Compared to Anatomy that produces about
14%-18% average error, the published data produced by our algorithm is significantly more useful. It is quite impressive to see that the error of our algorithm is consistently below 12 % despite of the growth of l.

Figure 3 investigates the influence of w and s on data utility. Figure 3(a), 3 (b) and 3 (c) plot the error of anatomy and permutation as a function of s with $w = 3, 5, 7$, respectively. Evidently, the query result becomes better if a higher s is given. This phenomenon is consistent with the existing understanding that both anonymization techniques provide better support for count queries when query results are larger. Apparently, our algorithm produces significantly more useful published data than anatomy.

To study the impact of w, Figure 3 (d), 3(e) and 3 (f) plot the error of anatomy and permutation as a function of w. The accuracy incurs less error as w increases. To explain this, recall that all queries have the same (expected) selectivity $s = 0.1$. When w becomes larger, the values of $b_i(1 \leq i \leq w)$ queried on each attribute increases considerably, leading to a more sizable search region, which in return reduces error.

Figure 5 examines relationship between the accuracy of each method and the cardinality of the data set. As expected, the accuracy descends as n grows. Again, permutation achieves significantly lower error in all cases.

5.2 Efficiency

Finally, we evaluate the overhead of performing anonymization. Figure 6 illustrates the cost of computing the publishable tables by two anonymization techniques, respectively, when the cardinality n linearly grows from 100k to 500k. The cost grows as n increases. This is expected, since longer time needs to be paid for larger number of tuples that participated in the anonymization.

From figure 6, we also can see that the advantages of our method in anonymization quality does not come for free. However, in all tested cases our algorithm can finish in less than 1 minute, which is acceptable especially for those cases where query accuracy is the critical concern.

6   Conclusion

The weakness of anatomy motivates us to develop a novel anonymization technique called permutation anonymization, which provides better privacy preservation than anatomy without sacrificing data utility. We systematically investigate the theoretic properties of this new technique and propose a corresponding algorithm to implement it. As verified by extensive experiments, our method allows significantly more effective data analysis, and simultaneously providing enough privacy preservation.

References

1. Sweeney, L.: k-anonymity: a model for protecting privacy. Int. J. Uncertain. Fuzziness Knowl.-Based Syst. 10(5), 557–570 (2002)
2. Samarati, P.: Protecting respondents' identities in microdata release. IEEE Trans. on Knowl. and Data Eng. 13(6), 1010–1027 (2001)
3. Xiao, X., Tao, Y.: Anatomy: simple and effective privacy preservation. In: VLDB 2006. VLDB Endowment, pp. 139–150 (2006)
4. Tao, Y., Chen, H., Xiao, X., Zhou, S., Zhang, D.: Angel: Enhancing the utility of generalization for privacy preserving publication. IEEE Transactions on Knowledge and Data Engineering 21, 1073–1087 (2009)
5. Kalnis, P., Ghinita, G., Mouratidis, K., Papadias, D.: Preventing location-based identity inference in anonymous spatial queries. IEEE Trans. on Knowl. and Data Eng. 19(12), 1719–1733 (2007)
6. Mokbel, M.F., Chow, C.-Y., Aref, W.G.: The new casper: query processing for location services without compromising privacy. In: VLDB 2006: Proceedings of the 32nd International Conference on Very Large Data Bases, pp. 763–774 (2006)

7. Machanavajjhala, A., Gehrke, J., Kifer, D., Venkitasubramaniam, M.: l-diversity: Privacy beyond k-anonymity. In: ICDE 2006: Proceedings of the 2008 IEEE 21st International Conference on Data Engineering, p. 24 (2006)

8. Xu, J., Wang, W., Pei, J., Wang, X., Shi, B., Fu, A.W.-C.: Utility-based anonymization using local recoding. In: KDD 2006: Proceedings of the 12th ACM SIGKDD International Conference on Knowledge Discovery and Data Mining, pp. 785–790. ACM, New York (2006)

9. Agrawal, R., Srikant, R.: Privacy-preserving data mining. SIGMOD Rec. 29(2), 439–450 (2000)

10. Aggarwal, C.C., Yu, P.S.: On static and dynamic methods for condensation-based privacy-preserving data mining. ACM Trans. Database Syst. 33(1), 1–39 (2008)

11. Aggarwal, G., Feder, T., Kenthapadi, K., Khuller, S., Panigrahy, R., Thomas, D., Zhu, A.: Achieving anonymity via clustering, pp. 153–162 (2006)

12. McIntyre, J., Fienberg, S.E.: Data Swapping: Variations on a Theme by Dalenius and Reiss. In: Domingo-Ferrer, J., Torra, V. (eds.) PSD 2004. LNCS, vol. 3050, pp. 14–29. Springer, Heidelberg (2004)

13. Zhang, Q., Koudas, N., Srivastava, D., Yu, T.: Aggregate query answering on anonymized tables. In: ICDE 2007: Proceedings of the 23nd International International Conference on Data Engineering, vol. 1, pp. 116–125 (2007)

Efficient Mining Top-k Regular-Frequent Itemset Using Compressed Tidsets

Komate Amphawan[1,2,3], Philippe Lenca[2,3], and Athasit Surarerks[1]

[1] Chulalongkorn University, ELITE Laboratory, 10330 Bangkok, Thailand
komate@live.com, Athasit.S@chula.ac.th
[2] Institut Telecom, Telecom Bretagne, UMR CNRS 3192 Lab-STICC, France
philippe.lenca@telecom-bretagne.eu
[3] Université européenne de Bretagne

Abstract. Association rule discovery based on support-confidence framework is an important task in data mining. However, the occurrence frequency (support) of a pattern (itemset) may not be a sufficient criterion for discovering interesting patterns. Temporal regularity, which can be a trace of behavior, with frequency behavior can be revealed as an important key in several applications. A pattern can be regarded as a regular pattern if it occurs regularly in a user-given period. In this paper, we consider the problem of mining top-k regular-frequent itemsets from transactional databases without support threshold. A new concise representation, called *compressed transaction-ids set (compressed tidset)*, and a single pass algorithm, called *TR-CT (Top-k Regular frequent itemset mining based on Compressed Tidsets)*, are proposed to maintain occurrence information of patterns and discover k regular itemsets with highest supports, respectively. Experimental results show that the use of the compressed tidset representation achieves highly efficiency in terms of execution time and memory consumption, especially on dense datasets.

1 Introduction

The significance of regular-frequent itemsets with temporal regularity can be revealed in a wide range of applications. Regularity is a trace of behavior and as pointed out by [1], behaviors can be seen everywhere in business and social life. For example in commercial web site analysis, one can be interested to detect such frequent regular access sequences in order to assist in browsing the Web pages and to reduce the access time [2,3]. In a marketing point of view, managers will be interested in frequent regular behavior of customers to develop long-term relationships but also to detect changes in customer behavior [4].

Tanbeer et al. [5] proposed to consider the occurrence behavior of patterns *i.e.* whether they occurs regularly, irregularly or mostly in specific time period of a transactional database. A pattern is said regular-frequent if it is frequent (as defined in [6] thanks to the support measure) and if it appears regularly (thanks to a measure of regularity/periodicity which considers the maximum compressed at which the pattern occurs).

L. Cao et al. (Eds.): PAKDD 2011 Workshops, LNAI 7104, pp. 124–135, 2012.

To discover a set of regular-frequent itemsets, the authors proposed a highly compact tree structure, named *PF-tree (Periodic Frequent patterns tree)*, to maintain the database content, and a pattern growth-based algorithm to mine a complete set of regular-frequent itemsets with the user-given support and regularity thresholds. This approach has been extended on incremental transactional databases [7], on data stream [8] and mining periodic-frequent patterns consisting of both frequent and rare items [9].

However, it is well-known that support-based approaches tend to produce a huge number of patterns and that it is not easy for the end-users to define a suitable support threshold. Thus, top-k patterns mining framework, which allows the user to control the number of patterns (k) to be mined (which is easy to specify) without support threshold, is an interesting approach [10].

In [11] we thus proposed to mine the top-k regular-frequent patterns and the algorithm MTKPP (Mining Top-K Periodic-frequent Patterns). MTKPP discovers the set of k regular patterns with highest support. It scans the database once to collects the set of transaction-ids where each item occurs in order to calculate their supports and regularities. Then, it requires an intersection operation on the transaction-ids set to calculate the support and the regularity of each itemset. This operation is the most memory and time consuming process.

In this paper, we thus propose a compressed tidset representation to maintain the occurrence information of itemsets to be mined. Indeed, compressed representation for intersection operation have shown their efficient like in Diffsets [12] and bit vector [13]. Moreover, an efficient single-pass algorithm, called *TR-CT (Top-k Regular-frequent itemsets mining based on Compressed Tidsets)* is proposed. The experimental results show that the proposed TR-CT algorithm achieves less memory usage and execution time, especially on dense datasets for whose the compressed tidset representation is very efficient.

The problem of top-k regular-frequent itemsets mining is presented in Section 2. The compressed tidset representation and the proposed algorithm are described in Section 3. In Section 4, we compare the performance of TR-CT algorithm with MTKPP. Finally, we conclude in Section 5.

2   Top-k Regular-Frequent Itemsets Mining

In this section, we introduce the basic definitions used to mine regular-frequent itemsets [5] and top-k regular-frequent itemsets [11].

Let $I = \{i_1, \ldots, i_n\}$ be a set of items. A set $X = \{i_{j_1}, \ldots, i_{j_l}\} \subseteq I$ is called an *itemset* or an *l-itemset (an itemset of size l)*. A transactional database $TDB = \{t_1, t_2, \ldots, t_m\}$ is a set of transactions in which each transaction $t_q = (q, Y)$ is a tuple containing a unique transaction identifier q (tid in the latter) and an itemset Y. If $X \subseteq Y$, it is said that t_q contains X (or X occurs in t_q) and is denoted as t_q^X. Therefore, $T^X = \{t_p^X, \ldots, t_q^X\}$, where $1 \le p \le q \le |TDB|$, is the set of all ordered tids (called *tidset*) where X occurs. The support of an itemset X, denoted as $s^X = |T^X|$, is the number of tids (transactions) in TDB where X appears.

Definition 1 (Regularity of an itemset X). *Let t_p^X and t_q^X be two consecutive tids in T^X, i.e. where $p < q$ and there is no transaction t_r, $p < r < q$, such that t_r contains X (note that p, q and r are indices). Then, $rtt_q^X = t_q^X - t_p^X$ represents the number of tids (transactions) not containing X between the two consecutive transactions t_p^X and t_q^X.*

To find the exact regularity of X, the first and the last regularities are also calculated : (i) the first regularity of X(fr^X) is the number of tids not containing X before it first occurs (i.e. $fr^X = t_1^X$), and (ii) the last regularity (lr^X) is the number of tids not containing X from the last occurring of X to the last tids of database (i.e. $lr^X = |TDB| - t_{|T^X|}^X$).

Thus, the regularity of X is defined as $r^X = max(fr^X, rtt_2^X, rtt_3^X, \ldots, rtt_{|T^X|}^X, lr^X)$ which is the maximum number of tids that X does not appear in database.

Definition 2 (Top-k regular-frequent itemsets). *Let us sort itemsets by descending support values, let S_k be the support of the kth itemset in the sorted list. The top-k regular-frequent itemsets are the set of first k itemsets having highest supports (their supports are greater or equal to S_k and their regularity are no greater than the user-given regularity threshold σ_r).*

Therefore, the top-k regular-frequent itemsets mining problem is to discover k regular-frequent itemsets with highest support from TDB with two user-given parameters: the number k of expected outputs and the regularity threshold (σ_s).

3   TR-CT: Top-k Regular-Frequent Itemsets Mining Based on Compressed Tidsets

We now introduce an efficient algorithm, called *TR-CT*, to mine the top-k regular-frequent itemset from a transactional database. It uses a concise representation, called *compressed transaction-ids set (compressed tidset)* to maintain the occurrence information of each itemset. It also uses an efficient data structure, named *top-k list* (as proposed in [11]) to maintain essential information about the top-k regular-frequent itemsets.

3.1   Compressed Tidset Representation

The compressed tidset representation is a concise representation used to store the occurrence information (tidset: a set of tids that each itemset appears) of the top-k regular-frequent itemsets during mining process. The main concept of the compressed tidset representation is to wrap up two or more consecutive continuous tids by maintaining only the first (with one positive integer) and the last tids (with one negative integer) of that group of tids. TR-CT can thus reduce time to compute support and regularity, and also memory to store occurrence information. In particular this representation is appropriate for dense datasets.

Definition 3 (Compressed tidset of an itemset X). *Let $T^X =$ $\{t_p^X, t_{p+1}^X, \ldots, t_q^X\}$ be the set of tids that itemset X occurs in transactions where $p < q$ and there are some consecutive tids $\{t_u^X, t_{u+1}^X, \ldots, t_v^X\}$ that are continuous between t_p^X and t_q^X (**where** $p \leq u$ **and** $q \geq v$). Thus, we define the compressed tidset of itemset X as:*

$$CT^X = \{t_p^X, t_{p+1}^X, \ldots, t_u^X, (t_u^X - t_v^X), t_{v+1}^X, \ldots, T_q^X\}$$

This representation is efficient as soon as there are three consecutive continuous transaction-ids in the tidsets. In the worst case, the compressed representation of a tidset is equal of the size of the tidset.

Table 1. A transactional database as a running example of TR-CT

tid	items
1	a b c d f
2	a b d e
3	a c d
4	a b
5	b c e f
6	a d e
7	a b c d e
8	a b d
9	a c d f
10	a b e
11	a b c d
12	a d f

From the TDB on the left side we have $T^a = \{t_1, t_2, t_3, t_4, t_6, t_7, t_8, t_9, t_{10}, t_{11}, t_{12}\}$ which is composed of two groups of consecutive continuous transactions. Thus, the compressed tidset of item a is $CT^a = \{1, -3, 6, -6\}$. For example, the first compressed tids $(1, -3)$ represents $\{t_1, t_2, t_3, t_4\}$ whereas $(6, -6)$ represents the last seven consecutive continuous tids. For the item a, the use of compressed tidset representation is efficient. It can reduce seven tids to be maintained comparing with the normal tidset representation. For items b and c, the sets of transactions that they occur are $T^b = \{t_1, t_2, t_4, t_5, t_7, t_8, t_{10}, t_{11}\}$ and $T^c = \{t_1, t_3, t_5, t_7, t_9, t_{11}\}$, respectively. Therefore, the compressed tidsets of the items b and c are $CT^b = \{1, -1, 4, -1, 7, -1, 10, -1\}$ and $CT^c = \{1, 3, 5, 7, 9, 11\}$ which are the examples of the worst cases of the compressed tidset representation.

With this representation a tidset of any itemset may contain some negative tids and the original Definition 1 is not suitable. Thus, we propose a new way to calculate the regularity of any itemset from the compressed tidset representation.

Definition 4 (Regularity of an itemset X from compressed tidset). *Let t_p^X and t_q^X be two consecutive tids in compressed tidset CT^X, i.e. where $p < q$ and there is no transaction t_r, $p < r < q$, such that t_r contains X (note that p, q and r are indices). Then, we denote rtt_q^X as the number of tids (transactions)* *between t_p^X and t_q^X that do not contain X. Obviously, rtt_1^X is t_1^X. Last, to find the exact regularity of X, we have to calculate the number of tids between the last tid of CT^X and the last tid of the database. This leads to the following cases:*

$$rtt_q^X = \begin{cases} t_q^X & \text{if } q = 1 \\ t_q^X - t_p^X & \text{if } t_p^X \text{ and } t_q^X > 0, 2 \leq q \leq |CT^X| \\ 1 & \text{if } t_p^X > 0 \text{ and } t_q^X < 0, 2 \leq q \leq |CT^X| \\ t_q^X + (t_p^X - t_{p-1}^X) & \text{if } t_p^X < 0 \text{ and } t_q^X > 0, 2 \leq q \leq |CT^X| \\ |TDB| - t_{|CT^X|}^X & \text{if } t_{|CT^X|}^X > 0, \text{ (i.e. } q = |CT^X| + 1) \\ |TDB| + (t_{|CT^X|}^X - t_{|CT^X|-1}^X) & \text{if } t_{|CT^X|}^X < 0, \text{ (i.e. } q = |CT^X| + 1) \end{cases}$$

Finally, we define the regularity of X as $r^X = max(rtt_1^X, rtt_2^X, \ldots, rtt_{m+1}^X)$.

For example, consider the compressed tidset $CT^a = \{1, -3, 6, -6\}$ of item a. The set of regularities between each pair of two consecutive tids is $\{1, 1, 6 + (-3 - 1), 1, 12 - (-6 - 6)\} = \{1,1,2,1,0\}$ and the regularity of item a is 2.

3.2   Top-k List Structure

As in [11], TR-CT is based on the use of a top-k list, which is an ordinary linked-list, to maintain the top-k regular-frequent itemsets. A hash table is also used with the top-k list in order to quickly access each entry in the top-k list. As shown in Fig. 1, each entry in a top-k list consists of 4 fields: (i) an item or itemset name (I), (ii) a total support (s^I), (iii) a regularity (r^I) and $(iiii)$ an compressed tidset where I occurs (CT^I). For example, an item a has a support of 11, a regularity of 2 and its compressed tidset is $CT^a = \{1, -3, 6, -6\}$ (Fig. 1(d)).

3.3   TR-CT Algorithm Description

The TR-CT algorithm consists of two steps: (i) Top-k list initialization: scan database once to obtain and collect the all regular items (with highest support) into the top-k list; (ii) Top-k mining: use the best-first search strategy to cut down the search space, merge each pair of entries in the top-k list and then intersect their compressed tidsets in order to calculate the support and the regularity of a new generated regular itemset.

Top-k Initialization. To create the top-k list, TR-CT scans the database once transaction per transaction. Each item of the current transaction is then considered. Thanks to the help of the hash table we know quickly if the current item is already in the top-k list or not. In the first case we just have to update its support, regularity and compressed tidset. If it is its first occurrence then a new entry is created and we initialize its support, regularity and compressed tidset.

To update the compressed tidset CT^X of an itemset X, TR-CT has to compare the last tid (t_i) of CT^X with the new coming tid (t_j). Thanks to the compressed representation (see Definition 3) it simply consists into the following cases:

- if $t_i < 0$, *i.e.* there are former consecutive continuous tids occur with the exact tid of t_i. TR-CT calculates the exact tid of $t_i < 0$ (*i.e.* $t_{i-1} - t_i$) and compares it with t_j to check whether they are continuous. If they are consecutive continuous tids (*i.e.* $t_j - t_{i-1} + t_i = 1$), TR-CT has to extend the compressed tidset CT^X (it consists only of adding -1 to t_i). Otherwise, TR-CT adds t_j after t_i in CT^X.
- if $t_i > 0$, *i.e.* there is no former consecutive continuous tid occurs with t_i. TR-CT compared t_i with t_j to check whether they are continuous. If they are consecutive continuous tids (i.e. $t_j - t_i = 1$), TR-CT creates a new tid in CT^X (it consists of adding -1 after t_i in CT^X). Otherwise, TR-CT adds t_j after t_i in CT^X.

After scanning all transactions, the top-k list is trimmed by removing all the entries (items) with regularity greater than the regularity threshold σ_r, and the remaining entries are sorted in descending order of support. Lastly, TR-CT removes the entries after the kth entry in the top-k list.

Top-k Mining. A best-first search strategy (from the most frequent itemsets to the least frequent itemsets) is adopted to quickly generate the regular itemsets with highest supports from the top-k list.

Two candidates X and Y in the top-k list are merged if both itemsets have the same prefix (*i.e.* each item from both itemsets is the same, except the last item). This way of doing will help our algorithm to avoid the repetition of generating larger itemsets and can help to prune the search space. After that, the compressed tidsets of the two elements are sequentially intersected in order to calculate the support, the regularity and the compressed tidset of the new generated itemset. To sequentially intersect compressed tidsets CT^X and CT^Y of X and Y, one has to consider four cases when comparing tids t_i^X and t_j^Y in order to construct CT^{XY} (see Definition 3):

(*1*) if $t_i^X = t_j^Y > 0$ add t_i^X at the end of CT^{XY}
(*2*) if $t_i^X > 0, t_j^Y < 0, t_i^X \leq t_{j-1}^Y - t_j^Y$, add t_i^X at the end of CT^{XY}
(*3*) if $t_i^X < 0, t_j^Y > 0, t_j^Y \leq t_{i-1}^X - t_i^X$, add t_j^Y at the end of CT^{XY}
(*4*) if $t_i^X, t_j^X < 0$, add $t_{|CT^{XY}|}^{XY} - (t_{i-1}^X - t_i^X)$ at the end of CT^{XY} if $t_{i-1}^X - t_i^X <$ $t_{j-1}^Y - t_j^Y$ otherwise add $t_{|CT^{XY}|}^{XY} - (t_{j-1}^Y - t_j^Y)$ at the end of CT^{XY}

From CT^{XY} we can easily compute the support s^{XY} and regularity r^{XY} of XY (see definition 4). TR-CT then removes the kth entry and inserts itemset XY into the top-k list if s^{XY} is greater than the support of the kth itemset in the top-k list and if r^{XY} is not greater than the regularity threshold σ_r.

3.4   An Example

Consider the TDB of Table 1, a regularity threshold σ_r of 4 and the number of desired results k of 5.

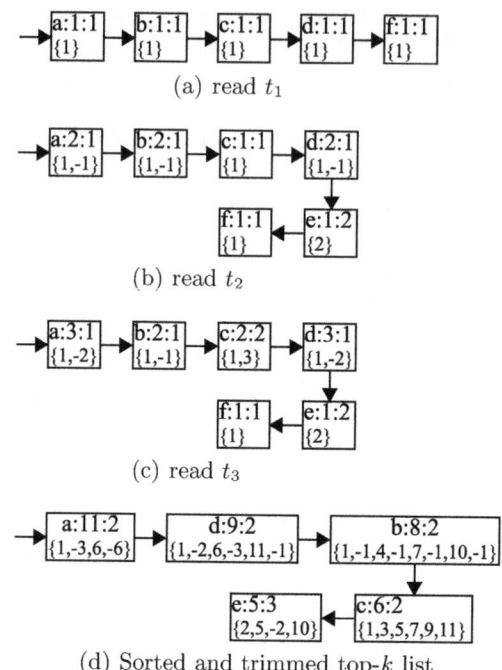

(a) read t_1

(b) read t_2

(c) read t_3

(d) Sorted and trimmed top-k list

Fig. 1. Top-k list initialization

After scanning the first transaction $(t_1 = \{a, b, c, d, f\})$, the entries for items a, b, c, d and f are created and their supports, regularities and compressed tidsets are initialized as $(1 : 1 : \{1\})$ (see Fig. 1(a)). With the second transaction $(t_2 = \{a, b, d, e\})$, TR-CT adds -1 at the end of the compressed tidsets of a, b and d, since these items occur in two consecutive continuous transactions. Then, the entry for item e is created and initialized (Fig. 1(b)). For the third transaction $(t_3 = \{a, c, d\})$, as shown in Fig. 1(c), the last tids of item a and d are changed to -2 (they occur in three consecutive continuous transactions t_1, t_2 and t_3) and the compressed tidset of item c is updated by adding t_3 as the last tid. After scanning all the transactions, the top-k list is sorted by support descending order and item f is removed (Fig. 1(d)).

In the mining process, item d is firstly merged with the former item a. The compressed tidsets CT^a and CT^d are sequentially intersected to calculate the support $s^{ad} = 9$, the regularity $r^{ad} = 3$ and to collect the compressed tidset $CT^{ad} = \{1, -2, 6, -3, 11, -1\}$ of itemset ad. Since the support s^{ad} is greater than $s^e = 5$ and the regularity r^{ad} is less than $\sigma_r = 4$, the item e is removed and

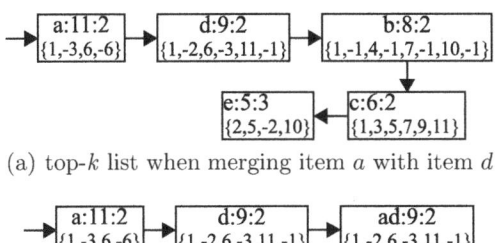

(a) top-k list when merging item a with item d

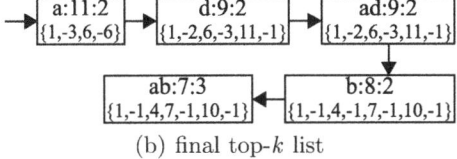

(b) final top-k list

Fig. 2. Top-k during mining process

ad is inserted into the top-k list as shown in Fig. 2(a). Next, the third itemset *i.e.* itemset ad is compared to the former itemsets a and b. These itemsets do not share the same prefix and thus are not merged. TR-CT then considers item b which is merged with a and d ($s^{ab} = 7$, $r^{ab} = 3$, $CT^{ab} = \{1, -2, 7, -1, 10, -1\}$; $s^{bd} = 5$, $r^{bd} = 5$, $CT^{bd} = \{1, -1, 7, -1, 11\}$). The itemset ab is thus added to the list and itemset c is removed and the itemset bd is eliminated. Lastly, the itemsets ab and ad are considered and we finally obtain the top-k regular-frequent itemsets as shown in Fig. 2(b).

4   Performance Evaluation

To validate the effectiveness of our proposed TR-CT algorithm, several experiments were conducted to compare the performance of TR-CT with the MTKPP algorithm[11] which is the first algorithm to mine top-k regular-frequent itemsets. All experiments were performed on an Intel®Xeon 2.33 GHz with 4 GB main memory, running Linux platform. Programs are in C. To measure the performance of the both algorithms, we focus on processing time (included top-k list construction and mining processes) and memory consumption (*i.e.* the maximum memory usage of the top-k list during mining process).

4.1   Test Environment and Datasets

The experiments were performed on several real datasets from the UCI Machine Learning Repository [14]. Table 2 shows some statistical information about the datasets used for experimental analysis. Accidents, connect, and pumsb are dense datasets (with long frequent itemsets) whereas retail is a sparse dataset (with short itemsets). These datasets are used to evaluate the computational performance of our algorithm. The regular and frequent patterns may have no sense (in particular with connect). We are here only interested in the efficiency evaluation (for k between 0 and 10000 and for σ_r between 0.5 and 10% of the total number of transactions in database).

Table 2. Database characteristics

Database	#items	Avg.length	#Transactions	Type
accidents	468	338	340, 183	dense
connect	129	43	67, 557	dense
pumsb	2, 113	74	49, 046	dense
retail	16, 469	10.3	88, 162	sparse

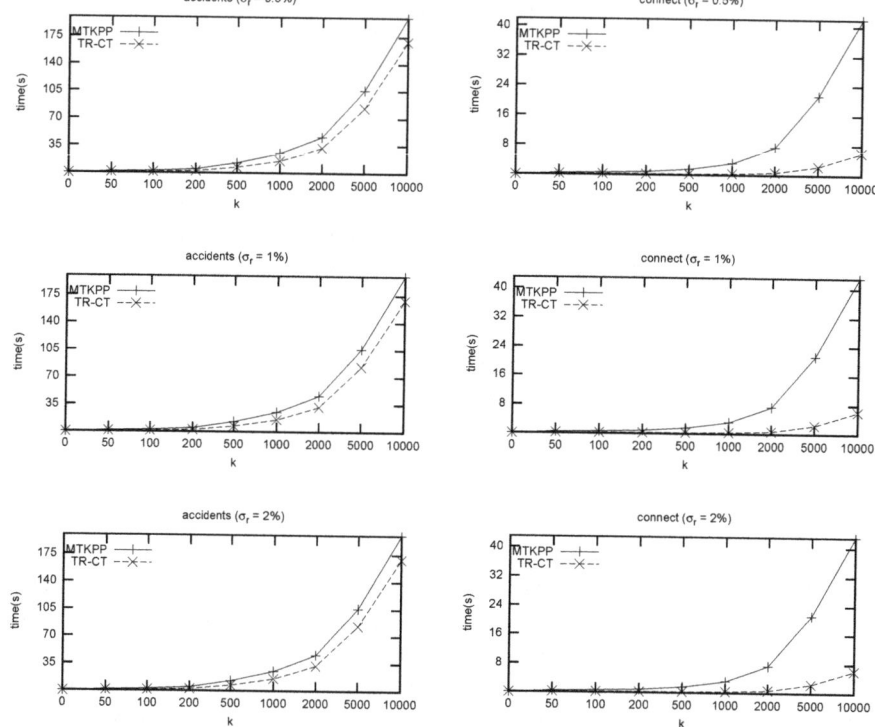

Fig. 3. Performance on accidents

Fig. 4. Performance on Connect

4.2   Execution Time

Figures 3, 4, and 5 give the processing time of dense datasets which are accidents, connect, and pumsb, respectively. From these figures, we can see that the proposed TR-CT algorithm runs faster than MKTPP algorithm using normal tids set under various value of k and regularity threshold σ_r. Since the characteristic of dense datasets, TR-CT can take the advantage of the compressed tidset representation which groups consecutive continuous tids together. Meanwhile, the execution time on sparse dataset retail is shown in Figure 6. Note that the performance of TR-CT is similar with MKTPP as with sparse dataset TR-CT can only take the advantage of grouping very few consecutive continuous tids.

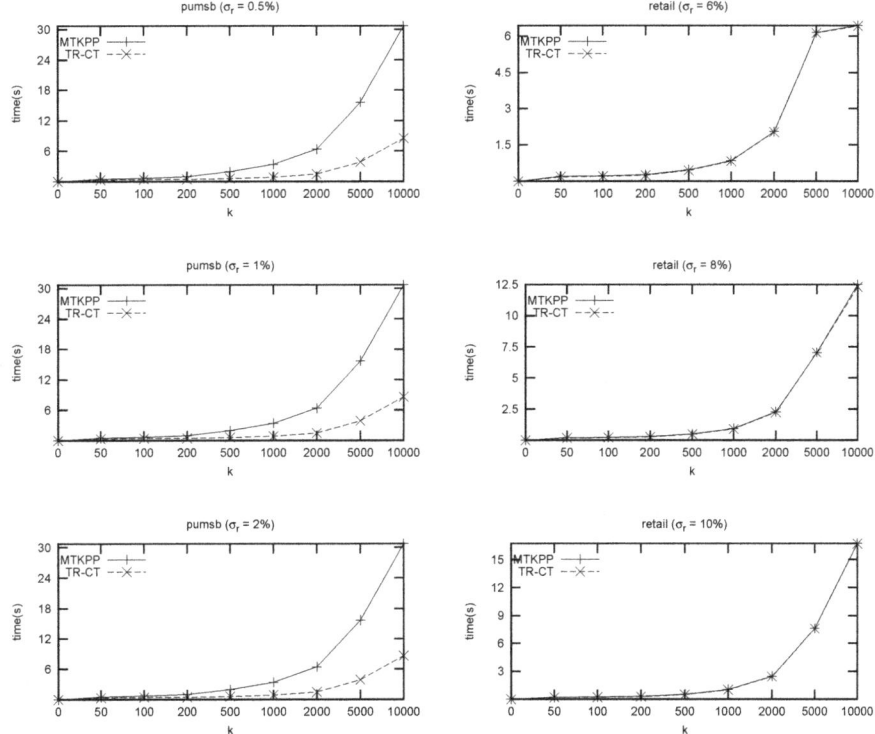

Fig. 5. Performance on Pumsb   **Fig. 6.** Performance on Retail

4.3 Space Usage

Based on the use of top-k list and compressed tidset representation, the memory usage and the number of maintained tids during mining process are examined. To evaluate the space usage, the regularity threshold σ_r is set to be the highest value (used in previous subsection) for each dataset. The first experiment compare the memory consumption of TR-CT and MTKPP algorithm. As shown in Fig. 7, TR-CT uses less memory than that of MTKPP on dense datasets (*i.e. accidents, connect* and *pumsb*) whereas the memory consumption of TR-CT is quite similar as MTKPP on sparse database *retail*. In some cases, the use of the compressed tidset representation may generate more concise tidsets than the original tidsets (used in MTKPP) since the former maintains only the first and last tids of the two or more consecutive continuous tids by using only one positive and one negative integer, respectively. That is why TR-CT has a good performance especially on dense datasets.

In the second experiment, the number of maintained tids is considered (see Fig. 8). The use of the compressed tidset representation may generate more concise tidsets than the original tidsets (used in MTKPP) since the former maintains only the first and last tids of the two or more consecutive continuous tids by using only one positive and one negative integer, respectively. The numbers of

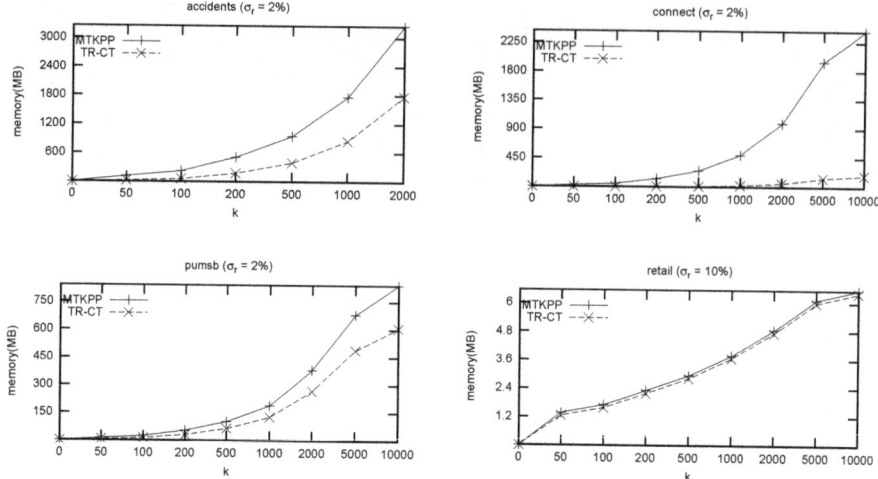

Fig. 7. Memory consumption of TR-CT

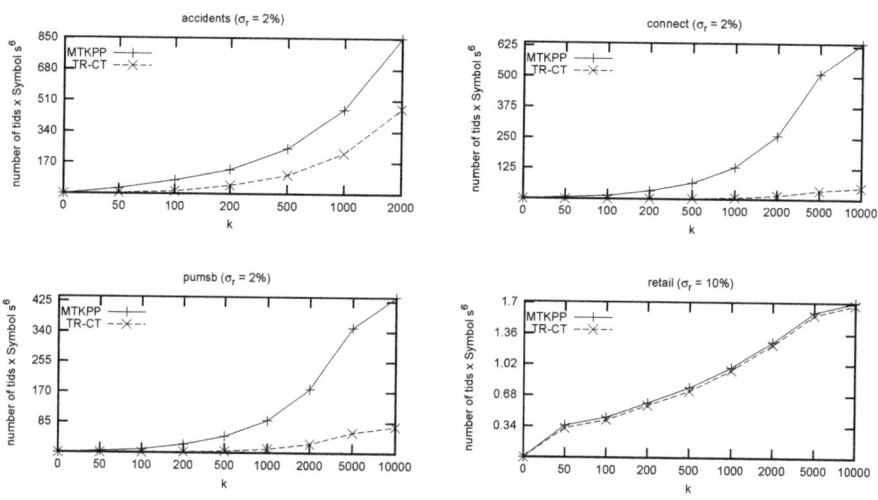

Fig. 8. Number of maintained transaction-ids

maintained tids between the two representations (algorithms) are shown in
Fig. 8. It is observed from the figure that the TR-CT maintained nearly the
same number of tids as the MTKPP when dataset are sparse. Meanwhile, TR-
CT significantly reduces the number of tids on dense datasets.

5 Conclusion

In this paper, we have studied the problem of mining top-k regular-frequent
itemsets mining without support threshold. We propose a new algorithm called

TR-CT (Top-k Regular-frequent itemset mining based on Compressed Tidsets) based on a *compressed tidset* representation. By using this representation, a set of tids that each itemset occurs consecutively continuous is transformed and compressed into two tids by using only one positive and negative integer. Then, the top-k regular-frequent itemsets are found by intersection compressed tidsets along the order of top-k list.

Our performance studies on both sparse and dense datasets show that the proposed algorithm achieves high performance, delivers competitive performance, and outperforms MTKPP algorithm. TR-CT is clearly superior to MTKPP on both the small and large values of k when the datasets are dense.

References

1. Cao, L.: In-depth behavior understanding and use: The behavior informatics approach. Inf. Sci. 180(17), 3067–3085 (2010)
2. Shyu, M.L., Haruechaiyasak, C., Chen, S.C., Zhao, N.: Collaborative filtering by mining association rules from user access sequences. In: Int. Workshop on Challenges in Web Information Retrieval and Integration, pp. 128–135. IEEE Computer Society (2005)
3. Zhou, B., Hui, S.C., Chang, K.: Enhancing mobile web access using intelligent recommendations. IEEE Intelligent Systems 21(1), 28–34 (2006)
4. Chen, M.C., Chiu, A.L., Chang, H.H.: Mining changes in customer behavior in retail marketing. Expert Syst. Appl. 28(4), 773–781 (2005)
5. Tanbeer, S.K., Ahmed, C.F., Jeong, B.S., Lee, Y.K.: Discovering Periodic-Frequent Patterns in Transactional Databases. In: Theeramunkong, T., Kijsirikul, B., Cercone, N., Ho, T.-B. (eds.) PAKDD 2009. LNCS, vol. 5476, pp. 242–253. Springer, Heidelberg (2009)
6. Agrawal, R., Srikant, R.: Fast algorithms for mining association rules in large databases. In: VLDB, pp. 487–499 (1994)
7. Tanbeer, S.K., Ahmed, C.F., Jeong, B.S.: Mining regular patterns in incremental transactional databases. In: Int. Asia-Pacific Web Conference, pp. 375–377. IEEE Computer Society (2010)
8. Tanbeer, S.K., Ahmed, C.F., Jeong, B.-S.: Mining Regular Patterns in Data Streams. In: Kitagawa, H., Ishikawa, Y., Li, Q., Watanabe, C. (eds.) DASFAA 2010. LNCS, vol. 5981, pp. 399–413. Springer, Heidelberg (2010)
9. Uday Kiran, R., Krishna Reddy, P.: Towards Efficient Mining of Periodic-Frequent Patterns in Transactional Databases. In: Bringas, P.G., Hameurlain, A., Quirchmayr, G. (eds.) DEXA 2010. LNCS, vol. 6262, pp. 194–208. Springer, Heidelberg (2010)
10. Han, J., Wang, J., Lu, Y., Tzvetkov, P.: Mining top-k frequent closed patterns without minimum support. In: IEEE ICDM, pp. 211–218 (2002)
11. Amphawan, K., Lenca, P., Surarerks, A.: Mining Top-K Periodic-Frequent Patterns without Support Threshold. In: Papasratorn, B., Chutimaskul, W., Porkaew, K., Vanijja, V. (eds.) IAIT 2009. CCIS, vol. 55, pp. 18–29. Springer, Heidelberg (2009)
12. Zaki, M.J., Gouda, K.: Fast vertical mining using diffsets. In: ACM SIGKDD KDD International Conference, pp. 326–335 (2003)
13. Shenoy, P., Haritsa, J.R., Sudarshan, S., Bhalotia, G., Bawa, M., Shah, D.: Turbocharging vertical mining of large databases. SIGMOD Rec. 29(2), 22–33 (2000)
14. Asuncion, A., Newman, D.: UCI machine learning repository (2007)

A Method of Similarity Measure and Visualization for Long Time Series Using Binary Patterns

Hailin Li[1], Chonghui Guo[1], and Libin Yang[2]

[1] Institute of Systems Engineering, Dalian University of Technology,
Dalian 116024, China
[2] College of Mathematics and Computer Science, Longyan University,
Longyan 364012, China
hailin@mail.dlut.edu.cn, guochonghui@tsinghua.org.cn,
ylib1982@163.com

Abstract. Similarity measure and visualization are two of the most interesting tasks in time series data mining and attract much attention in the last decade. Some representations have been proposed to reduce high dimensionality of time series and the corresponding distance functions have been used to measure their similarity. Moreover, visualization techniques are often based on such representations. One of the most popular time series visualization is time series bitmaps using chaos-game algorithm. In this paper, we propose an alternative version of the long time series bitmaps of which the number of the alphabets is not restricted to four. Simultaneously, the corresponding distance function is also proposed to measure the similarity between long time series. Our approach transforms long time series into SAX symbolic strings and constructs a non-sparse matrix which stores the frequency of binary patterns. The matrix can be used to calculate the similarity and visualize the long time series. The experiments demonstrate that our approach not only can measure the long time series as well as the "bag of pattern" (BOP), but also can obtain better visual effects of the long time series visualization than the chaos-game based time series bitmaps (CGB). Especially, the computation cost of pattern matrix construction in our approach is lower than that in CGB.

Keywords: Time series visualization, Binary patterns, Symbol representation, Similarity measure.

1 Introduction

Time series similarity measure is an interesting topic and also is a basic task in time series mining, which is an important tool for behavior informatics [3]. In the last decade, most of the studies have focused on the mining tasks based on similarity measure, including frequent patterns discovery, abnormal detection, classification, clustering, indexing and query. A common way to compare two

L. Cao et al. (Eds.): PAKDD 2011 Workshops, LNAI 7104, pp. 136–147, 2012.
© Springer-Verlag Berlin Heidelberg 2012

time series is Euclidean distance measure. Since Euclidean distance treats time series elements independently and is sensitive to outliers [1], it is not suitable for calculating the distance between two long time series, let alone compare the time series whose length is different. Another popular method to compare time series is dynamic time warping (DTW)[2,4,5], which is "warping" time axis to make a good alignment between time series points. The distance of DTW is the minimum value and obtained by dynamic programming. DTW is more robust against noise than Euclidean distance and provides scaling along the time axis, but it is not suitable to measure the similarity of long time series because of its heavy computation cost, i.e. $O(mn)$, where m and n are the length of the two time series respectively.

Besides similarity measure in time series, another issue concerns the high dimensionality of time series. Two basic approaches are used to reduce the dimensionality, i.e. a piecewise discontinuous function and a low-order continuous function. The former mainly includes discrete wavelet transform (DWT)[6], piecewise linear approximation (PLA)[7], piecewise aggregate approximation (PAA) [8], symbolic aggregate approximation (SAX)[9,10] and their extended versions. The latter mainly includes non-linear regression, singular value decomposition (SVD) [11] and discrete fourier transforms (DFT)[12]. After dimensionality reduction by the above techniques, Euclidean distance can be used to measure the similarity of long time series.

Recently, SAX is a popular method used to represent time series, which transforms time series into a symbolic string and provides the corresponding distance function to measure the similarity, which is applied widely in many behavior-related cases. Especially, the bag-of-words representation [13] based SAX constructs the "bag of patterns" (BOP) matrix and uses it to measure the similarity of long time series mining including clustering, classification and anomaly detection. In addition, there are more and more visualization tools based on SAX appearing in the filed of time series data mining. Two of the most popular tools are the VizTree proposed by Lin [14,15] and chaos-game based time series bitmaps (CGB) [16]. They first convert each time series into a symbolic string with SAX, compute the frequency of the substrings (patterns) which are extracted from the string according to a sliding window with a fixed length, map the frequency to the color bar and finally construct the visualization with suffix tree [17] or quadtree [16]. The visualizations entitles a user to discover clusters, anomalies and other regularities easily and fast. Especially, the chaos-game time series bitmaps (CGB) can arrange for the icons for time series files to appear as their bitmap representations. However, the number of alphabets used to divide normal distribution space in CGB is constricted to 4, which limits its applications. Moreover, the construction of pattern matrix demands a heavy computation cost. In most cases, the frequency matrix of patterns used to build the bitmaps is sparse [13], which means that little information on the time series is retained in the matrix and much extra memory space is used to store the data of which the value is 0. Furthermore, visual effects of time series bitmaps are not good for differentiating the long time series.

In this paper, we propose another alternative version to measure similarity of long time series and visualize them. It also transforms long time series into a symbolic string by SAX, easily constructs the matrix binary pattern which is a substring of length 2, calculates the frequency of each binary pattern, and form a binary pattern frequency matrix which can be used to measure the similarity and build the long time series bitmaps. The experiments demonstrate that our approach has the same clustering results as good as the method of "bag of patterns" (BOP), can also obtain non-sparse matrix of binary patterns and better visual effect with a lower time consumption of pattern matrix construction than that produced by CGB.

The rest of this paper is organized as follows. In section 2 we introduce the background and related work. In section 3 we propose our approach to measure similarity of long time series and visualize them. The empirical evaluation on clustering, visualization and computation cost comparison on long time series dataset is presented in section 4. In the last section we conclude our work.

2   Background and Related Work

Symbolic aggregate approximation (SAX) is a good method to simply represent time series. Time series is transformed into a symbolic string by SAX and the string can be used to calculate the similarity between two time series. Furthermore, the string also can be applied to draw bitmaps for the visualization of time series.

SAX represents time series by some fixed alphabets (words). It initially normalizes time series into a new sequence with the mean of 0 and the variance of 1, and then divides it into equal-sized w sections. The mean of each section is calculated and further represented by an alphabet whose precinct includes the mean. Fig. 1 shows one time series transformed into one string "DACCADADBB" by SAX.

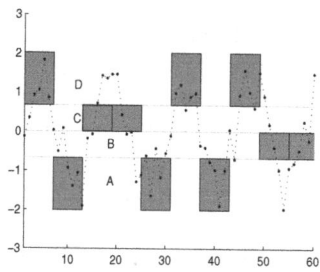

Fig. 1. Time series is transform into a string "DACCADADBB" by SAX

SAX representation can be used to find structural similarity in long time series. The paper [13] presented a histogram-based representation for the long time series, called it "bag of patterns" (BOP), presented in paper [13]. The word-sequence (pattern) matrix can be constructed after obtaining a set of strings for each time series. The matrix is used to calculate the similarity by any applicable distance measure or dimensionality reduction techniques. This approach is widely accepted by the text mining and information retrieval communities.

Furthermore, SAX representation can also be used to create the time series bitmaps. The authors [16] let SAX representation form a time series bitmaps by the algorithm of "chaos game" [18], which can produce a representation of DNA sequences. We call this Chaos-Game based time series Bitmaps CGB for short. CGB are defined for sequences with an alphabet size of 4. The four possible SAX symbolics are mapped to four quadrants of a square and each quadrant can be recursively substituted by the organized pattern in the next level. The method computes the frequencies of the patterns in the final square, linearly maps those frequencies to colors, and finally constructs a square time series bitmaps. Fig. 2(a) illustrates this process. The detail algorithm can be referred to the paper [16].

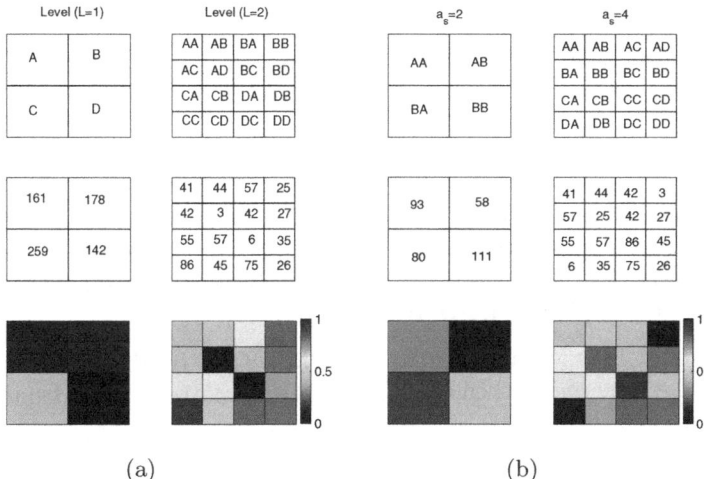

(a)   (b)

Fig. 2. (a) The process of time series bitmap construction by CGB. (b) The process of time series bitmap construction by our approach using binary patterns.

3   Binary Patterns Based Similarity and Visualization

Although BOP [13] is superior to many existing methods to measure similarity of long time series in the tasks of clustering, classification and anomaly detection, one of the disadvantages is that the size of possible SAX strings (the resulting dictionary size) is very large, which causes the word-sequence (pattern) matrix be sparse so that only few information on the original time series is retained. For example, with $\alpha_s = 4$ and $w = 10$, the size of the possible SAX strings is $\alpha_s^w = 4^{10} = 1048576$. Moreover, such large size needs a great of the memory space to store the data including many data points of which the value is 0. Even if we use a compress format [19] to store the data by compress column storage, the extra computation cost is likely to increase.

The chaos-game based time series bitmaps (CGB) [16] are only constructed by four symbolics, which is often used in DNA analysis. However, its applications are confined by the four symbolics because different number of the symbolics used to mine time series is demanded in most cases. With the high level in the square, the possibility of the frequent patterns in the long time series is quite low, which affects the visual effect of time series bitmaps. Moreover, it also needs a great of time consumption to construct the pattern matrix because of the recursive operation. Therefore, it is not reasonable to construct the bitmaps for long time series in the high level.

In this paper, we propose an alternative method to measure similarity and construct the bitmaps for long time series. It at least can overcome the above mentioned problems. For convenience, we give several definitions used to design our approach for long time series.

Definition 1. Alphabet set. The normalized time series is subject to the standard normal distribution $N(0,1)$. Alphabet set $S = \{s_1, s_2, \ldots, s_{a_s}\}$ is used to equiprobably divide the distribution space into a_s regions.

If the position of one point in time series locates the ith region which is represented by a alphabet s_i, then the point can be denoted as s_i.

Definition 2. The symbolic string of time series. A symbolic sequence $Q' = \{q'_1, q'_2, \ldots, q'_w\}$ is obtained from a time series $Q = \{q_1, q_2, \ldots, q_m\}$ by SAX.

Definition 3. Binary pattern. If the length of a substring is equal to 2, we call the substring a binary pattern.

For example, if a string obtained by SAX is $BAABCAC$, then the set of the binary patterns is $\{BA, AA, AB, BC, CA, AC\}$.

Definition 4. Binary pattern matrix (BPM). The alphabet set S is used to represent the time series, then the binary pattern matrix can be defined as

$$\begin{pmatrix} s_1 s_1 & s_1 s_2 & \cdots & s_1 s_{a_s} \\ s_2 s_1 & s_2 s_2 & \cdots & s_2 s_{a_s} \\ \vdots & \vdots & \vdots & \vdots \\ s_{a_s} s_1 & s_{a_s} s_2 & \cdots & s_{a_s} s_{a_s} \end{pmatrix}.$$

For example, if the alphabets $S = \{A, B, C\}$, then the binary pattern matrix is

$$\begin{pmatrix} AA & AB & AC \\ BA & BB & BC \\ CA & CB & CC \end{pmatrix}.$$

Definition 5. Binary pattern frequency matrix (BM). Suppose there is a string and the binary pattern matrix, the frequency of the binary pattern in the string can be computed. Those frequencies of the binary patterns constitute the Binary frequency matrix.

For example, if there is a string $ababaabccbacbaa$, then the binary pattern frequency matrix (BM) is

$$\begin{pmatrix} 2 & 3 & 1 \\ 4 & 0 & 1 \\ 0 & 2 & 1 \end{pmatrix}.$$

A long time series Q, of which the length is m, is transformed into $m - N + 1$ strings according to a sliding window of which the length is N by SAX. Each string produces a binary pattern frequency matrix BM_i, where $1 \leq i \leq m-N+1$. we can sum all the BM_i to obtain the total BM for the long time series, i.e. $BM = BM_1 + BM_2 + \ldots + BM_{m-N+1}$.

The inspired mind of our approach is something like the BOF method, which uses the word-sequence matrix to measure the similarity of long time series. However, it is obvious that the matrix of our approach is not sparse even though the alphabet size a_s is large. Moreover, since the number of binary patterns is fewer than that of the patterns produced by BOF and CGB in the high level, the memory used to store the binary patterns is lower.

Since the size of binary pattern matrix is small, the Euclidean distance function is a good choice to measure the similarity, i.e.

$$D(Q, C) = \sqrt{\sum_{j=1}^{a_s} \sum_{i=1}^{a_s} (BM_Q(i,j) - BM_C(i,j))^2}, \tag{1}$$

where BM_Q and BM_C are the binary frequency matrix of the long time series Q and C respectively.

After obtaining the BM, we also can apply it to construct the time series bitmaps. We can normalize the elements of BM and let every element's value locate into $[0, 1]$. The normalization function is not unique and can be formed by various ways. In our approach, the normalization function is

$$BFM = \frac{BM - min(BM)}{max(BM) - min(BM)} \in [0, 1], \tag{2}$$

where $min(BM)$ and $max(BM)$ can return the minimum and the maximum of the values in the BM.

Each normalized element in the matrix can be mapped to the color of which the value also locates in $[0,1]$. In this way, the matrix can be transform a bitmaps as shown in Fig. 2(b). We need to point out that the Matlab color bar is provided to construct all time series bitmaps in this paper. We know that when the alphabet size a_s in our approach is equal to 4 and the level L of quad-tree in the CGB approach is equal to 2, i.e. $a_s = 4$ and $L = 2$, both of the approaches have the same elements of sequence matrix so that their corresponding frequencies and the colors of the binary patterns are also equal as shown in Fig. 2(a) and 2(b). For example, The frequency of the binary pattern AB is equal to 41 in both approaches and their colors are also identical when there is a $a_s = 4$ and

$L = 2$. It demonstrates that the proposed method is available to measure and visualize time series as well as the traditional approaches.

For larger value of the parameter in the two respective methods, a_s in our method and L in the CGB approach, the number of binary patterns produced by our method is much less than that of other non-binary patterns produced by CGB and the binary patterns in our method can provide more sufficient information to approximate the long time series. Our approach using the matrix BM not only can conveniently measure the similarity of the long time series but also can construct the time series bitmaps with a good visual effect. Moreover, the computation cost for the pattern matrix construction is lower than the CGB. All those information can be demonstrated in next section. It is necessary to point out that, unlike the CGB method, the length of patterns in our approach is equal to 2, it can't deal with the patterns with different length except for the length 2. That is the reason why we call our approach an alternative version of long time series bitmaps.

4   Empirical Evaluation

In this section, we perform the hierarchical clustering to demonstrate that our approach can produce the same result as BOP [16] does, which also means that our approach is available for time series mining as good as BOP. We also compare the visual effects between our approach and CGB in the setting of different parameters to show that our approach has higher quality of visual effect of the long time series. The computation cost comparison between our approach and CGB for the pattern matrix construction is also discussed in the last subsection.

4.1   Hierarchical Clustering

It is well known that hierarchical clustering is one of the most widely used approaches to produce a nested hierarchy of the clusters. It can offer a great visualization power to analyze the relationships among the different time series. Moreover, unlike other clustering method, it does not require any input parameters.

We experiment on the long time series dataset derived from the paper [16] , of which the length is 1000. Fig. 3(a) and Fig. 3(b) show the resulting dendrogram for our approach and BOP when the size of the binary pattern matrix in our approach is equal to that of the word-sequence matrix in the BOP, i.e. 8×8. we can find that our method has the same clustering results as BOP does. Especially, when the alphabets size a_s is equal to 4 for our approach and the level is equal to 2, then the clustering results of two approach are identical as shown in Fig. 4. The reason is that in that case the binary pattern matrix (BPM) of our approach is corresponding to the word-sequence matrix of BOP.

From the clustering results we know that the performance of similarity measure in our approach is same to BOP. It also means that our approach is available for long time series data mining as BOP had done.

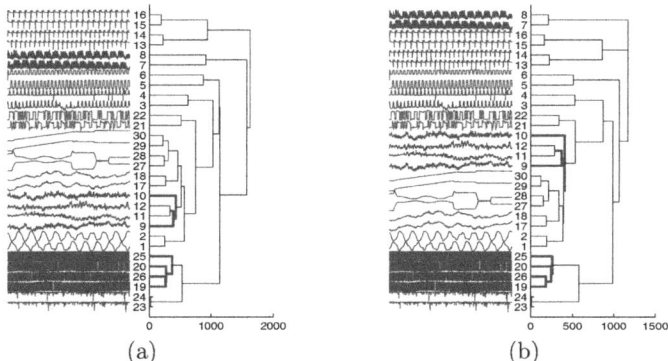

(a)   (b)

Fig. 3. (a) The clustering result of our approach is obtained when the alphabet size is 8 and , i.e. ($a_s = 8, N = 100, w = 10$). (b) The clustering result of BOP is obtained when the level of quad-tree is 3, i.e. ($L = 3, N = 100, w = 10$). N is the length of sliding window and w is the length of substring. Bold red lines denote incorrect substree.

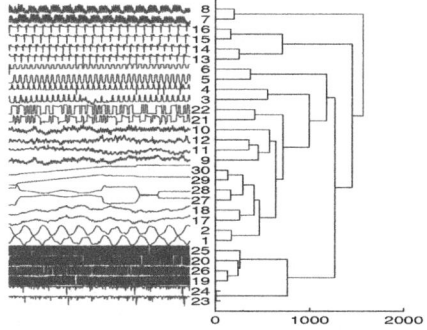

Fig. 4. The identical clustering result is obtained for our approach and BOP when $a_s = 4$ and $L = 2$ for the two approaches and ($N = 100, w = 10$)

4.2   Visual Effects

We also use the previous 30 long time series to produce the bitmaps by our approach and CGB. We compare their visual effects under the same size of the binary pattern matrix and word-sequence matrix. We provide two groups to make the experiments. One is under the matrix size of 4×4, that is, the alphabet size of our approach is 4 (i.e. $a_s = 4$) and the level of the quad-tree is 2 (i.e. $L = 2$). The results of the two approaches are shown in Fig. 5(a) and Fig. 5(b) respectively. The other is under the matrix size of 8×8, that is, the alphabet size of our approach is 8 (i.e. $a_s = 8$) and the level of the quad-tree is 3 (i.e. $L = 3$). The results of the two approaches are shown in Fig. 6(a) and Fig. 6(b) respectively.

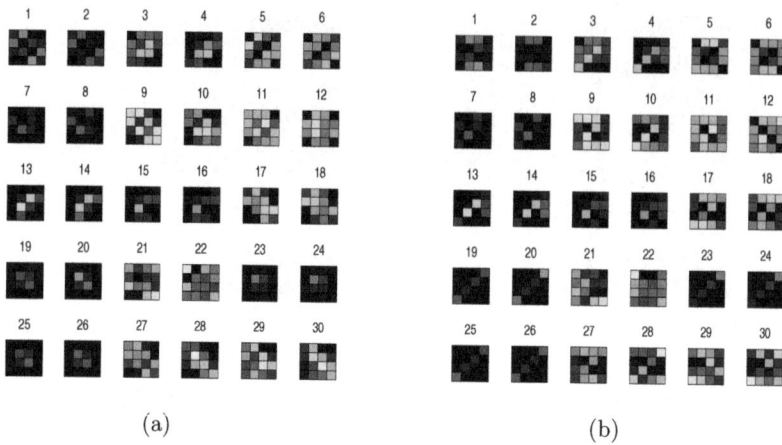

(a)   (b)

Fig. 5. (a) The bitmaps are constructed by our approach under $a_s = 4$. (b) The bitmaps are constructed by the CGB under $L = 2$.

It is easy to find that in Fig. 5(a) the color of each grid of one bitmap can be mapped to the color of each grid of the corresponding bitmap in Fig. 5(b) . As shown in Fig. 7(a) , the color of the grid is mapped to each other in that case, i.e. $a_s = 4$ in our approach and $L = 2$ in CGB. In other word, our alternative method to visualize long time series is available as CGB does.

In Fig. 6(a) and Fig. 6(b), it is obvious that the visual effect of our approach is better than the CGB. The hierarchical clustering results in previous subsection tell us that the group of time series 11 and 12 and the group of time series 21 and 22 are in the different clusters. However, Fig. 6(b) produced by CGB regard the four bitmaps as the members of the same cluster. But in Fig. 6(a) produced by our approach it is easy to distinguish the two groups. Other bitmaps of long time series also have such cases. Therefore, the visual effects of long time series bitmaps produced by our approach are better than those produced by CGB.

4.3   Comparison of Computation Cost

Since the computation cost of the pattern matrix construction is one of the most important differentia between our approach and the CGB, we make the experiments about the computation cost of the pattern matrix construction 6 times. Every time we truncate 50 time series with the same length from the long stock time series [20]. The average result of the computation cost is shown in Fig. 7(b) when the alphabet size a_s in our approach is equal to 4 and the quad-tree level in CGB is equal to 2. In other word, the comparison of the time computation is carried out under the same size of the pattern matrix for the two methods. It is obvious that the time consumption in our approach is lower than that in CBG.

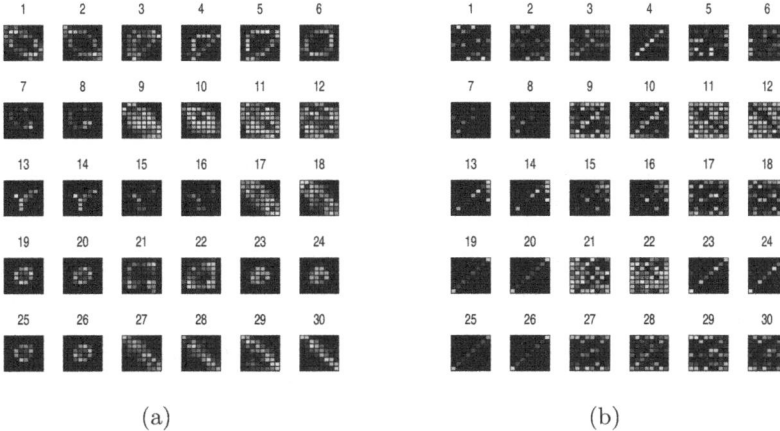

Fig. 6. (a) The bitmaps are constructed by our approach under $a_s = 8$. (b) The bitmaps are constructed by the CGB under $L = 3$.

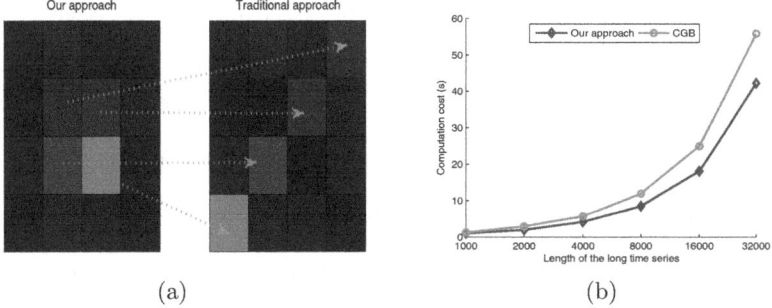

Fig. 7. (a) The color of the grid is mapped to each other under the matrix size of 4×4. (b)The time consumption of pattern matrix construction is compared between our approach and the CGB.

5   Conclusions

In this work, we have proposed an alterative approach to measure the similarity measure and visualize the long time series, which is based on binary patterns. Our approach counts the frequency of the occurrences of each binary patterns and forms a binary pattern frequency matrix. According to this matrix, we can measure the similarity of the long time series by Euclidean distance function and construct the bitmaps for the visualization of the long time series. In the experiments we demonstrated the effectiveness of our approach on the hierarchical clustering for the long time series, which shows that our approach can measure the similarity well as BOP does. Furthermore, the visual effect of time series bitmaps produced by our approach is better than the traditional one (CGB). From the comparison of the computation cost of pattern matrix construction, we conclude that our approach is also more efficient.

Acknowledgments. This work has been partly supported by the Natural Science Foundation of China under Grant Nos. 70871015 and 71031002.

References

1. Agrawal, R., Lin, K.I., Sawhney, H.S., Shim, K.: Fast similarity search in the presence of noise, scaling and translation in time-series databases. In: Proceedings of Very Large DataBase (VLDB), pp. 490–501 (1995)
2. Berndt, D.J., Clifford, J.: Finding patterns in time series: A dynmaic programming approach. In: Advances in Knowledge Discovery and Data Mining, pp. 229–248 (1996)
3. Cao, L.: In-depth Behavior Understanding and Use: the Behavior Informatics Approach. Information Science 180(17), 3067–3085 (2010)
4. Rabiner, L., Juang, B.H.: Fundamentals of speech recognition, Englewood Cliffs, N.J (1993)
5. Keogh, E.: Exact indexing of dynamic time warping. In: Proceedings of the 28th VLDB Conference, Hong Kong, China, pp. 1–12 (2002)
6. Popivanov, I., Miller, R.J.: Similarity search over time-series data using wavelets. In: Proceedings of the 18th International Conference on Data Engineering, pp. 212–221 (2002)
7. Iyer, M.A., Harris, M.M., Watson, L.T., Berry, M.W.: A performance comparison of piecewise linear estimation methods. In: Proceedings of the 2008 Spring Simulation Multi-Conference, pp. 273–278 (2008)
8. Lin, J., Keogh, E., Wei, L., Lonardi, S.: Experiencing SAX: a novel symbolic representation of time series. Data Mining and Knowledge Discovery 15, 107–144 (2007)
9. Lin, J., Keogh, E., Lonardi, S., Chiu, B.: A symbolic representation of time series with implications for streaming algorithms. In: Proceedings of the 8th ACM SIGMOD Workshop on Research Issues in Data Mining and Knowledge Discovery, pp. 2–11 (2003)
10. Keogh, E., Lin, J., Fu, A.: Hot SAX: efficiently finding the most unusual time series subsequence. In: Proceedings of the 5th IEEE International Conference on Data Mining, pp. 226–233 (2005)
11. Theodoridis, S., Koutroumbas, K.: Pattern Recognition, 4th edn., pp. 323–409 (2009)
12. Faloutsos, C., Ranganathan, M., Manolopoulos, Y.: Fast subsequence matching in time series databases. In: Proceedings of the ACM SIGMOD International Conference on Management of Data, pp. 419–429 (1994)
13. Lin, J., Li, Y.: Finding Structural Similarity in Time Series Data using Bag of Patterns Representation. In: Winslett, M. (ed.) SSDBM 2009. LNCS, vol. 5566, pp. 461–477. Springer, Heidelberg (2009)
14. Lin, J., Keogh, E., et al.: VizTree: a tool for visually mining and monitoring massive time series databases. In: Proceedings 2004 VLDB Conference, pp. 1269–1272. Morgan Kaufmann, St Louis (2004)
15. Lin, J., Keogh, E., et al.: Visually mining and monitoring massive time series. In: Proceedings of the 10th ACM SIGKDD International Conference on Knowledge Discovery and Data Mining, Seattle, WA, USA, pp. 460–469 (2004)
16. Kumar, N., Lolla, V.N., et al.: Time-series bitmaps: a practical visualization tool for working with large time series databases. In: SIAM 2005 Data Mining Conference, pp. 531–535 (2005)

17. Fu, T.C., Chung, F.L., Kwok, K., Ng, C.M.: Stock time series visualization based on data point importance. Engineering Applications of Artificial Intelligence 21(8), 1217–1232 (2008)
18. Barnsley, M.F.: Fractals everywhere, 2nd edn. Academic Press (1993)
19. Ekambaram, A., Montagne, E.: An Alternative Compress Storage Format for Sparse Matrices. In: Yazıcı, A., Şener, C. (eds.) ISCIS 2003. LNCS, vol. 2869, pp. 196–203. Springer, Heidelberg (2003)
20. Stock.: Stock data web page (2005), http://www.cs.ucr.edu/~wli/FilteringData/stock.zip

A BIRCH-Based Clustering Method for Large Time Series Databases

Vo Le Quy Nhon and Duong Tuan Anh

Faculty of Computer Science and Engineering,
Ho Chi Minh City University of Technology, Vietnam
dtanh@cse.hcmut.edu.vn

Abstract. This paper presents a novel approach for time series clustering which is based on BIRCH algorithm. Our BIRCH-based approach
performs clustering of time series data with a multi-resolution transform
used as feature extraction technique. Our approach hinges on the use of
cluster feature (CF) tree that helps to resolve the dilemma associated
with the choices of initial centers and significantly improves the execution time and clustering quality. Our BIRCH-based approach not only
takes full advantages of BIRCH algorithm in the capacity of handling
large databases but also can be viewed as a flexible clustering framework
in which we can apply any selected clustering algorithm in Phase 3 of
the framework. Experimental results show that our proposed approach
performs better than k-Means in terms of clustering quality and running time, and better than I-k-Means in terms of clustering quality with
nearly the same running time.

Keywords: time series clustering, cluster feature, DWT, BIRCH-based.

1 Introduction

Time series data arise in so many applications of various areas ranging from science, engineering, business, finance, economic, medicine to government. Because
of this fact, there has been an explosion of research effort devoted to time series
data mining in the last decade, in particular, in behavior-related data analytics
[8]. Beside similarity search, another crucial task in time series data mining which
has received an increasing amount of attention lately is time series clustering.
Given a set of unlabeled time series, it is often desirable to determine groups of
similar time series in such a way that time series belonging to the same group
are more "similar" to each other rather than time series from different groups.

Although there have been much research on clustering in general, most classic
machine learning and data mining algorithms do not work well for time series
due to their unique characteristics. In particular, the high dimensionality, very
high feature correlation and the large amount of noise in time series data present
a difficult challenge. Although a few time series clustering algorithms have been
proposed, most of them do not scale well to large datasets and work only in a
batch fashion.

L. Cao et al. (Eds.): PAKDD 2011 Workshops, LNAI 7104, pp. 148–159, 2012.

This paper proposes a novel approach for time series clustering which is based on BIRCH algorithm [13]. We adopt BIRCH method in the context of time series data due to its three inherent benefits. First, BIRCH incrementally and dynamically clusters incoming multi-dimensional metric data points to try to produce the best quality clustering with the available resources (i.e. available memory and time constraints). Second, BIRCH can find a good clustering with a single scan of the data and improve the quality further with a few additional scans. Third, BIRCH can scale well to very large datasets. To deal with the characteristics of time series data, we propose a BIRCH-based clustering approach that works by first using a multi-resolution transform to perform feature extraction on all time series in the database and then applying BIRCH algorithm to cluster the transformed time series data. Our BIRCH-based approach hinges on the use of cluster feature (CF) tree that helps to resolve the dilemma associated with the choices of initial centers and thus significantly improves the execution time and clustering quality. Our BIRCH-based approach not only takes full advantages of BIRCH method in the capacity of handling very large databases but also can be viewed as a flexible clustering framework in which we can apply any selected clustering algorithm in Phase 3 of the framework. Particularly, our BIRCH-based approach with an anytime clustering algorithm used in Phase 3 will also be an anytime algorithm. Experimental results show that our proposed approach performs better than k-Means in terms of clustering quality and running time, and better than I-k-Means in terms of clustering quality with nearly the same running time.

The rest of the paper is organized as follows. In Section 2 we review related work, and introduce the necessary background on multi-resolution transforms and BIRCH clustering algorithm. Section 3 describes our proposed approach for time series clustering. Section 4 presents our experimental evaluation which compares our method to classic k-Means and I-k-Means on real datasets. In section 5 we include some conclusions and suggestions for future work.

2 Background and Related Work

2.1 Dimensionality Reduction Using Multi-resolution Transforms

Discrete wavelet transform (DWT) is a typical case of time series dimensionality reduction method with multi-resolution property. This property is critical to our proposed framework. The Haar wavelet is the simplest and most popular wavelet proposed by Haar. The Haar wavelet transform can be seen as a sequence of averaging and difference operations on a time sequence. The transform is achieved by averaging two adjacent values on the time series at a given resolution to form a smoothed, lower-dimensional signal and the resulting coefficients at this given resolution are simply the differences between the values and their averages [1].

MPAA time series representation, proposed by Lin et al., 2005 [7], is another case of dimensionality reduction with multi-resolution property. MPAA divides time series X_i of length n into a series of lower dimensional signal with different resolution N where $N \in \{1, \ldots, n\}$. In the finest level, the data is divided into N segments of equal size. The mean value of the data falling within a segment is calculated and a vector of these values becomes the data reduced representation. Then recursively applying the above pairwise averaging process on the lower-resolution array containing the averages, we get a multi-resolution representation of time series.

2.2   Related Work on Time Series Clustering

The most promising approach for time series clustering is to cluster time series indirectly with features extracted from the raw data since feature extraction always help to reduce the dimensionality of time series. Some typical previous works on time series clustering based on this approach can be reviewed as follows.

Gavrilov et al., 2000, in [2] described a method that reduces the dimensionality of the stock time series by the PCA (Principal Component Analysis) feature extraction and then uses Hierarchical Agglomerative Clustering (HAC) algorithm to cluster the transformed data. The method was tested with the dataset consisting of 500 stocks from the Standard & Poor (S & P) index for the year and each stock is a series of 252 points.

Zhang et al., 2005 [12] put forward a k-Means clustering method for time series data that are reduced in dimensionality by the Haar wavelet transform. The interesting feature in this method is the unsupervised feature extraction using wavelet transform for automatically choosing the dimensionality of features. The method was tested with eight small datasets with cardinalities ranging from 20 to 128.

Lin et al., 2004 [7] presented an approach, called I-k-Means, to perform iterative clustering of time series at various resolutions using the Haar wavelet transform. First, the Haar wavelet decomposition is computed for all time series. Then, the k-Means clustering algorithm is applied, starting at the coarse level and gradually progressing to finer levels. The final centers at the end of each resolution are used as the initial centers for the next level of resolution. One important advantage of I-k-Means is that it is an *anytime* algorithm. It allows the user to monitor the quality of clustering results as the program executes. The user can interrupt the program at any level, or wait until the execution terminates once the clustering stabilize. In general, one can use any combination of iterative refining clustering algorithm and multi-resolution decomposition methods in I-k-Means. In comparison to k-Means, I-k-Means can produce a better clustering quality in a shorter running time since it can mitigate the problem associated with the choice of initial centers.

2.3   CF Tree and BIRCH Algorithm

BIRCH is designed for clustering a large amount of numerical data by integration of hierarchical clustering at the initial stage and other clustering methods, such as iterative partitioning at the later stage [13]. This method is based on two main concepts, clustering feature and clustering feature tree (CF tree), which are used to summarize cluster representations. These structures help the clustering method achieve good speed and scalability in large databases. BIRCH is also effective for incremental and dynamic clustering of incoming objects.

Given N d-dimensional points or objects $\vec{x_i}$ in a cluster, we can define the centroid $\vec{x_0}$, the radius R, and the diameter D of the cluster as follows:

$$\vec{x_0} = \frac{\sum_{i=1}^{N} \vec{x_i}}{N}$$

$$R = \left(\frac{\sum_{i=1}^{N} (\vec{x_i} - \vec{x_0})^2}{N}\right)^{\frac{1}{2}}$$

$$D = \left(\frac{\sum_{i=1}^{N} \sum_{i=j}^{N} (\vec{x_i} - \vec{x_j})^2}{N(N-1)}\right)^{\frac{1}{2}}$$

where R is the average distance from member objects to the centroid, and D is the average pairwise distance within a cluster. Both R and D reflect the compactness of the cluster around the centroid. A clustering feature (CF) is a triplet summarizing information about clusters of objects. Given N d-dimensional points or objects in a cluster, then the CF of the cluster is defined as

$$CF = \left(N, \vec{LS}, SS\right)$$

where N is the number of points in the subcluster, $\vec{LS}$ is the linear sum on N points and SS is the square sum of data points.

$$\vec{LS} = \sum_{i=1}^{N} \vec{x_i}$$

$$SS = \sum_{i=1}^{N} \vec{x_i}^2$$

A clustering feature is essentially a summary of the statistics for the given cluster. Clustering features are *additive*. Clustering features are sufficient for calculating all of the measurements that are needed for making clustering decisions in BIRCH. BIRCH thus utilizes storage efficiently by employing the clustering features to summarize information about the clusters of objects rather than storing all objects.

A CF tree is a height-balanced tree that stores the clustering features for a hierarchical clustering. An example is shown in Figure 1. By definition, a nonleaf node in the tree has descendents or "children". The nonleaf nodes store sums of the CFs of their children, and thus summarize clustering information about their children. Each entry in a leaf node is not a single data objects but a *subcluster*. A CF tree has two parameters: *branching factor* (B for nonleaf node and L for leaf node) and *threshold* T. The branching factor specifies the maximum number of children in each nonleaf or leaf node. The threshold parameter specifies the maximum diameter of the subcluster stored at the leaf nodes of the tree. The two parameters influence the size of the resulting tree.

BIRCH tries to produce the best clusters with the available resources. Given a limited amount of main memory, one of main concerns is to minimize the time required for I/O. BIRCH applies a multiphase clustering technique: a single scan of the data set yields a basic good clustering, and one or more additional scans can (optionally) be used to further refine the clustering quality. The BIRCH algorithm consists of four phases as follows.

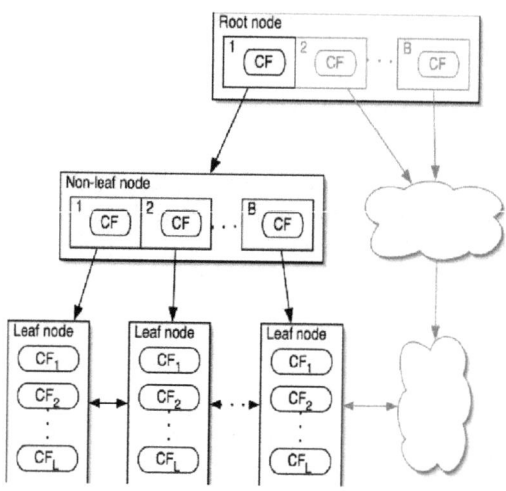

Fig. 1. A CF tree structure

Phase 1: (*Building CF tree*) BIRCH scans the database to build an initial in-memory CF tree, which can be viewed as a multilevel compression of the data that tries to preserve the intrinsic clustering structure of the data.

Phase 2: [optional] (*Condensing data*) Condense into desirable range by building a smaller CF tree.

Phase 3: (*Global Clustering*) BIRCH applies a selected clustering algorithm to cluster the leaf nodes of the CF tree.

Phase 4: [optional] *Cluster refining*

For Phase 1, the CF tree is build dynamically as objects are inserted. Thus, the method is incremental. An object is inserted to the closest leaf entry (subcluster). If the diameter of the subcluster stored in the leaf node after insertion is larger than the threshold value, the subcluster can not contain the object and the object can form a new entry and the leaf node in this case possibly is split. Node splitting is done by choosing the farthest pair of entries as seeds, and redistributing the remaining entries based on the closest criteria.

The size of the CF tree can be changed by modifying the threshold. If the size of the memory that is needed for storing the CF tree is larger than the size of the main memory, the CF tree should be made smaller, and a larger threshold value can be specified and the CF tree is rebuilt. The rebuild process (Phase 2) is performed by building a new tree from the leaf nodes of the old tree. Thus, the process of rebuilding the tree is done without the necessity of rereading all the objects or points. Therefore, for building the tree, data has to be read just once.

After the CF tree is built, any clustering algorithm, such as a typical partitioning algorithm, can be used in Phase 3 with the CF tree built in the previous phase. Phase 4, which is optional, uses the centroids of the clusters produced by Phase 3 as seeds and redistributes the data objects to its closest seed to obtain a set of new clusters.

3   The Proposed Approach – Combination of a Multi-resolution Transform and BIRCH

We can view our BIRCH-based approach as a flexible clustering framework in which we can apply any multi-resolution transform for time series dimensionality reduction and select any clustering algorithm in Phase 3 of the framework. The selected clustering algorithm can be a traditional algorithm (e.g. k-Means, k-Medoids, EM) that works in batch mode, or an *anytime* algorithm such as I-k-Means [7]. These major features are what we want to exploit in our BIRCH-based method for time series clustering.

Our BIRCH-based approach to perform clustering of time series is outlined as follows.

Step 1: A multi-resolution transform is computed for all time series data in the database. The process of feature extraction can be performed off-line, and needs to be done only once.

Step 2: We build and maintain the CF tree (that is Phase 1 of BIRCH algorithm) during the process of gathering the features of the multi-resolution transform. The CF tree stores a compact summarization of the original dataset.

Step 3: We use the CF tree to determine the initial cluster centers and start the k-Means, k-Medoids or I-k-Means algorithm that is used at Phase 3 of BIRCH algorithm. (Thus, if I-k-Means algorithm is used in the Phase 3, our clustering method can offer the capability of an *anytime* algorithm since it inherits the anytime property from I-k-Means).

Notice that if we apply a time series dimensionality reduction method without multi-resolution property in Step 1, we can not apply I-k-Means in Step 3 since this iterative clustering algorithm requires a dimensionality reduction with multi-resolution property. In other words, if we employ a dimensionality reduction with multi-resolution property in Step 1, we have more options to choose for the clustering algorithm in Step 3.

3.1   How to Determine the Appropriate Scale of a Multi-resolution Transform

One of the challenging problems in time series dimensionality reduction by a multi-resolution transform is how to choose the appropriate scale of the transform. In this work, we apply the same method proposed by Zhang et al. [12]

which is outlined as follows. The feature selection algorithm selects the feature dimensionality by leveraging two conflicting requirements, i.e., lower dimensionality and lower sum of squared errors between the features and the original time series. The main benefit of this method is that dimensionality of the features is chosen automatically.

3.2   Clustering with k-Means or I-k-Means in Phase 3

The algorithm works with the subclusters in leaf nodes (diameters of these subclusters are smaller than the threshold T) to perform clustering rather than working with the time series themselves. Notice that each subcluster is represented by its center. Given a set of time series, we have to partition them into K clusters. First, we build the CF tree. Then, we traverse the CF tree downwards from the root node to determine K initial clusters and their K associated cluster centers. After obtaining these K initial centers, we can use them to start a k-Means, k-Medoids or I-k-Means to perform Phase 3 of the BIRCH algorithm. The determination of initial centers through the traversal on CF tree helps the K-Means or I-k-Means not to be trapped into local minima and thus speeds up the execution time and improves the clustering quality. The outline of the algorithm for Phase 3 in the proposed clustering framework is described as in Figure 2.

/* The CF tree has been built in the maim memory */
1. From the root, recursively descend the CF tree by the following way:
    Put the root node of CF tree to queue Q
    while $(Q \neq \phi$ OR the number of nodes in Q is less than $K)$ do
    begin
        Get a node D from the queue;
        Put all the child nodes, if any, of D to Q;
    end
2. Determine K clusters:
    Now queue Q holds the root nodes of L subtrees T_i $(i \leq L)$. Each subtree T_i is exactly a cluster.
    if $(L \succ K)$ then
        Perform the merging of the pair of closest subclusters until $L \equiv K$.
    if $(L \prec K)$ then
        Perform repeatedly the split of some subclusters which diameter is largest until $L \equiv K$.
3. Apply k-Means, k-Medoids or I-k-Means to perform clustering with the initial centers obtained from the centers of K clusters determined in step 3. The data for clustering are the subclusters held in the leaf nodes of CF tree.

Fig. 2. The algorithm for Phase 3 in the proposed clustering framework

The algorithm can be further sped up by applying the heuristic that helps to avoid the merging and the split of subclusters in step 2. The main idea of the heuristic is that the selection of subclusters to be initial centers should be based on their density estimation and their distances from the selected initial centers. The further a subcluster is from its nearest existing initial center and the larger its density is, the more likely a candidate it is to be an initial center. This heuristic reflects the same spirit of the technique for initialization the k-Means clustering proposed by Redmond and Heneghan, 2005 [10].

4   Experimental Evaluation

In this experiment, we compare the performance of our proposed approach with k-Means and I-k-Means in term of clustering quality and running time. We applied Haar wavelet transform for time series dimensionality reduction. We conducted the experiment on two main datasets.

4.1   Clustering Quality Evaluation Criteria

The heading should be treated as a subsubsection heading and should not be assigned a number. We can use classified datasets and compare how good the clustered results fit with the data labels, which is the popular clustering evaluation method [3]. Five objective clustering evaluation criteria were used in our experiments: Jaccard, Rand, FM [3], CSM used for evaluating time series clustering algorithms [5] and NMI used recently for validating clustering algorithms [11].

Besides, since k-Means seeks to optimize the objective function by minimizing the sum of squared intra-cluster error [4], we evaluate the quality of clustering by using the following objective function:

$$F = \sum_{m=1}^{k} \sum_{i=1}^{N} ||x_i - c_m||$$

where x are the objects and c are the cluster centers.

4.2   Data Description

We tested on two publicly available datasets : Heterogeneous and Stock. The Heterogeneous, which is obtained from the UCR Time Series Data Mining Archive [6], is the classified dataset, therefore we can use it to evaluate the accuracy of the clustering algorithms. The Stock dataset is the stock data of year 1998 from Historical Data for S&P 500 Stocks [14] and each stock time series consists of 252 points representing the open prices of a particular stock.

We conduct the experiments on the two datasets with cardinalities ranging from 1000 to 10000 for each dataset. In the Heterogeneous dataset, each time series consists of 1024 points. In the Stock dataset, the length of each time series has been set to 252.

The Heterogeneous dataset is generated from a mixture of 10 real time series data from the UCR Time Series Data Mining Archive. Using the 10 time-series as seeds, we produced variation of the original patterns by adding small time shifting (2-3 % of the series length), and interpolated Gaussian noise. Gaussian noisy peaks are interpolated using splines to create smooth random variations.

4.3 Experimental Results

· **Heterogeneous:** For this dataset, we know that the number of clusters is 10 (K =10). We took the widely-used Euclidean distance for clustering algorithm. We experimented four clustering algorithms: k-Means, I-k-Means, BIRCH with k-Means in Phase 3 (called BIRCH-k-Means) and BIRCH with I-k-Means in Phase 3 (called BIRCH-I-k-Means) on the Heterogeneous dataset.

Here we use the CF tree with the parameter setting: $B = 2$, $L = 4$, $T = 0$. The threshold T is set to 0 since we want each entry (subcluster) in leaf node holds only one time series. Since we generated the heterogeneous dataset from a set of given time series data, we have the knowledge of correct clustering results in advance. In this case, we can compute the evaluation criteria such as Jaccard, Rand, FM, CSM, and NMI to assess the clustering quality of the competing algorithms. Table 1 shows the criteria values obtained from the experiments on the four clustering algorithms: k-Means, I-k-Means, BIRCH-k-Means and BIRCH-I-k-Means. As we can see from Table 1, our new algorithms (BIRCH-k-Means and BIRCH-I-k-Means) results in better criteria values than the classical k-Means and I-k-Means. Especially, BIRCH-I-k-Means is the best performer in terms of clustering quality. In Figure 3 we show the running time, and the objective

Table 1. Evaluation criteria values obtained from four clustering algorithms with Heterogeneous dataset

Criteria	k-Means	I-k-Means	BIRCH-k-Means	BIRCH-I-k-Means
Jaccard	0.6421	0.7124	0.7210	0.7341
Rand	0.5909	0.6934	0.7002	0.7112
FM	0.5542	0.7091	0.6991	0.7091
CSM	0.5095	0.6954	0.6542	0.6908
NMI	0.0541	0.6841	0.6431	0.6501

function (see 4.1) respectively from the experiments on the three clustering algorithms: k-Means, I-k-Means, and BIRCH-I-k-Means. In Figure 3.a we can see that the running time of I-k-Means and BIRCH-I-k-Means is much less than that of k-Means. Besides, the running time of BIRCH-I-k-Means is a bit more than that of I-k-Means due to the overhead of building CF tree. In Figure 3.b, we can observe that BIRCH-I-k-Means is also the best performer in term of the objective function, another criterion for clustering quality evaluation.

· **Stock:** For the Stock dataset, we do not have prior knowledge of correct clustering results. Lacking this information, we cannot evaluation criteria such as Jaccard, Rand, FM, CSM, and NMI to assess the clustering quality. So, we have to use the objective function values to compare the quality of the three clustering algorithms. We conduct the clustering with $K = 10$ on the Stock dataset. In Figure 4 we show the running time and the objective function, respectively from the experiments on the Stock dataset and with the three clustering algorithms: k-Means, I-k-Means, and BIRCH-I-k-Means. In Figure 4.a we can see that the running time of I-k-Means and BIRCH-I-k-Means is much less than that of k-Means. In Figure 4.b, again we can observe that BIRCH-I-k-Means is the best performer in term of the objective function.

The experimental results from both the datasets exhibit that when using BIRCH-I-k-Means, the quality of clustering as well as running time are both improved. We attribute this improvement to good choices of initial centers by employing CF tree in the selection of K initial clusters. The experiments also reveal that the number of iterations in BIRCH-I-k-Means is much smaller than that in I-k-Means.

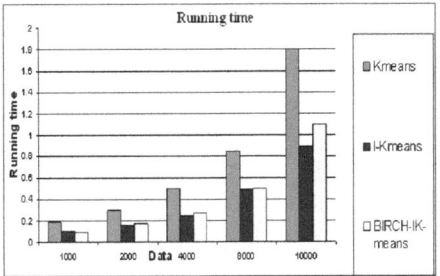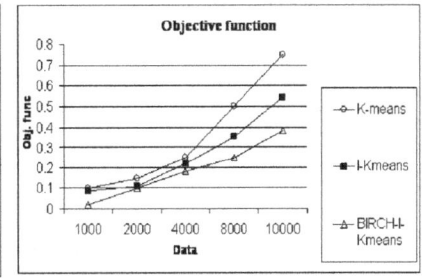

Fig. 3. BIRCH-I-k-Means vs. k-Means and I-k-Means in terms of (a) Running time for different data sizes. (b) Objection function for different data sizes (Heterogeneous dataset).

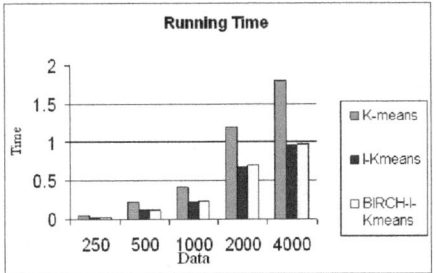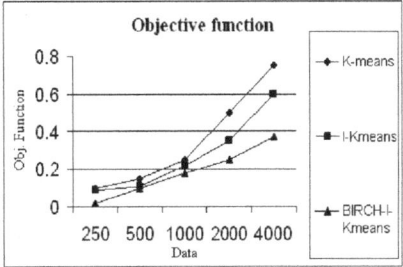

Fig. 4. BIRCH-I-k-Means vs. k-Means and I-k-Means in terms of (a) Running time for different data sizes. (b) Objective function for different data sizes (Stock dataset).

5   Conclusion

We have presented our BIRCH-based approach to perform clustering of time series data with a multi-resolution transform used as feature extraction method. This approach resolves the dilemma associated with the choices of initial centers and significantly improves the execution time and clustering quality. Our BIRCH-based approach can be viewed as a flexible clustering framework in which we can apply any selected clustering algorithm in Phase 3 of the framework. We empirically compared the performances of three clustering algorithms: k-Means, I-k-Means and BIRCH-I-k-Means and found out that BIRCH-I-k-Means performs better than k-Means in terms of clustering quality and running time, and better than I-k-Means in terms of clustering quality with nearly the same running time. Besides, our BIRCH-based approach has three other advantages. First, it can produce the best quality clustering in the environment with limited memory and time resources. Second, it can scale well to very large datasets where new time series can arrive dynamically. Thirdly, the user can perform clustering with several different values of K without having to start from the scratch since the CF tree is built only once.

At the moment, we focus our approach on the k-Means or I-k-Means employed at Phase 3 of the BIRCH algorithm. For future work, we plan to investigate: (i) extending our experiments to larger datasets and with time series of longer lengths and (ii) comparing the performances of our BIRCH-based approach over different multi-resolution feature extraction methods.

References

1. Chan, K., Fu, W.: Efficient time series matching by wavelets. In: Proceedings of the 15th IEEE Intl. Conf. on Data Engineering (ICDE 1999), March 23-26, pp. 126–133 (1999)
2. Gavrilov, M., Anguelov, M., Indyk, P., Motwani, R.: Mining The Stock Market: Which Measure is Best? In: Proc. of 6th ACM Conf. on Knowledge Discovery and Data Mining, Boston, MA, August 20-23, pp. 487–496 (2000)
3. Halkdi, M., Batistakis, Y., Vizirgiannis, M.: On Clustering Validation Techniques. J. Intelligent Information Systems 17(2-3), 107–145 (2001)
4. Han, J., Kamber, M.: Data Mining: Concepts and Techniques, 2nd edn. Morgan Kaufmann (2006)
5. Kalpakis, K., Gada, D., Puttagunta, V.: Distance Measures for Effective Clustering of ARIMA Time Series. In: Proc. of 2001 IEEE Int. Conf. on Data Mining, pp. 273–280 (2001)
6. Keogh, E., Folias, T.: The UCR Time Series Data Mining Archive (2002), http://www.cs.ucr.edu/~eamonn/TSDMA/index.html
7. Lin, J., Vlachos, M., Keogh, E.J., Gunopulos, D.: Iterative Incremental Clustering of Time Series. In: Hwang, J., Christodoulakis, S., Plexousakis, D., Christophides, V., Koubarakis, M., Böhm, K. (eds.) EDBT 2004. LNCS, vol. 2992, pp. 106–122. Springer, Heidelberg (2004)
8. Cao, L.: In-depth Behavior Understanding and Use: the Behavior Informatics Approach. Information Science 180(17), 3067–3085 (2010)

9. May, P., Ehrlich, H.-C., Steinke, T.: ZIB Structure Prediction Pipeline: Composing a Complex Biological Workflow Through Web Services. In: Nagel, W.E., Walter, W.V., Lehner, W. (eds.) Euro-Par 2006. LNCS, vol. 4128, pp. 1148–1158. Springer, Heidelberg (2006)

10. Redmond, S., Heneghan, C.: A Method for Initialization the k-Means Clustering Algorithm Using kd-Trees. Pattern Recognition Letters (2007)

11. Strehl, A., Ghosh, J.: Cluster Ensembles – A Knowledge Reuse Framework for Combining Multiple Partitions. J. of Machine Learning Research 3(3), 583–617 (2002)

12. Zhang, H., Ho, T.B., Zhang, Y., Lin, M.S.: Unsupervised Feature Extraction for Time Series Clustering Using Orthogonal Wavelet Transform. Journal Informatica 30(3), 305–319 (2006)

13. Zhang, T., Ramakrishnan, R., Livny, M.: BIRCH: A new data clustering algorithm and its applications. Journal of Data Mining and Knowledge Discovery 1(2), 141–182 (1997)

14. Historical Data for S&P 500 Stocks, http://kumo.swcp.com/stocks/

Visualizing Cluster Structures and Their Changes over Time by Two-Step Application of Self-Organizing Maps

Masahiro Ishikawa

Department of Media and Communication Studies, Tsukuba Int'l University, Japan
Manabe 6-20-1, Tsuchiura, Ibaraki, 300-0051 Japan
mi@tius.ac.jp

Abstract. In this paper, a novel method for visualizing cluster structures and their changes over time is proposed. Clustering is achieved by two-step application of self-organizing maps (SOMs). By two-step application of SOMs, each cluster is assigned an angle and a color. Similar clusters are assigned similar ones. By using colors and angles, cluster structures are visualized in several fashions. In those visualizations, it is easy to identify similar clusters and to see degrees of cluster separations. Thus, we can visually decide whether some clusters should be grouped or separated. Colors and angles are also used to make clusters in multiple datasets from different time periods comparable. Even if they belong to different periods, similar clusters are assigned similar colors and angles, thus it is easy to recognize that which cluster has grown or which one has diminished in time. As an example, the proposed method is applied to a collection of Japanese news articles. Experimental results show that the proposed method can clearly visualize cluster structures and their changes over time, even when multiple datasets from different time periods are concerned.

Keywords: Clustering, Visualization, Self-Organizing Map, Cluster Changes over Time.

1   Introduction

We are in the midst of the electric era where various kinds of data, in particular, behavioral data, are originally produced electrically or converted to electric forms from original forms. For example, sensored data are produced from scientific observation equipments in electric forms minute by minute and second by second. Everyday stock prices or some socio-economic indices are also produced and recorded day to day. In social media, for example in blogs, huge amount of text data has already been produced and accumulated, and more and more texts will be continuously produced and accumulated in the future too. Not limited to these types of data, variety of data are accumulated in electric forms.

Electrically accumulated data is easy to access and process by computers. However, huge amount of data is intractable if they are not organized systematically. Clustering is one of the most fundamental processes to organize them.

L. Cao et al. (Eds.): PAKDD 2011 Workshops, LNAI 7104, pp. 160–170, 2012.

It is used to grasp the overview of the accumulated data. Though clustering results might be used to navigate users to explore the whole data, numeric expressions are not enough. Even if accurate information is contained in the numeric expressions of clustering results, it is hard to recognize them for domain experts such as market analysts who are interested in utilizing the data in their application fields. Hence, visualization is very important for human beings. In many environments, including blogs on the web, data are continuously produced and accumulated, thus cluster structures will also change over time. Therefore, visualization methods should accommodate cluster behavior changes, an important task for behavior informatics [12]. It should visualize not only cluster structures in a time period, but also changes of cluster structures over time.

In this paper, a novel method for visualizing cluster structures and their changes over time is proposed. For the visualizations, two-step application of Batch Map, which is a batch learning version of self-organizing maps[1], is used for clustering and assigning angles and colors to clusters. Then colors and angles are used in visualizations. As an example, a collection of news articles is visualized by the proposed method. The example shows that cluster structures and their changes over time are clearly visualized by the proposed method.

2   Related Work

Visualization is important task especially for interactive data mining, and self-organizing map (SOM) is a popular tool for visualization. Thus, there are many studies on SOMs for visualization. There are popular methods for visualizing a single SOM, including U-matrix[1], P-matrix[15], and U*matrix[14]. These are applied after a SOM is constructed, and SOM map is used for visualization. However, even when a single map is concerned, cluster shape is obscure in those visualizations, so we cannot clearly "see" cluster structures.

In [13], Denny et al. proposed a SOM-based method for visualizing cluster changes between two datasets. Two SOMs are constructed from the first and the second datasets. Then, cells in the first SOM is clustered by k-means clustering method producing k clusters, and clusters are assigned different colors arbitrarily. Each cell in the second SOM is assigned the color of the most similar cell in the first SOM. Thus, clusters in two SOM maps are linked by colors. However, it can be applied to only two datasets. And the coloring scheme is not systematic, thus colors do not reflect relationship among clusters systematically. Moreover, even when new clusters emerge in the second SOM, they are forced to be assigned colors of clusters in the first SOM in spite of non-existence of correctly corresponding clusters in the first SOM.

In the second step of the method proposed in this paper, ring topology one-dimensional SOM is used and plays important roles. It makes systematic color assignment and application to arbitrary number of datasets possible.

3   The Proposed Method

3.1   Batch Map

Self-organizing map (SOM)[1] is a vector quantization and clustering method which is usually used to visualize cluster structures in high dimensional numeric vectors in a lower dimensional space, typically on two-dimensional plane. Thus it is a potential candidate to use when one would like to visualize huge amount of high dimensional numeric vectors. The key property of SOM is that it "keeps" topology in the original data space on the two-dimensional plane too. Similar vectors are mapped to cells which are close to each other on the two-dimensional plane, thus the cluster structure in the original space is "reproduced" on the plane. By applying some post-processing, U-matrix construction for example, data visualization is accomplished.

Though "SOM" is usually used to refer to on-line learning version of SOMs, the word refers to the batch-learning version, called *Batch Map* in the literature[1], in this paper. Here, Batch Map is summarized. Let $D = \{d_0, d_1, \cdots, d_{n-1}\}$ $(d_i \in \mathcal{R}^m)$ be a dataset to be processed, $\phi : \mathcal{R}^m \times \mathcal{R}^m \to \mathcal{R}$ be the distance(dissimilarity) function on D, and $\{c_{x,y}\}$ be the cells in two-dimensional (typically hexagonal) SOM map of $X \times Y$ grids. Each cell is associated with a reference vector $v(c_{x,y}) \in \mathcal{R}^m$. In the learning process of a Batch Map, each datum d_i is assigned to the cell associated with the reference vector which is most similar to d_i. By gradually updating the reference vectors, the Batch Map reproduces the cluster structures in D on the SOM map. Batch Map learning procedure is summarized below.

1. Initialize the reference vectors $v(c_{x,y})$.
2. Assign each datum d_i to the cell $c_{s,t}$ where

$$\phi(d_i, v(c_{s,t})) = \min_{x,y}\{\phi(d_i, v(c_{x,y}))\}.$$

3. Update reference vectors according to the following formula:

$$v(c_{x,y}) \leftarrow \frac{1}{|\mathcal{N}_{x,y}^{(R)}|} \sum_{d \in \mathcal{N}_{x,y}^{(R)}} d \tag{1}$$

Here, $\mathcal{N}_{x,y}^{(R)}$ denotes the set of data assigned to cells in the circle centered at $c_{x,y}$ with radius R. Thus, a reference vector is updated by the average of all data assigned to its neighborhood with respect to R.

4. Repeat step 2 and 3 until it converges, gradually decreasing radius R. In the last steps, R should be zero.

Note that when R equals to zero, this procedure is identical to that of the well-known k-means clustering.

3.2   Two-Step Application of Self-Organizing Maps

In the proposed method, SOM is applied in two steps. In the first step, two-dimensional SOM of $X \times Y$ cells is applied to the input dataset. As the result, each datum is assigned to the most similar cell (i.e. the cell associated with the most similar reference vector). In a sense, $X \times Y$ clusters are formed. They are called *micro clusters* in this paper. Reference vectors are the centroids of micro clusters. One of important reasons why SOM is used in this step is that the distribution of centroids obtained by a SOM approximately follows that of the original dataset, thus the micro clusters can be used as a compact representation of the original dataset.

Second, one-dimensional SOM with p cells is applied to centroids of micro clusters obtained above. An important point is that, ring topology must be adopted for the one-dimensional SOM. Hereafter, ring-topology one-dimensional SOM is called *coloring SOM* (cSOM). As the result of cSOM application, each centroid of a micro cluster is assigned to a cSOM cell, forming *macro clusters*. Reference vectors of a cSOM are the centroids of macro clusters. In this step, p macro clusters are formed.

3.3   Assigning Angles and Colors to Clusters

Using the centroids of macro clusters, each macro cluster is assigned an angle and a color. Let c_i be the i-th cell of the cSOM. Then, each cell c_i (thus a macro cluster) is assigned an angle ω_i which is calculated by the formula below. In the formula, p is the number of cells in the cSOM.

$$\omega_i = 2\pi \frac{\sum_{k=0}^{i-1} \phi(v(c_k), v(c_{k+1}))}{\sum_{k=0}^{p-1} \phi(v(c_k), v(c_{(k+1) \bmod p}))} \tag{2}$$

Then, a triplet $\langle R_i, G_i, B_i \rangle$ is calculated from ω_i by the following formulae.

$$R_i = \frac{sin(\omega_i) + 1}{2}$$

$$G_i = \frac{sin(\omega_i + \frac{2\pi}{3}) + 1}{2}$$

$$B_i = \frac{sin(\omega_i + \frac{4\pi}{3}) + 1}{2}$$

This triplet represents a color in the RGB coloring space $[0, 1]^3$, and assigned to the cell c_i (thus, to the corresponding macro cluster, micro clusters, and cells in the first step SOM). This coloring scheme means picking up a color from the hue circle at the angle ω_i. See Figure 1. If the distance between adjacent cells' reference vectors is smaller, the difference of assigned angles will also be smaller. Thus closer adjacent clusters are assigned more similar angles and colors systematically.

The focus of this paper is how to visualize the cluster structures and their changes over time. Thus, we should think of multiple datasets from multiple

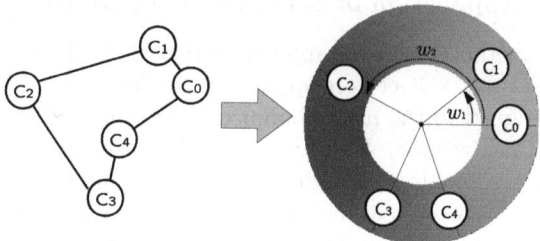

Fig. 1. Image of the Coloring Scheme: Left picture is the cSOM cells distributed in the input space. Then cells are placed on the hue circle according to the angles assigned to them. The colors under the cells are picked up.

periods of time. In such situations, each dataset is individually clustered by two dimensional SOMs, forming micro clusters in each dataset. To place similar clusters in two successive datasets at similar positions on two SOM maps, $i+1$-th SOM is initialized by the resultant reference vectors of the i-th SOM. The first SOM is randomly initialized.

Then all micro clusters of all datasets are used as input to a coloring SOM. As the result, even when they belong to different datasets, similar clusters are assigned similar angles and colors, which can be used to link similar clusters among multiple datasets.

3.4   Visualization by Colors and Angles

By using the assigned colors and angles, cluster structures are visualized in several fashions. One is based on the traditional SOM map, i.e. two-dimensional grid. Indeed two-dimensional SOM map is appealing, but there is no need to stick to it. Thus, two more fashions are introduced. They are briefly summarized here and the details are described in the next section with examples.

SOM Map Visualization. First of all, the two-dimensional SOM map is used to visualize the cluster structures. By coloring each cell the assigned color, a cluster can be seen as a continuous group of cells filled with similar colors.

Area Chart Visualization. In the area chart visualization, sizes (populations) of clusters are used. By using x-axis for time periods (datasets) and y-axis for sizes of clusters, changes of cluster sizes are visualized. The area dominated by a cluster is filled with the assigned color, and the stacking order of clusters is determined by the angles. Thus, adjacent areas are filled with similar colors, making a gradation. We can easily see which cluster (or group of similar clusters) has grown and which one has diminished.

Polar Chart Visualization. Cluster sizes can be visualized by histograms too. In a histogram, a cluster is represented by a bar, whose height reflects the cluster size. Again, by coloring the bars, relations among clusters can be visualized. However, the coloring scheme has some deficits. The most serious one is the perceived non-uniformity of color differences. The perceived color differences do not linearly reflect the differences of angles. Moreover they depend on the part of the hue circle. The other one is the device dependency of color reproduction. Thus, to link similar clusters, angles assigned to clusters are also used directly here. Instead of the orthogonal x-y axes, the polar axis is adopted. Each cluster bar is depicted at the angle assigned to it. Thus, similar clusters are placed on similar directions. Therefore, similar clusters in different datasets can be easily found in this visualization.

4 Example: Visualization of Clusters in News Articles

As an example, a collection of news articles are visualized here. Experiment was conducted on a PC Linux system with Intel Core i7 920 CPU and 12GB memory. Programs were implemented in Python language with the numpy[10] scientific computing extension.

4.1 Target Dataset

Japanese news articles released by Sankei digital corporation as "Sankei e-text" [9] in 2005 are used. It consists of 73388 articles.

4.2 Keyword Extraction and Matrix Representation

To apply SOMs, the dataset must be represented as a set of numeric vectors. First of all, each article is represented as a bag-of-words. To extract words from an article, a morphological analyzer is applied.[1] Then words tagged as noun are extracted, excluding some exceptions.[2] Then, according to the standard vector space model commonly used in information retrieval[2], news articles are represented as an $m \times n$ term-document matrix as follows:

$$D = \begin{pmatrix} d_{11} & d_{12} & \cdots & d_{1n} \\ d_{21} & d_{22} & \cdots & d_{2n} \\ \vdots & \vdots & \ddots & \vdots \\ d_{m1} & d_{m2} & \cdots & d_{mn} \end{pmatrix}$$

Where, m is the number of words in the collection (vocabulary size), n the number of news articles, and $d_{i,j}$ the weight of the i-th word in the j-th article. In this example, tf-idf values given by the formula below are used.

$$d_{i,j} = -f_{i,j} \log \frac{n_i}{n}$$

---

[1] Mecab[11] with IPA dictionary is used.
[2] Pronouns, adjuncts, suffixes, postfixes and numerals are excluded.

In the formula, $f_{i,j}$ is the frequency of i-th word in the j-th article, and n_i is the number of articles which contains the i-th word. In the matrix, a column vector represents an article.

In the target dataset, n is 73388, and m is 108750.

4.3   Dimension Reduction by Random Projection and LSI

Usually total number of articles, n, and vocabulary size, m, are very huge, thus memory and time costs are so high in the successive process that in many cases the problem becomes intractable. Thus, dimension reduction is applied here.

First, random projection[6][5] is applied to the original matrix D for reducing the dimensionality of the column vectors. Second, LSI (Latent Semantic Indexing)[8] is applied for further reduction. In the same time, LSI incorporates the latent semantics of words. This combination could achieve better clustering quality than sole application of random projection [7]. This aspect will not be further discussed in this paper since it is out of the scope. However, random projection and LSI are briefly described below to make this paper self-contained.

Random Projection. Random projection is a dimension reduction technique for high dimensional numeric vectors. To reduce the dimension of the column vectors of matrix D, randomly generated $m_1 \times m(m_1 \ll m)$ matrix R is used as follows.

$$D_1 = RD$$

If R satisfies some conditions, angles among column vectors are well preserved. In this example, elements of R are generated according to the following formula proposed in [4].

$$r_{ij} = \sqrt{3} \begin{cases} 1 & \text{(with probability } \frac{1}{6}) \\ 0 & \text{(with probability } \frac{2}{3}) \\ -1 & \text{(with probability } \frac{1}{6}) \end{cases}$$

Here, $\sqrt{3}$ is omitted since relative values are suffice for clustering purpose.

In this experiment, the dimensionality is reduced to 5000 by random projection.

Latent Semantic Indexing. LSI is one of popular techniques for dimension reduction, which, in the same time, incorporates latent semantics of words. Though LSI is the best technique for dimension reduction with respect to the least square error, it is intractable when a huge matrix is concerned due to its space and time complexities. However, by applying random projection beforehand, matrix size can be reduced to which LSI can be applied. By singular value decomposition, we have $D_1 = U \Sigma V^{\mathrm{T}}$. Or by eigen value decomposition, we get $D_1 D_1^{\mathrm{T}} = U \Lambda U^{\mathrm{T}}$. Then U_{m_2} is formed by taking the leftmost m_2 ($\ll m_1$) column of U. Dimension reduction is achieved by $D_2 = U_{m_2}^{\mathrm{T}} D_1$.

In this experiment, the dimensionality is reduce to 3600 by LSI. Accumulative contribution rate is 0.91.

Fig. 2. SOM map Visualization: Left map is the first week and the right is the second

4.4   Final Matrix and Distance Function

After dimension reduction, column vectors of D_2 are normalized to the unit length, obtaining 3600×73388 matrix X. Finally, according to their release dates, X is divided into 52 sub matrices $\{X_i\}$, each one contains articles released in each week of the year.

Similarity between articles is given by cosine measure which is popular in document retrieval and clustering. However, distance (dissimilarity) is needed in formula (2), thus the angle derived from the cosine measure is used as the distance function.

$$\phi(x, y) = \arccos(cos(x, y))$$

The range is restricted to $0 \leq \phi(x, y) \leq \pi$. Incidentally, learning procedure of Batch Map is modified in two points. First, in the step 1, reference vectors are normalized to the unit length. Second, in the step 3, updated reference vectors are also normalized to the unit length. As the result, Batch Map used in this example becomes the Dot Product SOM[1] which corresponds to the spherical k-means clustering[3]. Note that, these modifications are not mandatory though they are suitable for document clustering.

For the first step clustering, SOM of 10×10 cells with torus topology is used. Thus, 100 micro clusters are produced for each week and 5200 in total. For a coloring SOM, a ring of 64 cells is used, thus 64 macro clusters are produced by clustering 5200 micro clusters.

4.5   Visualization Results

SOM Maps. By successively applying SOMs, we got 52 SOM maps for 52 weeks. In Figure 2, the maps of the first and the second weeks are presented. We can easily find clusters as continuous regions of similar colors. In addition to separate maps, we can view all maps in a three-dimensional graphics at a glance. Figure 3 shows 3D views of all maps. Each cell is drawn as a cylinder. In the left view of Figure 3, 52 weeks are all visible. By controlling the transparency and the brightness of cells, we can view changes of specific clusters at a glance. In the

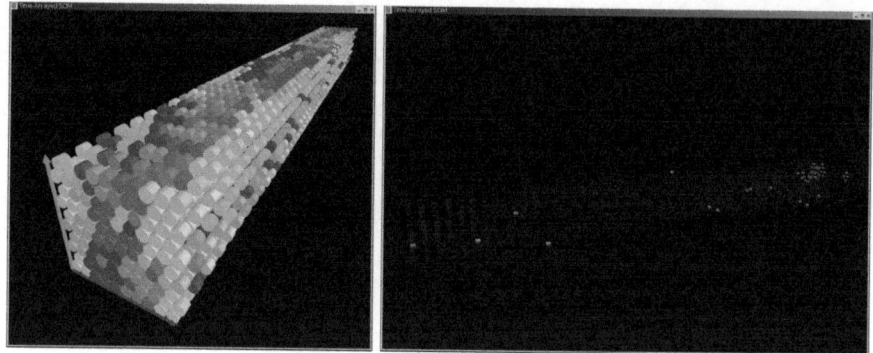

Fig. 3. 3D Visualization of SOM Maps: In the right view, only one cluster is emphasized

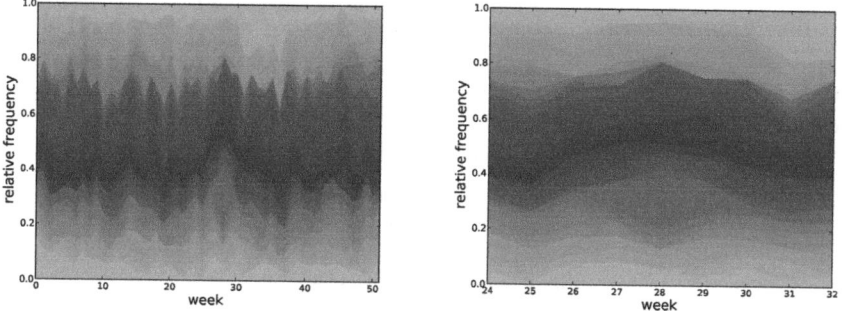

Fig. 4. Area Chart Visualization

right view of Figure 3, all clusters but one are made almost transparent. In the view, the rightmost is the first week and the leftmost is the most recent week. Recall that the distribution of reference vectors approximately follows that of the input dataset. Thus we can figure out that this cluster thrived at the beginning of the year, and rapidly diminished. This cluster mainly consists of articles about the giant tsunami in the Indian Ocean caused on the ending of the previous year.

Area Charts. Figure 4 shows area chart visualizations. In the figure, the stacking order of clusters are defined by the assigned angles. Thus, we can view not only single cluster but also a group of similar clusters as a unit. Note that the bottom and the top clusters are adjacent since a coloring SOM has ring topology.

In the right chart of Figure 4, the range from the 24th week to the 32nd week is magnified. It shows that a cluster filled with orange irregularly expanded in the 28th week. This cluster mainly consists of articles about the most popular Japanese high school baseball tournament held in August. Though we can see global structures at a glance in an area chart, it becomes intricate when the number of clusters got increased.

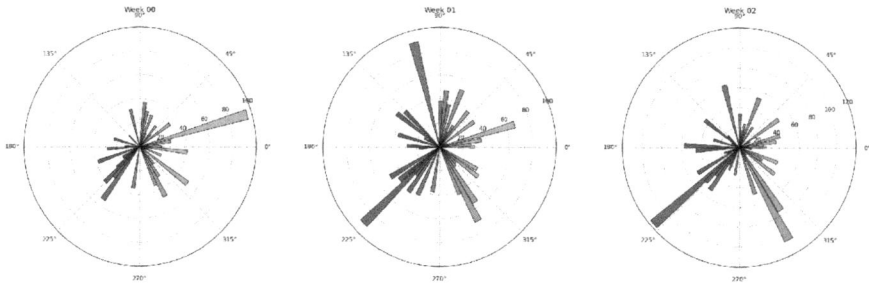

Fig. 5. Polar Chart Visualization

Polar Charts. Figure 5 shows polar charts for the first three weeks of the year. Each bar is a cluster and its length means the cluster size. Bars are filled with assigned colors. Note that a polar chart is not a pie chart. A cluster is placed at the angle assigned to it. Therefore, similar clusters are placed on similar directions, and angles between adjacent clusters reflects degrees of cluster separations. Crowded direction suggests that clusters on that direction might have to be merged into one cluster. The longest bar filled with green in the left chart of Figure 5 is the cluster about the giant tsunami in the Indian Ocean visualized in Figure 3. Even when it belongs to different weeks, the same cluster is placed on the same direction, thus we can easily find which cluster has grown and which one has diminished. These three charts show that the tsunami cluster is rapidly diminishing.

In many cases, it is difficult to determine how many clusters should be found in a dataset. Thus, for example in k-means clustering, sometimes several clustering trials with different values for k are needed. By forming sufficiently many micro and macro clusters, we might be able to figure out appropriate number of clusters by this visualization.

5   Conclusions and Future Work

In this paper, a novel method for visualizing cluster structures and their changes over time is proposed. A two-step SOM application method is introduced for clustering. Then, using the clustering results, each cluster is assigned an angle and a color systematically, which reflects relationship among clusters.

Visualization is achieved by using angles and colors assigned to clusters. In addition to the traditional two dimensional SOM map, area chart and polar chart are used for visualization. By experiments on a collection of Japanese news articles, effectiveness of the proposed method is demonstrated. Note that the proposed method can be applied to any kind of dataset as long as they are represented as numeric vectors and a distance function is defined. It should be pointed out that if SOM map visualization is not needed, the first step clustering could be done by one-dimensional SOM too. Moreover, it can be substituted for by any clustering method as long as the distribution of the resultant clusters' centroids follows that of the input dataset.

The same dataset is visualized in three fashions. Thus, cooperative work of them is important. Interactive user interface is also needed. Each fashion has merits and demerits, thus they should complement one another.

In this paper, cSOM is applied to the set of all micro clusters, thus every time a new dataset arrives cSOM must be applied anew to the incrementally augmented set of micro clusters. This is a drawback in dynamic environments. Therefore, incremental cSOM construction method should be invented in future work.

References

1. Kohonen, T.: Self-Organizing Maps, 3rd edn. Springer, Heidelberg (2001)
2. Baeza-Yates, R., Ribeiro-Neto, B.: Modern Information Retrieval. Addison Wesley (1999)
3. Dhillon, I.S., Modha, D.S.: Concept decompositions for large sparse text data using clustering. Machine Learning 42(1), 143–175 (2001)
4. Achlioptas, D.: Database-friendly Random Projections. In: Proc. of the 20th ACM SIGMOD-SIGACT-SIGART Symposium on Principles of Database Systems, pp. 274–281 (2001)
5. Bingham, E., Mannila, H.: Random projection in dimensionality reduction: applications to image and text data. In: Proc. of the 7th ACM SIGKDD International Conference on Knowledge Discovery and Data Mining, pp. 245–250 (2001)
6. Dasgupta, S.: Experiments with Random Projection. In: Proc. of the 16th Conference on Uncertainty in Artificial Intelligence, pp. 143–151 (2000)
7. Lin, J., Gunopulos, D.: Dimensionality reduction by random projection and latent semantic indexing. In: Proc. of SDM 2003 Conference, Text Mining Workshop (2003)
8. Papadimitriou, C.H., et al.: Latent Semantic Indexing: A Probabilistic Analysis. In: Proc. of the 17th ACM SIGACT-SIGMOD-SIGART Symposium on Principles of Database Systems, pp. 159–168 (1998)
9. Sankei e-text, https://webs.sankei.co.jp/sankei/about_etxt.html
10. Scientific Computing Tools for Python — numpy, http://numpy.scipy.org/
11. MeCab: Yet Another Part-of-Speech and Morphological Analyzer, http://mecab.sourceforge.net/
12. Cao, L.: In-depth Behavior Understanding and Use: the Behavior Informatics Approach. Information Science 180(17), 3067–3085 (2010)
13. Denny, Squire, D.M.: isualization for Cluster Changes by Comparing Self-organizing Maps. In: Ho, T.-B., Cheung, D., Liu, H. (eds.) PAKDD 2005. LNCS (LNAI), vol. 3518, pp. 410–419. Springer, Heidelberg (2005)
14. Ultsch, A.: U*-Matrix: A Tool to visualize Cluster in high-dimensional Data. In: Proc. of the 2008 Eighth IEEE International Conference on Data Mining, pp. 173–182 (2008)
15. Ultsch, A.: Maps for the Visualization of high-dimensional Data Spaces. In: Proc. of Workshop on Self-Organizing Maps 2003, pp. 225–230 (2003)

Analysis of Cluster Migrations Using Self-Organizing Maps

Denny[1,2], Peter Christen[1], and Graham J. Williams[1,3]

[1] Research School of Computer Science, The Australian National University,
Canberra, Australia
[2] Faculty of Computer Science, University of Indonesia, Indonesia
[3] Australian Taxation Office, Canberra, Australia
denny@cs.ui.ac.id, peter.christen@anu.edu.au, graham.williams@ato.gov.au

Abstract. Discovering cluster changes in real-life data is important in many contexts, such as fraud detection and customer attrition analysis. Organizations can use such knowledge of change to adapt business strategies in response to changing circumstances. This paper is aimed at the visual exploration of migrations of cluster entities over time using Self-Organizing Maps. The contribution is a method for analyzing and visualizing entity migration between clusters in two or more snapshot datasets. Existing research on temporal clustering primarily focuses on either time-series clustering, clustering of sequences, or data stream clustering. There is a lack of work on clustering snapshot datasets collected at different points in time. This paper explores cluster changes between such snapshot data. Besides analyzing structural cluster changes, analysts often desire deeper insight into changes at the entity level, such as identifying which attributes changed most significantly in the members of a disappearing cluster. This paper presents a method to visualize migration paths and a framework to rank attributes based on the extent of change among selected entities. The method is evaluated using synthetic and real-life datasets, including data from the World Bank.

Keywords: temporal cluster analysis, cluster migration analysis, visual data exploration, change analysis, Self-Organizing Map.

1 Introduction

Migration analysis tracks changing behaviors of a population and monitors the changes in clusters over time. This can, for example, be applied to group customers' purchase patterns to identify high-responsive customers for marketing promotions [10]. Moreover, it can be used to perform customer attrition analysis to prevent churning (customer turnover or loss of customers) from happening in the future. In industries such as telecommunications where getting new customers is costlier than retaining existing ones, knowledge about churning customers is important [10]. With this knowledge, strategies can be derived to prevent churn. For example, giving free incentives and getting feedback on customer satisfaction were recently used to target customers of a pay-TV company

L. Cao et al. (Eds.): PAKDD 2011 Workshops, LNAI 7104, pp. 171–182, 2012.

who had a high probability to churn [1]. Migration analysis can be used not only for analyzing churning customers, but also to analyze migrations from a cluster of highly profitable customers to a cluster of average customers.

While recent work in this area [3,5] focused on the analysis of changes in the structure of clustering, this paper focuses on analyzing cluster changes at the micro level (entity migration), i.e. changes in cluster membership. For example, analysts might be interested to know the attributes that have changed significantly for those entities that belong to a lost cluster, or they might want to know where entities from one cluster have migrated to in the next snapshot dataset.

This paper contributes a method for analyzing and visualizing entity migrations using Self-Organizing Maps (SOMs) between two snapshot datasets. The paper presents a visualization method to reveal the path of migration of entities, and a framework to rank attributes based on the extent of their change among selected entities. It is worth to note that this paper does not address time series clustering [11], where the aim is to cluster time series with similar patterns. The approach presented here clusters entities at points in time (snapshots), and analyzes cluster migrations between such snapshots.

SOMs are useful for visualizing clustering structure changes, and they have several advantages in temporal cluster analysis. First, they are able to relate clustering results by linking multiple visualizations, and they can detect various types of cluster changes, including emerging clusters and disappearing clusters [3,5]. Second, SOMs create a smaller but representative dataset that follows the distribution of the original dataset [7]. Third, SOMs topologically map a high-dimensional dataset into a two-dimensional map with similar entities being placed close to each other [7]. Importantly, SOMs offer various map visualizations [14] that allow non-technical users to explore high-dimensional data spaces through a non-linear projection onto a two-dimension plane.

This paper analyzes the migrations of entities given two observations for each entity in two datasets, $D(\tau_1)$ and $D(\tau_2)$, and two trained SOM maps, $M(\tau_1)$ and $M(\tau_2)$, with τ_1 and τ_2 being the time snapshots of the two datasets. The datasets should contain the same type of information on the same entities at different points in time, therefore all entities in dataset $D(\tau_1)$ should exists in dataset $D(\tau_2)$, and vice versa. This type of data is known as longitudinal data in statistics [6] which enables analyzing migrations and measuring change.

2   Self-Organizing Maps

A SOM is an artificial neural network that performs unsupervised competitive learning [7]. Artificial neurons are arranged on a low-dimensional grid, commonly a 2-D plane. Each neuron j has an d-dimensional prototype vector, m_j, where d is the dimensionality of dataset D. Each neuron is connected to neighboring neurons, with distances to its neighbours being equidistant in the map space. Larger maps generally have higher accuracy and generalization capability [15], but they also have higher computation costs.

Before training a map, the prototype vectors need to be initialized [7]. The preferred way is to use linear initialization which uses the first two largest principal components. At each training step t, the best matching unit b_i (BMU) for training data vector x_i from a dataset D, i.e. the prototype vector m_j closest to the training data vector x_i, is selected from the map. Using the batch training algorithm [7], the values of new prototype vectors $m_j(t+1)$ at training step $(t+1)$ are the weighted averages of the training data vectors x_i at training step t, where the weight is the neighbourhood kernel value $h_{b_i j}$ centered on the best matching unit b_i (commonly Gaussian).

A SOM can be visualized using various methods that allow non-technical users to explore a dataset. These methods include component plane visualizations and u-matrix visualization [14]. The prototype vectors of a trained SOM can be treated as 'proto-clusters' serving as an abstraction of the dataset. The prototype vectors are then clustered using a traditional clustering technique to form the final clusters [15].

3   Related Work

The similarity between two datasets can be shown by comparing two data hit histogram visualizations [14]. A data hit histogram shows the frequency with which data vectors are mapped to each node. This technique can indicate changes over time in the data mapping, but it is difficult to interpret these changes simply by comparing the data hit histograms, because this kind of visualization introduces inconsistencies in visualizing changes in datasets. If a vector is mapped into a dense area of the SOM, a small change in the data may cause it to be mapped to a different node. On the other hand, when a data vector is mapped into a sparse area, the same magnitude of change in the data vector might not cause this vector to be mapped to a different node. In addition, this visualization is not designed to show the migration paths of entities because it does not link the same entities on these two map visualizations.

Another way to visualize the movement of individual entities over time is by using the trajectory of an entity on a SOM. However, this technique can only show the movement of one entity. It would be very cluttered to show trajectories of many entities simultaneously. This visualization has been used to observe the evolution of a bank in terms of their financial data [13]. When an entity migrates to a new cluster in period τ_2 that is not represented on the original SOM (map $M(\tau_1)$), it leads to high quantization error in the visualization of the entity on map $M(\tau_2)$.

Changes in entities over time can be visualized using Trendalyzer developed by Gapminder[1]. This visualization shows multivariate time series data using an *animated bubble chart*. The bubble chart can plot three numerical attributes. The position of a bubble on the X and Y axes indicates values of two numerical attributes, and the size of the bubble represents the value of a third numerical attribute. The X and Y axes can use linear or log scale. The colour of the bubble can be used to represent a categorical attribute. The time dimension is controlled by a sliding bar. As the time advances, the position and the size of the bubbles

---

[1] http://www.gapminder.org/

change to reflect values at the time. The slider can be used to choose the time and animate the movement of the bubbles. While this visualization is suitable to illustrate changes over time, it is not effective for large datasets. A large number of bubbles would make the visualization too dense, thus making it difficult to discover patterns in large datasets. Furthermore, this visualization can only show three numerical attributes and one categorical attribute at one time.

Migration patterns can also be analyzed by finding changes in cluster composition. An approach to address changes in cluster composition over time and changes in cluster membership of individual data points has been proposed [8]. First, the data of the first period is clustered using k-means. Then, the resulting clusters are labeled based on descriptive statistics of the clusters. This analysis is repeated for subsequent periods. A bar chart is used to visualize changes in cluster size over time, and a pie chart to show the proportion of each cluster at a certain period of time. This method cannot detect new clusters, lost clusters, and the movement of cluster centroids, because it assumes that the structure of the subsequent clustering results would be the same as the first one.

Existing methods to analyze cluster migration have limitations. For example, the SOM trajectory method [13] is not suitable for large datasets. Other methods assume subsequent clustering structures in a snapshot dataset to be the same as in the previous one. The migration analysis method presented in this paper is suitable for analyzing migrations of large snapshot datasets and analyzing attribute interestingness based on changes in a selected group of migrants.

4   Visualizing Migrations

Because SOMs follow the distribution of the dataset it trained on, there is a problem with visualizing cluster migration paths on a single map. When a new region emerges in dataset $D(\tau_2)$, this region is not represented on map $M(\tau_1)$. As a result, when visualizing entities that migrate to a new region in period τ_2, this region is not represented by a prototype vector on map $M(\tau_1)$, as is illustrated in Figure 1. Similarly, when visualizing cluster migration paths only on map $M(\tau_2)$, entities which migrate from a lost cluster cannot be visualized properly, because the lost region is no longer represented on map $M(\tau_2)$. Therefore, migration paths that are involved with new clusters or lost clusters will not be shown properly by using one map only (either map $M(\tau_1)$ or map $M(\tau_2)$).

Migration paths should be visualized using both maps, because they follow the distribution of the dataset they were trained on. However, showing migrations using two 2D maps can be complicated. For example, a three-dimensional visualization could be used to show migrations between two 2D maps, but this can result in a cluttered visualization if many migration patterns are shown. As an alternative, migrations can be shown using two 1D maps. The challenge is to transform 2D maps into 1D maps.

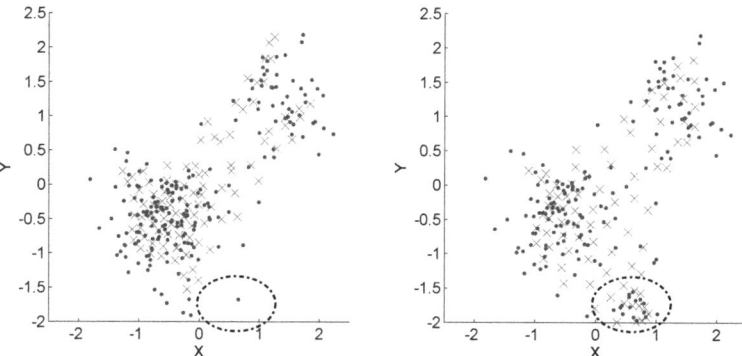

Fig. 1. Data (blue dots) and prototype vectors (red crosses) of the maps $M(\tau_1)$ *(left)* and $M(\tau_2)$ *(right)* trained using two synthetic datasets $D(\tau_1)$ and $D(\tau_2)$ respectively. A new cluster has emerged in dataset $D(\tau_2)$ (marked with the red circle in the right hand plot). It is not represented by any prototype vectors on map $M(\tau_1)$.

4.1   Transforming 2D Maps to 1D Maps

There are two approaches to transforming a SOM from 2D to 1D. In the first, a 1D map can be created by training the map using the prototype vectors of the 2D map as the training vectors. Experiments performed in this research show that this approach is not effective to visualize migrations, as discussed in the following section. In the second approach, a 1D map is created based on the clustering result of the 2D map.

Training a 1D Map Using Prototype Vectors of the 2D Map. In transforming a 2D map to a 1D map, it will be easier to analyze the 1D map if the topological order of the 2D map is preserved. One way to achieve this is by using a SOM to order the prototype vectors of the 2D map onto a 1D map. The 1D map (e.g. Figure 2(b)) with the size $|M| \times 1$ is trained with the prototype vectors of the 2D map (e.g. Figure 2(a)), where $|M|$ is the number of units of the 2D map. The trained 1D map will have different prototype vectors compared to the initial 2D map, even though it is topologically ordered. After that, the prototype vectors of the 2D map are sorted topologically with the previously trained 1D map as the reference.

However, map folding might occur in the 1D map as shown in Figures 2(b) and 2(c). This could happen because the SOM attempts to follow the distribution of a higher dimensional dataset. For example, the ordered 1D map shown in Figure 2(b) is then clustered using k-means with four cluster (chosen based on the plot of the Davies-Bouldin Index [2]). In this figure, there are two nodes of the green cluster that are separated from the main green cluster. This scattered green cluster will make it harder for an analyst to see migration patterns, even though the map is topologically ordered.

Showing migration from node to node between two 1D maps (such as the one shown in Figure 2(c)) can be difficult to analyze. The many migration paths

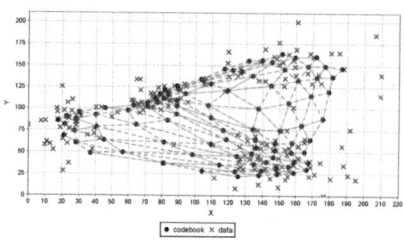 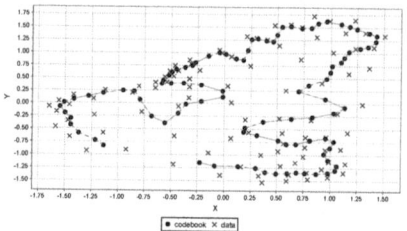

(a) Data vectors and prototype vectors of a 2D map trained using a synthetic dataset. The grey line shows connection between neighbouring units.

(b) Data vectors and prototype vectors of the 1D map trained using prototype vectors of the trained 2D map shown in Figure 2(a) as the training dataset.

(c) The clustering result of the 1D map shown in Figure 2(b) with four clusters. Notice that there are two folds (scattered nodes of the green cluster).

Fig. 2. Training a 1D map using prototype vectors of the 2D map

would result in a too crowded and too difficult to understand visualization. Arrows can come from most of the nodes on map $M(\tau_1)$ to the nodes on map $M(\tau_2)$. Alternatively, showing migration from cluster to cluster is easier to see and simpler, because the number of cluster is relatively smaller. In the next subsection, a migration visualization from cluster to cluster is developed.

Creating a 1D Map Based on the Clustering Result of a Trained 2D Map. In this approach, a 1D map is created based on a clustering result of a trained 2D map selected by a user. The user can use a plot of the Davies-Bouldin index [2] as a guide to choosing the optimal clustering result of the 2D map. The 1D map is visualized using stacked k bars where k is the number of clusters as shown in Figure 3(b). The height is in proportion to the size of the clusters of the 2D map. Colour is used to link the cluster on the 2D map and on the 1D map, as shown in Figure 3.

The migration paths are visualized (Figure 3(b)) using lines from the left hand map (period τ_1) to the right hand map (period τ_2). Using line arrows could clutter the visualization for large datasets with many clusters. Therefore, to avoid cluttered visualization of migration paths, a user can select interactively which migration paths from or to a cluster to show.

4.2   Visualizing Migrations

The number of entities of each migration path can be represented using colour or line thickness. Here, line thickness is used based on rank of number of migrants. To enable better visual data exploration, the visualization of migration paths is interactive. A user can acquire more details by right clicking on a migration

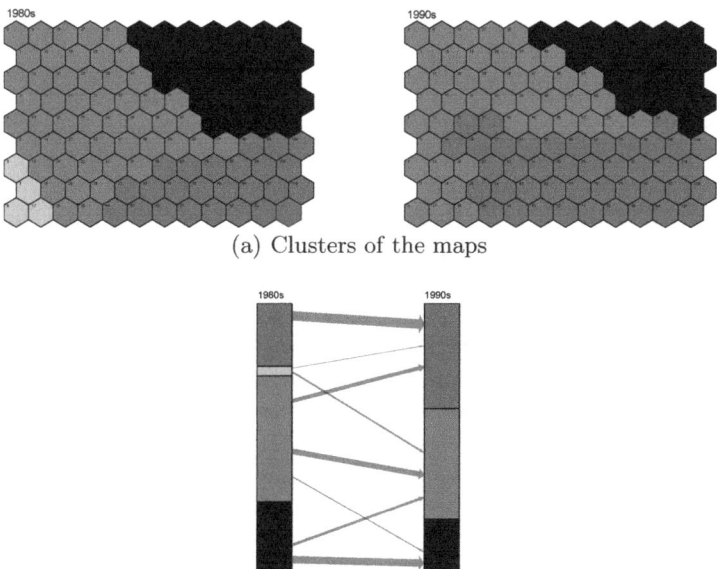

(a) Clusters of the maps

(b) Visualization of the migration paths

Fig. 3. The migration path visualization of the clusters on maps trained using World Development Indicator (WDI) datasets from the 1980s (*left*) and the 1990s (*right*). The green cluster located in the bottom left corner on the 1980s map is missing in the 1990s map.

path, a source cluster, or a destination cluster. The selected path or cluster is then analyzed as discussed in the next section.

5   Analyzing Attribute Interestingness of Migrants

As mentioned earlier, it can be of interest to know which attributes of a selected group of entities change more than others. A framework is developed to analyze attribute interestingness that can be applied to a variety of measures. Attribute interestingness related to migrations can be measured based on changes in attribute values for the selected entities between datasets $D(\tau_1)$ and $D(\tau_2)$. Then, the measurement results for each attribute are presented in a sorted table, as shown in Table 1.

Let $\mathcal{S}_{node}$ be the set of node indexes on map $M(\tau_1)$ that are selected by a user using brushing. The selection can be made based on ReDSOM visualizations [5], or clustering result visualizations (e.g. Figure 3(a)). Let $\mathcal{S}_{data}$ be the set of selected data indexes. $\mathcal{S}_{data}$ can be derived by mapping data vectors to nodes in $\mathcal{S}_{node}$ on map $M(\tau_1)$:

$$\mathcal{S}_{data} \leftarrow (\mathbf{x}_i \in D(\tau_1) \mid BMU\,(\mathbf{x}_i, M(\tau_1)) \in \mathcal{S}_{node}) \qquad (1)$$

Alternatively, entities can be selected based on their involvement in a selected migration path shown in migration visualization, such as in Figure 3(b).

In measuring change, each entity can be represented by its best matching unit (BMU) on the trained SOM, instead of using its actual value. This method is more efficient for large datasets, since it avoids accessing the whole dataset. The BMU of entities $\mathbf{x}_i(\tau_1)$ on map $M(\tau_1)$ is the first measurement, where $i \in \mathcal{S}_{data}$. Similarly, attribute values of the BMUs of the same entities $\mathbf{x}_i(\tau_2)$ on map $M(\tau_2)$ are the second repeated measurement.

There are a number of ways to measure change in order to analyze the migration patterns of the set of entities $\mathcal{S}_{data}$ from time period τ_1 to τ_2. Magnitude of change for one entity over time can be measured using difference, percent change, and percent difference. These individual values of a set of entities need to be aggregated to obtain a magnitude of change of a set of entities $\mathcal{S}_{data}$. The aggregates of $\mathcal{S}_{data}$ can be calculated by averaging the difference, or averaging the percent change. Comparing average differences between attributes does not make sense when the attributes have different magnitudes or scales. The selection of ways to average percentage change is dependent on the context of the application.

Statistical significance tests can also be used to measure whether changes in an attribute are significant for the chosen subpopulation. The use of these significance tests in this paper is merely for exploratory data analysis, because no treatments were provided to the entities that are mapped to selected nodes in $\mathcal{S}_{node}$, and most datasets are obtained from observational study rather than from experimental one. Observational data are not designed to make inferences of a larger population or to understand cause and effect [9]. Here, a P-value using the paired t-test for each attribute is used to evaluate the significance of changes in the attribute.

To calculate the t statistic for paired samples, one first calculates the difference between each matched pair to produce a single sample, then the one-sample t procedure is used [9]. It is important to keep in mind when analysing the P-value (e.g. Table 1) that the paired t-test calculates the t statistic without reference to the values of the first observations, such as in percentage change, so it can give the same t statistic value even though the differences may have different orders of magnitude for different attributes.

6   Application to the WDI datasets

The migration analysis developed in this paper has been applied to the World Development Indicator (WDI) datasets [16]. The datasets reflect different aspects of welfare of countries in the 1980s and 1990s.

Migration Path Visualization. Migration path visualizations in the WDI datasets between the 1980s and the 1990s is shown in Figure 3(b). With ReD-SOM visualization [5], a number of interesting clustering structure changes were identified in these datasets, such as a missing cluster (the light green cluster), a shrinking cluster (the dark red cluster), and a growing cluster (the blue cluster), as shown in Figure 3(a).

The migration patterns involving these clusters are shown in Figure 3(b). It is straightforward from this visualization to see that countries that belong to the light green (lost) cluster, which consists of four South American countries: Brazil, Argentina, Nicaragua, and Peru, migrated to the blue cluster and the orange cluster in the 1990s. Knowing that member countries of the light green cluster experienced an economic crisis in the 1980s, migration of Argentina to the blue cluster that contains OECD and developed countries is an interesting path. This change is likely to be a result of rapid reforms in the late 1980s and early 1990s performed by many Latin American countries [17].

Also from the visualization, a small portion of members of the shrinking cluster (the dark red cluster) migrated to the orange cluster. This shrinking cluster consists mostly of African countries that were characterized by low school enrolment, high mortality rate, and high illiteracy in the 1980s. The orange cluster has moderate values of secondary school enrolment, birthrate, and mortality under 5 years old compared to the dark red cluster in the 1980s. None of the countries from the shrinking cluster in the 1980s migrated to the blue cluster in the 1990s that contains OECD and developed countries. It would be a big step for countries from the dark red cluster to migrate to the blue cluster. The blue cluster received its new members from the orange cluster and the light green cluster.

Analyzing Attribute Interestingness. It can be of interest to know which attributes have a greater degree of change for the cluster changes identified previously. Results of analysis are presented in sorted tables (e.g., Tables 1 and 2). The attributes are sorted based on the P-value (the second column) which is the probability that the old values and the new values in the attribute have no significant difference. The evidence of change is stronger with lower P-value. The commonly used 5% confidence level ($\alpha \leq 0.05$) is used here. The 1980s and 1990s mean values including their normalized values (z-score) are shown in the last two columns.

Based on hot spot analysis [4] of the lost cluster, the distinguishing characteristics were food prices inflation and consumer prices inflation. This analysis has shown that there are some other attributes that have significantly changed for these countries, e.g., an increase in measles immunization, a decrease in children labour, a decrease in mortality under five year old, and a decrease in infant mortality, as shown in Table 1. It is also interesting to note that these improvements in health were achieved without a significant change in the number of physicians (see at the bottom of the Table 1). These improvements can be explained by a UNICEF program known as GOBI (Growth monitoring, Oral rehydration, Breast-feeding, and Immunization) which was targeting poor children [12].

The countries in the shrinking dark red cluster on the 1980s map (Figure 3(a)) demonstrated significant changes in many aspects of welfare (Table 2). Advances in education is indicated by the illiteracy decreasing significantly, while school enrolment in primary, secondary, and tertiary increased significantly. Even though these changes show development, the welfare of these countries is still far behind those in the blue cluster. For example, the mean tertiary school enrolment in the dark red cluster in the 1990s is 4.9% (Table 2) whereas tertiary school enrolment in the blue cluster in the 1990s is 45.1%.

Table 1. Attributes of the countries in the lost green cluster on the 1980s map in Figure 3 sorted by significance of change between 1980s and 1990s in each attribute measured using the paired t-test. The third column indicates direction of change. The line separates those attributes with significantly changed values between 1980s and 1990s with confidence level of 0.05.

Indicator Name	P-value	↕	1980s value (norm)	1990s value (norm)
Immunization Measles	0.0014	↑	66.91 (-0.21)	91.11 (0.98)
Labor Force Children	0.0087	↓	10.89 (-0.28)	6.7 (-0.53)
Inflation Consumer Prices	0.0095	↓	2,995.94 (4.72)	5.81 (-0.19)
Inflation Food Prices	0.0096	↓	2,556.06 (4.54)	4.5 (-0.21)
Mortality Rate Under 5	0.0478	↓	69.73 (-0.27)	49.03 (-0.53)
Mortality Rate Infant	0.0819	↓	48.83 (-0.18)	35.95 (-0.45)
:	:	:		
Daily Newspapers	0.8787	↓	68.47 (-0.27)	63.54 (-0.31)
Physicians	0.9926	↓	1.43 (0.23)	1.43 (0.23)

Table 2. Attributes of the countries in the shrinking dark red cluster on the 1980s map in Figure 3 sorted by significance of change between 1980s and 1990s in each attribute measured using the paired t-test

Indicator Name	P-value	↕	1980s value (norm)	1990s value (norm)
Labor Force Children	0.0000	↓	35.44 (1.19)	27.03 (0.69)
Birthrate	0.0000	↓	44.12 (1.05)	39.65 (0.71)
Television Sets	0.0000	↑	26.23 (-0.82)	55.69 (-0.65)
Illiteracy Rate Adult Female	0.0000	↓	67.36 (1.07)	54.22 (0.61)
Illiteracy Rate Adult Total	0.0000	↓	56.41 (1.08)	44.88 (0.60)
Mortality Rate Under 5	0.0000	↓	188.46 (1.21)	162.46 (0.89)
School Enrollment Secondary Female	0.0000	↑	14.18 (-1.03)	25.59 (-0.70)
School Enrollment Secondary	0.0000	↑	19.24 (-1.03)	29.36 (-0.72)
Mortality Rate Infant	0.0000	↓	114.86 (1.21)	101.81 (0.94)
:	:	:		

7   Conclusion

This paper has presented a method to visualize the migration paths of clusters from two snapshot datasets $D(\tau_1)$ and $D(\tau_2)$. Migration paths between clusters cannot be visualized properly on a single map, either on map $M(\tau_1)$ or on map $M(\tau_2)$, because there might be some regions that are not represented on the other map. Showing migrations using two 2D maps can be complicated as it requires 3D visualization with rotation. The most effective way to visualize the migration path is by representing clusters of 2D SOMs using 1D maps. Furthermore, experiments showed that transforming a 2D SOM into a 1D SOM can create folding in the 1D SOM cluttering visualization.

The paper has also presented a framework to understand which attributes have changed significantly in a set of selected entities from two snapshot datasets. This selection can be made though interactive brushing on the SOM visualizations, such as through ReDSOM visualization [5] or visualization of a clustering result of the map. Alternatively, entities can be selected based on their involvement in a user-selected migration path based on the migration visualization.

A number of measures can be used to assess whether or not these changes are likely to be significant, such as using differences, percent change, or paired t-test. The paired t-test evaluates the significance of changes, however it does not measure the magnitude of changes, the test only evaluates differences in an attribute relative to the variability of individual differences. This framework to analyze attribute interestingness can be applied using other measures as well.

Real-life datasets from the WDI were analyzed using the framework devised in the course of this research revealing interesting changes. As an example, the countries in the lost cluster who experienced high inflations in the 1980s achieved significant improvements in health in the 1990s without a significant change in number of physicians which can be explained by a UNICEF program known as GOBI. This hidden context was found after analysis using the method developed in this paper.

Evaluations of the visualizations and methodologies developed in this paper using real-life datasets demonstrated that they can help analysts discover hidden context by providing detailed insights into reasons for cluster changes between snapshot datasets.

7.1 Future Work

Since SOMs can handle only numerical data, the methodology and visualization methods developed in this paper cannot be applied to categorical data without scale conversion. Further research should consider other clustering approaches that can work with categorical data and also mixed types of data.

Because this research provides a framework to analyze migration, new measures can be developed or evaluated to analyze attribute interestingness in the framework. The paired t-test used in this paper does not take into account changes in the rest of the population. If a selected group of entities has similar changes to the rest of the population for a particular attribute, the attribute does not add value for an analyst in understanding the changes. In other words, changes that occur locally need to be normalized by changes that occur globally. Further work needs to be done to evaluate other methods, such as the two-sample t-test, to evaluate whether the magnitude of change in the selected group is similar to the magnitude of change in the rest of the population.

References

1. Burez, J., den Poel, D.V.: CRM at a pay-TV company: Using analytical models to reduce customer attrition by targeted marketing for subscription services. Expert Systems with Applications 32(2), 277–288 (2007)
2. Davies, D.L., Bouldin, D.W.: A cluster separation measure. IEEE Transactions on Pattern Analysis and Machine Intelligence (PAMI) 1(2), 224–227 (1979)
3. Denny, Squire, D.M.: Visualization of Cluster Changes by Comparing Self-Organizing Maps. In: Ho, T.-B., Cheung, D., Liu, H. (eds.) PAKDD 2005. LNCS (LNAI), vol. 3518, pp. 410–419. Springer, Heidelberg (2005)

4. Denny, Williams, G.J., Christen, P.: Exploratory Hot Spot Profile Analysis using Interactive Visual Drill-down Self-Organizing Maps. In: Washio, T., Suzuki, E., Ting, K.M., Inokuchi, A. (eds.) PAKDD 2008. LNCS (LNAI), vol. 5012, pp. 536–543. Springer, Heidelberg (2008)
5. Denny, Williams, G.J., Christen, P.: Visualizing Temporal Cluster Changes using Relative Density Self-Organizing Maps. Knowledge and Information Systems 25, 281–302 (2010)
6. Diggle, P.J., Liang, K.Y., Zeger, S.L.: Analysis of Longitudinal Data. Oxford University Press, New York (1994)
7. Kohonen, T.: Self-Organizing Maps, 3rd edn. Springer Series in Information Sciences, vol. 30. Springer, Heidelberg (2001)
8. Lingras, P., Hogo, M., Snorek, M.: Temporal cluster migration matrices for web usage mining. In: Proceedings of the 2004 IEEE/WIC/ACM International Conference on Web Intelligence, pp. 441–444. IEEE Computer Society (2004)
9. Moore, D.S.: The basic practice of Statistics. W H. Freemand and Company, New York (2000)
10. Reinartz, W.J., Kumar, V.: The impact of customer relationship characteristics on profitable lifetime duration. The Journal of Marketing 67(1), 77–99 (2003)
11. Roddick, J.F., Spiliopoulou, M.: A survey of temporal knowledge discovery paradigms and methods. IEEE Transactions on Knowledge and Data Engineering 14(4), 750–767 (2002)
12. Seear, M.: An Introduction to International Health. Canadian Scholars' Press Inc., Toronto (2007)
13. Serrano-Cinca, C.: Let financial data speak for themselves. In: Deboeck, G., Kohonen, T. (eds.) Visual Explorations in Finance with Self-Organizing Maps, pp. 3–23. Springer, London (1998)
14. Vesanto, J.: SOM-based data visualization methods. Intelligent Data Analysis 3(2), 111–126 (1999)
15. Vesanto, J., Alhoniemi, E.: Clustering of the Self-Organizing Map. IEEE Transactions on Neural Networks 11(3), 586–600 (2000)
16. World Bank: World Development Indicators 2003. The World Bank, Washington DC (2003)
17. Zagha, R., Nankani, G.T. (eds.): Economic Growth in the 1990s: Learning from a Decade of Reform. World Bank Publications, Washington, DC (2005)

Quality Issues,
Measures of Interestingness
and Evaluation
of Data Mining Models
Workshop (QIMIE 2011)

ClasSi: Measuring Ranking Quality
in the Presence of Object Classes
with Similarity Information*

Anca Maria Ivanescu, Marc Wichterich, and Thomas Seidl

Data Management and Data Exploration Group
Informatik 9, RWTH Aachen University, 52056 Aachen, Germany
{ivanescu,wichterich,seidl}@cs.rwth-aachen.de
http://www.dme.rwth-aachen.de/

Abstract. The quality of rankings can be evaluated by computing their correlation to an optimal ranking. State of the art ranking correlation coefficients like Kendall's τ and Spearman's ρ do not allow for the user to specify similarities between differing object classes and thus treat the transposition of objects from similar classes the same way as that of objects from dissimilar classes. We propose *ClasSi*, a new ranking correlation coefficient which deals with class label rankings and employs a class distance function to model the similarities between the classes. We also introduce a graphical representation of *ClasSi* akin to the *ROC curve* which describes how the correlation evolves throughout the ranking.

Keywords: ranking, quality measure, class similarity, ClasSi.

1 Introduction and Related Work

Evaluating the performance of an algorithm by comparing it against others is an important task in many fields such as in data mining and information retrieval. There are several evaluation methods developed for this purpose which can be integrated in the algorithm design process to improve effectiveness. Data mining and information retrieval models often return a ranking of the database objects. This ranking can be evaluated by checking if relevant documents are found before non relevant documents. Available measures for this evaluation are *precision* and *recall* as well as their weighted harmonic mean, known as the *F-measure* [9]. Related evaluation measures include the *mean average precision* [8], the *ROC curve* and the *area under the ROC curve (AUC)* [2]. These measures are all limited to binary class problems, distinguishing only between relevant and non relevant objects. Extensions of ROC to multi-class problems such as *generalized AUC* [4], the *volume under the curve* [1], and the *scalable multi-class ROC* [5] are combinations of two-class problems and do not consider class similarities.

* The authors gratefully acknowledge the financial support of the Deutsche Forschungsgemeinschaft (DFG) within the Collaborative Research Center (SFB) 686 "Model-Based Control of Homogenized Low-Temperature Combustion" and DFG grant SE 1039/1-3.

L. Cao et al. (Eds.): PAKDD 2011 Workshops, LNAI 7104, pp. 185–196, 2012.

When evaluating object rankings, statistical methods to measure the correlation between two rankings can also be employed (e.g. *Kendall's* τ [6] and *Spearman* rank correlation coefficient ρ [11]). A positive correlation coefficient indicates an agreement between two rankings while a negative value indicates a disagreement. Variants of τ such as τ_b, τ_c [7], gamma (Γ) [3], and Somers' asymmetric coefficients [10] address the case of tied objects through different normalizations. However, these rank correlation measures only take the order of objects into account and the degree of similarity between objects is ignored.

In this work we propose $ClasSi$, a rank correlation coefficient which is capable of handling rankings with an arbitrary number of class labels and an arbitrary number of occurrences for each label. The main advantage of $ClasSi$ is that it incorporates a <u>cla</u>ss <u>si</u>milarity function by which the user is able to define the degree of similarity between different classes.

In Section 2.1 we describe existing rank correlation coefficients τ and ρ. Section 2.2 defines $ClasSi$ and Section 2.3 examines its properties, showing that all requirements of a rank correlation coefficient are met. In Section 2.4 we discuss how to compute $ClasSi$ for the first k ranking positions, obtaining a graphical representation similar to the *ROC curve*. Section 3 analyzes the behavior of $ClasSi$ in an experimental setup. Finally, in Section 4 we conclude the paper.

2   Ranking Quality Measures for Objects in Classes

We propose a new ranking correlation coefficient which allows for a user defined class distance measure and also comes with a graphical representation.

As mentioned before, the state of the art evaluation measures cannot handle class label rankings where objects in the database are grouped into classes according to some property that confers a notion of group similarity. For example, in an image similarity search system, if an object from a query test set is assigned the class label "bonobos", it only matters that other objects from the class "bonobos" are retrieved early on, but it does not matter which particular bonobo appears early on. In addition, objects from the similar class "chimpanzees" should appear before objects from the dissimilar class "tigers".

The preferred order of object classes in a ranking then depends on the class of the object for which the other objects were ranked (e.g., a query object in a similarity search scenario). An optimal ranking in the presence of classes is one where objects from the same class as the query object come first, then objects from the neighboring classes and so on. For example, a query from the class "bonobos" may have following optimal ranking where the order of the objects within the classes is arbitrary (b = bonobo, c = chimpanzee, t = tiger):

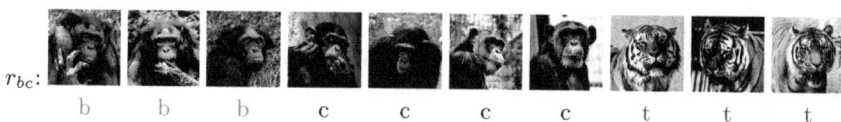

r_{bc}:

b   b   b   c   c   c   c   t   t   t

A worst-case ranking is one in which objects from the most dissimilar class come first, then objects from one class closer and so on. The objects from the class coinciding with the query are the last ones in this ranking. For the same example query, a worst case ranking is:

r_{wc}:

    t   t   t   c   c   c   c   b   b   b

To evaluate the quality of a computed ranking we look at its correlation to the optimal ranking. A positive correlation coefficient indicates a certain agreement between the two rankings. The higher the coefficient the more the ranking coincides with the optimal one, and if its value is 1 then it corresponds to it.

Additionally *ClasSi* accounts for the different degrees of dissimilarities between classes. For the example above, we can see that the classes "bonobos" and "chimpanzees" are more similar to each other, and both of them are dissimilar from the class "tigers". Consider the following two rankings:

$r1$:

  c   b   b   b   c   c   c   t   t   t

$r2$:

  b   b   b   c   t   c   c   c   t   t

Intuitively, the r_1 ranking coincides better with r_{bc} than r_2 does although both r_1 and r_2 have exactly one non bonobo object moved forward by three slots and r_1 has a non bonobo object at the very first position. The reason is that mistaking a tiger for a bonobo is a much bigger mistake than mistaking a chimpanzee for a bonobo.

2.1 Preliminaries: Measuring Ranking Quality

Throughout the paper we consider a database $DB = \{o_1, \ldots, o_m\}$ of cardinality m. A ranking of the objects in DB is defined as a bijective mapping r from DB to $\{1, \ldots m\}$ where $r(o)$ gives the position of object $o \in DB$ in the ranking and $r^{-1}(a)$ gives the ath object from DB according to the ranking.

Measures such as Kendall's τ [6] and Spearman's rank correlation coefficient [11] assess the correlation between two ranking. As they will serve as the basis for the measure proposed in Section 2.2, they shall be reviewed here shortly.

Definition 1. *Kendall's* τ. *Given a database* $DB = \{o_1, \ldots, o_m\}$ *and two ranking* r_1, r_2, *Kendall's correlation coefficient* τ *is defined as*

$$\tau = \frac{Con - Dis}{\frac{1}{2}m(m-1)} = 1 - \frac{2 \cdot Dis}{\frac{1}{2}m(m-1)}$$

with

$$Con = |\{(r_1^{-1}(a), r_1^{-1}(b)) \mid 1 \le a < b \le m \wedge r_1^{-1}(a) \prec_{r_2} r_1^{-1}(b)\}| \; and$$

$$Dis = |\{(r_1^{-1}(a), r_1^{-1}(b)) \mid 1 \le a < b \le m \wedge r_1^{-1}(b) \prec_{r_2} r_1^{-1}(a)\}|$$

where $o_a \prec_r o_b \Leftrightarrow r(o_a) < r(o_b)$.

Kendall's τ can be used to evaluate the quality of a ranking r by comparing r to an optimal ranking r^*. Kendall's τ then measures the correlation of these two rankings by counting the number of concordant object pairs (those, that are sorted the same way by both rankings) minus the number of discordant object pairs (those that are sorted differently by the two rankings). The result is then divided by the total number of pairs ($\frac{1}{2}m(m-1)$) to normalize the measure between -1 and 1 where 1 is reached for identical rankings and -1 for reversed rankings.

While Kendall's τ takes the number of discordant pairs into account Spearman's rank correlation coefficient ρ explicitly considers the difference in ranking positions when comparing two rankings.

Definition 2. *Spearman's* ρ. *Given a database* $DB = \{o_1, \ldots, o_m\}$ *and two ranking* r_1, r_2, *Spearman's rank correlation coefficient* ρ *is defined as*

$$\rho = 1 - \frac{6 \sum_{o \in DB} (r_1(o) - r_2(o))^2}{m(m^2 - 1)}.$$

Similar to Kendall's τ, Spearman's ρ is normalized between -1 and 1. Even though the difference in ranking position is considered by ρ, it is conceivable that two mismatches (each for example by ten ranking positions) may be of notably differing importance to the quality of the rankings. In the following we propose to incorporate knowledge on cross object (class) similarity into the evaluation of rankings. This allows for a more meaningful assessment of for instance similarity search results, which currently are mostly evaluated using either simple precision/recall measures or ranking quality measures that ignore class similarity information.

2.2 Class Similarity Ranking Correlation Coefficient ClasSi

We next introduce according class labeling and comparison functions that help us assess the quality of rankings in this scenario. The function $l : DB \to C$ assigns a class label from $C = \{c_1, \ldots c_n\}$ to each object $o \in DB$. The class distance function $d : C \times C \to \mathbb{R}$ conveys the notion of (dis)similarity between individual classes (e.g., $d(\text{bonobos}, \text{bonobos}) = 0$ and $d(\text{bonobos}, \text{chimpanzees}) < d(\text{bonobos}, \text{tigers})$). Based on the class distance function and a query object, the best case and worst case rankings are defined as follows.

Definition 3. *Optimal class ranking.* *For a query object q with class label c a ranking r_{bc} is said to be optimal iff*

$$d(c, l(r^{-1}(a))) \leq d(c, l(r^{-1}(a+1))) \qquad \forall a \in \{1, \ldots, m-1\}$$

where $l(r^{-1}(a))$ is the label of the ath object according to ranking r.

Definition 4. *Worst-case class ranking.* *For a query object q with class label c a ranking r_{wc} is said to be a worst-case ranking iff*

$$d(c, l(r^{-1}(a))) \geq d(c, l(r^{-1}(a+1))) \qquad \forall a \in \{1, \ldots, m-1\}.$$

The *ClaSi* ranking correlation coefficient not only takes into consideration the number of discordant pairs but also as their dissimilarities. The dissimilarity of a discordant pair of class labels c_i and c_j is appraised by looking at their distances to the query class label c_q.

$$cost(i, j) = \begin{cases} 0, & \text{if } d(c_q, c_i) \leq d(c_q, c_j), \\ d(c_q, c_i) - d(c_q, c_j), & \text{else.} \end{cases}$$

For a given class ranking r as defined above, we compute *ClaSi* by iterating through all positions and sum up the dissimilarity cost for each discordant pair.

Definition 5. *Given a database $DB = \{o_1, \ldots, o_m\}$, a class distance function $d : C \times C :\rightarrow \mathbb{R}$, a query object q which defines best and worst case rankings r_{bc} and r_{wc}, and the dissimilarity cost function $cost : C \times C :\rightarrow \mathbb{R}$, the ClaSi correlation coefficient between an arbitrary ranking r and r_{bc} is defined as*

$$ClaSi = 1 - \frac{2 \cdot DisCost_r}{DisCost_{r_{wc}}}$$

where $DisCost_r$ is the cost generated by the discordant pairs of r compared to r_{bc} and $DisCost_{r_{wc}}$ is the according cost generated by the worst case ranking:

$$DisCost_r = \sum_{a=1}^{m} \sum_{b=a+1}^{m} cost(l(r^{-1}(a)), l(r^{-1}(b)))$$

$$DisCost_{r_{wc}} = \sum_{a=1}^{m} \sum_{b=a+1}^{m} cost(l(r_{wc}^{-1}(a)), l(r_{wc}^{-1}(b)))$$

For the example above we define following class distance measure: $d(b, c) = 1$, $d(b, t) = 6$. The dissimilarity costs between classes in this case are: $cost(c, b) = 1$, $cost(t, b) = 6$, $cost(t, c) = 5$, and 0 for all other cases. To compute *ClaSi* we iterate through the ranking positions and sum up the cost of the discordant pairs. For *ClaSi* between r_1 and r_{bc} we count 3 discordant pairs: there are 3 labels b which occur after a c label, thus $DisCost_{r_1} = 3 \cdot cost(c, b) = 3$. Between r_{bc} and r_{wc} all possible discordant pairs occur. The corresponding cost

is $DisCost_{wc} = 4 \cdot 3 \cdot cost(t,c) + 3 \cdot 3 \cdot cost(t,b) + 3 \cdot 4 \cdot cost(c,b) = 126$. The
$ClasSi$ correlation coefficient is then $ClasSi_{r_1} = 1 - \frac{2 \cdot 3}{126} = 0.95$. For r_2 there are
3 labels c which occur after a t label, thus $DisCost_{r_2} = 3 \cdot cost(t,c) = 15$. We
obtain $ClasSi_{r_2} = 1 - \frac{2 \cdot 15}{126} = 0.76$ which is considerable smaller than $ClasSi_{r_1}$,
since the dissimilarity between a tiger and a bonobo is much higher than the
dissimilarity between a chimpanzee and a bonobo.

2.3   Properties of ClasSi

After introducing $ClasSi$ in the previous section, we now show that it has all the
properties of a correlation coefficient. With $ClasSi$ we measure the correlation
between an arbitrary ranking r and the optimal ranking r_{bc} of a set of class
labels. If r is also an optimal ranking then $ClasSi$ equals 1 as the two are
perfectly correlated. If r is a worst case ranking then $ClasSi$ equals -1 as they
perfectly disagree. Finally, if r is a random ranking then the expected value of
$ClasSi$ is 0 which means that the two rankings are independent.

Theorem 1. *The ClasSi correlation coefficient between ranking r and the optimal ranking r_{bc} is 1 if r corresponds to r_{bc}:*

$$l(r^{-1}(a)) = l(r_{bc}^{-1}(a))   \forall a \in 1,...,m$$

Proof. If r corresponds to the optimal ranking, then there are no discordant
pairs and thus no dissimilarity cost. In this case:

$$ClasSi = 1 - \frac{2 \cdot 0}{DisCost_{r_{wc}}} = 1$$

Theorem 2. *The ClasSi correlation coefficient between ranking r and the optimal ranking r_{bc} is -1 if r corresponds to r_{wc}:*

$$l(r^{-1}(a)) = l(r_{wc}^{-1}(a))   \forall a \in 1,...,m$$

Proof. If r corresponds to the worst case ranking, then $DisCost_r = DisCost_{r_{wc}}$
and in this case:

$$ClasSi = 1 - \frac{2 \cdot DisCost_{r_{wc}}}{DisCost_{r_{wc}}} = -1$$

Theorem 3. *The expected correlation coefficient $E(ClasSi)$ between a random ranking r and the optimal ranking r_{bc} is 0.*

Proof. Assume w.l.o.g. that there are m_i objects with label c_i. Then for each
object with label c_i there are Dis_i possible objects with a different label which
are more similar to the query object and would be discordant if they were to be
ranked after the c_i-labeled objects. More formally:

$$Dis_i = |\{o_a \mid d(c_q, c_i) > d(c_q, l(o_a)), \forall 1 \le a \le m\}|$$

The probability for the occurrence of a discordant pair can be modeled by means
of the *hypergeometric distribution*. For a sequence of s drawings without replacement from a statistical population with S entities, out of which M have a certain

property, the hypergeometric distribution describes the probability that k is the number of successful draws, i.e. the number of draws having that property:

$$P(X = k) = \frac{\binom{M}{k}\binom{S-M}{s-k}}{\binom{S}{s}} \qquad E(X) = s\frac{M}{S}$$

Let us consider position $m - e$ in the class ranking r which is followed by e entries and assume label $l(r^{-1}(m - e)) = c_i$ at this position. The probability that there are k discordant entries among the e following entries, is according to the hypergeometric distribution

$$P(Dis = k|c_i) = \frac{\binom{Dis_i}{k}\binom{m-1-Dis_i}{e-k}}{\binom{m-1}{e}} \tag{1}$$

The expected number of discordant entries occurring in the remaining e entries is then:

$$E_e(Dis|c_i) = \sum_{k=0}^{e} k\frac{\binom{Dis_i}{k}\binom{m-1-Dis_i}{e-k}}{\binom{m-1}{e}} = e\frac{Dis_i}{m-1} \tag{2}$$

For each object with label c_i we compute the average cost of a discordant pair:

$$\overline{cost}_i = \frac{1}{Dis_i} \cdot \sum_{\substack{1 \le j \le n \\ d(c_q,c_i) > d(c_q,c_j)}} m_j \cdot cost(i,j).$$

The expected associated cost at position $m - e$ is obtained by multiplying the expected number of discordant pairs with the average cost $\overline{cost}_i$ of a discordant pair for the label c_i:

$$E_e(DisCost_r|c_i) = e\frac{Dis_i \cdot \overline{cost}_i}{m-1} \tag{3}$$

For an arbitrary label and e entries left, the expected associated cost is:

$$E_e(DisCost_r) = \sum_{i=1}^{n} p_i e\frac{Dis_i \cdot \overline{cost}_i}{m-1} \tag{4}$$

where $p_i = \frac{m_i}{m}$ denotes the a priori class probability. If the expected costs at each position are summed up, then the expected costs generated by the expected number of discordant entries is obtained as

$$E(DisCost_r) = \sum_{e=1}^{m-1}\sum_{i=1}^{n} p_i \cdot e\frac{Dis_i \cdot \overline{cost}_i}{m-1}$$

$$= \frac{1}{m-1}\sum_{e=1}^{m-1} e\sum_{i=1}^{n} p_i \cdot Dis_i \cdot \overline{cost}_i$$

$$= \frac{1}{m-1}\frac{(m-1)m}{2}\sum_{i=1}^{n} p_i \cdot Dis_i \cdot \overline{cost}_i$$

$$= \frac{m \sum_{i=1}^{n} p_i \cdot Dis_i \cdot \overline{cost_i}}{2}$$

Knowing that m_i is the number of entries with label c_i

$$E(DisCost_r) = \frac{m \sum_{i=1}^{n} p_i \cdot Dis_i \cdot \overline{cost_i}}{2} = \frac{\sum_{i=1}^{n} m_i \cdot Dis_i \cdot \overline{cost_i}}{2}$$

At this point we obtained the following expected correlation coefficient:

$$E(ClasSi) = 1 - \frac{2 \cdot \frac{\sum_{i=1}^{n} m_i \cdot Dis_i \cdot \overline{cost_i}}{2}}{DisCost_{r_{wc}}}$$

Considering that for r_{wc} we have all possible discordant pairs, the associated dissimilarity cost can be computed by iterating over all n classes and taking all their objects, their possible discordant pairs, and their average cost of discordant pairs into account:

$$DisCost_{r_{wc}} = \sum_{i=1}^{n} m_i \cdot Dis_i \cdot \overline{cost_i}$$

Thus, the expected correlation coefficient between r and r_{bc} is

$$E(ClasSi) = 1 - \frac{2 \cdot \frac{\sum_{i=1}^{n} m_i \cdot Dis_i \cdot \overline{cost_i}}{2}}{\sum_{i=1}^{n} m_i \cdot Dis_i \cdot \overline{cost_i}} = 0.$$

Thus a ranking returned by a similarity measure can be also assessed by considering the $ClasSi$ correlation coefficient to the optimal ranking. Since for a random ranking the expected $ClasSi$ value is 0, a computed ranking should have a higher $ClasSi$ correlation coefficient to the optimal ranking.

Another important property of $ClasSi$ is that it not only considers the number of discordant pairs, but also the degree of their dissimilarity. By specifying the class distance function, the user specifies different degrees of dissimilarity for the discordant pairs. Nevertheless, only the relative differences matter.

Theorem 4. *Let d and d' be two class distance functions such that $d'(c_i, c_j) = \alpha \cdot d(c_i, c_j)$, and $ClasSi^{(d)}$ and $ClasSi^{(d')}$ be the corresponding rank correlation coefficients, then:*

$$ClasSi^{(d)} = ClasSi^{(d')}$$

Proof. From the relationship between the class distance functions we also obtain following relationship between the dissimilarity cost functions

$$cost'(c_i, c_j) = \alpha \cdot cost(c_i, c_j)$$

Thus the scaling of $DisCost'_r = \alpha \cdot DisCost_r$ and of $DisCost'_{r_{wc}} = \alpha \cdot DisCost_{r_{wc}}$ cancel each other.

2.4 ClasSi on Prefixes of Rankings

Up to now, the $ClasSi$ measure has been computed for a complete ranking of objects in a database yielding a single value that reflects the overall quality of the ranking. In some situations, it might be more interesting to have a quality score for a subset of the ranked objects. The first k positions of a ranking are of particular interest, since only these results might either be presented to a user (e.g., in a retrieval system) or be considered in a data mining process. The proposed $ClasSi$ measure can easily be adapted to suite this need in a meaningful manner. Instead of measuring the cost of misplaced objects for the whole ranking, the $ClasSi$ measure restricted to the top k positions measures the guaranteed cost of objects placed in the first k positions. That is, for each object o within the top k objects, it is checked how much cost will be generated due to objects o' appearing after o when they were supposed to appear before o in the ranking. Likewise, the cost of the worst case scenario is restricted to the cost guaranteed to be generated by the top k objects of the worst case ranking.

Definition 6. *Given a database $DB = \{o_1, \ldots, o_m\}$, a class distance function $d : C \times C :\rightarrow \mathbb{R}$, a query object q which defines best and worst case rankings r_{bc} and r_{wc}, and the dissimilarity cost function $cost : C \times C :\rightarrow \mathbb{R}$, the ClasSi correlation coefficient for the top k positions between a ranking r and r_{bc} is defined as*

$$ClasSi_k = 1 - \frac{2 \cdot DisCost_{r,k}}{DisCost_{r_{wc},k}}$$

where $DisCost_{r,k}$ is the cost generated by the discordant pairs of r rooted within the top k positions of r and $DisCost_{r_{wc},k}$ is the according cost generated by the worst case ranking:

$$DisCost_{r,k} = \sum_{a=1}^{k} \sum_{b=a+1}^{m} cost(l(r^{-1}(a)), l(r^{-1}(b)))$$

$$DisCost_{r_{wc},k} = \sum_{a=1}^{k} \sum_{b=a+1}^{m} cost(l(r_{wc}^{-1}(a)), l(r_{wc}^{-1}(b)))$$

Algorithms 1 and 2 return arrays filled with cumulative discordant pair costs and ClasSi values in $O(k * m)$ time. By keeping track of the number of objects seen for each class up to position k in $O(k * n)$ space, it is possible to speed up the computation to $O(k * n)$ if the number of objects per class is known a priori.

By plotting $ClasSi_k$ for all $k \in \{1, \ldots, m\}$ we obtain a curve, which describes how the correlation evolves. If the first $ClasSi_k$ values are small and the curve is growing, this means that most of the discordant pairs are at the beginning and towards the end the ranking agrees with the optimal one. If the curve is decreasing, this means that the quality of the ranking is better at the beginning and decreases towards the end of the ranking.

Algorithm 1. DisCost(Prefix k, Ranking r, Labeling l, Costs $cost$)

```
1  dc = Float[k];
2  for a = 1 to k do          // iterate over position pairs (a,b) in ranking r
3  |   dc[a] = 0;
4  |   for b = a+1 to m do
5  |   |   dc[a] += cost[l(r⁻¹(a))][l(r⁻¹(b))];       // sum discordant pair costs
6  |   end for
7  |   if (a > 1) then dc[a] += dc[a-1];
8  end for
9  return dc;
```

Algorithm 2. ClasSi(Prefix k, Rankings r, r_{wc}, Labeling l, Costs $cost$)

```
1  classi = Float[k];
2  rc = DistCost(k, r, l, cost);
3  wc = DistCost(k, rwc, l, cost);
4  for a= 1 to k do
5  |   classi[a] = 1 - (2 * rc[a] / wc[a]);
6  end for
7  return classi;
```

---

3 Examples

Using the values given by Algorithm 2, it is possible to track the progression of the *ClasSi* value for ascending values of k. Figure 1(a) shows graphs for four rankings as described in Section 2.2. The best case ranking results in a constant graph at value +1.0. Analogously, the $ClasSi_k$ values for the worst case ranking are constant at −1.0. Ranking r_1 with one c (i.e., chimpanzee) moved forward by 3 positions to the first position results in a graph that starts at 0.84 and then increases up to a value of 0.95 as all further positions do not include any more discordant pairs while the number (and cost) of potential discordant pairs grows. Ranking r_2 on the other hand starts with four objects ranked identically to the best case ranking, thus the resulting curve starts at 1. On the fifth position an object with high cost for discordant pairs appears and significantly reduces the quality of the ranking to 0.75.

The dissimilarity between "bonobos" and "tigers" is specified by the user through the class distance function. In Figure 1(b) we see how the *ClasSi*-curve for r_2 drops when the distance $d(b,t)$ is increased while $d(b,c)$ remains constant. The higher $d(b,t)$, the smaller *ClasSi* gets.

We further investigate the behavior of *ClasSi* for an increasing number of discordant pairs. We consider a synthetically generated optimal ranking and another one, which emerged from the optimal one by randomly choosing a pair of objects and switching their position. The rankings have 250 entries and are divided into 10 classes. The randomly generated class distance function from the target class to the other classes is plotted in Figure 2(b). In Figure 2(a) *ClasSi* curves are plotted for an increasing number of discordant pairs and we can see that the *ClasSi* values decrease with an increasing number of discordant pairs.

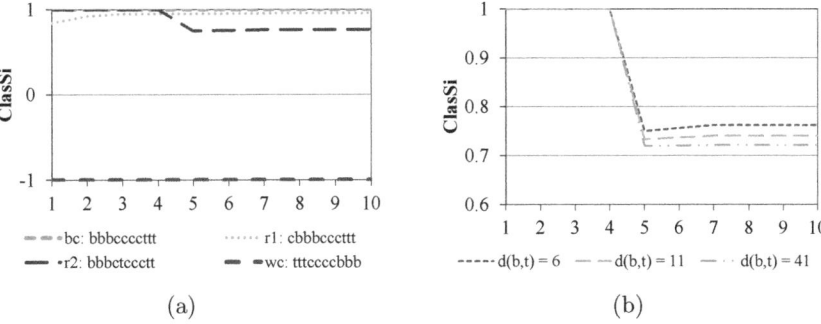

Fig. 1. *ClasSi* in our example from Section 2 for varying k

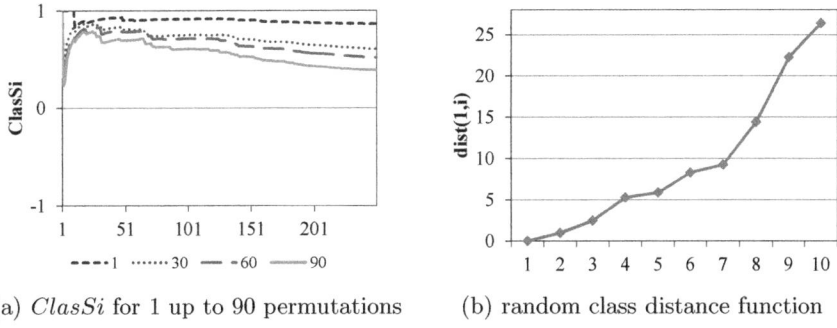

(a) *ClasSi* for 1 up to 90 permutations   (b) random class distance function

Fig. 2. *ClasSi* for an increasing number of object permutations

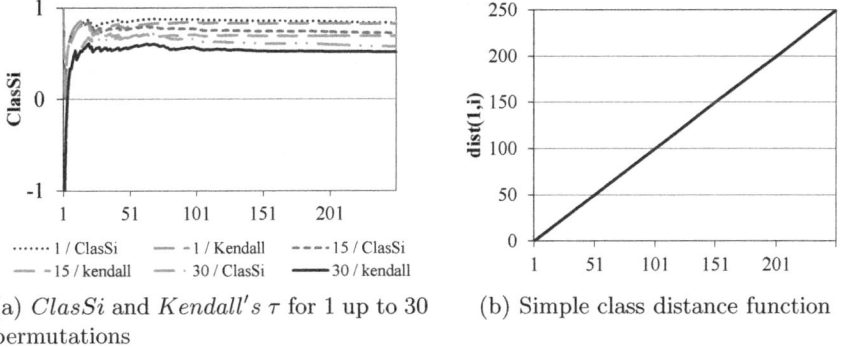

(a) *ClasSi* and *Kendall's* τ for 1 up to 30 permutations   (b) Simple class distance function

Fig. 3. *ClasSi* for the simple case

Note that although *ClasSi* can deal with multiple entries with the same label and allows the user to define the class similarities, this coefficient can be also used for rankings in which each entry occurs only once and/or the user only specifies the desired order of classes. In this particular case, *ClasSi* behaves similarly to

Kendall's τ as it can be seen in Figure 3(a). The simple class distance function resulted from the specified class ordering is plotted in Figure 3(b).

4   Conclusion

In this paper we introduced a new measure to evaluate rankings of class labels by computing the correlation to an optimal ranking. It also allows for the user to specify different similarities between different classes. We have also proven that $ClasSi$ has all the properties required for a correlation coefficient. $ClasSi$ can be computed by iterating through the ranking and can be stopped at every position k, delivering an intermediate result $ClasSi_k$. By plotting these values we obtain a representation akin to the ROC $curve$ from which we can recognize where the agreements and disagreements w.r.t. the optimal ranking occur.

References

1. Ferri, C., Hernández-Orallo, J., Salido, M.A.: Volume under the ROC surface for multi-class problems. In: Proceedings of 14th European Conference on Machine Learning, pp. 108–120 (2003)
2. Flach, P., Blockeel, H., Ferri, C., Hernández-Orallo, J., Struyf, J.: Decision support for data mining; introduction to ROC analysis and its applications. In: Data Mining and Decision Support: Integration and Collaboration, pp. 81–90. Kluwer Academic Publishers (2003)
3. Goodman, L.A., Kruskal, W.H.: Measures of association for cross classifications. Journal of the American Statistical Association 49(268), 732–764 (1954)
4. Hand, D.J., Till, R.J.: A simple generalisation of the area under the ROC curve for multiple class classification problems. Machine Learning 45, 171–186 (2001)
5. Hassan, M.R., Ramamohanarao, K., Karmakar, C.K., Hossain, M.M., Bailey, J.: A Novel Scalable Multi-Class ROC for Effective Visualization and Computation. In: Zaki, M.J., Yu, J.X., Ravindran, B., Pudi, V. (eds.) PAKDD 2010. LNCS, vol. 6118, pp. 107–120. Springer, Heidelberg (2010)
6. Kendall, M.: A new measure of rank correlation. Biometrika 30(1-2), 81–89 (1938)
7. Kendall, M., Gibbons, J.D.: Rank Correlation Methods. Edward Arnold (1990)
8. Manning, C.D., Raghavan, P., Schütze, H.: Introduction to Information Retrieval. Cambridge University Press (2008)
9. van Rijsbergen, C.J.: Information Retrieval, 2nd edn. Butterworths, London (1979)
10. Somers, R.H.: A new asymmetric measure of association for ordinal variables. American Sociological Review 27(6), 799–811 (1962)
11. Spearman, C.: The proof and measurement of association between two things. The American Journal of Psychology 100, 441–471 (1987)

The Instance Easiness of Supervised Learning for Cluster Validity

Vladimir Estivill-Castro*

Griffith University, QLD 4111, Australia
v.estivill-castro@griffith.edu.au
http://vladestivill-castro.net

Abstract. "The statistical problem of testing cluster validity is essentially unsolved" [5]. We translate the issue of gaining credibility on the output of un-supervised learning algorithms to the supervised learning case. We introduce a notion of instance easiness to supervised learning and link the validity of a clustering to how its output constitutes an easy instance for supervised learning. Our notion of instance easiness for supervised learning extends the notion of stability to perturbations (used earlier for measuring clusterability in the un-supervised setting). We follow the axiomatic and generic formulations for cluster-quality measures. As a result, we inform the trust we can place in a clustering result using standard validity methods for supervised learning, like cross validation.

Keywords: Cluster validity, Supervized Learning, Instance easiness.

1 Introduction

From its very beginning, the field of knowledge discovery and data mining considered validity as a core property of any outcome. The supervised case assumes a function $c(\boldsymbol{x})$ and attempt to fit a model $F(\boldsymbol{x})$ given the training set (or data points) $\{(\boldsymbol{x}_i, c(\boldsymbol{x}_i))\}_{i=1,\ldots,n}$. One can evaluate the quality of the fit with many solid alternatives [7,13]. However, in the un-supervised setting we are only presented with the set of cases $\{\boldsymbol{x}_i\}_{i=1,\ldots,n}$. Most likely we are performing such learning with no solid grounds for what is the actual (real-world) generator of these examples and any assumption on our part may actually constitute a far too large unjustified bias. What in fact constitutes learning and what is the goal?

How can we establish some confidence (or "credibility" in the language of Witten and Frank [13, Chapter 5]) on the result delivered by a clustering algorithm? This constitutes a fundamental question. The very well known distance-based clustering algorithm k-means is among the top 10 most used algorithms in knowledge discovery and data mining applications [14], however it is statistically inconsistent and statistically biased (converging to biased means even if the input is generated from k spherical multi-variate normal distributions with equal proportions). How do the users of such a method derive any trust in their results? Or in the credibility of any other clustering methods?

---

* Work performed while hosted by Universitat Popeu Fabra, Barcelona, Spain.

L. Cao et al. (Eds.): PAKDD 2011 Workshops, LNAI 7104, pp. 197–208, 2012.

This paper proposes two new measures of cluster quality. The fundamental idea is that no matter what clustering algorithms is used, in the end one desires to obtain a model that can accurately answer the question "are x_i and x_j in the same cluster?" When clusterings are partitions, this questions has only two disjoint answers (yes or no), and thus, the results of a clustering algorithm can be scrutinized by the facility by which supervised learning algorithms can discover a suitable classifier. We show that these two measures have mathematical properties that have been considered desirable by several authors. The measures are inspired by the intuition that if the clustering results does identify classes that are well-separated, these results constitute an easy problem in the supervized-learning sense. This implies formalizing a notion of "instance easiness". We measure how easy is to learn in the supervized-learning case by using a similar approach or easiness previously introduced for unsupervised learning. That is, we will draw on the notions of "clusterability" [2] to suggest the mechanisms to achieve this.

2   Instance Easiness

In the machine learning literature, instance easiness has been applied with the notion of *clusterability* [2] to the un-supervised learning (or clustering) case. That is, Ackerman and Ben-David introduced notions to measure how easy is to cluster a particular instance X into k clusters.

2.1   Generic Definitions

We now introduce formal definitions and nomenclature for the clustering problem (un-supervised learning) that follow the general formulations of Ackerman and Ben-David [1] since this is a generic form that is widely applicable.

Let X be some domain set (usually finite). A function $d : X \times X \to \Re$ is a *distance function* over X if

1. $d(x_i, x_i) \geq 0$ for all $x_i \in X$,
2. for any $x_i, x_j \in X$, $d(x_i, x_j) > 0$ if and only if $x_i \neq x_j$, and
3. for any $x_i, x_j \in X$, $d(x_i, x_j) = d(x_j, x_i)$ (symmetry).

Note that a distance function is more general than a metric; because the triangle inequality is not required[1].

A *k-clustering* of X is a k-partition, $C = \{C_1, C_2, \ldots, C_k\}$. That is, $\bigcup_{j=1}^{k} C_i = X$, $C_j \neq \emptyset$, for all $j \in \{1, \ldots, k\}$; and $C_i \cap C_j = \emptyset$ for all $i \neq j$. A *clustering* of X is a k-clustering of X for some $k \geq 1$. A clustering is *trivial* if $|C_j| = 1$ for all $j \in \{1, \ldots, k\}$ or $k = 1$. For $x_i, x_j \in X$ and a clustering C of X, we write $x_i \sim_C x_j$ if x_i and x_j are in the same cluster of clustering C, and we write $x_i \nsim_C x_j$ if they are in different clusters.

---

[1] Most authors prefer to call these functions *dissimilarities* and use *distance* as synonym to *metric*, but here we keep this earlier use of *distance* so our notation follows closely the notation in the clustering case [1,2].

A *clustering function* for some domain set X is a function (algorithm) that takes as inputs a distance function d over X, and produces as output a clustering of X. Typically, such clustering function is an algorithm that attempts to obtain the optimum of a loss function that formalizes some induction principle. In this form, a clustering problem is an optimization problem. For example, if we minimize the total squared error when we chose a set of k representatives in an Euclidean space (that is, d is the Euclidean metric $Eucl$) we obtain the problem that k-means attempts heuristically to solve. Given X with $|X| = n$ and $k > 1$, minimize $ErrorSQ_{Eucl}(R) = \sum_{i=1}^{n}[Eucl(\boldsymbol{x}_i, rep[\boldsymbol{x}_i, R])]^2$ where R is a set of k representatives and $rep[\boldsymbol{x}_i, R]$ is the nearest representative to $\boldsymbol{x}_i$ in R. This problem can be solved exactly by first enumerating all k-clusterings of X, and by taking the mean of each cluster as a representative and finally by evaluating the loss. However, the number of k-clusterings corresponds to the Stirling numbers of the second kind, and this approach has complexity at least exponential in n. A discrete version is usually referred as medoids where we also require that $R \subseteq X$. In this case, the problem remains NP-hard as it reduces to the p-median problem; however, we can now exhaustively test all subsets $R \subset X$ with $|R| = k$. The complexity of this exhaustive search algorithm is now at least proportional to $\binom{n}{k}$. This approach would have complexity $O(n^{k+1})$ and would now be polynomial in n. While Ackerman and Ben-David [1,2] refer to this as "polynomial" for some of their easiness results alluded earlier, it is perhaps more appropriate to refer to it as *polynomial in n for each fixed k* and thus our use of quotation marks (this class is also known as the class XP).

2.2   Instance Easiness for Supervised Learning

We introduce here a notion of instance easiness for the supervised learning problem. To the best of our knowledge, this is the first use of *instance easiness* applied to supervised learning. It also will be the building block for our presentation of cluster-quality measures. Our approach follows the notion of stability to perturbation of a problem (this approach was used for the unsupervised case by Ackerman and Ben-David [2]). Consider an instance of the supervised learning problem given by

1. a set of n pairs $\{(\boldsymbol{x}_i, c(\boldsymbol{x}_i))\}_{i=1}^{n}$, where $X = \{\boldsymbol{x}_1, \ldots, \boldsymbol{x}_n\}$ is the training set of labeled examples, and Y is a finite[2] set of labels (thus, $c(\boldsymbol{x}_i) \in Y$),
2. a family of models $\mathcal{F}$, so that if $F \in \mathcal{F}$, then $F : X \to Y$, and
3. a real valued loss function $\mathcal{L}$.

The goal is to find $F_O \in \mathcal{F}$ that optimizes the loss function. For brevity we will just write $[X, Y]$ for an instance of the supervised learning problem. For example, the family of models could be all artificial neural networks with a certain number of layers and neurons per layer and the loss function could be the total squared

---

[2]   In this paper we consider $|Y| \in \mathbb{N}$ and small. Thus we focus on classification and not on interpolation/regression.

error. Then, back-propagation can be seen as a gradient-descent approach to the corresponding optimization problem.

We also make some generic observations of what we require of a loss function. First, the loss function in the supervised learning setting is a function of the instance $[X, Y]$ and the classifier $F : X \rightarrow Y$, thus, we write $\mathcal{L}([X, Y], F)$. We expect F to always be a mathematical function (and not a relation). In the supervised learning setting, the instance $[X, Y]$ is typically formed by data points for a mathematical function c (that for each element in the domain, associates one and no more than one element in the codomain). However, in practice, it is not unusual to have contradictory examples; that is, it is not uncommon for data sets derived from practice to have two (or more) contradictory examples $(\boldsymbol{x}_i, c)$ and $(\boldsymbol{x}_i, c')$ with $c \neq c'$. Nevertheless, what we will require is that the loss function cannot be oblivious to the requirement that the classifier be a function in the following sense. Given an instance $[X, Y]$, at least for every classifier F_O that optimizes the loss function $\mathcal{L}([X, Y], F)$ the optimal value $\mathcal{L}([X, Y], F_O)$ cannot be the same to $\mathcal{L}([X', Y], F_O)$ when X' is the same as X except that X' contains one or more additional contradictory examples (perturbations can cause more contradictory examples, and that cannot improve the loss).

What we propose is that, if there is a distance function d over X, then we can consider an instance of supervised learning as easy if small perturbations of the set X result also in small perturbations of the loss. More formally, we say that two set X and X' are ϵ-close (with respect to a distance function d over $X \cup X'$) if there is a bijection[3] $\pi : X \rightarrow X'$ so that $d(\boldsymbol{x}_i, \pi(\boldsymbol{x}_i)) \leq \epsilon$, for all $i = 1, \ldots, n$. With these concepts we introduce our first fundamental definition.

Definition 1. *Let $[X, Y]$ and $[X', Y']$ be two instances of the supervised learning problem, we say they are ϵ-close if*

1. *$Y' \subseteq Y$ (no new classes are introduced),*
2. *X and X' are ϵ-close, and $c(\boldsymbol{x}_i) = c(\pi(\boldsymbol{x}_i))$ where $\pi : X \rightarrow X'$ provides the ϵ-closeness.*

That is, the training sets are ϵ-close and there are no more class labels.

Now, let $OPT_{\mathcal{L}}(X, Y)$ be the optimum value of the loss function $\mathcal{L}$ for the instance $[X, Y]$; that is, $OPT_{\mathcal{L}}(X, Y) = \min\{\mathcal{L}([X, Y], F) \mid F \in \mathcal{F}\} = \mathcal{L}(F_O)$.

Definition 2. *We say that the instance $[X, Y]$ is (ϵ, δ)-easy if*

1. *there is $F_0 : X \rightarrow Y$ a classifier that optimizes the loss, and*
2. *for all instances $[X', Y]$ that are ϵ-close to $[X, Y]$, we have*

$$\mathcal{L}([X', Y], F_0) \leq (1 + \delta)OPT_{\mathcal{L}}(X, Y).$$

The loss does not depend on any distance function on X. We also assume that the loss is based on the categorical/nominal nature of the set Y, and thus the loss value does not change if we rename the classes with any one-to-one function. We say such loss functions are isomorphism-invariant. Common loss functions are isomorphism-equivalent; that is, they do not depend on the name of the classes.

---

[3] Originally ϵ-closeness was defined [2] with a one-to-one mapping, but we ensure the relation "X is ϵ-close to X'" is symmetric, but this is not necessary for what follows.

2.3   Illustration

We here illustrate the concepts introduced earlier. For visualization purposes, we consider a two-dimensional data set, and for simplicity, we assume we are using a clustering algorithm like k-means and, because of some intuition, we are seeking two clusters[4], i.e. $k = 2$. The data set in Fig. 1 consist of 4 normal distributions in a mixture with equal proportions $(1/4)$. The respective means $\mu_1 = (0,0)$, $\mu_2 = (20,20)$, $\mu_2 = (-25,25)$, $\mu_2 = (-5,45)$. All have diagonal covariance matrices and all elements of the diagonal are equal to 10. While this data is not challenging for k-means, there are at least two local minima for the clustering loss function. Therefore, depending on its random initialization, k-means produces two clusterings. One with centers $M = \{(-15,35),(10,9)\}$ and another with centers $M' = \{(-12,11),(7,33)\}$. We used WEKA's SIMPLEKMEANS [8] and the first set is obtained with 7 iterations on average (see Fig. 1) while the second one required 29 iteration on average (see Fig. 1).

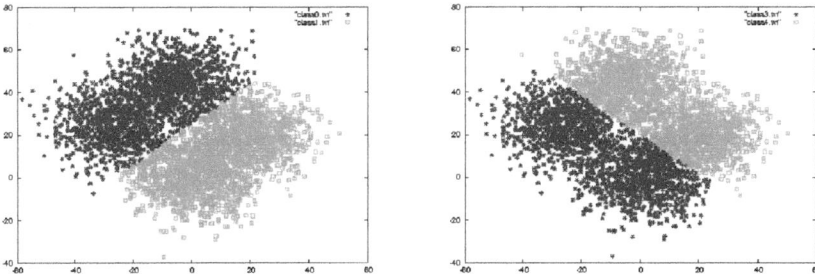

Fig. 1. Data with 4,000 points and two clusterings resulting from centers M and M'

For each of this clusterings we can draft a supervised learning problem. We argue that the corresponding supervised learning problem that results from this two clusters are different in terms of how easy they are. Note however, that Fig. 1 illustrates the corresponding supervised learning problems; and therefore, they correspond to linearly separable classes (if fact, k-means classes are a Voronoi partition of the universe and therefore, always separable by k-hyperplanes). This suggest that k-means always produces what can be regarded as an "easy" supervised-learning problem (we would expect linear discriminant, support vector machines, CART and many classifiers to do very well).

We suggest that if the clusters do reflect genuine structure and we have discovered meaningful classes, the corresponding job of using these concepts for learning a classifier should be "easier", and we measure "easier" by how stable the supervised learning instance is to perturbations. Obtaining very accurate classifiers for the two supervised learning problems in Fig. 1 can be achieved with the simplest of WEKA's algorithms. Using WEKA's stratified cross-validation to

---

[4] Although here we know the ground-truth, in a practical clustering exercise we would not know much about the data and identifying the value of k would be part of the challenge; one approach is to test if there are clusters by evaluating $k = 1$ vs. $k = 2$.

evaluate accuracy, NAIVEBAYES achieves 99.7% accuracy, only misclassifying 11 instances for M while it achieves 99.1% accuracy, only misclassifying 36 instances for M'. Similarly, WEKA's NNGE (Nearest-neighbor-like algorithm using non-nested generalized exemplars which are hyper-rectangles that can be viewed as if-then rules) achieves 99.2% accuracy, only misclassifying 32 instances for M while it achieves 99.2% accuracy, only misclassifying 33 instances for M'. However, if one takes the data set in Fig. 1 and associates all those points that have one or more negative coordinates to one class and those points that have both positive coordinates to another class, we obtain a supervised learning problem PROBLEM: $N = x_1 > 0 \wedge x_2 > 0$ that is also "easy" because the two classifiers above can also obtain high accuracy. In fact, (also evaluated by WEKA's stratified cross-validation) NNGE achieves 99.95% accuracy only missing 2 instances and NAIVEBAYES achieves 91% accuracy only missing 361 instances.

However, clearly the last supervised learning problem is not the result of a good clustering. What we do now is keep those classifiers learned with the unperturbed data, and use perturbed data as test data. We see that those classifiers that come from less quality clusterings degrade their accuracy more rapidly in proportion to the perturbation. We perturbed the 3 supervised learning problems by adding a uniformly distributed random number in $[-1, 1]$ to the attributes (but the class remains untouched). Now in M, NAIVEBAYES is 99.5% accurate (19 errors on average), NNGE is 99.6% accurate (14 errors on average). In M' NAIVEBAYES is 99% accurate (40 errors on average), NNGE is 99% accurate

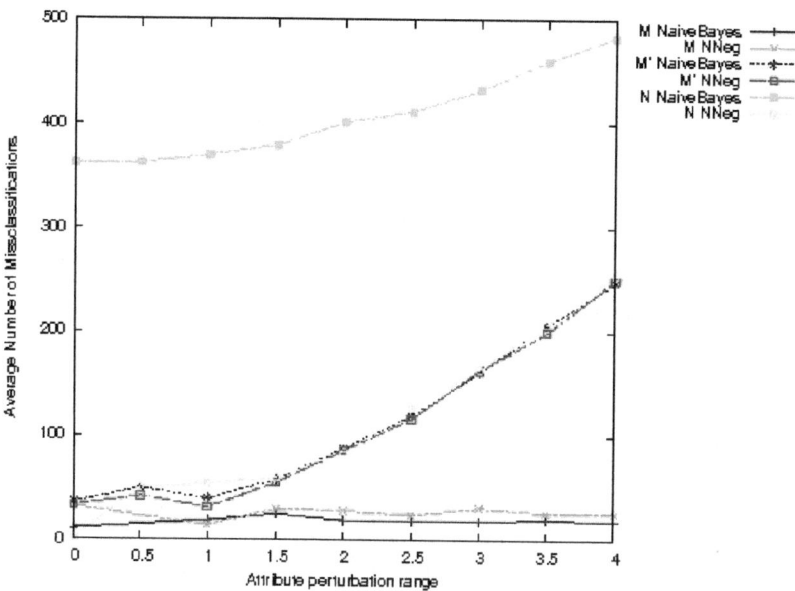

Fig. 2. Deterioration of performance of classifiers as perturbations on the attributes is larger. But classifiers derived from good clusters preserve accuracy.

(26 errors on average), In PROBLEM N: $x_1 > 0 \wedge x_2 > 0$ NAIVEBAYES is 90% accurate (369 errors on average), NNGE is 99% accurate (44 errors on average),

Fig. 2 shows points (ϵ, y) where misclassification rate y in the test-set corresponds to data that has suffered a perturbations with uniformly distributed random noise is $[-\epsilon, \epsilon]$. The behavior (effects on the loss) is essentially independent of the classifier, with the two classifiers for M remaining stable to perturbation. However, for the other supervised learning problems, the misclassification rate rapidly grows (classification accuracy deteriorates).

In fact, while the growth rate is important, we will be interested on the stability. That is, we focus on the largest $\epsilon > 0$ where the problem remains easy since in this region the loss suffers small impact. In Fig. 1, this correspond to how far right does the misclassification line remain flat before it starts to increase.

This experimental observation will be formalized in the next section to construct measures of cluster quality. Note that PROBLEM N has a simple boundary but not a good clustering. Although M' is a good clustering, the M clustering is the best because the groups are essentially the clouds at $(0,0)$ and $(20, 20)$ on one side and the clouds at $(-25, 25)$ and $(-5, 45)$ on the other. These pairing has more separation that the paring by M'.

We emphasize that this example is mainly for illustration purposes. By no reasonable standard data of 4 normal distributions with equal cylindrical covariance in two dimensions and equal proportions corresponds to a challenging clustering exercise. In fact, with 4, 000 points, it hardly corresponds to a challenging data mining setting. However, we believe this illustrates our point further. The most widely used clustering algorithm (k-means) even on this data set (which is supposed to be suitable for k-means since clusters are spherical and separated) can provide wrong answers. Clearly, part of the problem is the inappropriate value $k = 2$. Can the loss function for k-means indicate the better clustering between M and M'? This example is so simple that such is the case. For example, WEKA standard evaluation with SIMPLEKMEANS indicates a better loss function for M than for M'. But this example is for illustration only.

2.4   The Clustering-Quality Measure

A *clustering-quality measure* (*CQM*) is a function that is given a clustering C over (X, d) (where d is a distance function over X) and returns a non-negative real number. Many proposals of clustering-quality measures have been suggested for providing some confidence (or ensuring validity) of the results of a clustering algorithm. Before we introduce two new *CQM*, we need a bit of notation. If d is a distance function over X and $\lambda > 0$ with $\lambda \in \Re$, then the λ-*scaled version of* d is $d' = (\lambda d)$ and is defined by $(\lambda d)(\boldsymbol{x}_i, \boldsymbol{x}_j) = \lambda \cdot d(\boldsymbol{x}_i, \boldsymbol{x}_j)$. If d is a distance function over X, the *normalized* version of d is denoted by $nor(d)$ and is defined as the $1/\lambda$-scaled version of d when $\lambda = \max\{d(\boldsymbol{x}_i, \boldsymbol{x}_j) \mid \boldsymbol{x}_i, \boldsymbol{x}_j \in X\}$; that is, $nor(d) = (d/\lambda) = (d/\max\{d(\boldsymbol{x}_i, \boldsymbol{x}_j) \mid \boldsymbol{x}_i, \boldsymbol{x}_j \in X\})$.

Definition 3. *Given a clustering* $C = \{C_1, \ldots, C_k\}$ *of* (X, d), *the* CQM *by classification* m_c *is the largest* $\epsilon > 0$ *so that, if we construct a supervised learning* *instance* $[X, C]$ *derived from the clustering by* $Y = C$ *and* $c(\boldsymbol{x}_i) = C_j$ *such that* $\boldsymbol{x}_i \in C_j$, *then* $[X, C]$ *is* $(\epsilon, 0)$-*easy with respect to* $nor(d)$.

Definition 4. *Given a clustering* $C = \{C_1, \ldots, C_k\}$ *of* (X, d), *the* CQM *by pairing* m_p *is the largest* $\epsilon > 0$ *so that, if we construct a supervised learning instance* $[X \times X, \{\text{YES}, \text{NO}\}]$ *derived from the clustering* C *by*

$$c(\boldsymbol{x}_i, \boldsymbol{x}_j) = \begin{cases} \text{YES} & \text{if } \boldsymbol{x}_i \sim_C \boldsymbol{x}_j \\ \text{NO} & \text{if } \boldsymbol{x}_i \not\sim_C \boldsymbol{x}_j, \end{cases}$$

then the instance $[X \times X, \{\text{YES}, \text{NO}\}]$ *is* $(\epsilon, 0)$-*easy with respect to* $nor(d)$.

We are now in a position to prove the four (4) properties required by Ackerman and Ben-David of a *CQM*. These properties were inspired by the 3 axioms suggested by Kleinberg [9]. Kleinberg proved that although the axioms are desirable of all clustering functions, they were inconsistent. This is usually interpreted as the impossibility of defining what clustering is. However Ackerman and Ben-David properties are sound, thus suggesting it is feasible to describe what is good clustering. Scale invariance means that the output is invariant to uniform scaling of the input.

Definition 5. Scale invariance: *A* CQM *m satisfies* scale invariance *if* $\forall C$ *a clustering of* (X, d) *and* $\lambda > 0$, *we have* $m[C, X, d] = m[C, X, (\lambda d)]$.

Lemma 1. *On a bounded study region, The* CQM m_c *and the* CQM m_p *satisfy invariance.*

Since $nor(d) = nor(\lambda d)$ for all $\lambda > 0$ and the definitions of m_p and m_c use the normalized version of d the lemma follows. Thus, we now assume all distance functions are normalized to the largest ball that includes the study region.

For the next property we need to introduce the notion of isomorphic clusters, denoted $C \approx_d C'$. A distance-preserving isomorphism $\phi : X \to X$ satisfies that $\forall \boldsymbol{x}_i, \boldsymbol{x}_j \in X$, $d(\boldsymbol{x}_i, \boldsymbol{x}_j) = d(\phi(\boldsymbol{x}_i), \phi(\boldsymbol{x}_j))$. Two clusters C and C' of the same domain (X, d) are *isomorphic* if there exists a distance-preserving isomorphism such that $\forall \boldsymbol{x}_i, \boldsymbol{x}_j \in X$, we have $\boldsymbol{x}_i \sim_C \boldsymbol{x}_j$ if and only if $\phi(\boldsymbol{x}_i) \sim_{C'} \phi(\boldsymbol{x}_j)$.

Definition 6. Invariant under isomorphism: *A* CQM *is* invariant under isomorphism *(isomorphism-invariant) if* $\forall C, C'$ *non-trivial clusterings over* (X, d) *where* $C \approx_d C'$, *we have* $m[C, X, d] = m[C', X, d]$.

Lemma 2. *The* CQM m_c *and* m_p *are isomorphism-invariant.*

Definition 7. Richness: *A* CQM *satisfies* richness *if for each non-trivial partition* C *of* X *there is a distance function* $\hat{d}$ *such that* C *maximizes* $m[C, X, \hat{d}]$ *when considered as a function of* C.

Lemma 3. *The measures* m_c *and* m_p *satisfy richness.*

The final property of a *CQM* is *consistency*. Given a clustering C over (X, d) (that is, d is a distance function over X), we say that another distance function d' over X is a *C-consistent variant* of d if

1. $d'(\boldsymbol{x}_i, \boldsymbol{x}_j) \leq d(\boldsymbol{x}_i, \boldsymbol{x}_j) \; \forall \boldsymbol{x}_i \sim_C \boldsymbol{x}_j$, and
2. $d'(\boldsymbol{x}_i, \boldsymbol{x}_j) \geq d(\boldsymbol{x}_i, \boldsymbol{x}_j) \; \forall \boldsymbol{x}_i \nsim_C \boldsymbol{x}_j$.

Definition 8. Consistency: *A* CQM *satisfies* consistency *if* $\forall$ *C a clustering of* (X, d) *and* d' *that is* C*-consistent variant of* d, *we have* $m[C, X, d'] \geq m[C, X, d]$.

Lemma 4. *The* CQM m_c *and the* CQM m_p *satisfy consistency.*

3   Discussion

We have introduced two *CQM* and mathematically demonstrated fundamental properties that have several favorable implications. First, *CQM* that satisfy richness, scale invariance, consistency and isomorphism-invariance can be combined to produce new measures with also these properties [1]. Thus, we have not only enriched the set of *CQM* since m_c and m_p become generators to produce *CQM*.

Secondly, the methods to verify accuracy in supervised learning are now well established and many strong and solid implementations exist (like WEKA [8]). Therefore, the issue of cluster quality can now be simplified as we did in the earlier illustration. Before the proposal here, it is not surprising to find statements like: "Evaluation of clusterers is not as comprehensive as the evaluation of classifiers. Since clustering is unsupervised, it is a lot harder determining how *good* a model is" [4]. Our proposal here shows that the machinery for evaluating supervised learning can be useful to tackle the issue of cluster validity without the need of already classified (supervised) instances. We should aim for cluster validity methods that are as close as possible to the "comprehensive" landscape we have for supervised learning. Our proposal here suggests this direction.

Our proposal is applicable to the issue of alternative clusterings. The outputs from these algorithms are several alternative clusterings, because the data-miner believes there may be several meaningful ways to create such clusterings [3]. While this approach needs to resolve the issue of cluster similarity or dissimilarity (as in external cluster validity), it is also guided by a measure of cluster quality. That is, the approach also needs to provide some credibility for each of the multiple answers provided to an unsupervised learning problem.

Traditionally, cluster validity has taken three avenues: *internal cluster validity*, *external cluster validity* [7], and *experimental cluster validity*. A comprehensive discussion of the issues and challenges with each appears elsewhere. Approaches to cluster validity since then continue along these lines. But, typically there is an admission that evaluating a model built from a clustering algorithm is challenging [7]. Proposals like comparing a matrix of two clusterings [7] still face many problems, and lead to the challenges of measures of similarity between clusters. In the handbook of Data Mining and Knowledge Discovery, Chapter 15 [11] has a discussion of clustering methods, and some material on cluster evaluation and validity. M. Halkidi and M. Vazirgiannis [7, Chapter 30] also offer a discussion of cluster validity. The fact of the matter all remain variations of earlier methods. Our approach is perhaps most similar to other experimental approaches [6,15] which have justification in the intuition that the boundary of clusters should show sparsity. Since implicitly, support vector machines offer to find margins that are as wide as possible, one can hypothesize about outliers and boundaries and filter them out [6]. If removing few boundary items and repeating the clustering passes an external validity test that shows the clustering is robust to this change, then we can raise our confidence that the clustering has good quality. Otherwise the clustering is suspicious. A similar idea is derived from mathematical properties of proximity-graphs [15]. Here as well, a candidate clustering can be polished on the boundaries of clusters simply by removing those data points that the proximity graph suggests are on fringes of cluster. If repeating the alternation of polishing and clustering offers stable clusters (clusters do not change with respect to some external clustering validity measure), trust in the clustering is raised.

Both of these mechanisms [6,15] have a notion of robustness, and the foundations are derived from proximity structure reflected in the clusterer (clustering function) itself. However, they are computationally costly as clustering needs to be repeated, external cluster-validity functions need to be computed and the test can only be one of similarity between pairs of clusters. Our approach here is mathematically more formal, and it is easy to implement by the availability of supervised learning techniques and their implementations.

Consensus clustering or ensemble of clusters [10,12,16] is an extension of these earlier ideas [6,15] of agreement between clusterings. Although initially proposed for problems in bio-informatics, the concept seems quite natural, since in fact, many clustering algorithms will produce different clusters if initialized with different parameters. So, the same clustering approach leads to multiple answers. Proponents of consensus clustering argue it is sensible to produce a clustering that maximizes the agreement (a similar idea occurs with multiple classifiers or a classifier ensemble).

Our approach here also enables to give some assessment of the clustering participating in the ensemble as well as the resulting combined clustering. Our approach does not need to deal with the issues of cluster similarity. But, we illustrate we can apply our approach to consensus clustering with the data set made available by Dr. A. Strehl x8d5k.txt [12]. This is a mixture of 8 non-symmetrical Gaussians in 8 dimensions. The data consist of 1000 points, and the Original cluster labels from the mixture are provided. These original labels provide 200 data points from each cluster. Also, 5 clusterings V1, V2, V3, V4, and V5 are provided and the consensus clustering (Combined) of these five is the 6-th clustering. It corresponds to "the best known labels in terms of average normalized mutual information (ANMI)". We applied our approach to these 7 clustering and present our results in a similar way to our earlier illustration in Fig. 3. Because we have the Original clustering (sometimes referred as the *true* clustering), we can see that the alternative clusters are in fact weaker that the truth. However, the Combined cluster is extremely satisfactory and our approach shows that it is essentially equivalent for our CQM to the Original. These conclusions are also in agreement if the supervised learner is WEKA's NaiveBayes or NNge. This is what we would expect.

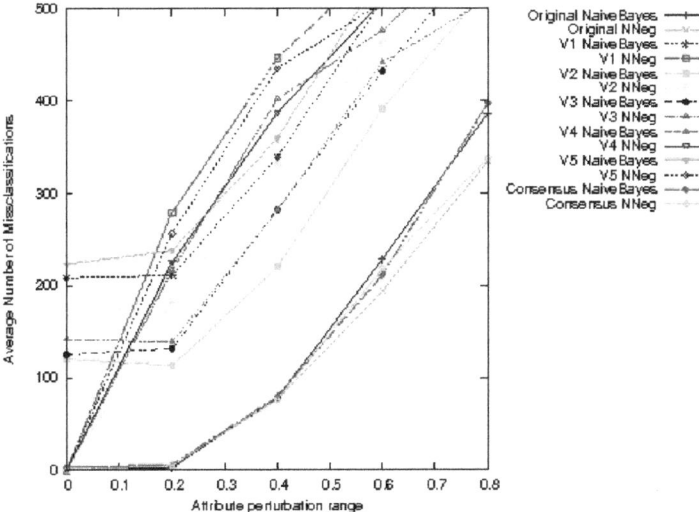

Fig. 3. Deterioration of performance of classifiers as perturbations on the attributes are larger (data set is `x8d5k.txt`[12]. But classifiers derived from true clusters and consensus cluster preserve accuracy.

4   Summary

Despite decades of research the fundamental issue of clustering validity has remained unsolved. So much so that widely accepted software like WEKA has minimal tools and resources for it. This contrasts with the large set of tools for validity for the supervised learning case. This paper enables to use the set of tools for the supervised case in the unsupervised case.

The intuition of our work is a simple idea. When we have to discover groups (classes) in a data set where we have no information regarding this, whatever results must be assessed for validity. A clustering algorithm's output must be evaluated and external validity approaches are out of the question, since if we had knowledge of the *true* clustering, why would be trying to find it? However, we would expect that the classes obtained by the clustering function are in some way separable and constitute meaningful concepts. They should be robust to small perturbations. A classifier obtained from corresponding supervised learning result should have performance that degrades rather slowly when presented with data that is close. Such data can be obtained by perturbations and then the robustness of the classifier measured by the now standard approaches of supervised learning.

We have provided illustrations that this idea is manifested in practical clustering scenarios including consensus clustering. We have also provided theoretical foundations by formalizing a notion of instance easiness for supervised clustering and then deriving measures of cluster quality. We have shown that these measures satisfy the generic properties of richness, scale invariance, isomorphism-invariance and consistency that are common to many measures (however, some of these other measures are very costly to compute). Thus, our approach enables a practical and theoretical useful mix.

References

1. Ackerman, M., Ben-David, S.: Measures of clustering quality: A working set of axioms for clustering. In: Advances in Neural Information Processing Systems 22 NIPS, Proceedings of the Twenty-Second Annual Conference on Neural Information Processing Systems, pp. 121–128. MIT Press, Vancouver (2008)
2. Ackerman, M., Ben-David, S.: Clusterability: A theoretical study. In: Proceedings of the Twelfth Int. Conf. on Artificial Intelligence and Statistics AISTATS, Clearwater Beach, Florida, USA, vol. 5, JMLR:W&CP (2009)
3. Bae, E., Bailey, J.: Coala: A novel approach for the extraction of an alternate clustering of high quality and high dissimilarity. In: Proceedings of the 6th IEEE Int. Conf. on Data Mining (ICDM), pp. 53–62. IEEE Computer Soc. (2006)
4. Bouckaert, R.R., Frank, E., Hall, M., Kirkby, R., Reutemann, P., Seewald, A., Scuse, D.: WEKA Manual for Version 3-6-2. The University of Waikato (2010)
5. Duda, R.O., Hart, P.E., Stork, D.G.: Pattern Classification, 2nd edn. John Wiley & Sons, NY (2001)
6. Estivill-Castro, V., Yang, J.: Cluster Validity using Support Vector Machines. In: Kambayashi, Y., Mohania, M., Wöß, W. (eds.) DaWaK 2003. LNCS, vol. 2737, pp. 244–256. Springer, Heidelberg (2003)
7. Halkidi, M., Vazirgiannis, M.: Chapter 30 — quality assessment approaches in data mining. In: The Data Mining and Knowledge Discovery Handbook, pp. 661–696. Springer, Heidelberg (2005)
8. Hall, M., Frank, E., Holmes, G., Pfahringer, B., Reutemann, P., Witten, I.H.: The WEKA data mining software: an update. SIGKDD Explorations 11(1), 10–18 (2009)
9. Kleinberg, J.: An impossibility theorem for clustering. In: The 16th conference on Neural Information Processing Systems (NIPS), pp. 446–453. MIT Press (2002)
10. Monti, S., Tamayo, P., Mesirov, J., Golub, T.: Consensus clustering: A resampling-based method for class discovery and visualization of gene expression microarray data. Machine Learning 52(1-2), 91–118 (2003)
11. Rokach, L., Maimon, O.: Chapter 15 — clustering methods. In: The Data Mining and Knowledge Discovery Handbook, pp. 321–352. Springer, Heidelberg (2005)
12. Strehl, A., Ghosh, J.: Cluster ensembles – a knowledge reuse framework for combining multiple partitions. J. on Machine Learning Research 3, 583–617 (2002)
13. Witten, I., Frank, E.: Data Mining — Practical Machine Learning Tools and Technologies with JAVA implementations (2000)
14. Wu, X., et al.: Top 10 algorithms in data mining. Knowledge and Information Systems 14(1), 1–37 (2008)
15. Yang, J., Lee, I.: Cluster validity through graph-based boundary analysis. In: Int. Conf. on Information and Knowledge Engineering, IKE, pp. 204–210. CSREA Press (2004)
16. Yu, Z., Wong, H.-S., Wang, H.: Graph-based consensus clustering for class discovery from gene expression data. Bioinformatics 23(21), 288–2896 (2007)

A New Efficient and Unbiased Approach for Clustering Quality Evaluation

Jean-Charles Lamirel[1], Pascal Cuxac[2], Raghvendra Mall[3], and Ghada Safi[4]

[1] LORIA, Campus Scientifique,
BP 239, Vandœuvre-lès-Nancy, France
jean-charles.lamirel@inria.fr
http://www.loria.fr
[2] INIST-CNRS, 2 allée du Parc de Brabois,
54500 Vandœuvre-lès-Nancy, France
pascal.cuxac@inist.fr
http://recherche.inist.fr
[3] Center of Data Engineering, IIIT Hyderabad,
NBH-61, Hyderabad, Andhra Pradesh, India
raghvendra.mall@research.iiit.ac.in
http://www.iiit.ac.in
[4] Department of Mathematics, Faculty of Science, Aleppo University,
Aleppo, Syria
ghada1@scs-net.org

Abstract. Traditional quality indexes (Inertia, DB, ...) are known to be method-dependent indexes that do not allow to properly estimate the quality of the clustering in several cases, as in that one of complex data, like textual data. We thus propose an alternative approach for clustering quality evaluation based on unsupervised measures of Recall, Precision and F-measure exploiting the descriptors of the data associated with the obtained clusters. Two categories of index are proposed, that are Macro and Micro indexes. This paper also focuses on the construction of a new cumulative Micro precision index that makes it possible to evaluate the overall quality of a clustering result while clearly distinguishing between homogeneous and heterogeneous, or degenerated results. The experimental comparison of the behavior of the classical indexes with our new approach is performed on a polythematic dataset of bibliographical references issued from the PASCAL database.

Keywords: clustering, quality indexes, unsupervised recall, unsupervised precision, labeling maximization.

1 Introduction

The use of classification methods is mandatory for analyzing large corpus of data as it is the case in the domain of scientific survey or in that of strategic analyzes of research. While carrying out a classification, one seeks to build homogeneous groups of data sharing a certain number of identical characteristics. Furthermore,

L. Cao et al. (Eds.): PAKDD 2011 Workshops, LNAI 7104, pp. 209–220, 2012.

the clustering, or unsupervised classification, makes it possible to highlight these groups without prior knowledge on the treated data. A central problem that then arises is to qualify the obtained results in terms of quality: a quality index is a criterion which indeed makes it possible altogether to decide which clustering method to use, to fix an optimal number of clusters, and to evaluate or to develop a new method. Even if there exist recent alternative approaches [2] [9] [10], the most usual indexes employed for the evaluation of the quality of clustering are mainly distance-based indexes relying on the concepts of intra-cluster inertia and inter-cluster inertia [15]:

- Intra-cluster inertia measures the degree of homogeneity between the data associated with a cluster. It calculates their distances compared to the reference point representing the profile of the cluster. It can be defined as:

$$Intra = \frac{1}{|C|} \sum_{c \in C} \frac{1}{|c|} \sum_{d \in c} \| p_c - p_d \|^2$$

where C represents the set of clusters associated to the clustering result, d represents a cluster associated data and p_x represents the profile vector associated to the element x.

- Inter-cluster inertia measures the degree of heterogeneity between the clusters. It calculates the distances between the reference points representing the profiles of the various clusters of the partition.

$$Inter = \frac{1}{|C|^2 - |C|} \sum_{c \in C} \sum_{c' \in C, c' \neq c} \| p_c - p_{c'} \|^2$$

Thanks to these two quality indexes or their adaptations, like the Dunn index [6], the Davies-Bouldin index [3], or the Silhouette index [19], a clustering result is considered as good if it possesses low intra-cluster distances as compared to its inter-cluster distances. However, it has been shown in [13] that the distance-based indexes are often strongly biased and highly dependent on the clustering method. They cannot thus be easily used for comparing different methods. Moreover, as Forest also pointed out [7], the experiments on these indexes in the literature are often performed on unrealistic test corpora constituted of low dimensional data and embedding a small number of potential classes. As an example, in their reference paper, Milligan and Cooper [18] compared 30 different methods for estimating the number of clusters relying only on simulated data described in a low dimensional Euclidean space. Nonetheless, using Reuters test collection, it has been shown by Kassab and Lamirel [11] that aforementioned indexes are often properly unable to identify an optimal clustering model whenever the dataset is constituted by complex data that must be represented in a both highly multidimensional and sparse description space, as it is often the case with textual data. To cope with such problems, our own approach takes its inspiration altogether from the behavior of symbolic classifiers and from the evaluation principles used in Information Retrieval (IR). Our Recall/Precision and

F-measure indexes exploit the properties of the data associated to each cluster after the clustering process without prior consideration of clusters profiles [13]. Their main advantage is thus to be independent of the clustering methods and of their operating mode. However, our last experiments highlighted that these new quality indexes did not make it possible to clearly distinguish between homogeneous results of clustering and heterogeneous, or degenerated ones [8]. After presenting our original quality indexes, we thus describe hereafter some of their extensions which make it possible to solve the aforementioned problem. We then experimentally show the effectiveness of our extended approach, as compared to classical distance-based approach, for discriminating between the results provided by three different clustering methods which have been applied on a polythematic documentary corpus containing various bibliographic records issued from the PASCAL CNRS scientific database.

2   Unsupervised Recall Precision F-Measure Indexes

2.1   Overall Clustering Quality Estimation

In IR, the **Recall R** represents the ratio between the number of relevant documents which have been returned by an IR system for a given query and the total number of relevant documents which should have been found in the documentary database [20]. The **Precision P** represents the ratio between the number of relevant documents which have been returned by an IR system for a given query and the total number of documents returned for the said query. **Recall** and **Precision** generally behave in an antagonist way: as **Recall** increases, **Precision** decreases, and conversely. The **F** function has thus been proposed by Van Rijsbergen [21] in order to highlight the best compromise between these two values. It is given by:

$$F = \frac{2(R * P)}{R + P} \tag{1}$$

Based on the same principles, the *Recall* and *Precision* indexes which we introduce hereafter evaluate the quality of a clustering method in an unsupervised way[1] by measuring the relevance of the clusters content in terms of shared properties, or features. In our further descriptions, a cluster content is supposed to be represented by the data associated with this latter after the clustering process and the descriptors (i.e. the properties or features) of the data are supposed to be weighted by values within the range [0,1].

Let us consider a set of clusters C resulting from a clustering method applied on a set of data D, the local *Recall (Rec)* and *Precision (Prec)* indexes for a given property p of the cluster c can be expressed as:

$$\underset{c}{Rec}(p) = \frac{|c_p^*|}{|D_p^*|}, \underset{c}{Prec}(p) = \frac{|c_p^*|}{|c|}$$

---

[1] Conversely to classical Recall and Precision indexes that are supervised.

where the notation X_p^* represents the restriction of the set X to the set members having the property p.

Then, for estimating the overall clustering quality, the averaged *Macro-Recall* (R_M) and *Macro-Precision* (P_M) indexes can be expressed as:

$$R_M = \frac{1}{|\overline{C}|} \sum_{c \in \overline{C}} \frac{1}{|S_C|} \sum_{p \in S_C} \underset{c}{Rec(p)}, \quad P_M = \frac{1}{|\overline{C}|} \sum_{c \in \overline{C}} \frac{1}{|S_C|} \sum_{p \in S_C} \underset{c}{Prec(p)} \quad (2)$$

where S_c is the set of peculiar properties of the cluster c, which can be defined as:

$$S_c = \left\{ p \in d, d \in c \middle| \overline{W_c^p} = \underset{c' \in C}{Max}\left(\overline{W_{c'}^p}\right) \right\} \quad (3)$$

and where $\overline{C}$ represents the peculiar set of clusters extracted from the clusters of C, which verifies:

$$\overline{C} = \{c \in C | S_c \neq \emptyset\}$$

and, finally:

$$\overline{W}_c^p = \frac{\sum\limits_{d \in c} W_d^p}{\sum\limits_{c' \in C} \sum\limits_{d \in c'} W_d^p} \quad (4)$$

where W_x^p represents the weight of the property p for element x.

It can be demonstrated (see [13] for more details) that if both values of *Macro-Recall* and *Macro-Precision* (Eq. 2) reach the unity value, the peculiar set of clusters $\overline{C}$ represents a Galois lattice. Therefore, the combination of this two measures enables to evaluate to what extent a numerical clustering model can be assimilated to a Galois lattice natural classifier.

Macro-Recall and *Macro-Precision* indexes can be considered as cluster-oriented quality measures because they provide average values of *Recall* and *Precision* for each cluster. They have opposite behaviors according to the number of clusters. Thus, these indexes permit to estimate an optimal number of clusters for a given method and a given dataset. The best data partition, or clustering result, is in this case the one which minimizes the difference between their values (see Figure 1B). However, similarly to the classical distance-based indexes, their main defect is that they do not permit to detect degenerated clustering results, whenever those jointly include a small number of heterogeneous or "garbage" clusters with large size and a big number of "chunk" clusters with very small size [8]. To correct that, we propose to construct complementary property-oriented indexes of *Micro-Recall* and *Micro-Precision* by averaging the *Recall/Precision* values of the peculiar properties independently of the structure of the clusters:

$$R_m = \frac{1}{|L|} \sum_{c \in \overline{C}, p \in S_c} \underset{c}{Rec(p)}, \quad P_m = \frac{1}{|L|} \sum_{c \in \overline{C}, p \in S_c} \underset{c}{Prec(p)} \quad (5)$$

where L represents the size of the data description space.

It is possible to refer not only to the information provided by the indices *Micro-Precision* and *Micro-Recall*, but to the calculation of the *Micro-Precision* operated cumulatively. In the latter case, the idea is to give a major influence to large clusters which are the most likely to repatriate the heterogeneous information, and therefore, to significantly lower by their own the quality of the resulting partition. This calculation can be made as follows:

$$CP_m = \frac{\sum_{i=|c_{inf}|,|c_{sup}|} \frac{1}{|C_{i+}|^2} \sum_{c \in C_{i+}, p \in S_c} \frac{|c_p|}{|c|}}{\sum_{i=|c_{inf}|,|c_{sup}|} \frac{1}{C_{i+}}} \tag{6}$$

where C_{i+} represents the subset of clusters of C for which the number of associated data is greater than i, and:

$$inf = argmin_{c_i \in C}|c_i|, sup = argmax_{c_i \in C}|c_i| \tag{7}$$

2.2   Cluster Labeling and Content Validation

Complementary to overall clustering model evaluation, the role of clusters labeling is to highlight the peculiar characteristics or properties of the clusters associated to a clustering model at a given time. Labeling can be thus used both for visualizing or synthesizing clustering results [14] and for validating or optimizing learning of a clustering method [1]. It can rely on endogenous data properties or on exogenous ones. Endogenous data properties represent the ones being used during the clustering process. Exogenous data properties represent either complementary properties or specific validation properties. Some label relevance indexes can be derivated from our former quality indexes using a probabilistic approach. The *Label Recall L-R* derives directly from Eq. 4. It is expressed as:

$$L - R(p) = \overline{W}_c^p \tag{8}$$

The *Label Precision P-R* can be expressed as:

$$L - P(p) = \frac{\sum_{d \in c} W_d^p}{\sum_{p' \in d, d \in c} W_d^p} \tag{9}$$

Consequently, the set of labels L_c that can be attributed to a cluster c can be expressed as the set of endogenous or exogenous cluster data properties which maximize the *Label F-measure* that combines *Label Recall* (Eq. 8) and *Label Precision* (Eq. 9) in the same way than the supervised F-measure described by (Eq.1) would do. As soon as *Label Recall* is equivalent to the conditional probability $P(c|p)$ and *Label Precision* is equivalent to the conditional probability $P(p|c)$, this former labeling strategy can be classified as an expectation maximization approach with respect to the original definition given by Dempster and al. [4].

3   Experimentation and Results

To illustrate the behavior of our new quality indexes, and to compare it to the one of the classical inertia indexes, our test dataset is build up from is a set of bibliographic records resulting from the INIST PASCAL database and covering one year of research performed in the French Lorraine area. The structure of the records makes it possible to distinguish the titles, the summaries, the indexing keywords and the authors as representatives of the contents of the information published in the corresponding article. In our experiment, the research topics associated with the keywords field are solely considered. Our test dataset represents a dataset of 1341 records. A frequency threshold of 3 being finally applied on the index terms, it resulted in a data description set of 889 indexing keywords. These keywords cover themselves a large set of different topics (as far one to another as medicine from structural physics or forest cultivation ...). Moreover, they comprise a high ratio of polysemic forms (like age, stress, structure, ...) that are used in the context of many different topics. The resulting experimental dataset can thus be considered as a complex dataset for clustering.

To carry out the clustering, we exploited in parallel the SOM fixed topology neural method [12], the Neural Gas (NG) free topology neural method [17] and the classical K-means method [16]. For each method, we do many different experiments letting varying the number of clusters from 9 to 324 clusters, employing the size of an increasing square SOM grid as a basic stepping strategy. In the next paragraphs, for the sake of clarity, we dont specifically report the results of K-means because they are similar to those of NG.

3.1   Overall Analysis of the Results

The analysis of the results performed by an expert showed that only the SOM method provided homogeneous clustering results on this dataset. Hence, in the case of the NG (or K-means) method, the analyst highlighted the presence of "garbage" clusters attracting most of the data in parallel with "chunk" clusters representing either marginal groups or unformed topics. This behavior, which corresponds to the case of a degenerated clustering result due to the dataset clustering complexity, can also be confirmed when one looks to the endogeneous labels or data properties that can be extracted from the clusters in an unsupervised way using the expectation maximization methodology described in section 2.2. Hence, cluster label extraction based on cluster data properties permits to highlight that the NG method mainly produced a "garbage" cluster with very big size that collects more than 80% of the data (1179 data among a total of 1341) and attracts (i.e. maximizes) many kinds of different labels (720 labels or data properties among a total of 889) relating to multiple topics (see Figure 1). Conversely, the good results of the SOM method can be confirmed in the same way. Indeed, endogeneous label extraction also shows that this latter method produces different clusters of similar size attracting semantically homogeneous groups of labels or data properties (see Figure 2).

[1198]720 -- Roots Density Proteins Scaling laws Respiratory system diseases Boundary conditions Forests Milk protein Environmental protection Forecasting Data analysis Air pollution Fibers In situ DNA Alcohols Rock mechanics Soils Analytical method Cosmology Data processing Mapping Vegetation Rhizosphere Drinking water Application Monte Carlo methods Urban areas Grains Pediatrics Streams Computer software Above ground plant part Enzyme inhibitors Phase diagrams Photosynthesis Nervous system diseases Gene regulation Seasonal variations Nucleosynthesis Biological availability Organic matter Engineering Program verification Flow cytometry T-Lymphocyte Diffusion Flocculation Principal component analysis Pest Detection Contrast media Hardwood forest tree Tolerance Food industry Delay system Fault tolerance Real time systems NP hard problem Respiratory system Polymers Synchronization .....

Fig. 1. Clusters content overview (endogeneous labels or data properties maximized by the clusters data) of NG generated partition

[11]17 -- Ecology – Forestry – Forests – Mountain - Organic matter – Climate - Softwood forest tree – Regeneration – pH - Image analysis - Forest management - Herbaceous plant - Chemical properties - Plant juvenile growth stage – Sensitivity - Forest site – Methodology

[13]23 -- Amorphous material - Growth from vapor - Magnetron - Preparation - Inorganic compound - Radiofrequency sputtering - Adhesion - Electrical conductivity - Nanostructured materials - Nanoindentation - Voltage dependence - Tribology - X-ray photoelectron spectra - Hardness - Ion implantation - Spectroscopic ellipsometry - Current voltage characteristics - EXAFS spectrometry - Thick film - Friction coefficient - Pulsed laser - Laser ablation technique - Diamond lattices

[15]22 -- Immunopathology - Allergy - Skin test - Immunology - IgE - Skin disease - Wheat - Exploration - Digestive diseases - Contrast media - Association - Adult - Food - Dermatology - Medical screening - Patient - Bronchus disease - Cytokine - Consensus conference - Lung disease - In vitro - Respiratory system diseases

[14]18 -- Temperate zone - Hardwood forest tree - Deciduous forest - Woody plant - Forest tree - Forest stand - Vegetation type - Silviculture - High forest - Dendrometry - Environmental factor - Ecophysiology - Entomology - Plant pathology - Stand characteristics - Artificial forest stand - Resource management - Wood line

Fig. 2. Clusters content overview (endogeneous labels or data properties maximized by the clusters data) of SOM generated partition

In addition, for highlighting the clusters purity regarding to exogeneous labels we have used the PASCAL hierarchical classification codes which have been initially associated to the data by the analysts. The exploited measure is the corrected purity index, formerly proposed by Diarmuid et al. [5]. This measure presents the advantage to discard clusters containing single data that would bias the purity results. It is expressed as :

$$mpur = \frac{\sum_{n_{prevalent(c_i)}>2} n_{prevalent(c_i)}}{|C|} \qquad (10)$$

where each cluster is associated with its prevalent classification code. The number of data in a cluster c that take this class is denoted $n_{prevalent(c)}$.

For highligting the behaviour of the different methods in terms of clusters purity, we have also exploited two different coding accuracies. Coding accuracy is directly related to the number of selected levels in the PASCAL classification codes hierarchy, which condition itself altogether the number and the precision

	Nbr of labels	SOM (purity)	NG (purity)
PASCAL C-codes (3 levels)	355	0.4832	0.0753
PASCAL C-codes (4 levels)	672	0.3788	0.0402

Fig. 3. SOM and NG clusters purity results exploiting exogeneous labels (PASCAL classification codes) and using two different coding accuracy.

of the utilized codes. The results presented in Figure 3 clearly highlight that, whatever is the considered coding accuracy, the SOM method is solely abble to achieve satisfying cluster purity results.

3.2   Quality Indexes Validation

On the one hand, the results presented in Figure 4A illustrate the fact that the classical indexes of inertia have an unstable behavior which does not make it possible to clearly identify an optimal number of clusters in both contexts of SOM and NG methods. On the other hand, it also appears in Figure 4B that the behavior of the *Macro-Recall/Precision* indexes is stable and makes it possible to identify an optimal number of clusters in all cases. Indeed, this optimal clusters number can be found out at the break-even point between the *Macro-Recall* and the *Macro-Precision* values (i.e. 100 clusters for NG and 256 clusters for SOM in Figure 4B).

Nonetheless, none of these former groups of indexes, whenever it is solely considered, permits to correctly estimate the quality of the results. Those do not make it possible in particular to discriminate between homogeneous results of clustering (SOM) and degenerated ones (NG or K-means). They even present the important defect to privilege this last family of results, illustrating a contradictory behavior (i.e. the worst results are identified as the best, and conversely). In the case of degenerated results, one potential explanation of the better values of aforementioned indexes is that the joint presence of a big amount of "chunk"

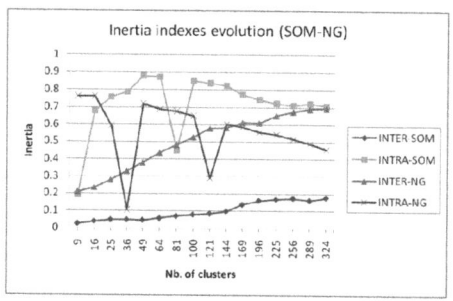
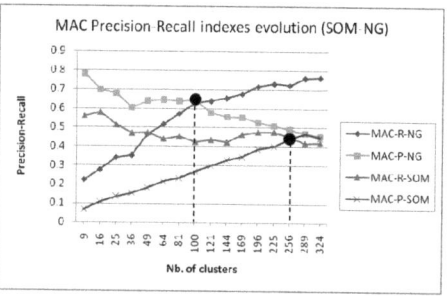

Fig. 4. Inertia (INTER, INTRA) (4A) and Macro-Recall/Precision (MAC-R, MAC-P) (4B) indexes evolution as regards to the number of clusters

clusters which are both coherent and necessarily distant of a small amount of "garbage" clusters can compensate, and even hide, the imprecision of these latter because of the cluster-based averaging process performed by those indexes. In the context of our approach, the detection of degenerated clustering results can although be achieved in two different ways:

The first way is the joint exploitation of the values provided by our *Macro-* and *Micro- Precision* indexes, as it is shown in Figure 5A and Figure 5B. The *Micro-Recall/Precision* indexes have general characteristics similar to the *Macro-Recall/Precision* ones. However, by comparing their values with those of these latter indexes, it becomes possible to identify heterogeneous results of clustering. Indeed, in this last case, the *"Precisions"* of the clusters with small size will not compensate for any more those of the clusters with big size, and the imprecise properties present in the latter, if they prove to be heterogeneous, will have a considerable effect on the *Micro-Precision*. Thus, in the case of NG the differences between the values of *Micro-* and *Macro-Precision* are increasingly more important than in the case of SOM, whatever the considered number of clusters (Figure 2A). It proves that the peculiar properties of the clusters in the partitions generated by NG (or K-means) are largely less precise than those of the clusters produced by SOM. The analysis of the evolution of the *Micro-Precision* curves of the two methods according to the size of the clusters (Figure 5B) permits to clearly highlight that this phenomenon affects more particularly the NG clusters with big size.

A second way to appropriately estimate the quality of clustering results is thus to directly exploit the results provided by the index of *Cumulated Micro-Precision* (CP_m) that focuses on the imprecision of big sized clusters (Eq. 10). In the case of NG, the value of *Cumulated Micro-Precision* remains very low, regardless to the expected number of clusters (Figure 6A). This is mainly due to the influence of the *Micro-Precision* of "garbage" clusters with significant size that can never be split into smaller groups by the method. In a complementary way, whatever the method considered, the index of *Cumulated Micro-Precision* ensures accurate monitoring of the quality depending on the chosen configuration in terms of number of clusters. In the case of SOM, the quality loss occurring for some grid sizes (eg. 244 clusters model, corresponding to a 12x12 grid, in Figure 6A) that induces the formation of large heterogeneous clusters is accurately characterized by highly decreasing values of this index. Figure 6B finally illustrates the interest of correcting the *Cumulated Micro-Precision* index (Eq. 10) by factorizing it with the ratio of non empty clusters for detecting the optimal number of clusters. The curve associated with this corrected index shows a plateau whose starting point (eg. 256 clusters model in Figure 6B) permits to identify the most efficient partition. In this case, such point highlights that no quality progress can be obtained with higher number of clusters. This information is compliant with the one obtained with *Macro-Recall* and *Macro-Precision* indexes (see Figure 4B and associated comments) and thus validates the choice of such point as the characteristic value of the clustering results for a given method.

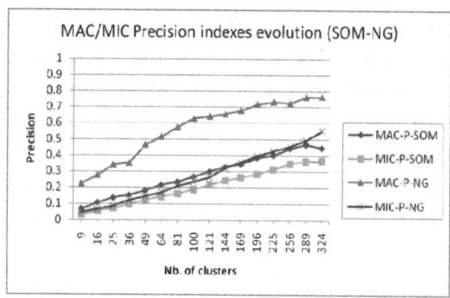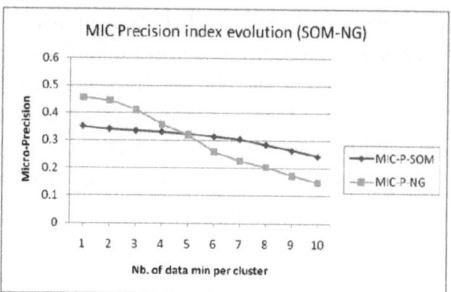

Fig. 5. Evolution of the values of the Micro-Precision (MIC-P) and Macro-Precision (MAC-P) indexes according to the number of clusters (5A) and their size (5B)

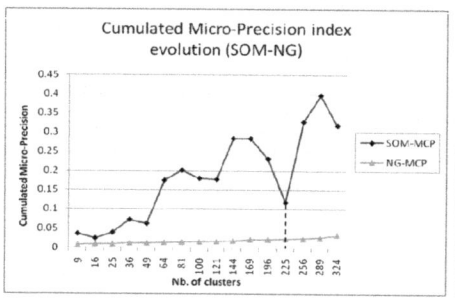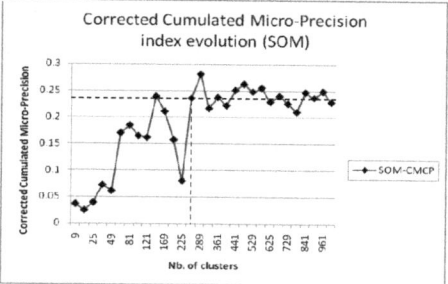

Fig. 6. Evolution of the values of Cumulated Micro-Precision (MCP) (6A) and Corrected Cumulated Micro-Precision (CMCP) (6B) indexes according to the number of clusters

4   Conclusion

We have proposed a new approach for the evaluation of the quality of clustering based on the exploitation of the properties associated with the clusters through the indexes of *Macro-* and *Micro- Recall/Precision* and their extensions. We have shown the advantages of this approach with respect to traditional approaches of evaluation of clustering quality based on distances, at the same time, by justifying its theoretical basis through its relationship with the symbolic classification approaches, and by showing practical results for the optimization of the number of clusters of a given method. Our experiments have been achieved in a realistic context constituted by a complex textual dataset. In such context, we have shown that our new indexes can accurately assess the global quality of a clustering result while giving the additional possibility to distinguish clearly between homogeneous and degenerated clustering results. We have also shown that our approach can apply to the comparison of the results issued from different methods, as well as to the fine-grained analysis of the results provided by a given method, avoiding in both cases to lead to clustering quality misinterpretation.

We have finally shown, through our experiments, the additional capabilities of our approach for synthesizing and labeling the clusters content and we have yet proved their usefulness for a better understanding of the nature of the clustering results. We more specifically tried out our methodology on textual data, but it proves sufficiently general to be naturally applicable on any other type of data, whatever is their nature.

References

1. Attik, M., Al Shehabi, S., Lamirel, J.-.C.: Clustering Quality Measures for Data Samples with Multiple Labels. In: IASTED International Conference on Artificial on Databases and Applications (DBA), Innsbruck, Austria, pp. 50–57 (February 2006)
2. Bock, H.-H.: Probability model and hypothese testing in partitionning cluster analysis. In: Arabie, P., Hubert, L.J., De Soete, G. (eds.) Clustering and Classification, pp. 377–453. World Scientific, Singapore (1996)
3. Davies, D., Bouldin, W.: A cluster separation measure. IEEE Transaction on Pattern Analysis and Machine Intelligence 1, 224–227 (1979)
4. Dempster, A., Laird, N., Rubin, D.: Maximum likelihood for incomplete data via the em algorithm. Journal of the Royal Statistical Society B-39, 1–38 (1977)
5. Diarmuid, Ó.S., Copestake, A.: Semantic classification with distributional kernels. In: Proceedings of COLING 2008, pp. 649–656 (2008)
6. Dunn, J.: Well Separated clusters and optimal fuzzy partitions. Journal of Cybernetics 4, 95–104
7. Forest, D.: Application de techniques de forage de textes de nature prédictive et exploratoire à des fins de gestion et danalyse thématique de documents textuels non structurés, PhD Thesis, Quebec University, Montreal, Canada (2007)
8. Ghribi, M., Cuxac, P., Lamirel, J.-C., Lelu, A.: Mesures de qualité de clustering de documents: Prise en compte de la distribution des mots-clés. In: Atelier EvalECD 2010, Hamamet, Tunisie (January 2010)
9. Gordon, A.D.: External validation in cluster analysis. Bulletin of the International Statistical Institute 51(2), 353–356 (1997); Response to comments. Bulletin of the International Statistical Institute 51(3), 414–415 (1998)
10. Halkidi, M., Batistakis, Y., Vazirgiannis, M.: On clustering validation techniques. Journal of Intelligent Information Systems 17(2/3), 147–155 (2001)
11. Kassab, R., Lamirel, J.-C.: Feature Based Cluster Validation for High Dimensional Data. In: IASTED International Conference on Artificial Intelligence and Applications (AIA), Innsbruck, Austria, pp. 97–103 (February 2008)
12. Kohonen, T.: Self-organized formation of topologically correct feature maps. Biological Cybernetics 43, 56–59 (1982)
13. Lamirel, J.-C., Al-Shehabi, S., Francois, C., Hofmann, M.: New classification quality estimators for analysis of documentary information: application to patent analysis and web mapping. Scientometrics 60, 445–562 (2004)
14. Lamirel, J.-C., Attik, M.: Novel labeling strategies for hierarchical representation of multidimensional data analysis results. In: IASTED International Conference on Artificial Intelligence and Applications (AIA), Innsbruck, Austria (February 2008)
15. Lebart, L., Morineau, A., Fenelon, J.P.: Traitement des données statistiques, Dunod, Paris (1979)

16. MacQueen, J.: Some methods of classification and analysis of multivariate observations. In: Proc. 5th Berkeley Symposium in Mathematics, Statistics and Probability, vol. 1, pp. 281–297. Univ. of California, Berkeley (1967)
17. Martinetz, T., Schulten, K.: A neural gas network learns topologies. Artificial Neural Networks, 397–402 (1991)
18. Milligan, G.W., Cooper, M.C.: An Examination of Procedures for Determining the Number of Clusters in a Data Set. Psychometrika 50, 159–179
19. Rousseeuw, P.J.: Silhouettes: a graphical aid to the interpretation and validation of cluster analysis. Journal of Computational and Applied Mathematics 20, 53–65
20. Salton, G.: The SMART Retrieval System: Experiments in Automatic Document Processing. Prentice Hall Inc., Englewood Cliffs (1971)
21. Van Rijsbergen, C.J.: Information Retrieval. Butterworths, London (1979)

A Structure Preserving Flat Data Format Representation for Tree-Structured Data

Fedja Hadzic

DEBII, Curtin University, Perth, Australia
fedja.hadzic@curtin.edu.au

Abstract. Mining of semi-structured data such as XML is a popular research topic due to many useful applications. The initial work focused mainly on values associated with tags, while most of recent developments focus on discovering association rules among tree structured data objects to preserve the structural information. Other data mining techniques have had limited use in tree-structured data analysis as they were mainly designed to process flat data format with no need to capture the structural properties of data objects. This paper proposes a novel structure-preserving way for representing tree-structured document instances as records in a standard flat data structure to enable applicability of a wider range of data analysis techniques. The experiments using synthetic and real world data demonstrate the effectiveness of the proposed approach.

Keywords: XML mining, tree mining, decision tree learning from XML data.

1   Introduction

Semi-structured documents such as XML possess a hierarchical document structure, where an element may contain further embedded elements, and each element can be attached with a number of attributes. It is therefore frequently modeled using a rooted ordered labeled tree. Many frequent subtree mining algorithms have been developed that driven by different application aims mine different subtree types [1-7]. Even though they are well scalable for large datasets, if the instances in the tree database are characterized by complex tree structures the approach will fail due to enormous number of candidate subtrees that need to be enumerated, as was analyzed mathematically in [5]. For an overview of the current state-of-the-art in the field of tree-structured data mining, please refer to [3, 8, 9].

There has been limited work in classification methods in tree-structured data. Some initial work was mainly based on tree-structured association rules and queries [10, 11], while very recently work presented in [12] employs a mathematical programming method to directly mine discriminative patterns as numerical feature. The XRules approach [13] is a rule based classification system based on the frequent subtree patterns discovered by the TreeMiner algorithm [6], and hence the structural information is taken into account during model learning. However, frequent subtree

L. Cao et al. (Eds.): PAKDD 2011 Workshops, LNAI 7104, pp. 221–233, 2012.
© Springer-Verlag Berlin Heidelberg 2012

patterns can be very large in number, many of which may not be useful for the classification task at hand, and one often needs to filter out many of irrelevant/uninteresting patterns using a variety of statistical and heuristic measures. Several methods using neural networks and kernel methods for processing tree-structured data have been proposed [14], where the approaches are typically tailored towards the use of a particular machine learning method. In this work, an alternative approach is taken with the focus on the data conversion process not tailored to any specific method, yet enabling the application of available methods for data in tabular form in general.

A unique and effective way of representing tree-structured data into a flat data structure format is proposed. The structural information is preserved in form of additional attributes, but the results can differ to those obtained using frequent subtree mining based approaches. A discussion is therefore provided regarding the scenarios in which the proposed approach can be particularly useful and can extract knowledge when other frequent subtree based approaches would fail due to inherent complexity. This is supported with an example case in which all closed/maximal subtrees can be obtained using a closed/maximal itemset mining method on the proposed representation, which as experimentally shown can significantly improve the performance. In the remainder of the experiments, the focus is on the application of a decision tree learning method, to evaluate its usefulness when applied on a synthetic and real-world tree-database converted into the proposed format. The preliminary results indicate the great potential in enabling the application of well established and theoretically/practically proven data analysis techniques that otherwise could not be applied to tree-structured data directly. Hence, many of general data mining techniques as well as quality and interestingness measures can be directly applied to tree-structured data by adopting the conversion approach described in this paper. This will indirectly improve the quality and diversity of knowledge models that can be discovered from tree-structured data.

2   Background of the Problem

A graph consists of a set of nodes (or vertices) that are connected by edges. Each edge has two nodes associated with it. A path is defined as a finite sequence of edges. A rooted tree has its top-most node defined as the root that has no incoming edges. In a tree there is a single unique path between any two nodes. A node u is said to be a parent of node v, if there is a directed edge from u to v. Node v is then said to be a child of node u. Nodes with the same parent are called siblings. The *fan-out/degree* of a node is the number of children of that node. The *level/depth* of a node is the length of the path from root to that node. The *Height* of a tree is the greatest level of its nodes. A *rooted ordered labelled tree* can be denoted as $T = (v_0, V, L, E)$, where (1) $v_0 \in V$ is the root vertex; (2) V is the set of vertices or nodes; (3) L is a labelling function that assigns a label $L(v)$ to every vertex $v \in V$; (4) $E = \{(v_1, v_2) | v_1, v_2 \in V \text{ AND } v_1 \neq v_2\}$ is the set of edges in the tree, and (4) for each internal node the children are ordered form left to right. The problem of *frequent subtree mining* can be generally stated as:

given a database of trees *Tdb* and minimum support threshold (σ), find all subtrees that occur at least σ times in *Tdb* [3, 9]. To reduce the number of candidate subtrees that need to be enumerated, some work has shifted toward the mining of maximal and closed subtrees [7]. A closed subtree is a subtree for which none of its proper supertrees has the same support, while for a maximal subtrees, no supertrees exist that are frequent. For formal definitions of frequent subtree mining related concepts and overview of the tree mining field, the interested reader is referred to [3, 8, 9].

The common way of representing trees is first discussed, to lay the necessary ground for explaining our tree-structured data to flat data format conversion process. The work described in this paper will be utilizing the pre-order (depth-first) string encoding (φ) as described in [6]. In the remainder of the paper the pre-order string encoding will be referred to simply as string encoding. A *pre-order traversal* can be defined as follows: If ordered tree *T* consists only of a root node *r*, then *r* is the pre-order traversal of *T*. Otherwise let $T_1, T_2, ..., T_n$ be the subtrees occurring at *r* from left to right in *T*. The pre-order traversal begins by visiting *r* and then traversing all the remaining subtrees in pre-order starting from T_1 and finishing with T_n. The pre-order string encoding [6] can be generated by adding vertex labels in a pre-order traversal of a tree $T = (v_0, V, L, E)$, and appending a backtrack symbol (for example '-1', '-1'$\notin L$) whenever we backtrack from a child node to its parent node. Fig. 1 shows an example tree database *Tdb* consisting of 6 tree instances (or transactions) $(T_0, ..., T_5)$ and the pre-order string encoding is shown below each tree (i.e. $\varphi(T_1) = $ 'k r -1 m -1').

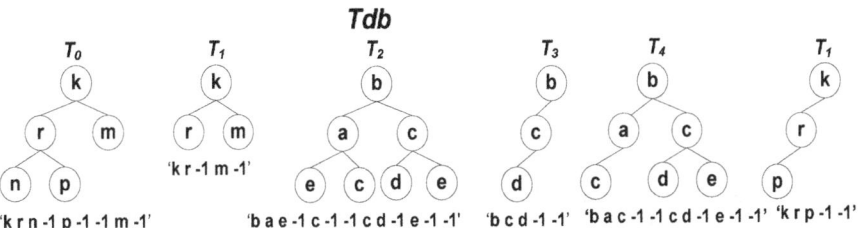

Fig. 1. Example of a tree-structured database consisting of 6 transactions

3   Proposed Tree-Structured to Flat Data Conversion

This section describes our technique to convert the string like presentation commonly used by frequent subtree mining algorithms [6] into a flat data structure format (henceforth referred simply as table) so that both structural and attribute-value information is preserved. The first row of a (relational) table consists of attribute names, which in a tree database are scattered through independent tree instances (transactions). One way to overcome this problem is to first assume a structure according to which all the instances/transactions are organized. Each of the transactions in a tree-structured document should be a valid subtree of this assumed structure, referred to as the *database structure model (DSM)*. This *DSM* will become the first row of the table, and while it does not contain the attribute names, it contains the most general structure where every instance from the tree database can be matched to. The *DSM* needs to ensure that when the labels of a particular transaction from the tree database are processed, they are placed in the correct column, corresponding to the position in the *DSM* where this label was matched to. Hence, the labels (attribute names) of this *DSM* will correspond to pre-order positions of the nodes of the *DSM* and sequential position of the backtrack ('-1') symbols from its string encoding.

The process of extracting a *DSM* from a tree database consists of traversing the tree database and expanding the current *DSM* as necessary so that every tree instance can be matched against *DSM*. Let the tree database consisting of n transactions be denoted as $Tdb = \{tid_0, tid_1, ..., tid_{n-1}\}$, and let the string encoding of the tree instance at transaction tid_i be denoted as $\varphi(tid_i)$. Further, let $|\varphi(tid_i)|$ denote the number of elements in $\varphi(tid_i)$, and $\varphi(tid_i)_k$ ($k = \{0, 1, ..., |\varphi(tid_i)|-1\}$) denote the k_{th} element (a label or a backtrack '-1') of $\varphi(tid_i)$. The same notation for the string encoding of the (current) *DSM* is used, i.e. $\varphi(DSM)$. However, rather than storing the actual labels in $\varphi(DSM)$, 'x' is always stored to represent a node in general, and 'b' to represent a backtrack ('-1'). The process of extracting the *DSM* from *Tdb* can be explained by the pseudo code below.

Database Structure Model Extraction from a Tree database *Tdb*
Input: *Tdb*
Output: *DSM*
inputNodeLevel = 0 // current level of $\varphi(tid_i)_k$
DSMNodeLevel = 0 // current level of $\varphi(T(h_{max}, d_{max}))_k$
$\varphi(DSM) = \varphi(tid_0)$ // set default DSM (use 'x' instead of labels)
for *i* = 1 to *n* − 1 *// n = |Tdb|*
  for each *$\varphi(tid_i)_k$ in $\varphi(tid_i)$*
    for each *p* = 0 to (|$\varphi(DSM)$|-1)
      if *$\varphi(tid_i)_k$ = -1* **then** *inputNodeLevel−−* **else** *inputNodeLevel++*
      if *$\varphi(DSM)_p$='b'* **then** *DSMNodeLevel−−* **else** *DSMNodeLevel++*
      if *inputNodeLevel ≠ DSMNodeLevel*
        if *$\varphi(tid_i)_k$ = -1* **then**
        while *inputNodeLevel ≠ DSMNodeLevel*
        p++
        if *$\varphi(DSM)_p$ = -1* **then** *DSMNodeLevel−−* **else** *DSMNodeLevel++*
        endwhile
      else
        while *inputNodeLevel ≠ DSMNodeLevel*
        if *$\varphi(tid_i)_k$ ≠ -1* **then** *$\varphi(DSM)_{p+1}$ = 'x'* **else** *$\varphi(DSM)_{p+1}$ = 'b'*
        k++
        p++
        if *$\varphi(tid_i)_k$ = -1* **then** *inputNodeLevel−−* **else** *inputNodeLevel++*
        endwhile
    endfor
   endfor
  endfor
return *DSM*

To illustrate the complete conversion process using *DSM* please refer back to Fig. 1. Using the string encoding format representation [6], the tree database *Tdb* from Fig. 1 would be represented as is shown in Table 1, where the left column corresponds to the transaction identifiers, and the right column is the string encoding of each subtree.

Table 1. Representation of *Tdb* (Fig. 1) in string encoding representation

	Tdb
0	k r n -1 p -1 -1 m -1
1	k r -1 m -1
2	b a e -1 c -1 -1 c d -1 e -1 -1
3	b c d -1 -1
4	b a c -1 -1 c d -1 e -1 -1
5	k r p -1 -1

In this example the *DSM* is reflected in the structure of T_2 in Fig. 1 and it becomes first row in Table 2 to reflect the attribute names as explained before. The string encoding is used to represent the *DSM* and since the order of the nodes (and backtracks ('-1')) is important the nodes and backtracks are labeled sequentially according to their occurrence in the string encoding. For nodes (labels in the string encoding), x_i is used as the attribute name, where i corresponds to the pre-order position of the node in the tree, while for backtracks, b_j is used as the attribute name, where j corresponds to the backtrack number in the string encoding. Hence, from our running example in Fig. 1 and Table 1, $\varphi(DSM) = $ '$x_0\ x_1\ x_2\ b_0\ x_3\ b_1\ b_2\ x_4\ x_5\ b_3\ x_6\ b_4\ b_5$'. To fill in the remaining rows every transaction from *Tdb* is scanned and when a label is encountered it is placed to the matching column (i.e. under the matching node (x_i) in the *DSM* structure), and when a backtrack ('-1') is encountered, a value '1' (or 'y') is placed to the matching column (i.e. matching backtrack (b_j) in *DSM*). The remaining entries are assigned values of '0' (or 'no', indicating non existence). The flat data format of *Tdb* from Table 1 (and Fig. 1) is illustrated in Table 2.

Table 2. Flat representation of *Tdb* in Fig.1 and Table 1

x_0	x_1	x_2	b_0	x_3	b_1	b_2	x_4	x_5	b_3	x_6	b_4	b_5
k	r	n	1	p	1	1	m	0	0	0	0	1
k	r	0	0	0	0	1	m	0	0	0	0	1
b	a	e	1	c	1	1	c	d	1	e	1	1
b	c	d	1	0	0	1	0	0	0	0	0	0
b	a	c	1	0	0	1	c	d	1	e	1	1
k	r	p	1	0	0	1	0	0	0	0	0	0

Implications of the Proposed Approach. The database structure model *DSM* governs what is to be considered as a valid instance of a particular tree characteristic, and the exact positions of the node within the *DSM* are taken into account. Hence if two subtrees have the same labels and are structurally the same, but their nodes have been matched against different nodes in *DSM* (i.e. they occur at different positions) they would be considered as different characteristic instances. Consider the subtree with encoding 'b c d' from the Tdb in Fig. 1. Within the current tree mining framework this subtree would be considered to occur in 3 transactions T_2, T_3 and T_4, while using the proposed approach the occurrence of 'b c d' in T_3 would be considered different than in T_2 and T_4, because nodes 'c' and 'd' occur at different (pre-order) positions, and hence the instance is represented differently in the table as can be seen in Table 2. This illustrates the key difference, and it is caused by the fact that not all the instances of a tree database may follow the same order or will have all the elements of the document input structure (e.g. XML schema) available. However, if all the instances in a tree database always follow the same structure and node layout as the input document structure, then there would be no such difference in the method. For example the proposed method would be more applicable in cases when there is an XML Schema available where each XML instance conforms to this schema and contains all the elements in the same order. This is demonstrated with an experiment at the end of this section. However, it is first worth mentioning that, in some applications, it may be the case that the different occurrence of a pattern in the document tree indicates that it is used in different context.

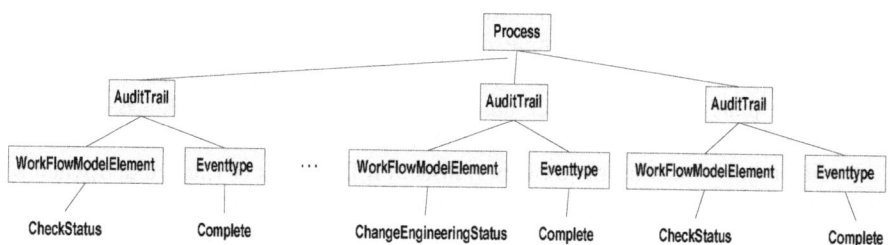

Fig. 2. An XML fragment from an instance of process log data

As an example, please consider Fig. 2 which shows a fragment of a process instance from a process log dataset obtained from http://prom.win.tue.nl/tools/prom. Let us say, that the user wants to find out whether the action 'CheckStatus' was performed at a particular stage in the workflow. As can be seen in Fig. 2, this action occurs multiple times within a process instance, and it needs to be distinguished based on its occurrence within the workflow model. In cases when the users are investigating the conformance of business process instances to a business process model, it may be important to know when an action or sets of actions occur in a phase of the workflow other than expected by the model, and hence the differing characteristic of the proposed method may be desired. Furthermore, the subtree patterns extracted would be more informative as they keep the exact positions of the nodes in the subtree with respect to the *DSM* (which can be seen as the general workflow model).

Another difficulty encountered in real world applications of frequent subtree mining, is related to the complexity issues caused by a number of attributes that are likely present in every instance or transaction. To provide an illustrative example, consider an application of credit risk assessment in banks for providing loans to small to medium enterprises (SMEs). Qualitative data on loan applications can usually be found from text databases while the quantitative information is usually stored in relational format. We have proposed our industry partner (bank) the use of XML to capture domain specific terms and effectively organize the available quantitative and qualitative information. The XML template was produced based on a small number of textual document instances provided, as was described in [15]. A simplified portion of the XML template is shown in Fig. 3 to show some of the information stored and the possible sets of values.

```
<XML>
<creditapplication>
<creditriskindicators>
<personalinfo>
  <industryrisk>very low,low,moderate,high,very high</industryrisk>
   <creditinformation>
       <principal>up to IDR 50 million,more than IDR 50 million up to IDR 100
       million</principal>
       <duration>up to 1 year,more than 1 year to 2 years,more than 2 years to 3 years,more
       than 3 years to 4 years,more than 4 years to 5 years</duration>
   </creditinformation>
  </personalinfo>
 ...
<customerprofile>
      ...
        <integrity>high, medium, low</integrity>
   <capacity>
       <salesperyear>below IDR 300 million, above IDR 300 million to IDR 2.5 billion,above
       IDR 2.5 billion to IDR 50 billion</salesperyear>
       <eatpermonth>less than 0.2 of sales, between 0.2 to 0.4 of sales,more than 0.4 of
       sales</eatpermonth>
   </capacity>
</customerprofile>
<financialreport>
<currentratio>below standard, above standard</currentratio>
 ....
<salesgrowthratio>below standard, above standard</currentratio>
</financialreport>
</creditriskindicators>
<creditassesment>accept,accept with considerations,reject</creditassessment>
 </creditapplication>
</XML>
```

Fig. 3. XML Template (part of) for credit risk assessment of SMEs

There are a total of 53 tags (attributes) which store additional and more specific information (eg, industry, sub-industry, etc.). At this stage we could not obtain sufficient amount of document instances from the bank as the rejected applications were not kept. We have conducted several interviews with the loan officers form the bank in order to obtain some domain knowledge in form of rules that reflect realistic credit parameters and the corresponding assessment decision. This resulted in obtaining 63 rules that can be applied to reflect a realistic assessment decision for the given criteria. A total of 100,000 XML instances were created based upon the pre-defined XML template, with random values for credit parameters. The 63 rules provided by domain experts were then incorporated on top of those instances to modify the class value realistically. We have removed each instance whose criteria values could not be found in a rule, which reduced the dataset to 53,081 instances. These rules were general with respect to the financial ratios, indicating that a ratio is above or below a standard for a particular industry/sub-industry. However, real ratios were used for the corresponding industry/sub-industry, giving a total of 41 possible variations where a general rule can apply. While on average a single rule would have modified 800 instances, the rule itself will not be present in 800 instances because specific ratios apply. Hence, low support thresholds would need to be used, but this has shown to be impractical because of the large amount of subtree patterns generated. This is caused by 13 nodes that are present in each instance and are mainly there to contextualize the information (e.g. creditriskindicators, customerprofile). As these nodes are present in patterns for any given support, combinatorial complexity problems occur at lower support thresholds. When running the IMB3-miner [4] algorithm for mining ordered induced subtrees, because of the just mentioned issues, enumerating all subtrees that occurred in less than half of the database was infeasible.

In this dataset the information in every instance is organized according to the pre-defined XML template and hence the key difference of mining the converted tree database in flat format will not occur. Due to the subset exclusion properties of closed and maximal pattern one would expect that the number of closed/maximal itemsets should be the same as the number of closed/maximal subtrees for a given support. To confirm this the results of the CMTreeMiner [7] algorithm that extracts closed and maximal ordered induced subtrees was compared with the results of Charm [16] and GenMax [17], that extract closed and maximal itemsets, respectively.

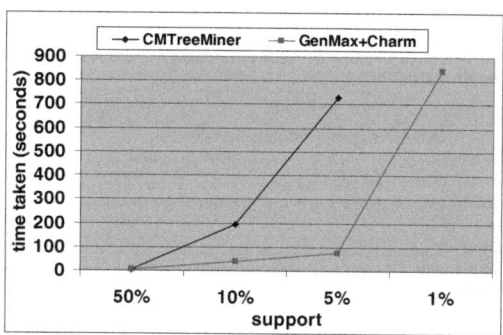

Fig. 4. Time performance of closed/maximal subtree and itemset mining

The results from Charm and GenMax required some post-processing, as nodes are not ordered and every backtrack ('-1) is present in the maximal/closed itemset. However, since the patterns contain attribute names, any node can be matched directly to the database structure model (*DSM*), and the right structural organization of the nodes worked out. The process consists of sequentially listing the values of each matched node in *DSM*, while keeping the level information of each current node in *DSM* and in the subtree pattern. This enables us to work out the exact number of edges between each parent-child node pair and store the right number of backtracks in the right places in the thus far generated subtree encoding. Since the *DSM* itself is ordered according to the pre-order traversal, this results in pre-order string encodings of the subtrees, giving the same results as CMTreeMiner. Fig. 4 shows the time performance comparison between CMTreeMiner and the sum of time taken by GenMax and Charm algorithms for varying support thresholds. When support threshold was set to 1% CMTreeMiner was taking over two hours and the run was terminated. As expected, enumerating frequent closed/maximal itemsets is faster than enumerating frequent closed/maximal subtrees, because in the latter the structural validity needs to be ensured during the candidate generation phase. In fact further performance gain could be achieved by removing all the backtrack attributes from the sets of items, as they are worked out based upon the occurrence of nodes in *DSM*. Hence, in scenarios where the schema is available, and all nodes are unique and the instances are organized in the same order, one can mine for maximal/closed itemsets from the converted flat representation to obtain patterns when approaches for mining closed/maximal subtrees would fail due to inherent structural complexity. Little post-processing is required but as experimentally shown the performance gain can be large. On the other hand, applying a C4.5 decision tree learning algorithm [18] on the proposed flat representation of the data could directly discover all class distinguishing rules. The flat data representation consisted of 107 attributes (54 nodes (including class with 3 possible values) and 53 backtracks) and the learned C4.5 decision tree had 100% accuracy (10-fold cross-validation), 804 nodes and 747 leaf nodes (rules).

4   Experimental Evaluation

The purpose of the experiments provided in this section is to demonstrate how using the proposed tree database transformation one can discover interesting knowledge using other data mining techniques than association rule mining. The C4.5 decision tree learning algorithm [18] is taken as case in point and Weka software [19] is used.

Synthetic Data. In this experiment several synthetic datasets are considered, that were specifically designed for the problem of classifying each tree instances according to its structural properties. Hence, the attribute values are not what can be used to classify each instance correctly, but rather it is the structural characteristics of each instance that determines its correct class. Hence, the used datasets and the classification problem mainly serves the purpose of evaluating the capability of the proposed representation approach to capture the structural properties of each tree instance and take them into account for classification.

The first dataset was created as follows. Each tree instance was accompanied with a class attribute, which indicates the maximum height and degree of that particular tree instance (transaction). For example, if the maximum height is 2, and the maximum degree is 3, then the class label will be 23_val. For the first test, the aim was to generate a database that contains every possible instance of a tree for a given maximum height of 2 and degree 4. One tree with height 2 was created, where all nodes have degree 4 and have the same label ('2'), from which all possible induced subtrees were enumerated. This will ensure that every instance of the class is present and there were a total of 9 classes (note that '00_val' is used for a tree consisting of a single node).

The database structure model (*DSM*) is therefore a tree with height 2 where all nodes have degree 4, as all class instances can be matched against this tree. Each attribute name is flagged on the node/edge of the *DSM* tree structure displayed in Fig. 5. Since the same label was chosen for all nodes, the node attributes will have the same predictive capability as the backtrack attributes. This would usually not be the case if the variety of labels are used, as it is through the backtrack attributes that the structural properties are preserved. Fig. 6 shows the learned C4.5 decision tree for this dataset. Two tests were performed so that in the second test (right side of Fig. 6) all of the attributes of node type ('x_i') were removed from learning, as in reality the node labels will be different and should not have predictive capability in this classification task. Model building took 0.03 seconds, and the 10-fold cross-validation test option was used and both decision trees had accuracy of 99.23%. Please note that in this test the class value distribution was very uneven, because of the way that the dataset was generated, as for higher degree the number of possible unique instances of a class is much larger. There were 695 instances of class 24_val (i.e. max height = 2 and max degree = 4), 71 of 23_val, 9 of 22_val, while the remaining 6 classes had only one representative instance. The test was repeated when the number of classes was made the same, and there was an increase in accuracy to 99.85%, but the decision tree was much larger, i.e. size 53 and 27 leaves. By comparing Figures 5 and 6 one can verify the extracted rules.

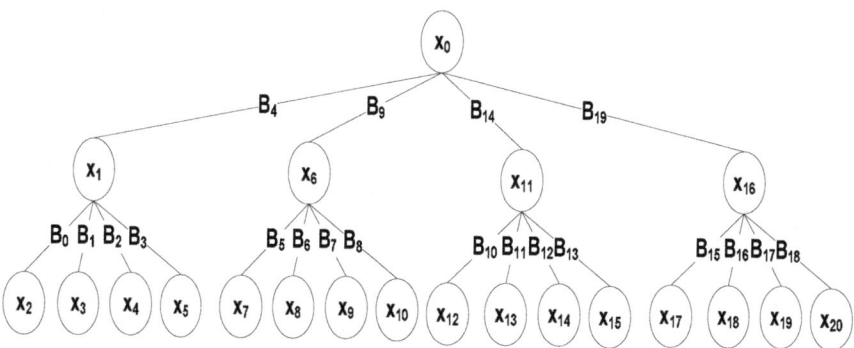

Fig. 5. Illustrating the attribute names for synthetic database

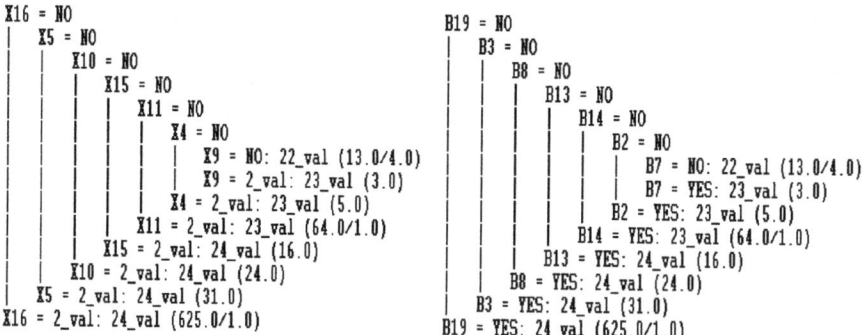

Fig. 6. C4.5 decision trees for synthetic dataset

The second experiment performed for this classification task has used the TreeGenerator software [6] to generate synthetic database of 50,000 records. The maximum height and degree were set to 5, giving us a total of 26 unique class values. Hence in this case each node was assigned a random label and it is not guaranteed that every instance of a particular class will be generated. The C4.5 decision tree learned form this data had 99.748% accuracy, it consisted of 1278 nodes and 1153 leaves and it took 34.57 seconds to obtain. This complex tree could be partly because the value for option n (number of items) was set to 1000. By analyzing the decision tree note that while some backtracks ('b_i') were used for instance splitting during decision tree construction, many constraints on labels for nodes ('x_i') occur at the leaf nodes. This indicates that particular labels happen to be present/absent at a particular node for many instances of a particular class. The experiment was repeated on a dataset generated with n option in TreeGenerator [6] set to 1 (i.e. all nodes share 1 label). The resulting decision tree reduced to only have 299 nodes and 150 leaves, with the classification accuracy of 99.786%, and the time taken was 24.88 seconds. These results demonstrate that the structural and attribute/value information is preserved and taken into account for the classification task at hand.

Web Log Data. This experiment set utilizes the publically available real-world server log data (US1924, US2430, US304) previously used in [13] to evaluate their XRules method. These datasets have a variety of tree structures, and each dataset contains a number of very deep and wide trees at the end. This kind of property is undesired in the proposed approach as the database structure model (*DSM*) from the database will be fairly large and many of the nodes and backtracks will be there because of a small number of tree instances that were out of the norm with respect to the general structure. Hence, two test sets are performed, one in which the dataset as originally found (e.g. US1924 in Table 3) is processed and one with only the first 7000 records as these do not contain those rare large instances (e.g. 7KUS1924 in Table 3). The accuracy (10-fold cross-validation) and size of each C4.5 decision tree learned from this data is shown in Table 3. This is a challenging classification problem for the proposed method. The dataset has a variety of small and large trees. The larger trees do not occur as frequently, but they still need to be captured by the *DSM*. This makes it much larger than necessary to capture the majority of trees in the database (hence the long running time when compared to 7K variants in Table 3). However, the results are comparable to the results of XRules presented in [13].

Table 3. C4.5 results for the Web log datasets

Datasets	US1924	7KUS1924	US2430	7KUS2430	US304	7KUS304
Accuracy(%)	76.68	84.4	76.75	81.91	83.18	77.29
Size	13222	12079	14882	16405	21460	20813
Leaves #	13195	12060	14862	16394	21413	20781
Time(seconds)	7044.16	53.14	4743.16	45.41	6825.56	37.66

5   Conclusion and Future Work

This paper has presented a unique and effective way of representing tree structured databases such as XML documents into a flat data format for which many more available data mining/analysis methods exist in comparison to available methods for tree-structured data format. The preliminary experimental results have demonstrated that the structural and attribute/value information is preserved in the proposed flat data representation. A standard decision tree learning algorithm performed very well in classifying tree instances according to their structural properties. Regarding, the real world dataset not particularly suitable for the proposed presentation, the decision tree learning algorithm still achieved comparable classification accuracy to previously reported results based on a frequent subtree mining approach. More studies will be performed in future and the necessary extensions for the approach to perform well in a wider range of cases. Generally speaking, enabling effective representation of all information from a tree-structured database in a flat data format, will give the opportunity to apply and evaluate well established and theoretically proven data analysis techniques that could not process tree-structured data directly.

References

1. Abe, K., Kawasoe, S., Asai, T., Arimura, H., Arikawa, S.: Optimized Substructure Discovery for Semistructured Data. In: Elomaa, T., Mannila, H., Toivonen, H. (eds.) PKDD 2002. LNCS (LNAI), vol. 2431, p. 1. Springer, Heidelberg (2002)
2. Tatikonda, S., Parthasarathy, S., Kurc, T.: TRIPS and TIDES: new algorithms for tree mining. In: ACM CIKM 2006, Arlington, Virginia, USA (2006)
3. Chi, Y., Nijssen, S., Muntz, R.R., Kok, J.N.: Frequent Subtree Mining - An Overview. Fundamenta Informaticae, Special Issue on Graph and Tree Mining 66, 1–2 (2005)
4. Tan, H., Dillon, T.S., Hadzic, F., Feng, L., Chang, E.: IMB3-Miner: Mining Induced/Embedded Subtrees by Constraining the Level of Embedding. In: 10th Pacific-Asia Conf. on Knowledge Discovery and Data Mining, Singapore, pp. 450–461 (2006)
5. Tan, H., Hadzic, F., Dillon, T.S., Feng, L., Chang, E.: Tree Model Guided Candidate Generation for Mining Frequent Subtrees from XML. ACM Transactions on Knowledge Discovery from Data (TKDD) 2(2) (2008)
6. Zaki, M.J.: Efficiently mining frequent trees in a forest: algorithms and applications. IEEE Transaction on Knowledge and Data Engineering 17(8), 1021–1035 (2005)
7. Chi, Y., Yang, Y., Xia, Y., Muntz, R.R.: CMTreeMiner: Mining Both Closed and Maximal Frequent Subtrees. In: Dai, H., Srikant, R., Zhang, C. (eds.) PAKDD 2004. LNCS (LNAI), vol. 3056, pp. 63–73. Springer, Heidelberg (2004)
8. Tan, H., Hadzic, F., Dillon, T.S., Chang, E.: State of the art of data mining of tree structured information. Int'l Journal Computer Systems Science and Eng. 23(2) (March 2008)
9. Hadzic, F., Tan, H., Dillon, T.S.: Mining of Data with Complex Structures. SCI, vol. 333. Springer, Heidelberg (2011)
10. Chen, L., Bhowmick, S.S., Chia, L.-T.: Mining Association Rules from Structural Deltas of Historical XML Documents. In: Dai, H., Srikant, R., Zhang, C. (eds.) PAKDD 2004. LNCS (LNAI), vol. 3056, pp. 452–457. Springer, Heidelberg (2004)

11. Braga, D., Campi, A., Ceri, S., Klemettinen, M., Lanzi, P.: Discovering interesting information in XML data with association rules. In: ACM Symposium on Applied Computing, Melbourne, Florida, pp. 450–454 (2003)
12. Kim, H., Kim, S., Weninger, T., Han, J., Abdelzaher, T.: DPMine: Efficiently Mining Discriminative Numerical Features for Pattern-Based Classification. In: Balcázar, J.L., Bonchi, F., Gionis, A., Sebag, M. (eds.) ECML PKDD 2010. LNCS, vol. 6322, pp. 35–50. Springer, Heidelberg (2010)
13. Zaki, M.J., Aggarwal, C.C.: XRules: An Effective Structural Classifier for XML Data. In: SIGKDD 2003, Washington DC, USA (2003)
14. Da San Martino, G., Sperduti, A.: Mining Structured Data. IEEE Computational Intelligence Magazine 5(1) (2010)
15. Ikasari, N., Hadzic, F., Dillon, T.S.: Incorporating Qualitative Information for Credit Risk Assessment through Frequent Subtree Mining for XML. In: Tagarelli, A. (ed.) XML Data Mining: Models, Method, and Applications. IGI Global (2011)
16. Zaki, M.J., Hsiao, C.-J.: CHARM: An Efficient Algorithm for Closed Itemsets Mining. In: 2nd SIAM Int'l Conf. on Data Mining, Arlington, VA, USA, April 11-13 (2002)
17. Gouda, K., Zaki, M.J.: GenMax: An Efficient Algorithm for Mining Maximal Frequent Itemsets. Data Mining and Knowledge Discovery 11(3), 223–242 (2005)
18. Quinlan, J.R.: C4.5: Programs for Machine Learning. Morgan Kaufmann Publishers, San Mateo (1993)
19. Holmes, G., Donkin, A., Witten, I.H.: Weka: A machine learning workbench. In: 2nd Australia and New Zealand Intelligent Info. Systems Conf. Brisbane, Australia (1994)

A Fusion of Algorithms in Near Duplicate Document Detection[*]

Jun Fan[1,2] and Tiejun Huang[1,2]

[1] National Engineering Laboratory for Video Technology, School of EE & CS,
Peking University, Beijing 100871, China
[2] Peking University Shenzhen Graduate School, Shenzhen 518055, China
{jfan,tjhuang}@jdl.ac.cn

Abstract. With the rapid development of the World Wide Web, there are a huge number of fully or fragmentally duplicated pages in the Internet. Return of these near duplicated results to the users greatly affects user experiences. In the process of deploying digital libraries, the protection of intellectual property and removal of duplicate contents needs to be considered. This paper fuses some "state of the art" algorithms to reach a better performance. We first introduce the three major algorithms (shingling, I-match, simhash) in duplicate document detection and their developments in the following days. We take sequences of words (shingles) as the feature of simhash algorithm. We then import the random lexicons based multi fingerprints generation method into shingling base simhash algorithm and named it shingling based multi fingerprints simhash algorithm. We did some preliminary experiments on the synthetic dataset based on the "China-US Million Book Digital Library Project"[1]. The experiment result proves the efficiency of these algorithms.

Keywords: duplicate document detection, digital library, web pages, near duplicate document.

1 Introduction

Duplicate and near duplicate documents detection plays an important role in both intellectual property protection and information retrieval. The definition of duplicate is unclear. The general notion is that files with minor edits of each other are also considered as duplicates.

The digital libraries provide users with on-line access to digitized news articles, book, and other information. This environment greatly simplifies the task of illegally retransmit or plagiarize the works of others which violates their copyrights.

In recent times, the dramatic development of the World Wide Web has led a proliferation of documents that are identical or almost identical. These copies of documents are same or only differ from each other in a very small portion. The appearances of duplicate and near duplicate documents in the search results annoy the users.

---

[*] The work is partially supported by National Natural Science Foundation of China under grant No. 90820003, the Important Scientific and Technological Engineering Projects of GAPP of China under grant No. GAPP-ZDKJ-BQ/15-6, and CADAL project.
[1] See http://www.ulib.org for more details.

L. Cao et al. (Eds.): PAKDD 2011 Workshops, LNAI 7104, pp. 234–242, 2012.

Brin et al. developed COPS [1] in the course of deploying a digital library system. COPS is a prototype of a document copy detection mechanism and dependents on sentence overlap. The registration based architecture of this prototype is widely used from then on. Shivakumar et al. [2] [3] proposed SCAM, which is based on comparing the word frequency occurrences of documents.

Andrei Broder et al.'s [7] [8] shingling algorithm and Charikar's [8] random projection based approach are considered "state of the art" algorithms for detecting near duplicate web documents. Henzinger [12] compared these two algorithms in a set of 1.6B distinct web pages and proposed a combined algorithm. The combined algorithm got a better precision compared to using the constituent algorithms individually. Another well known duplicate document detection algorithm called I-Match was proposed by Chowdhury et al. [4] and it was evaluated on multiple data collections. Kolcz et al. [5] studied the problem of enhancing the stability of I-Match algorithm with respect to small modifications on document content. They presented a general technique which makes use of multiple lexicons randomization to improve robustness.

This paper reports two attempts to improve the performance of simhash algorithm. In section 2, we described the three major algorithms (shingling [7], I-match [4], simhash [13]) in duplicate document detection and their developments in the following days. In section 3, we introduced our improvement attempts: take sequences of words (shingles) as the feature and fuse the random lexicons based multi fingerprints generation method with simhash. In section 4, we presented the experiments results in the "China-US Million Book Digital Library Project" dataset. Finally, Section 5 brings this paper to a conclusion.

2   Major Algorithms in Duplicate Document Detection

2.1   Shingling, Super Shingling, Mini-wise Independent Permutation Algorithms

Broder et al. [7] defined two concepts: resemblance and containment, to measure the similarity degree of two documents. Documents are represented by a set of shingles (or k-grams). The overlaps of shingle sets were calculated.

As there are too many shingles in a document, Broder et al. [7] [8] developed some sampling methods. Super shingling [7] and mini-wise independent permutation [8] are two kinds of the sampling methods. Super shingling method is shingling the shingles. The document is then represented by its super shingles.   Mini-wise independent permutation algorithm provide an elegant construction of a locality sensitive hashing schema for a collection of subsets with the set similarity measure of Jaccard Coefficient. [8]

2.2   I-Match, Multiple Random Lexicons Based I-Match Algorithms

Chowdhury et al. [4] extract a subset of terms from a document according to their NIDF (the normalized inverse document frequency) [4] values. They hashed the terms orderly and claimed that, if the terms are carefully chosen, near-duplicate documents are likely to have the same hash, whereas it is extremely unlikely that two dissimilar documents will hash to the same value. But the gathering of global collection statistic (NIDF) presents a significant challenge.

As the I-Match algorithm is based on the precondition of filtering out all the different words in near duplicate documents, its recall is relatively low. Kolcz et al. [5] proposed the multiple random lexicons based I-Match algorithm, which utilizes additional randomly created lexicons to generate multiple fingerprints. They claimed that, this method is also applicable to other single-signature schemes to improve recall. In our experiments about this algorithm, we set the parameters the same with theirs. This was also discussed in [5] [6].

2.3   Random Projection, Simhash Algorithms

Charikar [8] developed a locality sensitive hashing schema for a collection of vectors with the cosine similarity measure between two vectors, which is based on random projection of words. Henzinger [12] implemented this schema into the application of duplicate web page detection, and called it random projection.

Manku et al. [13] added the concept of feature weight to random projection, and named it simhash algorithm. Given feature vectors and corresponding feature weight, it generates a simhash fingerprint. The hamming distance of two vectors' simhash fingerprints is proportional to the cosine similarity of the two vectors. When the hamming distance of two simhash fingerprints is smaller than a threshold, the two documents of the two fingerprints are considered as duplicate. In our experiments, we set the size of the simhash fingerprint to 32 bits. And we examined the performance of each simhash based algorithm with a broad range of threshold from 0 to 31 bits.

3   Model Enhancements

In this paper, we did certain degree of fusion based on the characters of each class of algorithm mentioned above.

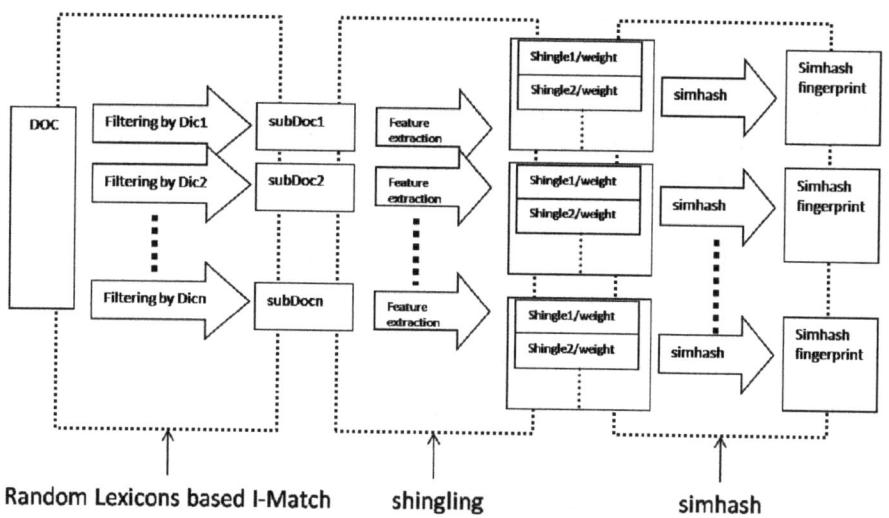

Fig. 1. Framework of our fusion algorithm

3.1   Shingling Based Simhash Algorithm

Henzinger [12] did a comparison between random projection and the shingling algorithm. At the end of that paper, the author proposed a combined algorithm which is in fact sequentially running random projection algorithm after the shingling algorithm to get a better performance. The author also proposed to study the performance of implement random projection algorithm on sequence of tokens, i.e., shingles, instead of individual tokens. Manku et al. [13] also mentioned to study how sensitive the simhash is to changes in features and weights of features as the future work.

Considering the simhash algorithm is independent of feature selection and assignment of weights to features. Intuitively, sequences of words (shingles) are more representative than individual words for a document. We use the k-shingles (word sequences of length k) as the features of the simhash algorithm, and the sum of IDF value of words in a k-shingle as the weight of the corresponding feature. This is shown in feature extraction part in Fig. 1. We named this as shingling based simhash algorithm, and regarded it as a fusion of shingling and simhash algorithms. It is different from simply running the two algorithms sequentially in [12].

3.2   Multiple Random Lexicons Based Simhash Algorithm

As mentioned by Kolcz et al. [5], randomly creating extra lexicons to generate additional fingerprints is applicable to other single-signature algorithm. We introduce this method into simhash algorithm. We filter documents by randomly created lexicons and generate multi simhash fingerprints as shown in Fig. 1. If the hamming distance between two fingerprints of two documents generated by the same extra random lexicon is smaller than the threshold, the two documents are reckoned as duplicate. We named this as multiple random lexicons based simhash algorithm.

We then fusion the two improvements into an integrated algorithm which is named shingling based multi fingerprints simhash algorithm.

4   Experiments

Although duplicate document detection has been studied for a long time and in a broad area, there isn't any widely accepted experiment dataset. Studies on duplicate document detection use their own datasets [4] [5] [6] [7] [8] [12] [13].

In our experiments, we randomly selected 1403 books (all in English) from the "China-US Million Book Digital Library Project". We then divided these books into 143,798 rough 4KB size texts. We propose these texts are unduplicated. We selected 5 texts randomly and modified (insert, delete, replace word) these 5 texts at random locations. We constructed 600 texts (120 texts for each source text) in this way and considered they are near duplicate documents. We calculated the fingerprints of these 144,403 texts, and used the 5 source texts as the queries. We calculated the precisions, recalls of these 5 queries and counted the macro-averages. P_i, R_i are the precision and recall value corresponding to each query. MacroP, MacroR are the macro-averages of precisions and recalls.

$$MacroP = \frac{1}{n}\sum_{i=1}^{n} P_i \qquad (1)$$

$$MacroR = \frac{1}{n}\sum_{i=1}^{n} R_i \qquad (2)$$

The F-measure is calculated as:

$$F\text{-measure} = 2 * MacroP * MacroR / (MacroP + MacroR) \qquad (3)$$

The experiments results we listed blew are all the macro-averages of the 5 queries. In order to clearly distinguish different curves, results of some parameter values aren't listed in the following figures.

Before the implementation of various copy detection algorithms, each document is first passed through a stopword-removal and stemming process, which removes all the stopwords and reduces every word to its stem.

4.1   Shingling Based Simhash Algorithm

As shown in Fig. 2, and the algorithm gets the best performance when shingle size equals 2 (words sequence of size 2). Shingle size of 1 is in fact the original simhash

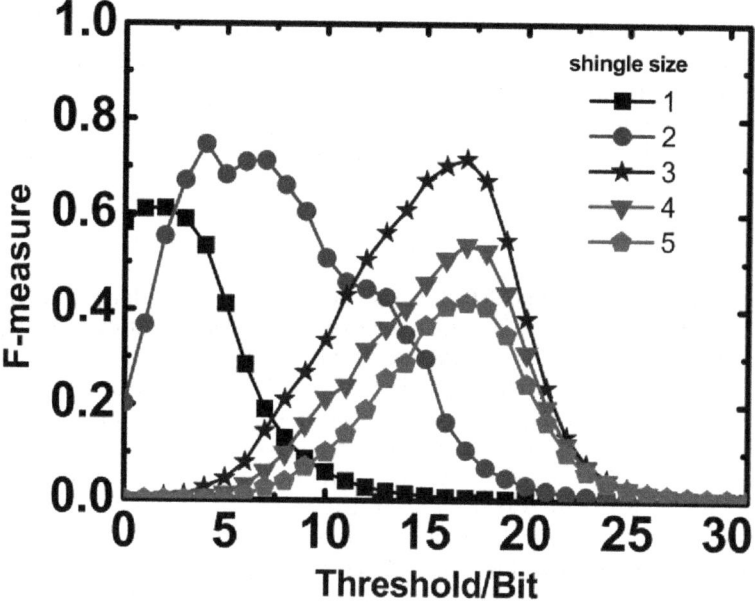

Fig. 2. The F-measure with shingle size range from 1 to 5, and the threshold of hamming distance range from 0 to 31

algorithm. The best F-measure value was improved from 0.6117 to 0.7469 as the shingle size grows from 1 to 2. In shingling based simhash algorithm with shingle size k, if we modified n words in random locations, the affected features range from n to n*k. With the increase of shingle size k, the affected features increase multiplied and the performance decreases. In the other side, if we only select single words as features, there maybe two document with roughly the same words, in different sequences and with different meanings are considered as duplicate. With larger shingle size k, we can reduce this kind of false positive obviously. By taking the k-shingles as the features, the effect of word order was considered. Therefore, there is a tradeoff of shingle size k to keep the balance.

4.2   Multiple Random Lexicons Based Simhash Algorithm

In this experiment, we set the features of simhash to be shingles with size 2. It is in fact the shingling based simhash algorithm with shingle size 2, when the random lexicon size is set to 1. With the increase of the random lexicon size and the threshold, the recall increases, this was showed in Fig. 3. Chowdhury et al. [4] shown the significant increase in recall of multiple random lexicons method, but didn't illustrate the precision in their paper. In Fig. 4, we can see the precision decreases slightly accordingly. The F-measure was shown in Fig. 5. From the F-measure's view, it doesn't mean that the larger the random lexicon size is, the better the performance will be. There exists a balance between precision and recall on the selection of random lexicon size.

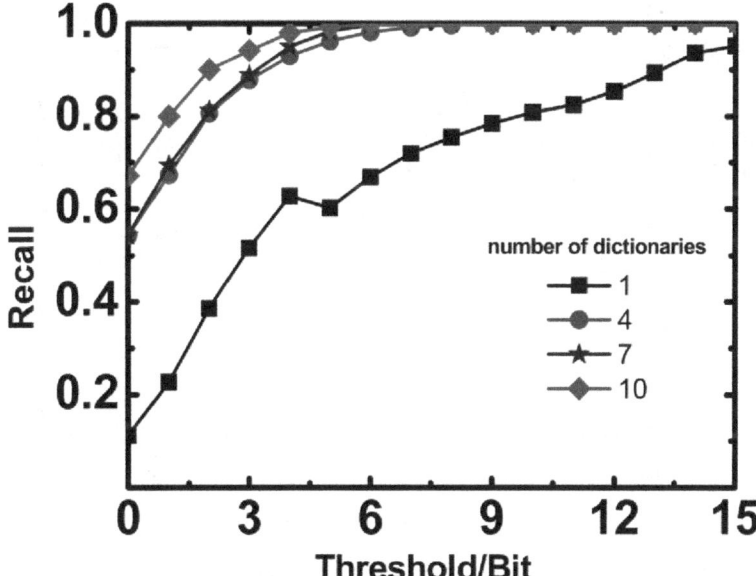

Fig. 3. The recall of the shingling based multi fingerprints simhash algorithm with random lexicon size 1, 4, 7, 10, and the threshold of hamming distance range from 0 to 15

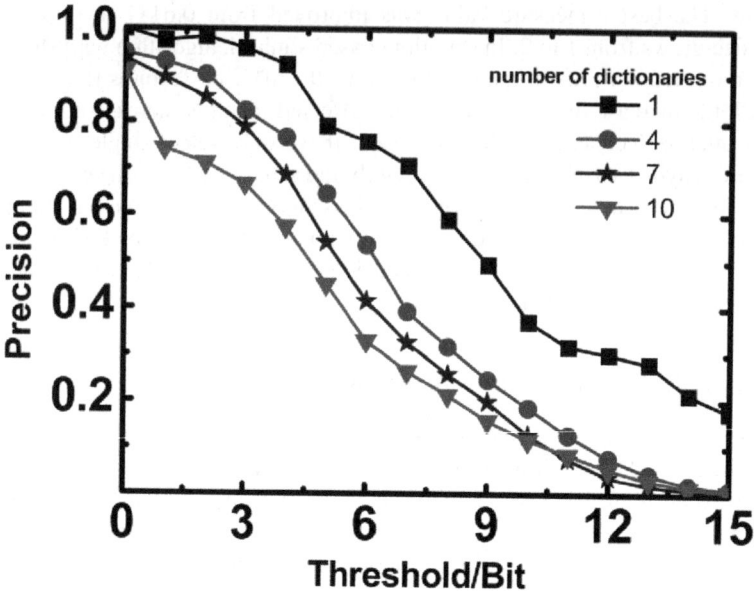

Fig. 4. The precision of the shingling based multi fingerprints simhash algorithm with random lexicon size 1, 4, 7, 10, and the threshold of hamming distance range from 0 to 15

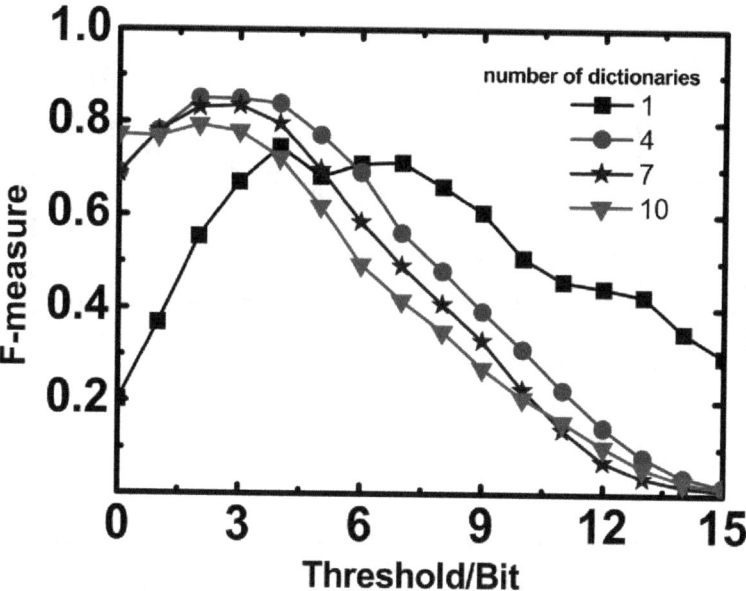

Fig. 5. The F-measure of the shingling based multi fingerprints simhash algorithm with random lexicon size 1, 4, 7, 10, and the threshold of hamming distance range from 0 to 15

We listed the best F-measure of each random lexicon size with their corresponding thresholds in table 1. We can see that, when random lexicon size equals 5 and threshold is 3, we get the best F-measure value 0.8805 in this experiment environment. There is about an 18% percent improvement compared with shingle based simhash algorithm with shingle size 2, and a 44% improvement compared with the original simhash algorithm.

Table 1. The best F-measure of each random lexicon size with their corresponding thresholds

random lexicon size	Best F-measure score	Threshold
1	0.7469	4
2	0.8071	6
3	0.8213	3
4	0.8515	2
5	0.8805	3
6	0.7855	3
7	0.8356	3
8	0.8077	1
9	0.8139	2
10	0.7942	2

We also tested shingling (we set the parameters the same with D. Fetterly et al. [7]), I-Match and multiple random lexicons based I-Match algorithms, the performances are roughly the same with most published papers. Especially, the performances of I-Match and multiple random lexicons based I-Match algorithms are similar with Theobald et al. [14] in our experiment environment. In our experiments, the two algorithms got even lower recalls, also with high precisions. Besides the character of the two algorithms themselves, the small size of the experiment dataset that yields a poor collection statistic may be another reason.

5   Conclusions

We described the three major algorithms (shingling, I-match, simhash) in duplicate document detection and their development in the following days. We introduced our idea of fusing these algorithms and presented the experiment results in the "China-US Million Book Digital Library Project" dataset. The performance of shingling based simhash algorithm was affected by test dataset in two sides. There exists a balance in selection of shingle size. Multiple random lexicons based simhash algorithm can improve recall but impair precision slightly. We should seek a balance when choose random lexicon size. As there is no conflict between feature selection and multi fingerprints generation, we implemented the combination which performances much better than the original simhash algorithm in our synthetic dataset. We are now constructing larger test dataset to validate our algorithm, and trying to implement our algorithm on other datasets.

References

1. Brin, S., Davis, J., Garcia-Molina, H.: Copy Detection Mechanisms for Digital Documents. In: Proceedings of the ACM SIGMOD Annual Conference (1995)
2. Shivakumar, N., Garcia-Molina, H.: SCAM: A copy detection mechanism for digital documents. In: Proceedings of the 2nd International Conference in Theory and Practice of Digital Libraries, DL 1995 (1995)
3. Shivakumar, N., Garcia-Molina, H.: Building a scalable and accurate copy detection mechanism. In: Proceedings of the 1st ACM Conference on Digital Libraries, DL 1996 (1996)
4. Chowdhury, A., Frieder, O., Grossman, D., Mccabe, M.C.: Collection statistics for fast duplicate document detection. ACM Transactions on Information Systems 20(2) (2002)
5. Kołcz, A., Chowdhury, A., Alspector, J.: Improved Robustness of Signature-Based Near-Replica Detection via Lexicon Randomization. In: Proceedings of the tenth ACM SIGKDD, Seattle, WA, USA (2004)
6. Conrad, J.G., Guo, X.S., Schriber, C.P.: Online Duplicate Document Detection: Signature Reliability in a Dynamic Retrieval Environment. In: Proceedings of the Twelfth International Conference on Information and Knowledge Management (2003)
7. Broder, A.Z., Glassman, S.C., Manasse, M.S.: Syntactic clustering of the Web. In: Proceedings of the 6th International Web Conference (1997)
8. Broder, A.Z., Charikar, M., Frieze, A., Mitzenmacher, M.: Min-Wise Independent Permutations. Journal of Computer and System Sciences, 630–659 (2000)
9. Fetterly, D., Manasse, M., Najork, M.: On the evolution of clusters of near-duplicate web pages. In: Proceedings of First Latin American Web Congress, pp. 37–45 (2003)
10. Fetterly, D., Manasse, M., Najork, M.: Detecting Phrase-level Duplication on the World Wide Web. In: The 28th ACM SIGIR, pp. 170–177 (2005)
11. Charikar, M.S.: Similarity estimation techniques from rounding algorithms. In: Proceedings of 34th Annual Symposium on Theory of Computing (2002)
12. Henzinger, M.: Finding near-duplicate web pages: a large-scale evaluation of algorithms. In: Proceedings of the 29th ACM SIGIR, pp. 284–291 (2006)
13. Manku, G.S., Jain, A., Sarma, A.D.: Detecting near-duplicates for web crawling. In: Proceedings of the 16th International Conference on World Wide Web, pp. 141–150 (2007)
14. Theobald, M., Siddharth, J., Paepcke, A.: SpotSigs: robust and efficient near duplicate detection in large web collections. In: Proceedings of ACM SIGIR (2008)

Searching Interesting Association Rules Based on Evolutionary Computation

Guangfei Yang[1], Yanzhong Dang[1], Shingo Mabu[2],
Kaoru Shimada[2], and Kotaro Hirasawa[2]

[1] Institute of Systems Engineering, Dalian University of Technology, China
[2] Graduate School of IPS, Waseda University, Japan

Abstract. In this paper, we propose an evolutionary method to search interesting association rules. Most of the association rule mining methods give a large number of rules, and it is difficult for human beings to deal with them. We study this problem by borrowing the style of a search engine, that is, searching association rules by keywords. Whether a rule is interesting or not is decided by its relation to the keywords, and we introduce both semantic and statistical methods to measure such relation. The mining process is built on an evolutionary approach, Genetic Network Programming (GNP). Different from the conventional GNP based association rule mining method, the proposed method pays more attention to generate the GNP individuals carefully, which will mine interesting association rules efficiently.

Keywords: Association Rule, Semantic Annotation, Genetic Network Programming.

1 Introduction

In association rule mining research, it remains an open problem to find interesting association rules because the conventional mining algorithms usually give a large number of rules. In this paper we propose an evolutionary method to search association rules, by borrowing the style of a search engine. In a search engine, a user expresses his demands by keywords. Similarly, we let the user input keywords to search rules.

Different from most of the previous studies to find interesting association rules which mainly consider objective statistical aspects, our definition of **interestingness** is also directly related to human beings' subjective opinions. So there are two kinds of criteria integrated in this paper: semantic measure and statistical measure.

As soon as the keywords are input and the semantic/statistical measures are selected, we start generating association rules. The association rule mining method based on Genetic Network Programming (GNP) [1] has been studied extensively recently [2] [3], and we will improve it to directly mine interesting association rules. Previously, the GNP individuals were usually generated randomly in the initial generation. In this paper, we generate GNP individuals more

L. Cao et al. (Eds.): PAKDD 2011 Workshops, LNAI 7104, pp. 243–253, 2012.

carefully, and develop a probability method to select potentially good attributes. The mined rules will be ranked according to their interestingness with respect to the keywords in descending order. In order to make it easier for the user to understand, we give some explanations or notations for each rule by adopting the idea of semantic annotation [4] which is learnt from the style of dictionary.

2   Related Work

Some methods are proposed to prune rules and give a smaller number of significant rules. In [5], they develop a technique to prune rules and then give a summary to the un-pruned rules. The idea of closed pattern is proposed to represent all the sub-patterns with the same support [6], which compresses the whole set of rules without loss of any information. In [7], an efficient method TFP is proposed to mine closed patterns without specifying the minimum support threshold, which removes the burden to set this subtle parameter. The TFP method also proposes another important concept: top-k pattern, that is, only a few most frequent and closed patterns are mined. Xin et al. [8] extends it by mining redundancy-aware top-k patterns by making a trade-off between the significance and redundancy of patterns.

There are also a lot of statistical measures developed to describe the significance of the rule from different aspects [9], and how to analyze the effects of these measures is of great importance. Tan et al. [10] study several important properties of different measures, and find that each single measure has its own advantages and disadvantages. They argue that whether a measure could contribute to the evaluation of interestingness depends on what kind of problem to be solved. Vaillant et al. [11] also find that different measures may give distinct evaluation results. Liu et al. [12] argue that a rule is only interesting in the context of other rules and each rule may not be interesting in many cases. Lenca et al. [13] analyze the interestingness measures based on Multiple Criteria Decision Aid (MCDA), and show a profitable way to select the right measure for specific problems.

Although the statistical measures could describe the usefulness or significance of rules effectively, the semantic knowledge hidden in the association rules still needs to be discovered further. The research in [14] discusses the subjective interestingness measures by focusing the actionable and unexpected rules, which is based on a belief system. In [15], they let the user input the expected patterns, and then a fuzzy matching technique is proposed to find and rank the patterns against the user's expectation. In the recent research [4], Mei et al. interpret the discovered frequent patterns by generating semantic annotations with a context model, which mimics the entities with semantic information in a dictionary.

3   Measuring the Similarity

In order to measure the semantic relations between different attributes, we first build an ontology, where the concepts in ontology are related to the attributes in databases.

Usually, there are three kinds of attributes: continuous, categorical and binomial. After discretization, the continuous attributes could be regarded as categorical ones. Then, each attribute has several categorical values, and each categorical value could be a binomial attribute. Finally, we build a new database, where all the attributes are binomial. Here is a simple example. Let us consider three attributes from a census dataset $\{age, job, gender\}$, which are continuous, categorical and binomial attributes, respectively. The continuous attribute age could be discretized into three categories: $\{young, middle, old\}$, the categorical attribute job could have three categories: $\{teacher, worker, farmer\}$, and the binomial attribute $gender$ has two categories: $\{male, female\}$. Now, we could build a new database, where the attributes are: $\{age = young, age = middle, age = old, job = teacher, job = worker, job = farmer, gender = male, gender = female\}$, and each attribute is binomial.

The columns in original database are **concepts** in the ontology, denoted as C. For example, $\{age, job, gender\}$ are concepts in $census$ ontology. The columns in transformed database are called **attributes**, denoted as A, and each attribute belongs to a certain concept. For example, $\{age = young, age = middle, age = old\}$ are attributes belonging to the concept age, $\{job = teacher, job = worker, job = farmer\}$ are attributes of the concept job, and $\{gender = male, gender = female\}$ are attributes of the concept $gender$. When measuring the semantic relations between attributes, we should consider their concepts. For example, the semantic relation between $gender = male$ and $age = middle$ is measured by the semantic relation between $gender$ and age.

If an attribute a_i belongs to a concept c_j, it is denoted as: $\Xi(a_i) = c_j$. If the set of all attributes belonging to c_j is A_k, then it is denoted as: $\Omega(c_j) = A_k$.

3.1   Measuring the Similarity

We mainly discuss three measures: semantic similarity, co-occurrence information and mutual information. We use the function, $f(o_1, o_2)$, to denote the similarity between object o_1 and object o_2, where an object could be an attribute, a concept, a rule or a set of attributes or concepts.

Semantic Similarity. We use the function, $f_s(o_1, o_2)$, to denote the semantic similarity. An ontology is built for measuring the semantic similarity between different concepts. Given two concepts c_i and c_j in the ontology, the semantic similarity between them is calculated by [16]:

$$f_s(c_i, c_j) = -\log[\frac{minlen(c_i, c_j)}{2 \times \text{MAX}}], \qquad (1)$$

where, $minlen(c_i, c_j)$ is the length of the shortest path between c_i and c_j, and MAX is the maximum depth of the ontology hierarchy.

The semantic similarity between two attributes a_i and a_j is calculated by:

$$f_s(a_i, a_j) = f_s(\Xi(a_i), \Xi(a_j)). \qquad (2)$$

The semantic similarity between two sets of concepts, $C^1 = \{c_1^1, c_2^1, \ldots, c_m^1\}$ and $C^2 = \{c_1^2, c_2^2, \ldots, c_n^2\}$, is calculated by:

$$f_s(C^1, C^2) = \frac{\sum_{i=1}^{m} \sum_{j=1}^{n} f_s(c_i^1, c_j^2)}{m \times n}. \tag{3}$$

It is easy to derive the semantic similarity calculation between sets of attributes based on Eq. (2) and Eq. (3). Since each rule could also be regarded as a set of attributes, we also know how to compute the semantic similarity between two rules. The normalized semantic similarity $\overline{f_s}$ is:

$$\overline{f_s}(o_i, o_j) = \frac{f_s(o_i, o_j) - f_s^{MIN}}{f_s^{MAX} - f_s^{MIN}}, \tag{4}$$

where, f_s^{MAX} is the maximum value of all semantic similarities between each pair of objects, and f_s^{MIN} is the minimum one.

Co-occurrence. We use the function, $f_c(o_1, o_2)$, to denote the co-occurrence information between object o_1 and object o_2. Given two attributes a_i and a_j, the co-occurrence between them is calculated by:

$$f_c(a_i, a_j) = \frac{N_{a_i, a_j}}{N}, \tag{5}$$

where, N_{a_i, a_j} is the number of records that contain both a_i and a_j, and N is the number of all the records in the database.

If we want to calculate the co-occurrence information between two concepts, we should consider all the attributes that belong to the concept. Given two concepts c_i and c_j, if one attribute $a_x \in \Omega(c_i)$ appears in a record, we say c_i appears in this record. Sometimes, more than one attributes in $\Omega(c_i)$ appear in a record at the same time, then this record is counted only for one time once. For example, if both attribute $a_{x1} \in \Omega(c_i)$ and attribute $a_{x2} \in \Omega(c_i)$ are in a record, we say c_i appears one time, not two times. Suppose the number of records containing at least one attribute in $\Omega(c_i)$ and at least one attribute in $\Omega(c_j)$ is N_{c_i, c_j}, and then we could calculate $f_c(c_i, c_j)$:

$$f_c(c_i, c_j) = \frac{N_{c_i, c_j}}{N}. \tag{6}$$

When calculating the co-occurrence information between sets of attributes or rules, the method is similar to Eq. (5). The normalized co-occurrence information $\overline{f_c}$ is:

$$\overline{f_c}(o_i, o_j) = \frac{f_c(o_i, o_j) - f_c^{MIN}}{f_c^{MAX} - f_c^{MIN}}, \tag{7}$$

where, f_c^{MAX} is the maximum value of all co-occurrence information between each pair of objects, and f_c^{MIN} is the minimum one.

Mutual Information. Co-occurrence could not describe the independence between attributes well. Mutual information is widely used to describe the independence of two variables in information theory, and it has been introduced into association rule mining successfully [4].

We use the function, $f_m(o_1, o_2)$, to denote the mutual information between object o_1 and object o_2. Given two attributes a_i and a_j, the mutual information between them is calculated by:

$$f_m(a_i, a_j) = \sum_{x_i \in \{0,1\}} \sum_{x_j \in \{0,1\}} p(x_i, x_j) \times log \frac{p(x_i, x_j)}{p(x_i) \times p(x_j)}, \tag{8}$$

where, x_i and x_j are variables indicating the appearance of attribute a_i and a_j, respectively, and $p(x)$ is the probability of x. $x_i = 1$ means attribute a_i appears, while $x_i = 0$ means attribute a_i is absent. By counting the number of records in the database containing attribute a_i and attribute a_j, we could calculate the mutual information by:

$$f_m(a_i, a_j) = p_{ij} \times log \frac{p_{ij}}{p_i \times p_j} + (p_j - p_{ij}) \times log \frac{p_j - p_{ij}}{(1 - p_i)p_j}$$
$$+ (p_i - p_{ij}) \times log \frac{p_i - p_{ij}}{(1 - p_j) \times p_i} + (1 - p_i - p_j + p_{ij})$$
$$\times log \frac{1 - p_i - p_j + p_{ij}}{(1 - p_i) \times (1 - p_j)}$$
$$\tag{9}$$

where, $p_{ij} = \frac{N_{a_i, a_j}}{N}$, $p_i = \frac{N_{a_i}}{N}$ and $p_j = \frac{N_{a_j}}{N}$ (N_{a_i} is the number of records containing attribute a_i and N_{a_j} is the number of records containing attribute a_j).

If we want to calculate the mutual information between two concepts, then we should consider all the attributes that belong to the concept. In other words, by considering the presence of $\Omega(c_i)$ and $\Omega(c_j)$ for two concepts c_i and c_j, we could calculate $f_m(c_i, c_j)$. When calculating the mutual information between rules, or sets of attributes, the method is the same as Eq. (8). The normalized mutual information $\overline{f_m}$ is:

$$\overline{f_m}(o_i, o_j) = \frac{f_m(o_i, o_j) - f_m^{MIN}}{f_m^{MAX} - f_m^{MIN}}, \tag{10}$$

where, f_m^{MAX} is the maximum value of all mutual information between each pair of objects, and f_m^{MIN} is the minimum one.

Combining Multiple Measures. We could combine the semantic and statistical information together to measure the relation between two objects, and there are two solutions: (1), combing semantic similarity and co-occurrence information; (2), combining semantic similarity and mutual information.

Given two objects o_i and o_j, the similarity measured by both semantic information and statistical information is calculated by:

$$f(o_i, o_j) = \alpha \times \overline{f_s}(o_i, o_j) + (1 - \alpha) \times \overline{f_c}(o_i, o_j), \tag{11}$$

or,

$$f(o_i, o_j) = \alpha \times \overline{f_s}(o_i, o_j) + (1 - \alpha) \times \overline{f_m}(o_i, o_j), \tag{12}$$

where, $\alpha \in [0,1]$. The user could specify the parameter α, and choose one of the above two methods. In the experiments in this paper, when measuring the similarity between attributes and keywords for generating rules, we use Eq. (11), and when measuring the similarity between rules and keywords for annotating rules, we use Eq. (12). The first method is a little easier for computation, making the generation of rules fast, while the second one could describe the relation between rules and keywords with some more information about independence.

4   Mining Association Rules by Genetic Network Programming

Genetic Network Programming (GNP) is an extension of GP, which uses directed graphs as genes for evolutionary computation [1]. As Fig. 1 shows, the basic structure of GNP has three kinds of nodes: a Start Node, some Judgement Nodes and some Processing Nodes. The start node is a special node, indicating the first position to be executed. Judgement nodes judge the information from the environment and determine the next node to be executed. Processing nodes represent some functional actions that could be taken.

There are some previous research about mining association rules by GNP [2] [3]. The initial GNP individuals are usually generated randomly. In this section, we will improve GNP based association rule miner by adopting the keyword-style searching. The association rules are generated by GNP individuals, and the attributes in the rules are associated with GNP judgement nodes. In order to mine interesting rules directly, the GNP judgement nodes should be generated more carefully. We develop a probabilistic method to select proper attributes for judgement nodes. Suppose a user has input several keywords, denoted as K, then he wants to find some interesting rules according to K. There are five steps to mine interesting association rules:

1. Generate GNP individuals. First, calculate the similarity between each attribute and the keywords K. Then, select the attributes by probability to generate judgement nodes in GNP individuals. The connections between nodes are decided by random. The probability of selecting attribute a, $p(a)$, is calculated by:

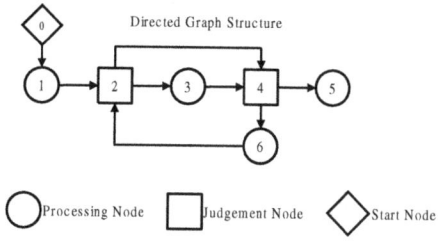

Fig. 1. Basic structure of GNP

$$p(a) = Z \times f(a, K), \tag{13}$$

where, Z is a parameter. It is true that:

$$\sum_{i=1}^{m} p(a_i) = 1, \tag{14}$$

where, m is the number of all attributes. So we could calculate the parameter Z by:

$$Z = \frac{1}{\sum_{i=1}^{m} f(a_i, K)}. \tag{15}$$

The more interesting attributes will have more probabilities to be selected into a rule.

2. Mine rules by GNP. The details has been explained in Section 3. Among all the rules mined in each individual, only new rules are saved and the number of new rules is the fitness value of this individual.

3. Evaluate rules. The similarity between rule r and the keyword K is $f(r, K)$, which has been introduced in Section 2, and the rules are displayed in the descending order of their similarity to K. We also calculate the support and confidence of each rule.

4. Rank rules. We design two kinds of policies to rank the rules.
   Policy 1. Given two rules r_i, r_j and keywords K, r_i is ranked higher than r_j if:
   (a) $f(r_i, K)$ is higher than $f(r_j, K)$, or
   (b) the similarities are the same, but the confidence of r_i is greater than that of r_j, or
   (c) their confidences are the same, but the support of r_i is greater than that of r_j.
   Policy 2. Each rule r is given a ranking value, $rank(r)$, and the rules are sorted in descending order depending on the ranking values, where, $rank(r) = \theta_1 \times f(r, K) + \theta_2 \times support(r) + \theta_3 \times confidence(r)$, where, $\theta_1 \in [0, 1]$, $\theta_2 \in [0, 1]$, $\theta_3 \in [0, 1]$, and $\theta_1 + \theta_2 + \theta_3 = 1$. In the following experiments, all of three parameters are $\frac{1}{3}$.

5. Annotate rules. Besides giving the basic statistics, such as support and confidence, we adopt the dictionary-like style to describe the rules, in order to give more meaningful explanations. We design two kinds of annotations: (1), similar rule; (2), representative transaction. (Here the meaning of a *transaction* is the same as a *record*.)
   Given two rules r_i and r_j, they are similar rules to each other if:

$$f(r_i, r_j) > \gamma, \tag{16}$$

where, $\gamma \in (0, 1)$ is a parameter defined by the user.
A transaction could be regarded as a set of attributes. Given a rule r and transaction t_i, t is the representative transaction of r if:

$$t = \arg\max_{t_i \in T} f(r, t_i), \tag{17}$$

where, T is the set of all transactions in the database.

The similar rule and representative transaction mimic the synonym and example sentences in a dictionary, respectively, and they give more information to help the user to understand the rule better.

5   Simulations

We select the census data set from UCI data repository [17] for the simulations, and demonstrate how to mine and analyze interesting rules by keywords. It is important to understand the data set, however, most of the data sets are hard to understand, because the understanding needs some additional domain knowledge. Census is a little easier to understand, but some concepts are still difficult to interpret. As a result, we mainly consider a small number of concepts to make the understanding easier in the following experiments. The data set has 199523 records, and we select a sample of 1000 records. There are 41 concepts in the census data set, and 334 attributes after preprocessing. We set the parameter α in Eq. (11) and Eq. (12) to be 0.5. The keyword is an attribute "income::<50000", which means that the concept is "income" and its value is "<50000". The resulted rules ranked by Policy 1 and Policy 2 are shown in Table 1 and Table 2, respectively. In each Table, we select top 3 rules.

In Table 1 and Table 2, for each rule, we list its antecedent attributes, consequent attributes, similarity to the keywords, support value, confidence value, ID of the similar rule and ID of the representative transaction. Although there are usually more than one similar rules, for simplicity, we only present the most similar one among all the similar rules. The rules in Table 1 are ranked by Policy 1, and we could see that some of the rules with large similarity but small support are ranked high. The measure of the support usually describes the usefulness of association rules, and most of the conventional association rule mining methods consider that the rules with large support are interesting. However, sometimes it is the similarity that attracts the user first. In other words, even if a rule has small support, it may provide some knowledge for the user since it is related to the user's keyword.

In this case, the ranking in Table 1 is useful. The user could see whether the closely related rules are really interesting or not from the subjective point of view.

Of course, if the user only wants to know the statistically strong rules, that is, the rules with high support and high confidence, then Table 2 will be better, because in Table 2, the rules with both large similarity to keywords and large statistical values are ranked high.

There are only IDs of the similar rule or representative transaction listed in Table 1 and Table 2, and each ID is corresponding to one rule. To make it more understandable for the readers, we give some short descriptions about the similar rules and representative transactions mentioned in Table 1. For RULE 1

Table 1. Three top rules ranked by policy 1 for keyword "income::<50000" ($\alpha = 0.5$)

RULE 1	[antecedent]	income::<50000
	[consequent]	own business or self employed::no
	[similarity]	0.71
	[support]	0.86
	[confidence]	0.92
	[rule]	361
	[transaction]	838
	[rank(r)]	0.83
RULE 2	[antecedent]	income::<50000
	[consequent]	citizenship::Native Born abroad Parent
	[similarity]	0.41
	[support]	0.01
	[confidence]	0.9
	[rule]	950
	[transaction]	576
	[rank(r)]	0.44
RULE 3	[antecedent]	country of birth father::Vietnam
	[consequent]	income::<50000
	[similarity]	0.36
	[support]	0.004
	[confidence]	1.0
	[rule]	882
	[transaction]	60
	[rank(r)]	0.45

in Table 1, the similar rule is about "own business or self employed" and "tax filer status", and the representative transaction is associated with a 51-years-old man, with 7th and 8th grade education, working in business and repair services, single, divorced and native born in US. For RULE 2 in Table 1, the similar rule is about "income" and "country of birth farther", and the representative transaction is about a 37-years-old man, with some colleges but no degree in education, working in public administration, single, divorced and native born in US. For RULE 3 in Table 1, the similar rule is about "citizenship" and "fill in questionnaire for veteran", and the representative transaction is corresponding to a 23-years-old woman, with education about some colleges but no degree, working in retail trade, married-civilian and spouse present, born in China and her father born in Vietnam.

The similar rules and representative transactions provide more semantic information to help the user to understand the rules, and the user could have some impressive information with such annotations. It costs too much to describe the contents of similar rules and representative transactions, that is why we skip the detailed descriptions in the tables and only present the IDs for conciseness.

Table 2. Three top rules ranked by policy 2 for keyword "income::<50000" ($\alpha = 0.5$)

RULE 1	[antecedent]	income::<50000
	[consequent]	own business or self employed::no
	[similarity]	0.71
	[support]	0.86
	[confidence]	0.92
	[rule]	361
	[transaction]	838
	[rank(r)]	0.83
RULE 2	[antecedent]	weeks worked in year::<49.899
	[consequent]	income::<50000
	[similarity]	0.34
	[support]	0.32
	[confidence]	0.85
	[rule]	849
	[transaction]	22
	[rank(r)]	0.50
RULE 3	[antecedent]	class of worker::private
	[consequent]	citizenship::native born
		household::Spouse of householder
	[similarity]	0.25
	[support]	0.18
	[confidence]	0.83
	[rule]	95
	[transaction]	4
	[rank(r)]	0.42

6   Conclusions

In this paper, we have proposed a novel evolutionary method to directly mine interesting association rules. We learn from the search engine and let the user describe his demands by some keywords. This style is intuitive, simple and effective. By defining some semantic and statistical methods to measure the relation, we could directly mine the interesting rules based on Genetic Network Programming.

Acknowledgments. This work is supported by the National Natural Science Foundation of China under Grant No. 71001016, the China Postdoctoral Science Foundation under Grant No. 20100471000, and the Key Program of National Natural Science Foundation of China under Grant No. 71031002.

References

1. Mabu, S., Hirasawa, K., Hu, J.: A Graph-Based Evolutionary Algorithm: Genetic Network Programming (GNP) and Its Extension Using Reinforcement Learning. Evolutionary Computation 15(3), 369–398 (2007)

2. Shimada, K., Hirasawa, K., Hu, J.: Class Association Rule Mining with Chi-Squared Test Using Genetic Network Programming. In: IEEE SMC (2006)
3. Yang, G., Shimada, K., Mabu, S., Hirasawa, K., Hu, J.: Mining Equalized Association Rules from Multi Concept Layers of Ontology Using Genetic Network Programming. In: IEEE CEC, Singapore, pp. 705–712 (2007)
4. Mei, Q., Xin, D., Cheng, H., Han, J., Zhai, C.: Generating Semantic Annotations for Frequent Patterns with Context Analysis. In: SIGKDD, pp. 337–346 (2006)
5. Liu, B., Hsu, W., Ma, Y.: Pruning and Summarizing the Discovered Associations. In: Proc. of the Fifth Int'l Conf. on Knowledge Discovery and Data Mining (KDD 1999), pp. 125–134 (1999)
6. Zaki, M.J., Hsiao, C.J.: CHARM: An Efficient Algorithm for Closed Itemset Mining. In: Proc. of the SIAM International Conference on Data Mining, pp. 457–473 (2002)
7. Han, J., Wang, J., Lu, Y., Tzvetkov, P.: Mining Top-k Frequent Closed Patterns without Minimum Support. In: Proc. of the 2002 IEEE Int'l Conf. on Data Mining (ICDM 2002), pp. 211–218 (2002)
8. Xin, D., Cheng, H., Yan, X., Han, J.: Extracting Redundancy-aware Top-k Patterns. In: Proc. of the Int'l Conf. on Knowledge Discovery and Data Mining(KDD 2006), pp. 444–453 (2006)
9. Suzuki, E.: Pitfalls for Categorizations of Objective Interestingness Measures for Rule Discovery. SCI, vol. 127, pp. 383–395 (2008)
10. Tan, P., Kumar, V., Srivastava, J.: Selecting the Right Interestingness Measure for Association Patterns. In: Proc. of the Eighth Int'l Conf. on Knowledge Discovery and Data Mining (KDD 2002), pp. 32–41 (2002)
11. Vaillant, B., Lenca, P., Lallich, S.: A Clustering of Interestingness Measures. Discovery Science, 290–297 (2004)
12. Liu, B., Zhao, K., Benkler, J., Xiao, W.: Rule Interestingness Analysis Using OLAP Operations. In: Proc. of the 12th Int'l Conf. on Knowledge Discovery and Data Mining (KDD 2006), pp. 297–306 (2006)
13. Lenca, P., Meyer, P., Vaillant, B., Lallich, S.: On Selecting Interestingness Measures for Association Rules: User Oriented Description and Multiple Criteria Decision Aid. European Journal of Operational Research 184(2), 610–626 (2008)
14. Silberschatz, A., Tuzhilin, A.: What Makes Patterns Interesting in Knowledge Discovery Systems. IEEE Trans. on Knowledge and Data Engineering 8(6), 970–974 (1996)
15. Liu, B., Hsu, W., Mun, L., Lee, H.: Finding Interesting Patterns Using User Expectations. IEEE Trans. on Knowledge and Data Engineering 11(6), 817–832 (1999)
16. Resnik, P.: Using information content to evaluate semantic similarity in a taxonomy. IJCAI (1995)
17. UC Irvine Machine Learning Repository, http://archive.ics.uci.edu/ml/

An Efficient Approach to Mine
Periodic-Frequent Patterns
in Transactional Databases

Akshat Surana, R. Uday Kiran, and P. Krishna Reddy

Center for Data Engineering,
International Institute of Information Technology-Hyderabad,
Hyderabad, Andhra Pradesh, India - 500032
{akshat.surana,uday_rage}@research.iiit.ac.in, pkreddy@iiit.ac.in

Abstract. Recently, temporal occurrences of the frequent patterns in
a transactional database has been exploited as an interestingness criterion to discover a class of user-interest-based frequent patterns, called
periodic-frequent patterns. Informally, a frequent pattern is said to be
periodic-frequent if it occurs at regular intervals specified by the user
throughout the database. The basic model of periodic-frequent patterns
is based on the notion of "single constraints." The use of this model to
mine periodic-frequent patterns containing both frequent and rare items
leads to a dilemma called the "rare item problem." To confront the problem, an alternative model based on the notion of "multiple constraints"
has been proposed in the literature. The periodic-frequent patterns discovered with this model do not satisfy *downward closure property*. As
a result, it is computationally expensive to mine periodic-frequent patterns with the model. Furthermore, it has been observed that this model
still generates some uninteresting patterns as periodic-frequent patterns.
With this motivation, we propose an efficient model based on the notion
of "multiple constraints." The periodic-frequent patterns discovered with
this model satisfy *downward closure property*. Hence, periodic-frequent
patterns can be efficiently discovered. A pattern-growth algorithm has
also been discussed for the proposed model. Experimental results show
that the proposed model is effective.

Keywords: Data mining, frequent patterns and rare item problem.

1 Introduction

Periodic-frequent patterns were introduced in [5]. A frequent pattern is said to
be periodic-frequent if it occurs at regular intervals specified by the user in the
database. Technically, a pattern (i.e. a set of items or itemset) is considered
periodic-frequent if it satisfies the user-specified minimum support ($minsup$)
and maximum periodicity ($maxprd$) constraints. $Minsup$ controls the minimum
number of transactions in which a pattern must appear in the database. $Maxprd$
controls the maximum time difference between two consecutive appearances of
a pattern in the database.

L. Cao et al. (Eds.): PAKDD 2011 Workshops, LNAI 7104, pp. 254–266, 2012.
© Springer-Verlag Berlin Heidelberg 2012

Since only single *minsup* and single *maxprd* values are used for the entire database, the basic model implicitly assumes that all items in a database have uniform frequencies and similar occurrence behavior. However, this is often not the case in many real-world databases. In a real-world database, some items reoccur frequently, while others reoccur relatively infrequent (or rarely). Furthermore, the rare items may have a larger reoccurrence interval than the frequent items. If the items' frequencies in a database vary widely, usage of single *minsup* and single *maxprd* framework to discover periodic-frequent patterns containing both frequent and rare items leads to the dilemma known as the "rare item problem" [6].

A model based on multiple *minsups* and multiple *maxprds* framework has been proposed in [4] to confront "rare item problem." However, this model is computationally expensive to implement because periodic-frequent patterns mined do not satisfy *downward closure property*, i.e., not all non-empty subsets of a periodic-frequent pattern are periodic-frequent.

In this paper, we have proposed an improved model to mine periodic-frequent patterns with multiple *minsups* and multiple *maxprds* framework. A pattern growth approach has also been proposed for efficient mining of periodic-frequent patterns. The periodic-frequent patterns mined with the proposed model satisfy *downward closure property*. As a result the proposed model is computationally efficient than the model discussed in [4]. Experimental results show that the proposed approach is efficient in mining periodic-frequent patterns containing both frequent and rare items.

The rest of the paper is organized as follows. In Section 2, we discuss the background on mining periodic-frequent patterns in transactional databases. In Section 3, we discuss the motivation and introduce the proposed model. A pattern-growth approach to mine periodic-frequent patterns is discussed in Section 4. Experimental results are reported in Section 5. The last section contains conclusions.

2   Background

2.1   Periodic-Frequent Pattern Model

Periodic-frequent patterns [5] are a class of user-interest based frequent patterns that exist in a database. The basic model of periodic-frequent pattern mining is as follows.

Let $I = \{i_1, i_2, \cdots, i_n\}$ be a set of items. A set of items X where $X \subseteq I$ is called a **pattern** (or an itemset). A transaction $t = (tid, Y)$ is a tuple, where tid represents a transaction-id (or a timestamp) and Y is a pattern. A transactional database T over I is a set of transactions, $T = \{t_1, \cdots, t_m\}$, $m = |T|$, where $|T|$ is the size of T in total number of transactions. If $X \subseteq Y$, it is said that t contains X or X occurs in t and such transaction-id is denoted as t_j^X, $j \in [1, m]$. Let $T^X = \{t_k^X, \cdots, t_l^X\} \subseteq T$, where $k \le l$ and $k, l \in [1, m]$ be the ordered set of transactions in which pattern X has occurred. Let t_j^X and t_{j+1}^X, where $j \in [k, (l-1)]$ be two consecutive transactions in T^X. The number of transactions or time difference between t_{j+1}^X and t_j^X can be defined as a **period** of X, say p^X. That is, $p^X = t_{j+1}^X - t_j^X$. Let $P^X = \{p_1^X, p_2^X, \cdots, p_r^X\}$, be the set of periods for pattern X. The **periodicity** of X, denoted as $Per(X) = max(p_1^X, p_2^X, \cdots, p_r^X)$. The **support** of X, denoted as $S(X) = |T^X|$. The pattern X is said to be periodic-frequent pattern, if $S(X) \geq minsup$ and $Per(X) \leq maxprd$, where $minsup$ and $maxprd$ are user-specified minimum support and maximum periodicity constraints. Both periodicity and support of a pattern can be described as a percentage of $|T|$.

Table 1. Transactions ordered based on times-tamp

Table 2. Periodic-frequent patterns mined for the database of Table 1. **S** and **P** represent the support count and periodicity, respectively. The columns entitled **I, II** and **III** represent the patterns mined using the basic, MCPF and MaxCPF models, respectively.

TID	Items
1	a, b, h
2	c, d
3	a, b, d
4	c, e, f
5	a, b

TID	Items
6	c, d, g
7	a, b, c, e
8	d, e, f
9	a, b, c
10	g, h

Pattern	S	P	I	II	III
{a}	5	2	✓	✓	✓
{b}	5	2	✓	✓	✓
{c}	5	2	✓	✓	✓
{d}	4	3	✓	✓	✓
{e}	3	4	✓	✓	✓

Pattern	S	P	I	II	III
{f}	2	4	✓	✓	✓
{a, b}	5	2	✓	✓	✓
{c, d}	2	4	✓	✗	✗
{c, e}	2	4	✓	✓	✗
{e, f}	2	4	✓	✓	✓

Example 1. Consider the transactional database shown in Table 1. Each transaction in it is uniquely identifiable with a transactional-id (*tid*) which is also a timestamp of that transaction. Timestamp indicates the time of occurrence of the transaction. The set of items, $I = \{a, b, c, d, e, f, g, h\}$. The set of a and b i.e., $\{a, b\}$ is a pattern. This pattern occurs in *tids* of $1, 3, 5, 7$ and 9. Therefore, $T^{\{a,b\}} = \{1, 3, 5, 7, 9\}$. Its support count (or support), $S(a, b) = |T^{\{a,b\}}| = 5$. The periods for this pattern are $1(= 1 - t_i)$, $2(= 3 - 1)$, $2(= 5 - 3)$, $2(= 7 - 5)$, $2(= 9 - 7)$ and $1(= t_l - 9)$, where $t_i = 0$ represents the initial transaction and $t_l = 10$ represents the last transaction in the transactional database. The periodicity of $\{a, b\}$, $Per(a, b) = maximum(1, 2, 2, 2, 2, 1) = 2$. If the user-specified $minsup = 4$ and $maxprd = 2$, the pattern $\{a, b\}$ is a periodic-frequent pattern because $S(a, b) \geq minsup$ and $Per(a, b) \leq maxprd$.

The periodic-frequent patterns discovered using this model satisfy *downward closure property*, i.e. all non-empty subsets of a periodic-frequent pattern must be periodic-frequent. A pattern-growth algorithm using a tree structure called periodic-frequent tree (PF-tree) has also been discussed to discover the complete set of periodic-frequent patterns. The algorithm uses *downward closure property* to efficiently discover the complete set of periodic-frequent patterns by reducing the search space.

2.2 Rare Item Problem

The basic model of periodic-frequent patterns suffers from the dilemma called the "rare item problem", which is as follows:

- If we specify a high *minsup* and/or low *maxprd* values, we miss the periodic-frequent patterns consisting of rare items.
- To mine patterns containing both frequent and rare items, a low *minsup* value and high *maxprd* value needs to be set. However, this results in combinatorial explosion, producing too many periodic-frequent patterns, because frequent items will associate with one another in all possible ways.

The rare item problem is illustrated in Example 2.

Example 2. Consider the transactional database shown in Table 1. On setting high *minsup* and/or low *maxprd* values, say *minsup* = 4 and *maxprd* = 2, we miss the patterns involving rare items, *e* and *f*. To mine periodic-frequent patterns involving the rare items, we have to set low *minsup* and high *maxprd* values. Let *minsup* = 2 and *maxprd* = 4. The columns in Table 2 with title 'I' represent the discovered periodic-frequent patterns. It can be observed that along with the interesting patterns {*a, b*} and {*e, f*}, the uninteresting patterns i.e., {*c, d*} and {*c, e*} (patterns represented in bold letters) have also been generated as periodic-frequent patterns. These patterns are uninteresting because they contain frequent items which are occurring together in very few transactions, and their reoccurrence interval is very high. These patterns can be considered interesting if they satisfy high *minsup* and low *maxprd*, say *minsup* = 4 and *maxprd* = 2.

2.3   Minimum Constraints Model of Periodic-Frequent Pattern

Since single *minsup* and single *maxprd* framework is unable to capture all the items' frequencies and occurrence behavior, an effort has been made in [4] to mine periodic-frequent patterns with multiple *minsups* and multiple *maxprds* framework, where each pattern can satisfy different *minsup* and *maxprd* values depending upon the items within it. The model proposed in [4] is known as Multi-Constraint Periodic-Frequent (MCPF) model. In this model, each item in the transactional database is specified with two types of constraints: (i) a support constraint, called *minimum item support* (*MinIS*), to capture each item's occurrence frequency, and (ii) a periodicity constraint, called *maximum item periodicity* (*MaxIP*), to capture each item's occurrence behavior. A pattern $X = \{i_1, i_2, \cdots, i_k\} \subseteq I$ is periodic-frequent if:

$$S(X) \geq minimum(MinIS(i_1), MinIS(i_2), \cdots, MinIS(i_k)) \quad and \quad (1)$$
$$Per(X) \leq maximum(MaxIP(i_1), MaxIP(i_2), \cdots, MaxIP(i_k))$$

Example 3. For the transactional database shown in Table 1, let the user-specified *MinIS* values for the items a, b, c, d, e, f, g and h be 4, 4, 4, 3, 2, 2, 2 and 2, respectively and the user-specified *MaxIP* values for these items be 2, 2, 2, 3, 4, 4, 4 and 4, respectively. The columns in Table 2 with title '**II**' represent the periodic-frequent patterns mined using this model. It can be seen that the uninteresting pattern {*c, d*} has been pruned out. However, the pattern {*c, e*} is still generated as a periodic-frequent pattern.

The periodic-frequent patterns mined using this model do not satisfy *downward closure property*. As a result, the pattern-growth algorithm designed for MCPF-model carries out exhaustive search on the constructed tree. The limitations of this model are discussed in the next section.

3   Proposed Model

The notion of "multiple *minsups* and multiple *maxprds*" framework suggested in [4] can efficiently address the "rare item problem." However, the MCPF-model faces the following limitations.

1. The periodic-frequent patterns mined with the MCPF model do not satisfy *downward closure property*. Hence, mining periodic-frequent patterns with this model is computationally expensive.
2. In real world applications, if a user specifies *MinIS* and *MaxIP* values for an item it can mean that any periodic-frequent pattern involving the respective item must have support no less than the specified *MinIS* value and periodicity no more than the *MaxIP* value. However, this notion of user's intuition is not captured by the MCPF-model.

   Example 4. Consider the item 'c' in the transactional database of Table 1. If the user specifies $MinIS(c) = 4$ and $MaxIP(c) = 2$, it can mean that any periodic-frequent pattern involving the item 'c' must appear at least 4 times in a database and all its reoccurrences must be within every 2 transactions. It can be observed that the MCPF-model does not capture such user requirement.

To confront the limitations of the MCPF-model, we propose another model based on "multiple *minsups* and multiple *maxprds*" framework.

   In the proposed model, the user specifies a support constraint, called *minimum item support* (*MinIS*) and a periodicity constraint, called *maximum item periodicity* (*MaxIP*) for each item in the transactional database. A pattern $X = \{i_1, i_2, \cdots, i_k\}$, $X \subseteq I$ is said to be periodic-frequent if:

$$S(X) \geq maximum(MinIS(i_1), MinIS(i_2), \cdots, MinIS(i_k)) \quad and \quad (2)$$
$$Per(X) \leq minimum(MaxIP(i_1), MaxIP(i_2), \cdots, MaxIP(i_k))$$

This model enables the user to specify high *minsup* and low *maxprd* for the patterns that involve frequent items, and specify low *minsup* and high *maxprd* for patterns involving only rare items. Thus, addressing the "rare item problem."

Property 1. If $X \subseteq I$, $Y \subseteq I$ and $Y \subseteq X$, then $S(X) \leq S(Y)$ and $maximum(\{MinIS(i); \forall i, i \in X\}) \geq maximum(\{MinIS(i); \forall i, i \in Y\})$.

Property 2. If $X \subseteq I$, $Y \subseteq I$ and $Y \subseteq X$, then $Per(X) \geq Per(Y)$ and $minimum(\{MaxIP(i); \forall i, i \in X\}) \leq minimum(\{MaxIP(i); \forall i, i \in Y\})$.

Theorem 1. *The periodic-frequent patterns generated using the proposed model satisfy* downward closure property.

Proof. If $X = \{i_1, i_2, \ldots, i_k\}$, $X \subseteq I$ is a periodic-frequent pattern, then $S(X) \geq maximum(\{MinIS(i); \forall i \in X\})$ and $Per(X) \leq minimum(\{MaxIP(i); \forall i \in X\})$. Let Y be a pattern such that $Y \subseteq X$. Using Property 1 and Property 2, it can be derived that $S(Y) \geq maximum(\{MinIS(i); \forall i, i \in Y\})$ and $Per(Y) \leq minimum(\{MaxIP(i); \forall i, i \in Y\})$. Thus, Y is also a periodic-frequent pattern.

4   MaxCPF-Tree: Design, Construction and Mining

In this section, we describe the structure, construction and mining of periodic-frequent patterns using a tree structure known as Maximum Constraints Periodic-Frequent-pattern-tree (MaxCPF-tree).

4.1   Structure of MaxCPF-Tree

The structure of the MaxCPF-tree consists of a prefix-tree and a MaxCPF-list. The MaxCPF-list is divided into the following five fields: item (i), support (S), periodicity (P), $MinIS$ (mis) and $MaxIP$ (mip). The items in the MaxCPF-list are sorted in decreasing order of their respective $MinIS$ values. We call this ordering the L-order. The structure of nodes in the prefix-tree of MaxCPF-tree are similar to that of the prefix-tree of PF-tree [5].

To maintain occurrence information for each transaction, the prefix-tree in MaxCPF-tree keeps an occurrence transaction-id list, called tid-list, only at the last node of every transaction. There are two types of nodes in the prefix-tree: ordinary node and $tail$ node. The ordinary node is similar to the nodes used in FP-tree [3], whereas the $tail$ node represents the last item of any transaction sorted in L-order. The structure of a $tail$-node is $N[t_1, t_2, ..., t_n]$, where N is the node's item name and $t_i, i \in [1, n]$, (n be the total number of transactions from the root up to the node) is a transaction-id where item N is the last item. To facilitate tree traversal, each node in the prefix-tree maintains parent, children and node traversal pointers. However, unlike FP-tree, no node in MaxCPF-tree maintains support count value in it. We now explain construction and mining of MaxCPF-tree.

4.2   Constructing MaxCPF-Tree

For construction of MaxCPF-list, a temporary array id_l is used to record the $tids$ of the last occurring transactions of all items. Let t_{cur} and p_{cur} respectively denote the tid of current transaction and the most recent period for an item $i \in I$. The MaxCPF-tree is, therefore, maintained according to the process given in Algorithm 1.

Consider the transactional database shown in Table 1. Let the user-specified $MinIS$ values for the items a, b, c, d, e, f, g and h be 4, 4, 4, 3, 2, 2, 2 and 2,

---

Algorithm 1 MaxCPF-tree (*T*: Transactional database, *I*: set of items, *MinIS*: items' minimum item support, *MaxIP*: items' maximum item periodicity)

1: Let *L* be the list of items sorted in descending order of their *MinIS* values.
2: In *L* order, insert each item $i \in I$ into the MaxCPF-list with $S(i) = 0$, $Per(i) = 0$, $mis(i) = MinIS(i)$ and $mip(i) = MaxIP(i)$.
3: **for** each transaction $t_{cur} \in T$ **do**
4:     **for** each $i \in t_{cur}$ **do**
5:         $S(i) + +$; $p_{cur} = t_{cur} - id_l(i)$;
6:         **if** $(p_{cur} > Per(i))$ **then**
7:             $Per(i) = p_{cur}$;
8:         **end if**
9:     **end for**
10: **end for**
11: Compute p_{cur} for each item by taking *tid* of the last transaction in *T* as t_{cur}. Update $Per(i) = p_{cur}$ for all items *i* having $p_{cur} > Per(i)$.
12: **for** each item *i* in MaxCPF-list **do**
13:     **if** $((S(i) < MinIS(i))$ or $(Per(i) > MaxIP(i)))$ **then**
14:         Remove *i* from the MaxCPF-list.
15:     **end if**
16: **end for**
17: Let L' be the sorted list of items in MaxCPF-list.
18: **for** each transaction $t \in T$ **do**
19:     After sorting items in L' order, create a branch in MaxCPF-tree as in PF-tree.
20: **end for**
21: Maintain node-links in MaxCPF-tree for tree-traversal.

Fig. 1. Algorithm for constructing MaxCPF-Tree

i	S	P	MIS	MIP	id1
a	0	0	4	2	0
b	0	0	4	2	0
c	0	0	4	2	0
d	0	0	3	3	0
e	0	0	2	4	0
f	0	0	2	4	0
g	0	0	2	4	0
h	0	0	2	4	0

(a) Before scanning the transactional dataset

i	S	P	MIS	MIP	id1
a	1	1	4	2	1
b	1	1	4	2	1
c	0	0	4	2	0
d	0	0	3	3	0
e	0	0	2	4	0
f	0	0	2	4	0
g	0	0	2	4	0
h	1	1	2	4	1

(b) After scanning the first transaction

i	S	P	MIS	MIP	id1
a	5	2	4	2	9
b	5	2	4	2	9
c	5	2	4	2	9
d	4	3	3	3	8
e	3	4	2	4	8
f	2	4	2	4	8
g	2	6	2	4	10
h	2	9	2	4	10

(c) After scanning the last transaction

i	S	P	MIS	MIP
a	5	2	4	2
b	5	2	4	2
c	5	2	4	2
d	4	3	3	3
e	3	4	2	4
f	2	4	2	4
g	**2**	**6**	**2**	**4**
h	**2**	**9**	**2**	**4**

(d) Updated MaxCPF-list. Items in bold are pruned to give the final MaxCPF-list.

Fig. 2. Construction of MaxCPF-list

respectively. Let the user-specified *MaxIP* values for these items be 2, 2, 2, 3, 4, 4, 4 and 4, respectively. Therefore, $L = \{a, b, c, d, e, f, g, h\}$.

In Fig. 2, we show how the MaxCPF-list is populated for the transactional database shown in Table 1. Fig. 2(a) shows the MaxCPF-list populated after inserting items in *L* order (Line 2 of Algorithm 1). Fig. 2(b) and Fig. 2(c) show the MaxCPF-list generated after scanning the first (i.e., $t_{cur} = 1$) and every transaction (i.e., $t_{cur} = 10$), respectively (lines 3 to 10 in Algorithm 1). To reflect the correct periodicity for each item in the MaxCPF-tree, the whole MaxCPF-list is refreshed as mentioned in line 11 of Algorithm 1. The resultant MaxCPF-list is shown in Fig. 2(d).

The fact that the periodic-frequent patterns mined using the proposed model follow *downward closure property*, helps in reducing the search space significantly. By using this property, any item that either has its support less than its $MinIS$ value or periodicity greater than its $MaxIP$ value is removed from the MaxCPF-list (Lines 12 to 16 in Algorithm 1). For example, items 'g' and 'h' are removed from the MaxCPF-list since their periodicities are greater than the respective $MaxIP$ values. The pruned items 'g' and 'h' are shown in bold in Fig. 2(d). Let L' be the new list of items in MaxCPF-list that are sorted in descending order of their $MinIS$ values.

The prefix-tree is constructed with the help of L', by performing another scan on the transactional database. The construction of prefix-tree in MaxCPF-tree is same as the construction of prefix-tree in PF-tree. Fig. 3(a) and (b) show the construction of MaxCPF-tree after scanning first and every transaction in the transactional database. In MaxCPF-tree, node-links are maintained as in FP-growth. For simplicity of figures, we are omitting them.

The structure of prefix-tree in MaxCPF-tree is similar to that of PF-tree [5]. The correctness of PF-tree (and thus for MaxCPF-tree) has already been proved in [5].

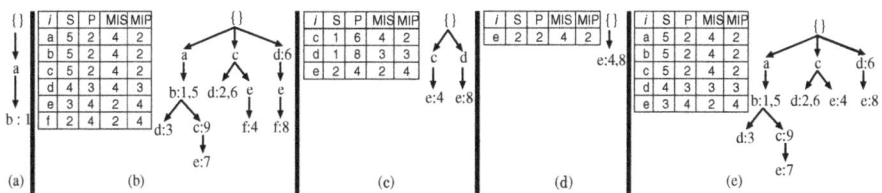

Fig. 3. Construction and mining of MaxCPF-tree. (a) Prefix-tree after scanning the first transaction (b) Prefix-tree after scanning every transaction (c) PT_f (d) CT_f and (e) MaxCPF-tree after pruning f.

4.3   Mining of *MaxCPF-Tree*

Consider the bottom-most item, say i, in the MaxCPF-list. For i, construct a prefix-tree, say PT_i by accumulating only the prefix sub-paths of nodes i. Since i is the bottom-most item in the MaxCPF-list, each node labeled i in the MaxCPF-tree must be a tail-node. While constructing PT_i, we map the tid-list of every node of i to all items in the respective path explicitly in a temporary array (one for each item). It facilitates the periodicity and support calculation for each item in the MaxCPF-list of PT_i. The contents of temporary array for the bottom item j in the MaxCPF-list of PT_i represents T^{ij} (i.e., set of all $tids$ where items i and j occur together). From T^{ij}, it is a simple calculation of $S(ij)$ and $Per(ij)$ for the pattern $\{i, j\}$.

Let $S_i(j)$ and $Per_i(j)$ be the support and periodicity, respectively of an item j in PT_i. To construct conditional tree CT_i from PT_i, the following two pruning strategies are employed:

(i) Using *downward closure property*, any item j having $S_i(j) < MinIS(j)$ cannot generate a periodic-frequent pattern.

(ii) Let I' be the set of items in PT_i having support greater than their corresponding $MinIS$ values. Thus, any item $j \in I'$ having $Per_i(j) > minimum\{MaxIP(j), MaxIP(i)\}$ cannot generate any periodic-frequent pattern in PT_i.

Using (i) and (ii), the conditional tree CT_i for PT_i is constructed by removing all those items j which have support less than their corresponding $MinIS$ value or periodicity greater than the minimum $MaxIP$ of $\{i, j\}$. If a deleted node is a *tail*-node, its *tid*-list is pushed-up to its parent node.

For every item j in CT_i, if $S(ij) \geq MinIS(j)$ and $Per(ij) \leq minimum(MaxIP(i), MaxIP(j))$, then $\{i, j\}$ is considered as a periodic-frequent pattern and the same process of creating prefix-tree and its corresponding conditional tree is repeated for further extensions of 'ij'. Else, based on *downward closure property*, we do not check for further extensions of 'ij'. Next, i is pruned from the original MaxCPF-tree. (The procedure for pruning i is same as pruning non-periodic-frequent items in earlier steps.) The whole process of mining each item is repeated until MaxCPF-list $\neq \emptyset$.

Consider the item 'f', which is the last item in the MaxCPF-list of Fig. 2(d). The PT_f generated from $MaxCPF$-tree is shown in Fig. 3(c). Out of the items 'c', 'd' and 'e' in PT_f, only 'e' satisfies its corresponding $MinIS$ value and $MaxIP_f (= minimum(MaxIP_f, MaxIP_e))$. The conditional tree CT_f after pruning items 'c' and 'd' is shown in Fig. 2(d). From $T^{\{e,f\}}$, the pattern $\{e, f\}$ will be generated as a periodic-frequent pattern. The MaxCPF-tree generated after pruning 'f' is shown in Fig. 3(e). Similar process is repeated for other items in the MaxCPF-tree.

The complete list of periodic-frequent patterns mined using the proposed model is shown in columns with title '**III**' in Table 2. Some patterns generated using MCPF-model like $\{c, e\}$ in which a highly frequent item 'c' combines with a rare item 'e', have failed to be periodic-frequent patterns in the proposed model. It is because $\{c, e\}$ failed to satisfy $minsup = 4$ and $maxprd = 2$. In this way the proposed model is able to efficiently address *rare item problem* while mining periodic-frequent patterns containing both rare and frequent items.

5   Experimental Results

In this section, we compare the performance of the proposed model with the basic model [5] and the MCPF-model [4]. All programs are written in C++ and run with Ubuntu 8.1 on a 2.66 GHz machine with 2 GB memory. The runtime specifies the total execution time. The experiments are pursued on the following datasets: synthetic dataset ($T10I4D100k$) [1] and real-world datasets (*Retail* [2] and *Mushroom*). $T10I4D100k$ and *Retail* are sparse datasets and *Mushroom* is a dense dataset. All of these datasets are widely used in frequent pattern mining literature.

To specify items' $MaxIP$ and $MinIS$ values, we use the methodologies provided in [4].

$$MaxIP(i) = maximum(\beta \times S(i) + Per_{max},\ Per_{min}) \tag{3}$$

$$MinIS(i) = maximum(\gamma \times S(i),\ LS) \tag{4}$$

Here $i \in I$ and $S(i)$ is the support of the item i. In Equation 3, Per_{max} and Per_{min} are the user-specified maximum and minimum periodicities such that $Per_{max} \geq Per_{min}$ and $\beta \in [-1, 0]$ is a user-specified constant. In Equation 4, LS is the user-specified lowest minimum item support allowed and $\gamma \in [0, 1]$ is a parameter that controls how the $MinIS$ values for items should be related to their supports.

5.1 Experiment 1

In this experiment, we analyze the generation of periodic-frequent patterns at fixed $MaxIP$ values and varying $MinIS$ values. For T10I4D100K, Retail and Mushroom datasets, both *minsup* and LS values are fixed at 0.1%, 0.1% and 25%, respectively. With $\gamma = \frac{1}{\alpha}$, and varying α, we present the number of periodic-frequent patterns generated in these databases at different *maxprd* ($maxprd = P_{min} = P_{max} = x\%$) values in Fig. 4(a)–(c). It can be observed that the number of periodic-frequent patterns are significantly reduced by our model when α is not too large. When α becomes larger, the number of periodic-frequent patterns found by our model gets closer to that found by the basic model (single *minsup-maxprd* model) and the MCPF-model. The reason is that when γ becomes larger, more and more items' $MinIS$ values reach LS.

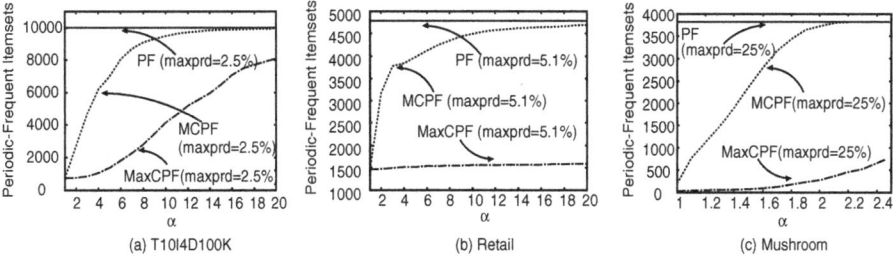

Fig. 4. Periodic-Frequent patterns generated at different $MinIS$ values

5.2   Experiment 2

In this experiment, we analyze the generation of periodic-frequent patterns at fixed $MinIS$ values and varying $MaxIP$ values. For T10I4D100K, Retail and Mushroom datasets, $minsup$ and LS values are fixed at 0.1%, 0.1% and 25%, respectively. Next, γ was set at 0.1. With $\beta = -\frac{1}{\alpha}$ and varying α, we present the number of periodic-frequent patterns generated in these databases at different $maxprd$ values in Fig. 5(a)–(c). It can be observed that the number of periodic-frequent patterns is significantly reduced by the proposed model when α is small. When α becomes larger, the number of periodic-frequent patterns found by the proposed model get closer to that found by the traditional model (single $minsup$-$maxprd$ model). The reason is because when α increases more and more items' $MaxIP$ values reach P_{max}. Fig. 6(a), (b) and (c) show the comparison between the runtime taken by the three models at different $MaxIP$ values. It can be observed that the runtime taken by the proposed approach to generate periodic-frequent patterns increases with increase in α value. It is because of the increase in the number of periodic-frequent patterns.

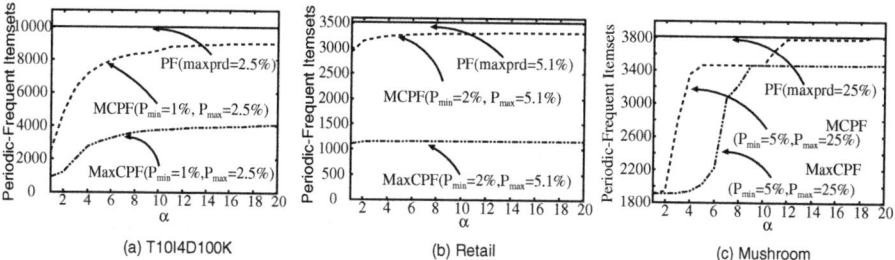

Fig. 5. Periodic-Frequent patterns generated at different $MaxIP$ values

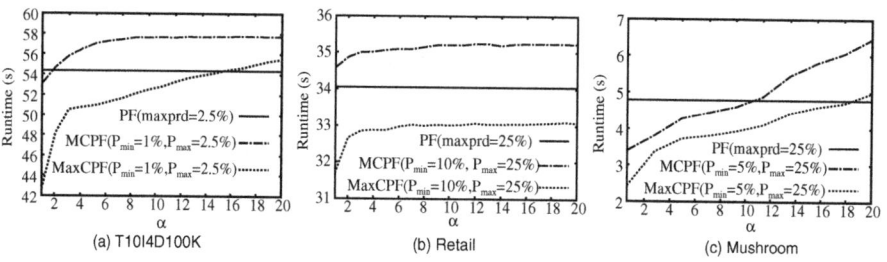

Fig. 6. Runtime taken for generating periodic-frequent patterns at different $MaxIP$ values

It can be observed that the proposed model (MaxCPF) takes less runtime than the the MCPF-model for all values of α. It is because the periodic-frequent patterns mined using the MCPF-model do not satisfy *downward closure property* whereas those mined using the proposed MaxCPF-model does, reducing the search space significantly and thus, resulting in less runtime. Also, it can be observed that the proposed model takes less runtime to find periodic-frequent patterns than the basic model (PF) for smaller values of α. It is because the number of periodic-frequent patterns mined using the proposed model is much less compared to that mined using the basic model for smaller values of α. At higher values of α, the proposed approach requires more runtime to find periodic-frequent patterns. The reason is that in the proposed model, we have to compute the minimum $MaxIP$ value among all items in a pattern. This computation needs to be done for all the patterns, thereby increasing the runtime.

Observations from Retail Dataset: To verify the nature of periodic-frequent patterns mined using MCPF and MaxCPF models, we extracted the set of periodic-frequent patterns from Retail dataset mined at $\alpha = 10$ in Figure 5(b). The MCPF-model discovered 3290 periodic-frequent patterns, whereas the proposed MaxCPF-model discovered 1162 periodic-frequent patterns (about 65% less than the MCPF-model). Here, we report some patterns which were discovered using MCPF-model but were pruned out by the proposed MaxCPF-model.

(i) The item '48' is a highly frequent item with support 42,135 and periodicity 19. The item '3714' is a rare item with support 129 and periodicity 3083. The MCPF-model has generated $\{48, 3714\}$ as a periodic-frequent pattern with support 89 and periodicity 4185. It can be observed that this pattern is uninteresting because a highly frequent item (i.e. '48') is combining with a rare item (i.e. '3714') giving a pattern having very low support and high periodicity. The proposed MaxCPF-model has pruned out this uninteresting pattern.

(ii) Similarly, the MCPF-model generated an uninteresting periodic-frequent pattern $\{39, 48, 98\}$ with support 89 and periodicity 3258. It is because the pattern consists of the highly frequent items '39' (with support 50,675 and periodicity 15), '48' (with support 42,135 and periodicity 19) and a rare item '98' (with support 232 and periodicity 2390). The proposed MaxCPF-model has pruned out this uninteresting pattern.

6   Conclusion

In this paper, we proposed an efficient model to address "rare item problem" while mining periodic-frequent patterns in real-world datasets. We have also developed a pattern-growth approach for mining periodic-frequent patterns using the proposed model. The experimental results on various datasets show that the proposed model can efficiently discover periodic-frequent patterns containing both frequent and rare items. Furthermore, the results also show that the pattern-growth approach is effective.

References

1. Agrawal, R., Srikant, R.: Fast algorithms for mining association rules in large databases. In: VLDB, pp. 487–499 (1994)
2. Brijs, T., Swinnen, G., Vanhoof, K., Wets, G.: Using association rules for product assortment decisions: A case study. In: Knowledge Discovery and Data Mining, pp. 254–260 (1999)
3. Han, J., Pei, J., Yin, Y., Mao, R.: Mining frequent patterns without candidate generation: A frequent-pattern tree approach. Data Min. Knowl. Discov. 8(1), 53–87 (2004)
4. Uday Kiran, R., Krishna Reddy, P.: Towards Efficient Mining of Periodic-Frequent Patterns in Transactional Databases. In: Bringas, P.G., Hameurlain, A., Quirchmayr, G. (eds.) DEXA 2010. LNCS, vol. 6262, pp. 194–208. Springer, Heidelberg (2010)
5. Tanbeer, S.K., Ahmed, C.F., Jeong, B.-S., Lee, Y.-K.: Discovering Periodic-Frequent Patterns in Transactional Databases. In: Theeramunkong, T., Kijsirikul, B., Cercone, N., Ho, T.-B. (eds.) PAKDD 2009. LNCS, vol. 5476, pp. 242–253. Springer, Heidelberg (2009)
6. Weiss, G.M.: Mining with rarity: a unifying framework. SIGKDD Explor. Newsl. 6(1), 7–19 (2004)

Algorithms to Discover Complete Frequent Episodes in Sequences

Jianjun Wu, Li Wan, and Zeren Xu

Chongqing University China
wjjalss@stu.xjtu.edu.cn, wanli@cqu.edu.cn, erbantou@hotmail.com

Abstract. Serial episode is a type of temporal frequent pattern in sequence data. In this paper we compare the performance of serial episode discovering algorithms. Many different algorithms have been proposed to discover different types of episodes for different applications. However, it is unclear which algorithm is more efficient for discovering different types of episodes. We compare Minepi and WinMiner which discover serial episodes defined by minimal occurrence of subsequence. We find Minepi cannot discover all minimal occurrences of serial episodes as the literature, which proposed it, claimed. We also propose an algorithm Ap-epi to discover minimal occurrences of serial episode, which is a complement of Minepi. We propose an algorithm NOE-WinMiner which discovers non-overlapping episodes and compare it with an existing algorithm. Extensive experiments demonstrate that Ap-epi outperforms Minepi(fixed) when the minimum support is large and NOE-WinMiner beats the existing algorithm which discovers non-overlapping episodes with constraints between the two adjacent events.

Keywords: Serial Episode, Non-overlapping Serial Episode, Sequence.

1   Introduction

Episodes are a common form of data that can contain important knowledge to be discovered. Finding the frequent episodes may provide interesting insight to experts in various domains. In particular, it has been shown to be very useful for alarm log analysis, financial events and stock trend relationship analysis [12]. There are a lot of algorithms which have been very successful in applications. One kind of those is to count the minimal occurrences of episode (shortly named Mo) such as Winepi[1],Minepi[2],WinMiner[3], while another kind is used to count the occurrences of non-overlapped episodes such as Nol [4]. But the minimal occurrences of episodes found by different algorithms are different, so that it's hard to compare the time performances of these algorithms with a same criterion.

Contributions:

1. We propose two lemmas to get the complete candidates; therefore we can discover all minimal occurrences of serial episodes.

2. We propose our algorithm named Ap-epi which can be used to find frequent episodes. This algorithm shows a better performance than Minepi(fixed) when the support is large.

L. Cao et al. (Eds.): PAKDD 2011 Workshops, LNAI 7104, pp. 267–278, 2012.

3. We have another algorithm that used to find the occurrences of non-overlapped episodes, which named NOE-WinMiner. The algorithm can find the frequent episodes with the interest-gapmax, while the existing algorithms cannot do.

4. Our algorithms are shown to outperform with different interests as well as the other algorithms.

The remainders of this paper are organized as follows. The next section presents a brief overview of the frequent episode discovery framework. Sec.3 presents the principle for discovering minimal occurrences of serial episodes. Sec.4 presents the new frequency counting algorithms. We compare the effectiveness and efficiency of our algorithms with other algorithms through some experiments in Sec.5. In Sec.6 we present the conclusion.

2   Overview of Frequent Episodes Mining Framework

The related research is to efficiently discover frequent serial episode in a *single* deterministic sequence [5, 6]. Both [2] and [3] present methods to discover frequent serial episode based on minimal occurrence of serial episode and suppose Aprior property holds for serial episode. However, neither of them discovered that Aprior property cannot be utilized to prune infrequent candidate serial episode based on the concept of minimal occurrence. Since [2] designed candidate generation method supposing Aprior property [7, 8, and 9] holds for serial episode, it cannot discover the complete set of frequent serial episodes. Of course, the QIMIE organization committee also makes a lot of contributions to interesting measures, such as [11], while we pay close attention to the limits in the frequent episodes mining.

Formally, we also follow the standard notions of event sequence, serial episode, minimal occurrences, and support used in [3] and [4] or give equivalent definition, when more appropriated to our presentation.

Definition 1. (event, ordered sequence of events) Let E be a set of event types. An event is defined by the pair (e, t) where $e \in E$ and $t \in N$. The value t denotes the time at which the event occurs. An ordered sequence of events s is a tuple $s = <(e_1,t_1),(e_2,t_2),...,(e_n,t_n)>$ such that $\forall i \in (1,...,n)$, $e_i \in E \wedge t_i \in N$ and $\forall i \in (1,...,n-1)$, $t_i \leq t_{i+1}$.

Definition 2. (event sequence) An event sequence S is a triple (s, T_s, T_e), where s is an ordered sequence of events of the form $<(e_1,t_1),(e_2,t_2),...,(e_n,t_n)>$ and T_s, T_e are natural numbers such that $T_s \leq t_1 \leq t_n \leq T_e$.

Definition 3. (episode) An episode is a tuple α of the form $\alpha = <e_1, e_2,..., e_k>$ with $e_i \in E$ for all $i \in \{1,...,k\}$. In this paper, we will use the notation $e_1 \rightarrow e_2 \rightarrow ... \rightarrow e_k$ to denote the episode $<e_1, e_2,..., e_k>$ where '$\rightarrow$' may be read as 'followed by'. We denote the empty episode by $\varnothing$.

Definition 4. (occurrence) An episode $\alpha = <e_1, e_2,..., e_k>$ occurs in an event sequence $S = (s, T_s, T_e)$, if there exists at least one ordered sequence of events $s' = <(e_1,t_1),(e_2,t_2),...,(e_k,t_k)>$ such that s' is a subsequence of s.

The interval $[t_1,t_k]$ is called an occurrence of α in S. The set of all the occurrences of α in S is denoted by $occ(\alpha,S)$.

These episodes and their occurrences correspond to the serial episodes of [6], up to the following restriction: the event types of an episode must occur at different time stamps in the event sequence. This restriction is imposed here for the sake of simplicity, and the definitions and algorithms can be extended to allow several event types to appear at the same time stamp. However, it should be noticed that this constraint applies on occurrences of the patterns, and not on the dataset (i.e., several events can occur at the same time stamp in the event sequence).

Gapmax is a user-defined threshold that represents the maximum time gap allowed between two consecutive events. That means in the occurrence, such as the time stamps of s, $\forall i \in (1,...,k-1)$, $0 < t_{i+1} - t_i < $gapmax.

Definition 5. (minimal occurrence) Let $[t_s, t_e]$ be an occurrence of an episode α in the event sequence s. If there is no other occurrence $[t_s', t_e']$ such that $(t_s < t_s' \wedge t_e' \leq t_e) \vee (t_s \leq t_s' \wedge t_e' < t_e)$ (i.e. $[t_s', t_e'] \subset [t_s, t_e]$),then the interval $[t_s, t_e]$ is called a minimal occurrence of α. The set of all minimal occurrences of α in S is denoted by mo(α,S).

Intuitively, a minimal occurrence is simply an occurrence that does not contain another occurrence of the same episode.

Definition 6. (non-overlapped occurrence) Consider an episode α =<e_1, e_2,..., e_k>. Two occurrences, h_1 = <t_1, t_2,..., t_k> and h_2 = <t_1', t_2',..., t_k'> of α are said to be non-overlapped if, either (i) $t_1 > t_k'$ or (ii) $t_k < t_1'$. A collection of occurrences of α is said to be non-overlapped if every pair of occurrences in it is non-overlapped. The corresponding frequency for episode α is defined as the cardinality of the largest set of non-overlapped occurrences of α in the given event sequence.

For example, in Fig.1, (A→B→A) has three minimal occurrences ([1, 2, 3], [3, 4, 6], [6, 7, 8]), while the non-overlapped episode (A→B→A) only occurs two ([1, 2, 3], [6, 7, 8]) in the sequence.

3   Principle for Discovering Minimum Occurrence of Serial Episode

In context of deterministic sequence, reference [2, 10] based their frequent serial episode discovery algorithm on concept of minimal occurrence and presented a lemma supposing Apriori property holds for serial episode.

Lemma 1. If a serial episode α is frequent in a sequence S, then all its sub-episodes α_s are frequent

We claim that "Lemma 1" does not hold for serial episode that is defined based on minimal occurrences.

Proof. Fig.1.(a) gives a counter example which demonstrates that the frequency of a serial episode might be larger than the frequency of its sub-episodes. As illustrated in Fig.1.(a), there is a sequence S and the frequency of serial episode α_1={A→B→B→A} is 1, mo(α_1,S)={[3,6]}. The frequency of α_2= {A→B→B→B} is 2, mo(α_2,S)={[1,5],[3,7]}. The frequency of α_3= {A→B→B→B→A} is 2, mo(α_3,S)={[1,6],[3,8]}. It is obvious that α_1 and α_2 are both sub-episodes of α_3. But the frequency of α is larger than that of α_1. If we define the minimum support threshold as 1, then α_3 is frequent but its sub- episode α_1 is not.

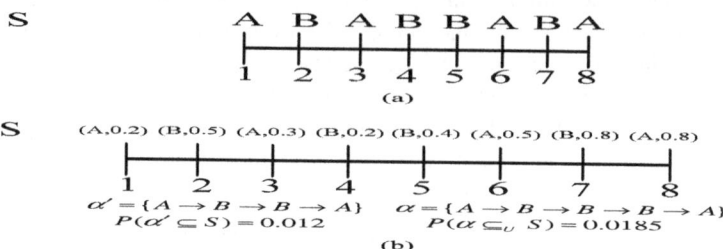

Fig. 1. (a)A counter example of Lemma 1 in determnistic sequence; (b) A counter example of Lemma 1 in uncertain sequence.

We present another lemma which holds for serial episode based on minimal occurrence. The necessary concepts are defined as follows.

Definition 7. (*Prefix of a serial episode*): Let α be a serial episode of size k in sequence S and $[t_s,t_e] \in$ mo(α,S). A sub-episode α_1 of size k-1 is the prefix of α such that $\exists t_1 \in [t_s,t_e]$, $[t_s,t_1] \in$ mo(α_1,S). α_1 is denoted as prefix(α).

Lemma 2. If a serial episode is frequent, its prefix must be frequent.

Proof. We prove this lemma by contradiction.
Assume that a serial episode α is frequent in S, but its prefix prefix(α) is not frequent, i.e. inequation (1) holds.

$$|mo(prefix(\alpha),S)| < |mo(\alpha,S)| \quad ^1$$
$$(1)$$

Therefore, there must be at least two minimal occurrences of α, $[t_{as},t_{ae}]$, $[t_{as}{}',t_{ae}{}'] \in$ mo(α,S) equation reference goes here and two occurrences of prefix(α),$[t_{as},t_1]$, $[t_{as}{}',t_2]$ ($t_1 < t_{\alpha e}$, $t_2 < t_{\alpha e}'$). By the assumption of inequation (1), $[t_{as},t_1]$ and $[t_{as}{}',t_2]$ could not be both minimal occurrence. Therefore, let us assume $[t_{as},t_1] \subset [t_{as}{}',t_2]$ and get the next inequation.

$$t_{as}{}' < t_{as} < t_1 \leq t_2 < t_{ae}{}'$$
$$(2)$$

By inequation (2):
   If $t_{ae} \leq t_{ae}{}'$, $[t_{as}{}',t_{ae}{}'] \in$ mo(α,S) does not hold.
   If $t_{ae} > t_{ae}{}'$, the type of tokens appear at t_{ae} and $t_{ae}{}'$ must be the same. Therefore, the occurrence $[t_{as},t_{ae}]$ includes another occurrence $[t_{as},t_{ae}{}']$ such that $[t_{as},t_{ae}{}'] \in$ mo(α,S) does not hold.
   We can conclude that, for each minimal occurrence of α there must be a corresponding minimal occurrence of its prefix prefix(α), i.e. $|mo(prefix(\alpha),S)| \geq |mo(\alpha,S)|$.Therefore, the assumption of $|mo(prefix(\alpha),S)| < |mo(\alpha,S)|$ does not hold.

---

[1] Prefix(α) < min_fr $\leq |mo(\alpha,S)|$.

4   Algorithms

In this section, we present two frequent discovery algorithms. In Sec 5.1, we first show how the algorithm to obtain the frequencies of serial episodes. Then in Sec 5.2, we present an algorithm for counting non-overlapped occurrences of serial episodes under constraint of gapmax.

4.1   An Apriori-Like Algorithm for Serial Episodes

Serial candidate episodes are recognized in an event sequence by using state automata that accept the candidate episodes and ignore all other input. Since we need to count the frequencies of all candidate episodes in one pass through the data, there are many automata that have to be simultaneously tracked. In order to access these automata efficiently they are indexed using a Waits (.) list. For each event type $A \in E$, the automata that accept A are linked together to a list Waits(A).The list contains entries of the form(a, x, init, arrive) that episode a is waiting for its *xth* event, and the first node of a comes at time init, the last arrived node of a comes at time arrive. This idea of efficiently indexing automata through a Waits (.) list was introduced in the windows-based frequency counting algorithm. The list was used only of the form (a, x), but we record the init time and the last node's arriving time so that we can add constraints between adjacent nodes.

The overall structure of the algorithm is as follows. The event sequence is scanned in time order. Given the current event, say (E_i, t_i), we consider all automata waiting for an event with event type E_i, i.e., the automata in the list waits (E_i). Automata transitions are effected and fresh automata for an episode are initialized by adding and removing elements from appropriate waits (.) lists. So we use a temporary storage called bag to store which is found necessary to add elements to the waits (.) list after the current loop. The idea of storing the necessary information was introduced in the Non-overlapped frequency counting algorithm. But as we want to count the minimal occurrences of episodes, we have to store more information in the bag, i.e., there always the automatons which wait for the first nodes of all candidate episodes needed to be put in the bag.

Since we want to count the minimal occurrences of the episodes, we may have N automata for N nodes of an N-Node episode. When the last node of an episode comes, several automatons of one episode may get their ends at the same time, then we have to record all the occurrences of the episodes, and calculate the minimal occurrence at last. Therefore, we use the other temporary storage named as *Occ* to store the occurrences of all the episodes which reach the end in this loop.

The strategy for counting minimal occurrences is very simple. At the beginning, we initialize all the automatons for every candidate episode, i.e. for an candidate episode a, it's first node is A, then add (a, 1, 0, 0) to the wait (A).As we go down the data sequence, this automation makes earliest possible transitions into each successive state. Once it reaches its finial state, an occurrence of the episode is recognized and stored in the Occ, and after the loop, we pick up the minimal occurrence from the Occ, and the frequency of this episode is increased by one. A fresh automation is initialized for this episode when an event corresponding to its first node appears in the data and the process of recognizing an occurrence is repeated. Algorithm 1 gives the pseudo code for counting minimal occurrences of serial episodes. In the description below we refer to the line numbers in the pseudo code.

Ap-epi:

Input: Set C of candidate N-node serial episodes, event stream $S=\{(E_1,t_1), \dots, (E_n,t_n)\}$, support threshold min_fr.

Output: The set F of frequent serial episodes in C

```
1.   for all event types A do
2.       Initialize Waits(A) = ∅
3.   for all a∈C do
4.       Add (a,1,0,0) to Waits(a[1])
5.       Initialize a.freq =0
6.   Initialize bag = ∅
7.   for i = 1 to n do
8.       /*n is the length of the data stream*/
9.       Initialize Occ = ∅
10.      for all (a, j, init, arrive) ∈ Waits(Eᵢ) do
11.          Record a,j,init,arrive
12.          Remove (a, j, init, arrive) from Waits(Eᵢ)
13.          Set j' = j + 1
14.          if j = 1 then
15.              init = i //Record the first element
16.              if j ≠ N
17.                  Add (a,1,0,0) to bag
18.          if j = N then
19.              Set j' = 1
20.              Add (a,1,0,0) to bag
21.              Add (a, init, i) to Occ
22.          else if a[j'] = Eᵢ  then
23.              Add (a, j', init, i) to bag
24.          else
25.              Add (a, j', init, i ) to Waits(a[j'])
26.      Empty bag into Waits(Eᵢ)
27.      for all a appears in   do
28.          pick up mo of all (a, init, arrive) ∈ Occ
29.          Update a. freq = a. freq + 1
30.Output F={ a∈C | a. freq ≥ min_fr}
```

For every init time, there is only one automaton of each episode which starts at that time; when an episode gets the final node, there may be several automatons of that episode reach end, and the init time of these automaton will not appear in any later occurrence, and the init time of the later automatons will greater than that of this one, that means the later occurrences will all overlap with the occurrences which ends at this time, and there must be an minimal occurrence that is one of the occurrences in the several automations. That means when the automatons of an episode reach the end in a loop, the frequency of the episode must be increased by one. If we do not want to calculate the *mo* of the episode, we can set a flag for each candidate episode to remark which the frequency of episode has been increased in the loop, and increase the frequency of the episode without *Occ* (delete line 27-29, Algorithm 1).

Space and Time Complexity. At any stage in the algorithm, there may be N active automaton per episode at most, which means that there are N|C| automatons being tracked simultaneously. The maximum possible number of elements in bag and Occ are both N|C|. Thus the space complexity of Algorithm 1 is O(N|C|).

The initialization time is O(|C|+|ε|), where |ε| denotes the size of the event type of the candidate episodes. The time required for actual data pass is linear in the length, n, of the data sequence. And in one loop, both the cost of time to record the occurrences of all episodes in C with the automations and to find minimal occurrence of the episodes from Occ are O(N|C|) at worst. Thus, the time complexity of Algorithm 1 is O (nN|C|).

Thus, Algorithm 1, both time-wise and space-wise, is an efficient procedure for obtaining frequencies of a set of serial episodes. In fact, the Aprior-like algorithm will even be better than the FP-Growth-like algorithms such as Minepi when we search the very frequency episodes. At that time, the projection database in FP-Growth-like algorithms is not much less than the original one, so for each episode, we have to scan the projection database of its prefix and suffix. The time complexity of FP-Growth-like is O (n^2|C|), where n is the length of projection database, so when length of projection database is not short enough, Algorithm 1 is better, like the situation above where the support threshold is so great that the iterations end earlier.

4.2   An FP-Growth-Like Algorithm for Non-overlapped Serial Episodes with Gapmax

The algorithm is based on WinMiner, that' the reason why we call it NOE-WinMiner. To get non-overlapping episode, we need to modify *join* function in WinMiner. More details of WinMiner can be found in Ref. [3]. Extend join function joins a non-overlapping frequent prefix X with a 1-length frequent episode Y and count its support. X.pattern denote the pattern template of X. X.Occ denotes the instance of X, and each instance is represented by a 2-tuple(t_{start}, T_{end}), where t_{start} is the starting time point of this instance of X, T_{end} is a set of time points which contains all the possible ending time points of this instance. Gapmax is a user specified threshold, and the time interval between any adjacent objects in an episode must not exceed it.

```
Function: extend_join
```

```
extend_join:
```
Input: x and y, two 2-tuples, containing an episode (e.g. x.Pattern)
and its set of minimal prefix occurrence (e.g. y.Occ), and where y
corresponds to an episode of size 1.
Output: z, a 2-tuple containing the non-overlapping episode
x.Pattern→y.Pattern and its set of minimal prefix occurrence.

```
1.Let z.Pattern = x.Pattern→y.Pattern
2.Let z.Occ = ∅, last = ∅
3.For all  (ts,T)∈x.Occ  do
4.   Let  L = ∅
5.   For all  t∈T  do
6.      Let EndingTimes = { ts'|∃(ts',T') ∈ y.Occ such that
7.      ts'> ts ∧ ts'-t<gapmax ∧ ∀(t1,T'')∈x.Occ, ts< t1⇒∀t2∈ T'', ts'≤ t2}
8.      Let  L=L∪ EndingTimes
9.   If  L ≠ ∅
10.      Let  z.Occ = z.Occ∪{( ts,L)}
11.      If ( last = ∅ || ∃te ∈ last ⇒ te< ts)
```

```
12.         z.spt++// increase non-overlapping support
13.         last = L
```

Compared to function *join* in Ref.[3], We add a variable *last* in *extend _ join* and line 10 to 13 are extended steps. *last* restores the set of all the possible ending time points of the most recently visited non-overlapping instance of *z.Pattern* . Steps from line 10 to 13 check whether the first object of current visiting instance occurs latter than at least one time point in set *last* . If so, the current visiting instance is a non-overlapping instance of episode *z.Pattern* , then we increase non-overlapping support *z.spt* and update *last* by the set of all the possible ending time points of the current visiting instance.

```
exploreLevelN:
```
Input: x a E/O-pair, and L₁ the set of E/O-pairs of frequent episodes
of size 1.

```
1.For all y∈L₁ do
2.   Let  =extend_join(x,y)
3.   If  |z.Occ|≥σ
4.         exploreLevelN(z,L₁)
```

By combining exploreLevelN and extend_join, we get the complete algorithm of NOE-WinMiner.

```
NOE-WinMiner:
Input: x and y, two E/O-pairs, containing an episode and its set of mpo
and where y corresponds to an episode of size 1, support threshold σ, a
maximum time gap constraint gapmax.
Output: The set F of frequent serial episodes
```

```
1.Initialize F = ∅
2.Scan database D to get all frequent sequential patterns of size 1 and
   their minimal occurrences in complete implicated database of D, L₁ₗ
3.For all x∈ L₁ do
4.        exploreLevelN(x, L₁)
```

The space complexity and time complexity of NOE-WinMiner are both as same as those of WinMiner.

5   Experiments

We have run a series of experiments using Ap-epi, Minepi, Nol and NOE-WinMiner. Of course, in this part Minepi is fixed by getting candidates with lemma 2, so that can find all frequent episodes. The general performance of the methods, the effect of the various parameters, and the scalability of the methods are considered is this section.

The test databases are all from UCI Machine Learning Repository. There are three database been used in the experiment. The title of the first database is Primate splice-junction gene sequence (DNA) with associated imperfect domain theory. We shortly named it "splice". There are 3190 sequences in it, and every sequence only has 61 events, and the kinds of characters in the data are 8.The title of the second database is E. coli promoter gene sequences (DNA) with associated imperfect domain theory. We shortly named it "promoter". There are 106 sequences in it. The third database is named as "Web-log" by us. There are 989818 sequences in it.

The experiments have been run on a PC with Intel(R) Core(TM)2 Duo CPU T5250 1.50GHz and 1GB main memory, under Windows XP system. The sequences resided in a flat text file.

5.1   Comparison of Algorithms Ap-epi and Minepi

We experimented with support threshold between 12 and 23. We make the max bound be 60 in Minepi to get all frequent episodes.

The frequent episodes found by Ap-epi and Minepi are the same, we also observe that the same general tendency for a rapidly decreasing number of episodes as the support threshold increases. As there are only a few kinds of characters in the sequences, there must be a lot of over-lapped episodes. That's also the reason why we get so many frequent episodes.

Both Fig.2 and Fig.3 show the time requirement for finding frequent episodes with Ap-epi and Minepi. We can easily find that Ap-epi has a good performance with the greater support threshold, while Minepi will be better with the support threshold decreasing.

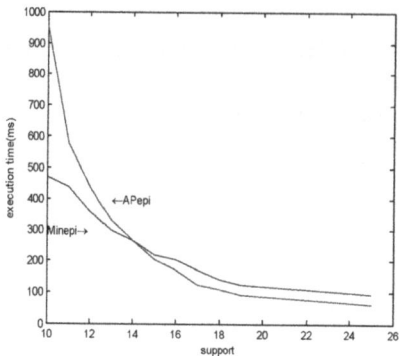

Fig. 2. Processing time for serial episodes with Ap-epi(solid line) and Minepi(dotted line):promoter

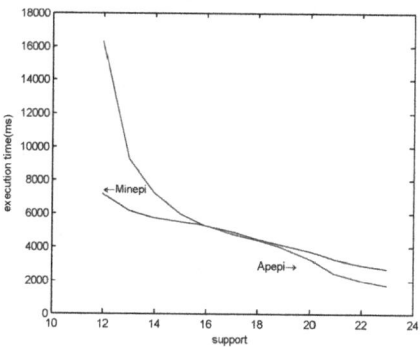

Fig. 3. Processing time for serial episodes with Ap-epi(solid line) and Minepi(dotted line):splice

On one hand, The Aprior-like algorithms such as Ap-epi, do not need projection database and related comparison operations. When the support threshold is great, with less iteration and converges faster, the cost of projection database and comparison operations is more expensive than scan the whole database. (The projection database is not smaller enough than the original database).

On the other hand, when the support threshold is smaller, the iterations will be more and the converge will be slower, then the projection database will be much smaller than the original one. At this time the FP-Growth-like algorithms will show the better performance.

5.2   Performance of NOE-WinMiner with Gapmax

Secondly, we want to see the performance of the non-overlapped algorithms with gapmax.

At first, we let support threshold=5, use 3000-gene data to do the experiment, get Fig.4. Then we also use the web-log data to do the experiment, and get the Fig.5.

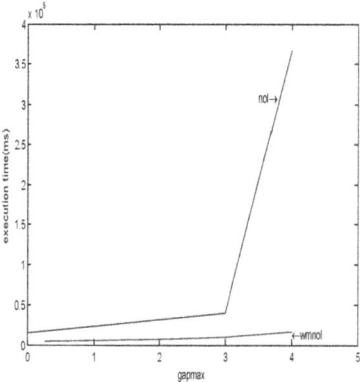

Fig. 4. Processing time for serial episodes as a function of gapmax with Nol(solid line) and NOE-WinMiner (dotted line):splice, support threshold=5

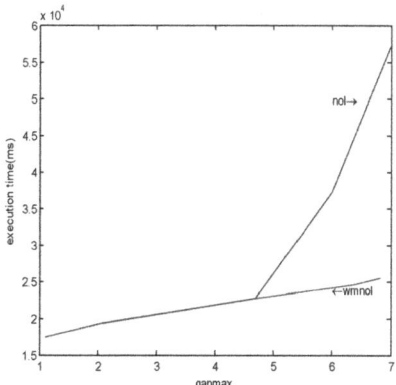

Fig. 5. Processing time for serial episodes as a function of gapmax with Nol(solid line) and NOE-WinMiner (dotted line):web-log, support threshold=30

All the figures show that the NOE-WinMiner algorithm with gapmax is more efficient than Nol. If we add the constraints to Nol, the strategy of pruning has changed. It will have much more Waits. If there're M candidates and each candidate is N-len, it can have M(N-1) Waits at most in a moment. While without gpmax, it only has M Waits. So as long as the episodes are, Nol with gapmax is more inefficient.

As NOE-WinMiner record every occurrence of any episode, when we want to add the constraints-gapmax, the information of adjacent tokens can easily get from the Mos. So NOE-WinMiner is suitable for mining data screams with gapmax.

The NOE-WinMiner algorithm has the important advantage: the relationship between adjacents the can be restricted and this constraint is helpful for the minging.

6   Conclusions

In this paper, we present the lemma that helps to discover all candidates, and an algorithm-Ap-epi based on it. We also present an algorithm named NOE-WinMiner, which discover non-overlapping episodes with gapmax. The experiments show that Ap-epi outperforms Minepi(fixed) when the minimum support is large and NOE-WinMiner beats the existing algorithm which discovers non-overlapping episodes with constraints between the two adjacent events. Furthermore, the study of the non-overlapped episodes is just beginning, and the algorithm of non-overlapped episodes can be extended to the applications with limited resources.

References

1. Mannila, H., Toivonen, H., Verkamo, A.I.: Discovering frequent episodes in sequences. In: Proceedings of the First International Conference on Knowledge Discovery and Data Mining (KDD 1995), Montréal, Canada, pp. 210–215 (1995)
2. Mannila, H., Toivonen, H.: Discovering generalized episodes using minimal occurrences. In: Proceedings of the Second International Conference on Knowledge Discovery and Data Mining (KDD 1996), Portland, OR, pp. 146–151 (1996)
3. Méger, N., Rigotti, C.: Constraint-Based Mining of Episode Rules and Optimal Window Sizes. In: Boulicaut, J.-F., Esposito, F., Giannotti, F., Pedreschi, D. (eds.) PKDD 2004. LNCS (LNAI), vol. 3202, pp. 313–324. Springer, Heidelberg (2004)
4. Laxman, S., Sastry, P.S., Unnikrishnan, K.P.: A fast algorithm for finding frequent episodes in event streams. In: Proceedings of the 13th ACM SIGKDD International Conference on Knowledge Discovery and Data Mining, San Jose, California, USA, August 12-15 (2007)
5. Han, J., Pei, J.: Mining Frequent Patterns by Pattern-Growth: Methodology and Implications. ACM SIGKDD Explorations (Special Issue on Scaleble Data Mining Algorithms) 2(2) (December 2000)
6. Han, J., Pei, J., Yin, Y.: Mining Frequent Patterns without Candidate Generation (PDF). In: Proc. 2000 ACM-SIGMOD Int. Conf. on Management of Data (SIGMOD 2000), Dallas, TX (May 2000)
7. Agrawal, R., Imieliński, T., Swami, A.: Mining association rules between sets of items in large databases. In: Proceedings of the 1993 ACM SIGMOD International Conference on Management of Data, Washington, D.C., United States, May 25-28, pp. 207–216 (1993)
8. Agrawal, R., Srikant, R.: Mining Sequential Patterns. In: Proceedings of the Eleventh International Conference on Data Engineering, March 06-10, pp. 3–14 (1995)
9. Srikant, R., Agrawal, R.: Mining Sequential Patterns: Generalizations and Performance Improvements. In: Apers, P.M.G., Bouzeghoub, M., Gardarin, G. (eds.) EDBT 1996. LNCS, vol. 1057, pp. 3–17. Springer, Heidelberg (1996)
10. Mannila, H., Toivonen, H., Verkamo, A.I.: Discovery of frequent episodes in event sequences. Data Mining and Knowledge Discovery 1(3), 259–289 (1997)
11. Suzuki, E.: Interestingness Measures-Limits, Desiderata, and Recent Results. In: QIMIE 2009 (2009)
12. Ng, A., Fu, A.: Mining Frequent Episodes for Relating Financial Events and Stock Trend. In: Whang, K.-Y., Jeon, J., Shim, K., Srivastava, J. (eds.) PAKDD 2003. LNCS (LNAI), vol. 2637, pp. 27–39. Springer, Heidelberg (2003)

Certainty upon Empirical Distributions

Joan Garriga

Dptmnt. de Llenguatges i Sistemes Informàtics,
Universitat Politècnica de Catalunya,
jgarriga@ceab.csic.es
http://www.lsi.upc.edu/~jgarriga/

Abstract. We address the problem of assessing the information conveyed by a finite discrete probability distribution, within the context of knowledge discovery. Our approach is based on two main axiomatic intuitions: (i) the minimum information is given in the case of a uniform distribution, and (ii) knowledge is akin to a notion of richness, related to the dimension of the distribution. From this perspective, we define a statistic that has a clear interpretation in terms of a *measure of certainty*, and we build up a plausible hypothesis, which offers a comprehensible insight of knowledge, with a consistent algebraic structure. This includes a native value for the uncertainty related to unseen events. Our approach is then faced up with entropy based measures. Finally, by implementing our measure in a decision tree induction algorithm, we show an empirical validation of the behavior of our measure with respect to entropy. Our conclusion is that the contributions of our measure are significant, and should definitely lead to more robust models.

Keywords: knowledge discovery, measures of information, entropy.

1 Introduction

Many data mining tasks for knowledge discovery rely on the use of the so called *information measures*. Such measures are intended in order to select an optimal model (statistical model selection, graphical modeling), an optimal set of rules (classification rule mining), an optimal split at each node of a tree (induction of decision trees), an optimal discretization of a continuous variable, among other examples. In any case, all of them are particular forms of expressing knowledge learned from data which in this context, means *the degree of certainty with respect to the outcome of a random variable*. But, regardless to the final objective of the mining process - let's suppose that no prior knowledge is available -, knowledge is invariably and uniquely expressed by occurrences and co-ocurrences of values observed in the sample. Therefore, such measures intend to assess the amount of information conveyed by any finite discrete probability distribution observed in the data.

Among others, Shannon's entropy [11] is the most widely known measure of uncertainty associated to a probability distribution. Some attractive properties

L. Cao et al. (Eds.): PAKDD 2011 Workshops, LNAI 7104, pp. 279–290, 2012.

hold for this measure and make it uniquely characterized [2]. A nice correspondence can be established between these properties and what is commonly accepted as a plausible axiomatic definition of knowledge [1]. This is the reason of its success, and the basis of a comprehensive later work.

Although entropy is strongly rooted within the information theory community, entropy's characterization does not properly cover some aspects of knowledge. It is well known for instance that, when applied to the induction of decision trees, entropy shows a certain bias for attributes with greater cardinality. It also yields some undesired results, when the attributes to be used and/or the classes to be learned have highly imbalanced frequencies. Thus, further generalizations have been developed, like Rényi's entropies of type α [10], and Daróczy's entropies of type β [3]. Furthermore, other solutions have been suggested in the form of combined entropies, (entropic gain [8], the u coefficient of Theil [13], the gain-ratio [9], the Kvalseth coefficient [6], or more recently the off-centered entropies [7]).

We propose alternatively a new measure of certainty [1] derived from a slightly different axiomatization of knowledge which takes into account these aspects. As we will show, when applied to practical problems such as induction of decision trees, this perspective leads to some different results.

In section 2, we describe the aim and major contributions of our approach. In section 3, we briefly review the groundwork of the proposed measure. In section 4, we focus on entropy based measures and how they relate to our contributions. In section 5, we give some experimental results. Finally, in section 6, we comment the results and summarize our conclusions.

2   Contributions

In a few words, we introduce a measure of *Certainty* upon empirical probability distributions. With this measure, we aim at overcoming two important aspects of knowledge described below.

2.1   The Cardinality Scaling of Knowledge

Let's figure the problem of modeling the price of a house from a set of features (city, square meters, garage, ...). We can surely deal with it as a regression problem. But we could also try to discretize the price variable and treat it as a classification problem. In this case, it is clear that - overfitting issues apart - the higher the cardinality of the class variable in our final model, the more information it expresses. This is a direct illustration of our intuition that knowledge is akin to a notion of richness related to the dimension of the distribution. But we would like to depict a deeper idea of this concept.

---

[1] Certainty and uncertainty are indeed quite the same thing: just different degrees of knowledge. But we want to emphasize this aspect of certainty, as opposed to entropy based measures of uncertainty.

So, let's figure the following horse race betting example. We have a sample of races in which two horses (Tomcat and Apache) show equal number of victories. Apparently, there exist no compelling reasons to favour either of them in future races. Now, let's suppose that we have an equal size second sample of races in which Tomcat is running against two competitors and we observe the same 0.5 chance of winning for Tomcat. In this case, it could make sense to bet. Going further, if more competitors are involved in the race and the sample keeps yielding the same 0.5 chance for Tomcat, the odds for each one of the other competitors are bound to decrease and our confidence on Tomcat's victory could increase. Indeed, we are always bound to loose 0.5 of our bets in the long run but it is clear that our epistemic state is quite different and we are increasingly compelled to bet on those more populated races, though Tomcat's chances remain the same.

The aim of this example is to show that knowledge about random events is conditioned not only by the observed frequencies but also by the dimension of the probability distribution under consideration - or the cardinality of the set of possible outcomes. Therefore, we call this effect the *cardinality scaling of knowledge*, which we may regard as different levels of *quality* of knowledge. We must seek the roots of this effect not in the decreasing chances of each competitor but rather in the curse of dimensionality: given a finite fixed sample size, the higher the dimension of the sample space - the number of competitors -, the lower the prior probability of observing a 0.5 chance for Tomcat. Thus, allthough the frequencies observed are equal, their statistical significance is not the same.

Now let's think of it as it is the case in data mining processes: given a single fixed sample, models of different complexity are taken under consideration. For each model considered, the statistical significance of the corresponding frequencies is different and so must be the amount of knowledge they convey.

In fig.1 we show how entropy and certainty capture this aspect of knowledge. This is a piecewise depiction of each event's elementary contribution to the total certainty/uncertainty of the random process for a growing number of uniform competitors - given by s - while Tomcat's chances remain the same. Tomcat's 0.5 chance contribution is represented at the bottom.

In essence, what this figure shows is the evolution of our epistemic state with respect to the outcome of the race, in relation with the growing number of runners. Both of them express the obvious increase of uncertainty with the number of competitors. But we can observe some differences:

1. the increasing rate of uncertainty is different, so we should assume that, at a global level, each approach expresses a different scaling of knowledge;
2. the bounding is also different, $\lim_{s \to \infty}(H_s) = \infty$, while $\lim_{s \to \infty}(K_s) = 0.25$;
3. the single event *Tomcat wins* contributes a fixed (cardinality independent) amount to the total uncertainty, while its contribution to the total certainty is dependent on the number of runners (cardinality dependent);
4. as the number of runners increases, the competitors contribute together an increasing amount of uncertainty, while their joint contribution to the total certainty tends to zero.

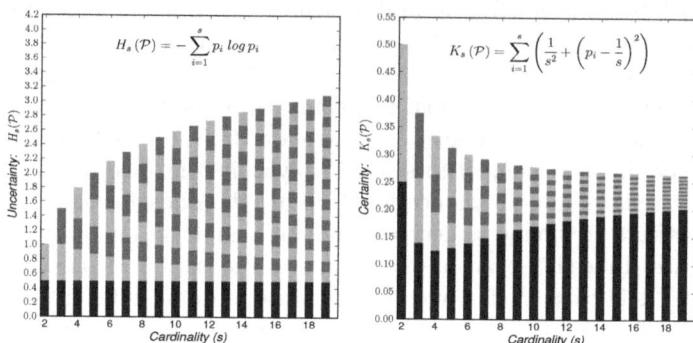

Fig. 1. Given $\mathcal{P}\left(0.5, \frac{0.5}{(s-1)}, ..., \frac{0.5}{(s-1)}\right)$, depiction of the elementary contributions to: (left) uncertainty as expressed by entropy; (right) certainty as expressed by equation 3, (promptly introduced in section 3). Please note the different scaling.

Now, given an ideal, large enough, sample size, the question is: does our initial state of knowledge (with two runners) evolve to an unbound uncertainty, as the number of runners grows (ideally) indefinitely? This can be argued in many different ways, but our point of view is that the answer to this question is definitely no. Otherwise, this assumption would leave no room to our certainty that Tomcat is going to win half of the races.

Our reasoning comes from the certainty side. As the cardinality increases, our certainty decreases, and each competitor's contribution is less because their chances are lower. Up to here, this is equally expressed by both measures. The difference in our approach is that, while the global certainty decreases, an increasing part of it is due to Tomcat's chances. At the limit, we reach an ideal situation in which we just have the amount contributed by Tomcat. This is because the competitors are (tending to) infinite and uniform, two reasons that explain a null contribution to the final certainty. At the same time, as each competitor's chances has almost vanished, Tomcat's victory seems to be amazingly guaranteed. But our certainty can not be one because Tomcat's chances are less than one, leading us to the maximum possible certainty, which is bounded by the value contributed by Tomcat in our initial state. We judge this as a more comprehensible description of our epistemic state, than a state of unbounded uncertainty.

2.2 Uncertainty about Unseen Events

The former example is an illustration of what we refer as the *algebra of knowledge*, that is, the way we measure the pieces of knowledge contributed by each one of the elementary events of the sample space, and the way these pieces should be combined in order to yield a global measure of the information conveyed by their empirical joint distribution. Obviously, this must include the fraction of the input space that is not represented in the sample, what we call the *unseen events*. This is again, a well known consequence of the curse of dimensionality.

Given a set of data, we can consider different models with different sets of input features. For each one of the models, the set of all joint configurations of the input features defines a particular input space, the dimension of which grows geometrically with the cardinalities of the input features. Thus, as soon as the complexity of the model involves just a few number of features, the sample will become very sparse and a great number of configurations will not be present in the sample. The joint configurations not observed in the sample are the *unseen events* and represent a fraction of the input space in which the model will inevitably fail. Of course, one may assume that they will rarely occur, but it is clear that they give rise to a certain amount of uncertainty that should not be obviated.

With regard to such basic concept, entropy yields a puzzling result: the uncertainty contributed by an unseen event is zero!, as expressed by the weird mathematical artifact $H(0) = 0 \log \infty = 0$. This result is somewhat clashing, and according to this, we are lead to believe that some issues seem to be sneaking through entropy's axiomatization of knowledge. The evidence is that, at this point, one has to rely on *ad-hoc* smoothing procedures, in order to estimate a complete probability distribution from the frequencies observed in the sample.

Conversely, we start up with a slightly different axiomatic approach to knowledge. From there, we derive a measure of certainty which achieves both: it takes into account the cardinality scaling of knowledge, and it yields a natively smoothed piece of certainty about each single event, wether observed or not in the sample. This measure is characterized by an analog set of properties to those holding for entropy. But, in our case, the algebra of knowledge is more clearly stated and offers a quite comprehensible insight of knowledge.

3   A Measure of Certainty

In the following, we denote by $\mathcal{P} = (p_1, p_2, \ldots, p_s)$ a finite discrete probability distribution, where $\mathcal{P}$ is a vector of observed frequencies over the set of disjoint dependent events $\Omega = \{e_1, e_2, \ldots, e_s\}$ observed in a sample. Also, we denote by $C_s = (s-1)/s$ and $U_s = 1/s$ what we call the *certainty* and *uncertainty* factors associated to the cardinality s of the distribution.

Our starting point is to measure the deviation of any such distribution, with respect to uniformity. Uniformity means equiprobability, which is the most uninformative distribution about the outcome of a random variable. Thus, our interpretation follows straightforward: the larger the deviation, the greater the amount of knowledge expressed by the distribution. As a measure of such deviation we take the L_2 norm to the uniform distribution:

$$L_2(\mathcal{P}, \mathcal{U}) \equiv \sum_{j=1}^{s} (p_j - U_s)^2 \quad .$$

Now, let's introduce our simple axiomatic requests: (i) the minimum knowledge we can have is that given in case of uniformity, and (ii) knowledge is akin to a notion of richness, related to the cardinality of the distribution. From the combination of both we may derive that, in the worst case, knowledge should not be zero. A more consequent alternative is to consider that the minimum is expressed by the uncertainty factor. That is, at the point of minimum information we have U_s and it increases as the square deviations increase. The most direct expression of this idea is,

$$K_s\left(\mathcal{P}\right) = U_s + \sum_{j=1}^{s} \left(p_j - U_s\right)^2 = \sum_{j=1}^{s} p_j^2 \tag{1}$$

It is straightforward to show that the following properties hold: (i) normalization, (ii) monotonicity (with respect to deviation), (iii) symmetry and (iv) expansibility.

And yet a fifth property holds, in relation to the composition of two successive random variables: given $\mathcal{P} = (p_1, p_2, \ldots, p_s)$ and $\mathcal{T} = (t, 1 - t)$, and their composition $\mathcal{Q} = (t\, p_1, (1 - t)\, p_1, p_2, \ldots, p_s)$, we have,

$$K_{s+1}\left(\mathcal{Q}\right) = K_s\left(\mathcal{P}\right) - p_1^2 \left(1 - K_2\left(\mathcal{T}\right)\right) \tag{2}$$

This looks quite natural: our knowledge about the final outcome of the successive composition of two distributions, is the certainty of the first distribution except for the additional uncertainty contributed by the second distribution.

3.1   Disjoint Dependent Events

One may think that there is nothing new in eq.1: it just looks like a simple translation of the euclidean distance to the uniform distribution, leading us to the well known *indexes of diversity/concentration*, long ago defined by [4], [5] or [12], among others. Furthermore, eq.1 is apparently independent of cardinality and explicitly expresses that any unseen event has a null contribution to the total certainty.

The point comes with the algebra of knowledge that underlies this expression. In fact, for each elementary event we have that,

$$p_j^2 = \left(U_s + \left(p_j - U_s\right)\right)^2 = U_s^2 + \left(p_j - U_s\right)^2 + 2\, U_s \left(p_j - U_s\right)$$

Therefore, we see that the certainty offset U_s is equally distributed among all possible outcomes of the distribution, yielding a term U_s^2. Thus, the amount contributed by each e_j is limited to $\left(p_j - U_s\right)^2 + 2\, U_s \left(p_j - U_s\right)$, from which the second term globally cancels out.

Consequently, if p_j is the observed probability of occurrence of event e_j, our knowledge about the outcome of e_j is the composition of two terms: a cardinality dependent offset and a deviation with respect to the uniform distribution,

$$K_s\left(p_j\right) = U_s^2 + \left(p_j - U_s\right)^2 \tag{3}$$

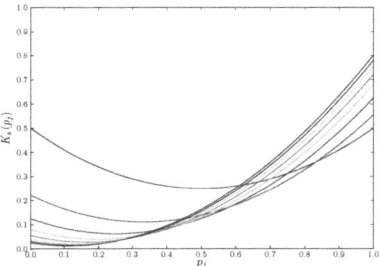

Fig. 2. Single event's Certainty for $s = \{2, 4, 8, 16, 32, 64, 128, \infty\}$

While remaining consequent with eq.1, that is, $K_s(\mathcal{P}) = \sum_{j=1}^{s} K_s(p_j)$, this expression (shown in fig. 2) offers a quite different picture of the elementary contributions to global certainty:

- It is explicitly dependent on s. In the limit, where this measure would hardly apply, certainty meets (square) probabilities,

$$\lim_{s \to \infty} K_s(p_j) = p_j^2 \quad .$$

- The minimum value is coherently given at the point of equiprobability, where we have $K_s(U_s) = U_s^2$.
- It is continuous at zero, yielding a value greater than zero for any unseen event, that is, $K_s(0) = U_s^2 + U_s^2$. This value is the expression of a conservative attitude with respect to future coming examples, while at the same time, it is an assertion of the certainties with which the rest of events have been observed.
- Being consequent with the previous, for any event with an observed probability of one, the measure yields a value lower than one, $K_s(1) = U_s^2 + C_s^2$
- In case of uniformity, or minimum information, we have,

$$K_s(\mathcal{P}) = \sum_{j=1}^{s} K(p_j) = U_s \quad .$$

- In the case of observing only one event, we have maximum certainty, (which does not mean absolute certainty),

$$K_s(\mathcal{P}) = K_s(1) + (s-1) K_s(0) = 1 \quad .$$

Together with eq.2, these features synthesize the additive algebra of knowledge that is implicit by the measure of certainty. Knowledge is defined as the sum of the pieces contributed by disjoint dependent events, and as the square weighted sum of knowledge about combined events.

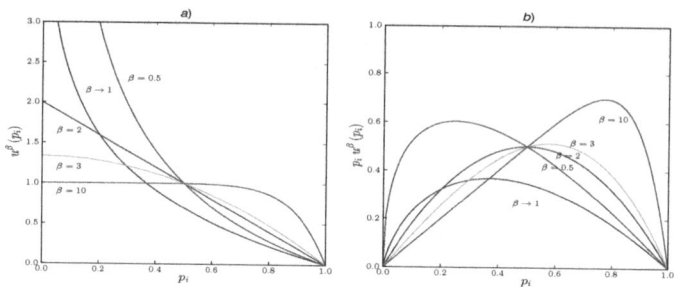

Fig. 3. a) Single event's uncertainty; b) Single event's weighted contribution

4   Entropy Based Measures

Though initially not conceived as such, Shannon's entropy [11] is, by far, the most widely used measure of information,

$$H\left(\mathcal{P}\right) = H\left(p_1, p_2, \ldots, p_n\right) = \sum_{i}^{n} p_i log_2 \frac{1}{p_i} \tag{4}$$

Further generalizations of entropy have been defined, first by Rényi [10],

$$H_\alpha\left(\mathcal{P}\right) = H_\alpha\left(p_1, p_2, \ldots, p_n\right) = \frac{1}{1-\alpha} log_2 \left(\sum_{i=1}^{n} p_i^\alpha\right), \tag{5}$$

(with, $\alpha > 0$, and $\alpha \neq 1$; in the limiting case of $\alpha \to 1$, Rényi's entropy tends to Shannon's entropy), and later by Daróczy [3],

$$H^\beta\left(\mathcal{P}\right) = \sum_{i=1}^{n} p_i\, u^\beta\left(p_i\right) \text{ , where, } u^\beta\left(p_i\right) = \frac{2^{\beta-1}}{2^{\beta-1}-1}\left(1 - p_i^{\beta-1}\right)$$

what yields the so called entropies of type β, (with $\beta > 0$, and $\beta \neq 1$),

$$H^\beta\left(\mathcal{P}\right) = H^\beta\left(p_1, p_2, \ldots, p_n\right) = \frac{2^{\beta-1}}{2^{\beta-1}-1}\left[1 - \sum_{i=1}^{n} p_i^\beta\right] \tag{6}$$

(in the limiting case of $\beta \to 1$, Daróczy generalization tends to Shannon's entropy, and setting $\beta = 2$, yields the quadratic entropy, $H^2 = 2\left(1 - \sum_i^n p_i^2\right) = 2\sum_i^n p_i\left(1 - p_i\right)$, identical to the so called Gini index [4]).

The contribution of Rényi's extended notion of entropy is that the term $-log\left(p_i\right)$ is interpreted as the entropy of the generalized distribution consisting of the single probability p_i, becoming thus evident that equation 4 is, indeed, a mean value [10]. Daróczy generalization is even more explicit by introducing the function of uncertainty $u^\beta\left(p_i\right)$ for a single event e_i.

The second contribution refers to the shape of the curve, determined by the β factor. As to our concern, this curve resolves a particular value of uncertainty for unseen events - undetermined in Shannon's entropy - given by $u^\beta(0) = \frac{2^{\beta-1}}{2^{\beta-1}-1}$. In any case, the global uncertainty remains as a mean value and therefore this particular value is meaningless by itself.

This is what we depict in fig.3: at the left (fig.3a), we plot the values of uncertainty $u^\beta(p_i)$ for different values of β, with special emphasis on Rényi's uncertainty, $u^{\beta \to 1} = -log(p_i)$, and on the quadratic entropy, $u^2 = 2(1 - p_i)$; at the right (fig.3b), we plot the corresponding weighted contribution of a single event, showing how the contribution of unseen events, inevitably, vanishes.

As formerly stated, there are reasons to believe that entropy does not cover a proper axiomatization of knowledge. If we do believe that unseen events are to be taken under consideration, we easily come to the conclusion that the algebra of knowledge does not fit well with the concept of a weighted mean measure. Furthermore, if we do believe that knowledge is akin to a notion of quality, we may yet find a reason to understand entropy's bias. It is argued that entropy is cardinality dependent - being its maximum given by $log(s)$ - but as long as there is no room for unseen events, this argument becomes worthless.

Conversely, we are giving a different answer to each one of these questions: certainty's algebra of knowledge is just additive (not a weighted mean), cardinality dependent, and yields not null values for unseen events. Beyond this, expansibility is still implicit in this algebra. With respect to entropy, expansibility follows straightforward from the fact of being a weighted mean. Herein, the connotations of certainty's expansibility are much stronger.

5   Empirical Validation

We have run some experiments to empirically study the behavior of our measure, by way of implementing it in a decision tree algorithm, and comparing it with the landmark decision tree C4.5 algorithm [8].

The classical implementation of an ID3 algorithm has the drawback of expanding the tree until all leaf nodes are pure, or no more attributes are left to split on. Therefore, the C4.5 algorithm was developed, with two special enhancements: subtree replacement and subtree raising. Both of them are postpruning operations, at the cost of some accuracy on the training set. These operations are based on some weak statistical reasoning [15], and they involve some parameters. However, they seem to work well in practice, even with the default values suggested for the parameters.

Thus, our challenge is clear: if certainty expresses a proper cardinality scaling of knowledge, not only it should be able to choose the optimal attribute to split on at each node, but also it should stop expanding the tree, whenever further splits are unnecessary. Therefore, the algorithm stands parameter free, and no postpruning operations are needed.

In order to carry out our evaluation, we have used the publicly available machine learning tool Weka [15], implementing our measure in the ID3 decision tree algorithm. Implementing certainty is straightforward: at each node we compare the class marginal distribution certainty, $K_s(class)$, with the class conditional distribution certainty, $K_s(class \mid att.)$, given each one of the pending attributes to split on. If there is no candidate yielding $K_s(class \mid att.) > K_s(class)$, the node is not expanded. If some exist, we choose the one with the lowest marginal distribution certainty, $K_r(att)$, (r refers to the attribute's cardinality). A low $K_r(att)$ means a more balanced marginal distribution. Thus, we split on the attribute which ensures a better coverage of all of its branches, eventually decreasing the chance of getting empty leaves in succesive splits. Let's mention that empty leaves are the realization of unseen events in decission trees, so reducing the number of empty leaves is equivalent to reducing the unseen fraction of the input space.

The experimental setup includes some heterogeneous data sets from the UCI repository [14]. The ID3 algorithm does not deal with continuous or missing values. Therefore, examples with missing values have been discarded, and continuous values have been discretized to a reasonable number of equal width intervals. In all runs, we use a 10 fold cross validation method. Whenever independent train and test sets are available, we also perform an independent train/test classification. For the C4.5 algorithm we always use the default parameters.

The results are shown in table 1. We also show the results yielded by the entropic-gain ID3 original algorithm. Thus, it is easier to figure out the way

Table 1. Comparison of trees induced by entropy (ID3, C4.5) and certainty (Crt.)

DataBase	setSize	attr.	Clssf.	tree	treeSize	nodes	leaves	nullLvs.	%uncovered	%correct
BreastCancer	683	10	10fld	ID3	211	21	190	95	50.00	91.65
			10fld	C4.5	61	6	55	14	25.45	93.41
			10fld	Crt.	51	5	46	4	8.70	95.46
SegmentChallenge	1500	20	10fld	ID3	390	44	346	193	55.78	93.92
			10fld	C4.5	213	23	190	102	53.68	94.93
			10fld	Crt.	171	27	144	48	33.33	91.73
OpticalDigits	5620	65	10fld	ID3	11493	676	10817	7582	70.09	44.11
			10fld	C4.5	4023	241	3782	2334	61.71	63.02
			10fld	Crt.	1769	104	1665	333	20.00	54.02
	1797		testSet	C4.5	3010	177	2833	1737	61.31	56.82
	1797		testSet	Crt.	1225	72	1153	198	17.17	54.26
penDigits	10992	17	10fld	ID3	5798	527	5271	2955	56.06	86.69
			10fld	C4.5	2366	215	2151	1068	49.65	89.16
			10fld	Crt.	1805	164	1641	342	20.84	86.85
	3498		testSet	C4.5	1915	174	1741	910	52.27	84.08
	3498		testSet	Crt.	1288	117	1171	227	19.39	81.76
letterRecognition	20000	17	10fld	ID3	30561	1910	28651	21832	76.20	73.53
			10fld	C4.5	13409	838	12571	9033	71.86	77.73
			10fld	Crt.	4657	291	4366	2060	47.18	71.65
Soybean	562	36	10fld	ID3	50	51	116	31	26.72	83.77
			10fld	C4.5	69	22	47	10	21.28	91.81
			10fld	Crt.	161	63	98	3	3.06	87.72
CarEvaluation	1728	7	10fld	ID3	408	112	296	0	0.00	89.35
			10fld	C4.5	182	51	131	0	0.00	92.36
			10fld	Crt.	213	58	155	0	0.00	94.21
			trainSet	C4.5	182	51	131	0	0.00	96.30
			trainSet	Crt.	213	58	155	0	0.00	96.30
Nursery	12960	9	10fld	ID3	1159	320	839	0	0.00	98.19
			10fld	C4.5	511	152	359	0	0.00	97.05
			10fld	Crt.	1031	274	757	0	0.00	96.37
			trainSet	C4.5	511	152	359	0	0.00	98.13
			trainSet	Crt.	1031	274	757	0	0.00	98.59

entropy tends to cover the sample space, and the posterior effect of the pruning phase of the C4.5 algorithm.[2]

6   Conclusions

Let's note the trade off between the complexity of the model, (column labeled *treeSize*), the dimension of the input space, (column *leaves*), the number of empty leaves, that is, the unobserved part of that input space, (column *nullLvs*), and the accuracy (column *%correct*). To better appreciate this tradeoff, the column labeled *%uncovered* specifies the ratio of empty leaves with respect to the total number of leaves.[3]

Useless to deny it: the C4.5 algorithm does yield somewhat better accuracies. This is just due to the greedy behavior of entropy with respect to the conservative attitude of certainty. But, at a little cost in accuracy, we get dramatic reductions in the complexity of the models and dramatic reductions in the unobserved part of the input space, along with significant reductions in computational cost. This means that, beyond the accuracies yielded by a ten fold cross validation, we can be much more confident on the behavior of certainty models with respect to future coming examples.

In the OpticalDigits, PenDigits and LetterRecognition data bases, the features are vectors of integers ranging from 0 to 16. At this level of cardinalities, entropy begins to show some undesired behavior, and the differences between both measures become evident: C4.5 yields an extremely large uncovered part of the input space, and we should be cautious about relying on the accuracy figures given.

Soybean, CarEvaluation and Nursery, are exceptional cases in which certainty yields more complex models, with larger input spaces. The reason is that there exists an extra sample subspace sufficiently covered by the sample, and all this extra information can be efficiently exploited. For the Soybean case, though yielding a more complex model, the unobserved part of its input space is still much smaller. On the other hand, the CarEvaluation and Nursery examples are two special data bases, in which all attributes are perfectly balanced, and the sample space is completely covered, thus, no unseen events exist. Under such conditions, certainty tends to exploit all the information available and expands the tree over the whole sample space. Again, that extra information seems to be efficiently used and does not turn into excessive overfitting. Furthermore, the classification results over the training set (also included in table 1) show that, once all information about the sample space is available, the accuracies achieved are exactly the same.

In summary, certainty's algebra of knowledge seems to work well. The cardinality scaling of the measure, along with the uncertainty of unseen events, seems

---

[2] Regarding some of the data sets used, better accuracies have been reported using other methods. Please keep in mind, that the aim of the experiment is just to compare the behavior of certainty and entropy as measures of information.

[3] The tree size values refer to the basic model build upon the whole training set.

to guarantee a proper comparison of the information conveyed by distributions of different dimensions. In this case, it allows implementing a parameter free algorithm that stops the expansion of the tree at a reasonable level. This is a significant contribution, since pruning is the most computationally expensive part of tree induction. Thus, although we should not expect the best results in a validation phase, it looks like a promising tool whenever the goal is to get some fast, robust and reliable knowledge.

References

1. Aczél, J., Forte, B., Ng, C.T.: Why Shannon and Hartley entropies are natural. Adv. Appl. Probab. 6, 131–146 (1974)
2. Aczél, J., Daróczy, Z.: On Measures of Information and Their Characterizations. Academic Press, New York (1975)
3. Daróczy, Z.: Generalized information functions. Information and Control 16, 36–51 (1970)
4. Gini, C.W.: Variability and Mutability, contribution to the study of statistical distributions and relations. In: Studi Economico-Giuricici della R. Universita de Cagliari (1912)
5. Herfindahl, O.C.: Concentration in the U.S. Steel Industry. Unpublished doctoral dissertation. Columbia University (1950)
6. Kvalseth, T.O.: Entropy and correlation: some comments. IEEE transactions on Systems, Man and Cybernetics 17(3), 517–519 (1987)
7. Lenca, P., Lallich, S., Vaillant, B.: Construction of an Off-Centered Entropy for the Supervised Learning of Imbalanced Classes: Some First Results. Communications in Statistics- Theory and Methods 39(3), 493–507 (2010)
8. Quinlan, J.R.: Induction of decision trees. Machine Learning 1(1), 81–106 (1986)
9. Quinlan, J.R.: C4.5: Programs for Machine Learning. Morgan Kaufmann, San Mateo (1993)
10. Rényi, A.: On Measures of Entropy and Information. In: Proceedings of the Fourth Berkeley Symposium on Mathematical Statistics and Probability, vol. 1, pp. 547–561. University of California Press (1961)
11. Shannon, C.E.: A Mathematical Theory of Communication. The Bell System Technical Journal 27, 379–423, 623-656 (1948)
12. Simpson, E.H.: Measurement of Diversity. Nature 163, 688 (1949)
13. Theil, H.: On the estimation of relationships involving qualitative variables. The American Journal of Sociology 76(1), 103–154 (1970)
14. Frank, A., Asuncion, A.: UCI Machine Learning Repository. University of California, School of Information and Computer Science, Irvine (2010), http://archive.ics.uci.edu/ml
15. Hall, M., Frank, E., Holmes, G., Pfahringer, B., Reutemann, P., Witten, I.H.: The WEKA Data Mining Software: An Update. SIGKDD Explorations 11(1) (2009)

Workshop
on Biologically Inspired
Techniques for Data Mining
(BDM 2011)

A Measure Oriented Training Scheme for Imbalanced Classification Problems

Bo Yuan and Wenhuang Liu

Intelligent Computing Lab, Division of Informatics, Graduate School at Shenzhen,
Tsinghua University, Shenzhen 518055, P.R. China
{yuanb,liuwh}@sz.tsinghua.edu.cn

Abstract. Since the overall prediction error of a classifier on imbalanced problems can be potentially misleading and biased, it is commonly evaluated by measures such as G-mean and ROC (Receiver Operating Characteristic) curves. However, for many classifiers, the learning process is still largely driven by error based objective functions. As a result, there is clearly a gap between the measure according to which the classifier is to be evaluated and how the classifier is trained. This paper investigates the possibility of directly using the measure itself to search the hypothesis space to improve the performance of classifiers. Experimental results on three standard benchmark problems and a real-world problem show that the proposed method is effective in comparison with commonly used sampling techniques.

Keywords: Imbalanced Datasets, Neural Networks, ROC, G-Mean, SMOTE.

1 Introduction

The challenging issue of imbalanced datasets is inevitable in many real-world data mining applications, such as network intrusion detection, video surveillance, oil spills detection in satellite radar images, diagnoses of rare medical conditions and text categorization [2, 11]. These applications share a common characteristic: samples from one class are rare (referred to as minority or positive samples), compared to the number of samples in other classes (referred to as majority or negative samples).

For example, in medical diagnosis applications, it is important to build a predictive model that can reliably identify people with high risk of acquiring certain disease in the earliest stage [18]. However, abnormal samples typically only account for a small fraction of all subjects under test, resulting in a highly imbalanced dataset. Note that a naïve model that simply classifies all subjects as being negative can still achieve high overall predication accuracies but is otherwise useless as it is incapable of identifying positive samples.

The major challenge comes from the fact that the rarely occurring samples are usually overwhelmed by the majority class samples so that they are much harder to be identified. In the meantime, traditional learning algorithms usually aim at achieving the lowest overall misclassification rate (i.e., use an error based objective function to search the hypothesis space), which creates an inherent bias in favor of the majority classes because the rare class has less impact on accuracy.

L. Cao et al. (Eds.): PAKDD 2011 Workshops, LNAI 7104, pp. 293–303, 2012.
© Springer-Verlag Berlin Heidelberg 2012

Strictly speaking, almost all real-world datasets are imbalanced and how to train and evaluate a classifier taking into account all classes is an important research question. In recent years, this topic has attracted more and more attention from the research community, focusing on mainly two aspects: informative performance measures and how to improve the performance of classifiers [2]. Consequently, some more appropriate measures such as G-mean, ROC, Lift analysis and F-measure have been employed. For example, ROC is very flexible as it assumes no fixed threshold values, and can be used to project a complete image of the classifier in face of the tradeoff between true positive and false positive rates.

In the meantime, various sampling techniques have been proposed, aiming at directly manipulating the original dataset to modify the class distributions. The original dataset can be either over-sampled or under-sampled to increase the influence of positive samples or to reduce the dominance of negative samples. Although these methods do alleviate the challenge to some extent, they also bring in new issues. For instance, the under-sampling methods may unintentionally remove important samples and are not economic when samples are expensive to acquire; the over-sampling methods, on the other hand, may introduce samples that are infeasible in the specific domain and/or lead to overfitting. After all, there is still an open question on the optimal class distributions, which are likely to be domain and classifier dependent.

There is another branch of research on applying Ensemble methods [8] to solving imbalanced problems [4, 7, 12]. For example, misclassification costs can be incorporated into the procedure of weight updating or over-sampling techniques can be embedded to increase the sampling weights for minority samples. This topic is beyond the scope of our current study, which focuses on using a single classifier.

Apart from the many successful applications of sampling methods on solving imbalanced classification problems, there is still a gap between the measure according to which the classifier is to be evaluated and how the classifier is trained. Regardless of what sampling methods are in use, in many situations, the search in the hypothesis space is still driven by error based objective functions. For instance, MSE (Mean Square Error) is commonly used in training Neural Networks. Unfortunately, the relationship between the training error and the measures for after-training evaluation is generally non-trivial. In idealized situations where the dataset can be perfectly separated, the classifier will have zero misclassification error with a possibly very small MSE value. In this situation, commonly used measures such as G-mean, Lift analysis, AUC (Area Under Curve, a performance metric for ROC curves) will also reach their maximum values. However, in many cases, such classifiers may not exist or, due to the local optima in the search space, cannot be found in practice.

In order to bridge the gap between the objective function and the performance measure, in this paper, we conduct an investigation on the possibility and effect of directly using performance measures as the objective functions in the training of classifiers. It is expected that the training process will become more targeted and efficient, with the guidance from the more informative objective functions.

Section 2 gives a brief review of the sampling techniques and performance measures for imbalanced classification problems. Section 3 shows how to train a classifier with measure based objective functions. Experimental results are presented in Section 4 and this paper is concluded in Section 5 with some discussions and a number of directions for future work.

2   Techniques for Imbalanced Problems

Existing techniques for imbalanced classification problems can be roughly grouped into two topics: how to train a classifier properly and how to evaluate a classifier in a meaningful way.

2.1   Sampling Methods

Sampling is one of the common data preprocessing techniques. The idea of sampling is to purposefully manipulate the class distributions so that positive samples can be well represented in the training set. Its major advantage is that the classifier and the training algorithm do not need to be changed. The basic version of sampling is to randomly remove some negative samples, called under-sampling, and/or make copies of positive samples, called over-sampling. In under-sampling, some important samples (e.g., samples along the class boundary) may be discarded, resulting in information loss and a less than optimal model. In the meantime, since over-sampling makes exact copies of positive samples, adding no new information to the dataset, it may cause the overfitting problem.

Since the basic version of sampling does not work well in practice, a series of studies have been conducted most of which focused on developing smart heuristic sampling methods [15, 16]. A widely used over-sampling technique is called SMOTE (Synthetic Minority Over-sampling Technique), which creates synthetic samples between each positive sample and one of its neighbors [3]. It can introduce new samples to enrich the dataset and counter the sparsity in the distribution and create larger and more general decision regions compared to over-sampling with replication.

2.2   Performance Measures

For binary classification problems, classifiers are normally evaluated based on the confusion matrix, as shown in Table 1. Given a specific threshold (e.g., 0.5 for continuous outputs within [0, 1]), samples are classified as being either positive or negative and the overall prediction accuracy is defined as (TP+TN)/(TP+FP+TN+FN). The major issue is that, for imbalanced problems, a classifier can still achieve a high prediction accuracy by simply marking all samples as being negative. Instead, a good classifier should be able to achieve high accuracies on predicting both positive samples (i.e., high TP values) and negative samples (i.e., high TN values).

Table 1. A confusion matrix for binary classification problems. TP: True Positive, TN: True Negative, FP: False Positive, FN: False Negative.

	Actual Positive	Actual Negative
Predicted Positive	TP	FP
Predicted Negative	FN	TN

Based on the confusion matrix, two popular measures have been proposed: G-mean and F-measure, defined as below:

$$G\text{-}mean = (Acc^+ \times Acc^-)^{1/2} \tag{1}$$

$$where \quad Acc^+ = \frac{TP}{TP + FN}; \quad Acc^- = \frac{TN}{TN + FP}$$

$$F\text{-}measure = \frac{2 \times Precision \times Recall}{Precision + Recall} \tag{2}$$

$$where \quad Precision = \frac{TP}{TP + FP}; \quad Recall = \frac{TP}{TP + FN} = Acc^+$$

In Eq.1, Acc^+ and Acc^- are the true positive rate and true negative rate respectively, and G-mean represents a tradeoff between the accuracies on both classes. In Eq. 2, *Precision* refers to the proportion of true positive samples among all samples that are predicted as being positive while *Recall* is the proportion of true positive samples that are correctly identified by the classifier, which is the same as Acc^+.

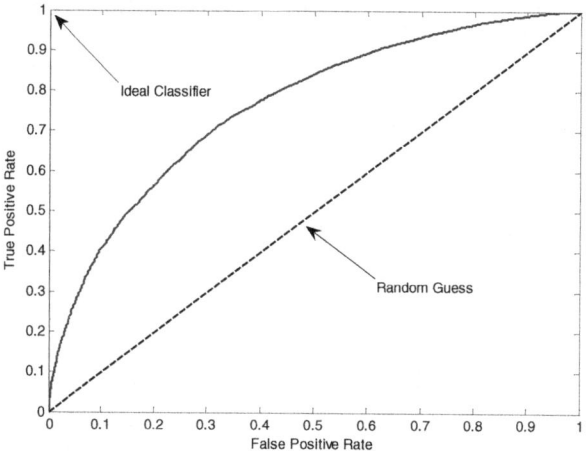

Fig. 1. An illustration of ROC curves. The dashed diagonal line represents the performance of a classifier making random guess. AUC is defined as the area under the curve with a maximum value of 1, corresponding to a classifier that perfectly separates the two classes.

The major advantage of ROC is that it can project a complete image on the behavior of a classifier in face of the tradeoff between true positive rate and true negative rate. The horizontal axis represents the false positive rate ($1\text{-}Acc^-$) while the vertical axis indicates the corresponding true positive rate. With the maximum threshold value, all samples are classified as being negative, corresponding to the origin. By contrast, with the minimum threshold, all samples are classified as being positive, corresponding to the upper right corner.

3   Measure Oriented Training Scheme

Given a hypothesis space containing all candidate classifiers that can be possibly reached, each measure creates a different fitness landscape (i.e., the hypothesis space plus an extra dimension for the measure) with its own structural properties (e.g., the locations of optima). It is unlikely that this fitness landscape is precisely consistent with the one implied by the objective function used in the training of classifiers. As a result, it is conceptually plausible and appealing to directly use the measure of interest as the objective function, in order to bridge the gap between the two landscapes.

In the meantime, many classifiers feature a localized learning pattern in that each time the classifier is updated only a subset of the samples or even a single sample takes effect. For instance, when training a neural network, for each sample, the expected and real outputs are compared and the error information is used to modify the weights and thresholds. When constructing a decision tree, how to split a certain node is only dependent on the subset of samples belonging to that node. In both cases, there is no consideration of the overall performance of the classifier. It is likely that some samples may need to be sacrificed to achieve better global performance.

However, most measures cannot be easily applied in a straightforward manner as it is difficult to derive analytical solutions based on them. Instead, since the measures describe the overall performance of classifiers, it is more appropriate to evaluate and update the classifier as a whole. We must admit that there is no readily available solution to each type of classifiers but for some classifiers, such as neural networks, there is a well developed solution: learning by evolution [20].

The parameters of a neural network are typically encoded into a real-valued vector called chromosome or individual. A population of such individuals represents a set of candidate solutions, which are to be evaluated according to the measure in use and evolved in parallel by evolutionary techniques such as Genetic Algorithms [10]. Each individual is evaluated as a black box and the training process does not require the objective function to have analytical solutions or to be differentiable. Traditionally, the major motivation of using evolutionary techniques over gradient based learning algorithms for training neural networks is to alleviate the curse of local optima. Also, it is possible to evolve the network structure at the same time, solving another well known dilemma. By contrast, the reason that we choose to evolve a neural network is to take its advantage of being flexible with objective functions.

The next question is which measure to choose? Theoretically, all measures can be incorporated into this learning by evolution framework. Since all real-world problems are literally imbalanced to some extent, without loss of generality, we assume that both classes are equally important. For problems where the positive class has significantly higher values, we regard them as falling into another category called minority mining problems, which is beyond the scope of this paper.

Furthermore, some measures such as ROC curves do create another issue: the fitness landscapes are not search friendly. Since ROC only considers the order of samples in terms of the classifier's outputs, there are potentially an infinite number of classifiers with the same AUC value, which creates a search space with many flat areas, a nightmare for all optimization techniques. Also, a classifier with high AUC values may have exceptionally large MSE values (e.g., all samples have output values close to 1 or 0). As a result, G-mean was selected as the measure in our studies.

4   Experiment

To validate the proposed measure oriented training (MOT) scheme, a series of experiments were conducted with standard feedforward neural networks and G-mean as the classifier and the measure respectively. The objective was not to perform a comprehensive and competitive test against existing state-of-the-art techniques. Instead, the major motivation was to demonstrate its general effectiveness and explore its performance with regard to the property of datasets.

4.1   Specification

In experimental studies, there are many factors that can put an impact on the final outcomes. In order to ensure a comparison that is as fair as possible and improve the replicability of the results, in our studies, all parameters were chosen without any specific tuning and were kept unchanged as we were not interested in finding out the best setting for each specific dataset. Also, the standard GA routines implemented in Matlab 2009 were used to reduce any coding related effects.

Table 2 gives a summary of the key experimental settings. Four datasets were used as benchmarks three of which were from the UCI Repository [19] and the last one contained real customer data from a major national bank in China. Note that, in imbalanced datasets, there are often only a handful of positive samples available and running a 10-fold cross validation will result in test sets with few positive samples. As a result, all datasets were randomly divided into the training set and the test set. Some of the properties of the datasets are shown in Table 3.

Table 2. Experimental settings used in the following studies

Parameters	Values
NN: Number of Input Nodes	The dimension of dataset
NN: Number of Hidden Nodes	5
GA: Population Size	200
GA: Initial Range	[-5, 5]
GA: Other Parameters	As default
SMOTE Sampling Ratio	100% for Cancer; 500% for others

Table 3. The four datasets used in the experimental studies

Datasets	Number of Attributes	Number of Instances	Proportion of Positive Samples
Cancer	9	683	34.99%
Yeast	8	1484	3.44%
Wine	11	4898	3.67%
Churn	27	1524	4.79%

4.2   Data Preprocessing

All datasets were normalized so that the attribute values were within the range of [0, 1] and all samples with missing values were removed. The positive and negative samples were labeled by "1" and "0" respectively. The Cancer dataset was created from the Wisconsin Breast Cancer Database [17] by removing its 16 samples with missing values (from totally 699 samples). The original Yeast Dataset [13] is a multiclass problem and the class named "ME2" was arbitrarily chosen to be the positive class while all other 9 classes were merged as the negative class, creating a highly imbalanced binary classification problem. The Wine dataset was created from the Wine Quality Dataset (white wine) [5] by marking all samples with scores no less than 8 out of 10 as the positive samples and all other samples as the negative ones. The Churn dataset consisted of various attributes of bank customers such as age, gender, profession, income and so on and was used to predict whether a customer was going to opt out of the service. From Table 3, it is clear that the last three datasets are expected to create much greater challenge for classifiers without appropriate techniques for handling imbalanced class distributions.

4.3   Results

The neural network was evolved by a GA with three training schemes: training based on MSE (**Baseline**), training based on MSE with oversampled training data (**SMOTE**) and training based on G-mean (**MOT**). Each scheme was tested on each dataset for 10 independent trials (the GA is a stochastic optimization algorithm) and the true positive rate, the true negative rate and the G-mean value on the test set were recorded. The average results (with standard deviations for G-mean) are shown in Tables 4-7. It is clear that the Cancer dataset was everyone's game and it was a two-horse race on the Yeast dataset while MOT stood out of the crowd on the Wine dataset and the Churn dataset. The three schemes were also tested on a few different pairs of training set and test set for each dataset, which produced similar performance patterns.

Table 4. Experimental results on the Cancer dataset

Methods	True Positive Rate	True Negative Rate	G-mean
Baseline	0.97984	0.96146	0.97058±0.0090
SMOTE	0.98488	0.95688	0.97068±0.0072
MOT	0.98152	0.95778	0.96952±0.0096

Table 5. Experimental results on the Yeast dataset

Methods	True Positive Rate	True Negative Rate	G-mean
Baseline	0.20716	0.99064	0.39388±0.25
SMOTE	0.62142	0.94086	**0.76390±0.036**
MOT	0.69288	0.91416	**0.79568±0.020**

Table 6. Experimental results on the Wine dataset

Methods	True Positive Rate	True Negative Rate	G-mean
Baseline	0	1	0±0.00
SMOTE	0.19766	0.95628	0.42042±0.12
MOT	0.71164	0.7330	**0.72192±0.011**

Table 7. Experimental results on the Churn dataset

Methods	True Positive Rate	True Negative Rate	G-mean
Baseline	0.06154	0.99084	0.18968±0.18
SMOTE	0.44104	0.90214	0.62926±0.046
MOT	0.63076	0.87196	**0.74064±0.040**

4.4   Analysis

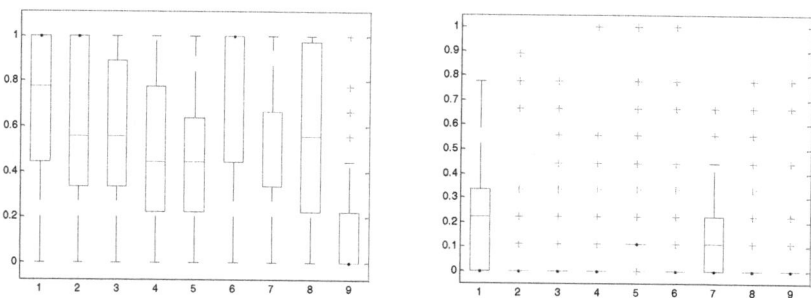

Fig. 2. The box plots of the Cancer dataset: Positive Class (left) and Negative Class (right). The horizontal axis represents the 9 attributes and the vertical axis shows the attribute values.

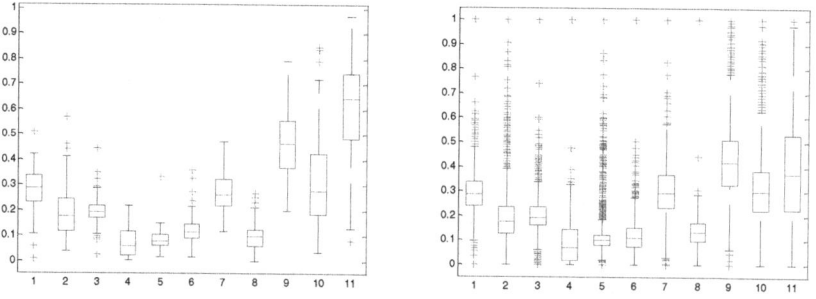

Fig. 3. The box plots of the Wine dataset: Positive Class (left) and Negative Class (right). The horizontal axis represents the 11 attributes and the vertical axis shows the attribute values.

In addition to the quantitative results, it would be interesting to move one step further to address the *why* part of the story. Certainly, for high dimensional datasets, it is difficult to visualize the distributions of data and the decision boundaries. Here, we show the box plots of the Cancer dataset on which Baseline performed very well and the Wine dataset on which Baseline performed extremely badly. Fig. 2 shows that the positive samples and negative samples of the Cancer dataset are reasonably well separated (the horizontal axis shows the attributes). It is reasonable to speculate that the decision boundaries were somewhere between the two classes, as evidenced by the small MSE values (Baseline: 0.02752; SMOTE: 0.03146; MOT: 0.04264).

There is a totally different situation on the Wine dataset where the two classes overlap significantly. In order to achieve a small MSE value, it is tempting to classify all samples as being negative as the positive samples only account for less than 4% of the dataset. On the other hand, in order to achieve a high G-mean value, a large portion of positive samples must be classified correctly, even at the cost of misclassifying some negative samples. In fact, the average MSE value was 0.03310 for Baseline and 0.25648 for MOT. This is a clear example where the MSE based objective function does not agree with the G-mean measure.

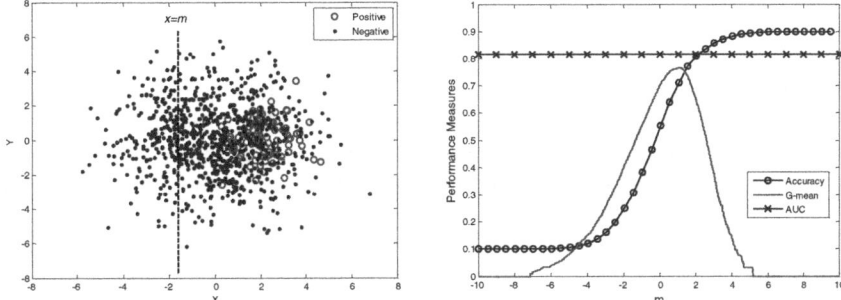

Fig. 4. Illustration of a 2D dataset where the positive class and the negative class overlap significantly (left) and the values of overall accuracy, G-mean and AUC as the decision boundary (*x=m*) moves horizontally (right)

To better demonstrate the property of imbalanced datasets and the relationship among the measures, a 2D dataset was created with 1000 positive samples and 9000 negative samples. The two classes overlapped significantly as shown in Fig. 4 (left). Note that only 10% samples were plotted for better visual effect. For simplicity, the classifier was assumed to be a vertical line (its output was defined as *x-m*) and samples on the right hand side of the line *x=m* were classified as being positive while samples on the left hand side were classified as being negative.

As the value of *m* changed from -10 to +10 continuously, the decision boundary moved horizontally from left to right and, at each position, the corresponding values of overall accuracy, G-mean and AUC were recorded as shown in Fig. 4 (right). The patterns of these measures were quite different. The AUC value was constant as the order of samples along the horizontal axis did not change with *m*. In the meantime, the overall accuracy monotonously increased from 0.1 to 0.9 (the positive samples accounted for 10% of the dataset). By contrast, G-mean reached its peak with $m \approx 1$ when there was a good balance between TP and TN. Note that the value of G-mean reduced to zero when the overall accuracy was at its top.

5   Conclusion

In this paper, we approached the imbalanced classification problems from a new angle. Instead of trying to manipulate the datasets through sampling to change the class distributions or assigning different costs to classes, we proposed to explicitly use the measure itself as the objective function when searching the hypothesis space. This scheme is conceptually plausible as the learning process will become more targeted and efficient by bridging the gap between the traditional error based objective functions and the measures of interest.

The results on three benchmark datasets as well as a real world dataset suggested that, as the challenge of the datasets went up, the advantage of the proposed scheme became more distinctive. Certainly, it is too early to make any conclusive claim on the comparison between this measure oriented training scheme and existing techniques for classifying imbalanced datasets, which requires more extensive and rigorous empirical and theoretical studies. Nevertheless, it offers a new perspective for developing more effective approaches to imbalanced datasets. In fact, since the performance measures in data mining problems are often conflicting with each other, the idea of applying multiobjective evolutionary techniques has become increasingly popular in recent years [14], with some interesting applications in the domain of imbalanced classification problems [1, 6, 9].

As to future work, for classifiers that cannot be evolved in the straightforward manner, other strategies need to be developed to incorporate this measure oriented objective function. In the meantime, since Ensemble methods can adaptively modify the weights of samples, influencing the class distributions in a more informative way, it is also interesting to investigate the possibility of combining measure oriented training with Ensemble methods.

Acknowledgement. This work was supported by the Scientific Research Foundation for Returned Overseas Scholars, Ministry of Education, P.R. China and National Natural Science Foundation of China (No. 60905030 and No. 61003100). The authors are also grateful to the anonymous reviewers for their very helpful comments.

References

1. Bhowan, U., Zhang, M.J., Johnston, M.: Multi-Objective Genetic Programming for Classification with Unbalanced Data. In: Twenty-Second Australasian Conference on Artificial Intelligence, pp. 370–380 (2009)
2. Chawla, N.V.: Data Mining for Imbalanced Datasets: An Overview. In: Data Mining and Knowledge Discovery Handbook: A Complete Guide for Practitioners and Researchers, pp. 853–867. Springer, Heidelberg (2005)

3. Chawla, N.V., Bowyer, K.W., Hall, L.O., Kegelmeyer, W.P.: SMOTE: Synthetic Minority Oversampling Technique. Journal of Artificial Intelligence Research 16, 321–357 (2002)
4. Chawla, N.V., Lazarevic, A., Hall, L.O., Bowyer, K.W.: SMOTEBoost: Improving Prediction of the Minority Class in Boosting. In: Lavrač, N., Gamberger, D., Todorovski, L., Blockeel, H. (eds.) PKDD 2003. LNCS (LNAI), vol. 2838, pp. 107–119. Springer, Heidelberg (2003)
5. Cortez, P., Cerdeira, A., Almeida, F., Matos, T., Reis, J.: Modeling Wine Preferences by Data Mining from Physicochemical Properties. Decision Support Systems 47(4), 547–553 (2009)
6. Ducange, P., Lazzerini, B., Marcelloni, F.: Multi-Objective Genetic Fuzzy Classifiers for Imbalanced and Cost-Sensitive Datasets. Soft Computing 14(7), 713–728 (2010)
7. Fan, W., Stolfo, S.J., Zhang, J., Chan, P.K.: AdaCost: Misclassification Cost-Sensitive Boosting. In: Sixteenth International Conference on Machine Learning, pp. 97–105. Morgan Kaufmann (1999)
8. Freund, Y., Schapire, R.E.: Experiments with a New Boosting Algorithm. In: Thirteenth International Conference on Machine Learning, pp. 148–156 (1996)
9. García, S., Aler, R., Galván, I.M.: Using Evolutionary Multiobjective Techniques for Imbalanced Classification Data. In: Diamantaras, K., Duch, W., Iliadis, L.S. (eds.) ICANN 2010. LNCS, vol. 6352, pp. 422–427. Springer, Heidelberg (2010)
10. Goldberg, D.E.: Genetic Algorithms in Search, Optimization, and Machine Learning. Addison Wesley (1989)
11. Han, S.L., Yuan, B., Liu, W.H.: Rare Class Mining: Progress and Prospect. In: 2009 Chinese Conference on Pattern Recognition, pp. 137–141. IEEE Press (2009)
12. Hoens, T.R., Chawla, N.V.: Generating Diverse Ensembles to Counter the Problem of Class Imbalance. In: Zaki, M.J., Yu, J.X., Ravindran, B., Pudi, V. (eds.) PAKDD 2010. LNCS, vol. 6119, pp. 488–499. Springer, Heidelberg (2010)
13. Horton, P., Nakai, K.: A Probabilistic Classification System for Predicting the Cellular Localization Sites of Proteins. In: Fourth International Conference on Intelligent Systems for Molecular Biology, pp. 109–115 (1996)
14. Jin, Y.C., Sendhoff, B.: Pareto-Based Multiobjective Machine Learning: An Overview and Case Studies. IEEE Transactions on Systems, Man and Cybernetics - Part C: Applications and Reviews 38(3), 397–415 (2008)
15. Kubat, M., Matwin, S.: Addressing the Curse of Imbalanced Training Sets: One Sided Selection. In: Fourteenth International Conference on Machine Learning, pp. 179–186. Morgan Kaufmann (1997)
16. Liu, X.Y., Wu, J., Zhou, Z.H.: Exploratory Under-Sampling for Class-Imbalance Learning. In: Sixth International Conference on Data Mining, pp. 965–969 (2006)
17. Mangasarian, O.L., Setiono, R., Wolberg, W.H.: Pattern Recognition via Linear Programming: Theory and Application to Medical Diagnosis. In: Coleman, T.F., Li, Y. (eds.) Large-Scale Numerical Optimization, pp. 22–30. SIAM Publications (1990)
18. Qu, X.Y., Yuan, B., Liu, W.H.: A Predictive Model for Identifying Possible MCI to AD Conversions in the ADNI Database. In: Second International Symposium on Knowledge Acquisition and Modeling, vol. 3, pp. 102–105. IEEE Press (2009)
19. UCI Machine Learning Repository, http://archive.ics.uci.edu/ml
20. Yao, X.: Evolving Artificial Neural Networks. Proceedings of the IEEE 87(9), 1423–1447 (1999)

An SVM-Based Approach to Discover MicroRNA Precursors in Plant Genomes

Yi Wang, Cheqing Jin, Minqi Zhou, and Aoying Zhou

Shanghai Key Laboratory of Trustworthy Computing,
Software Engineering Institute, East China Normal University, China
yiwang@ecnu.cn, {cqjin,mqzhou,ayzhou}@sei.ecnu.edu.cn

Abstract. MicroRNAs (miRNAs) are noncoding RNAs of ∼22 nucleotides that play versatile regulatory roles in multiculleler organisms. Since the cloning methods for miRNAs identification are biased towards abundant miRNAs, the computational approaches provide useful complements to identify miRNAs which are highly constrained by tissue- and time-specifically expression manners. In this paper, we propose a novel Support Vector Machine (SVM) based detector, named MiR-PD, to identify pre-miRNAs in plants. The classifier is constructed based on twelve features of pre-miRNAs, inclusive of five global features and seven sub-structure features. Trained on 790 plant pre-miRNAs and 7,900 pseudo pre-miRNAs, MiR-PD achieves 96.43% five-fold cross-validation accuracy. Tested on the newly identified 441 plant pre-miRNAs and 62,883 pseudo pre-miRNAs, MiR-PD reports an accuracy of 99.71% with 77.55% sensitivity[1] and 99.87% specificity[2], suggesting a feasible genome-wide application of this miRNAs detector so as to identify novel miRNAs (especially for those species-specific miRNAs) in plants without relying on phylogenetical conservation.

Keywords: MicroRNAs, plant, support vector machine, MiR-PD.

1 Introduction

MicroRNAs (miRNAs) are endogenous, non-protein-coding, small (∼22 nt) RNAs which represent an abundant class of gene regulatory molecules in multicellular organisms by targeting mRNAs for cleavage or translational repression [5]. The discovery of lin-4 miRNA, as a key regulator of larval development in Caenorhabditis elegans in 1993 and another miRNA(let-7) in 2000, initiated the revolution of miRNAs' world [17,22]. Since then, large amounts of new miRNA genes have been uncovered across plants, animals and viruses. Nowadays, there are 15,172 miRNAs spanning 142 species deposited in the miRBase (Release 16.0) [10].

There are two methods to identify miRNAs: the cloning method by experimental biology and the bioinformatics prediction after experimental validation.

---

[1] sensitivity = true_positive/(true_positive + false_negative)
[2] specificity = true_negative/(true_negative + false_positive)

L. Cao et al. (Eds.): PAKDD 2011 Workshops, LNAI 7104, pp. 304–315, 2012.
© Springer-Verlag Berlin Heidelberg 2012

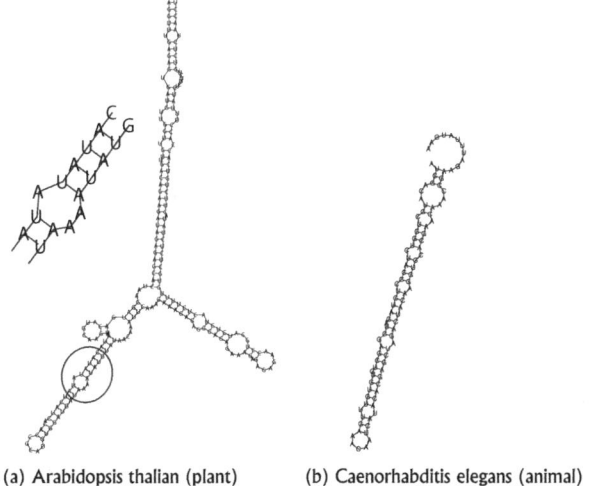

(a) Arabidopsis thalian (plant)   (b) Caenorhabditis elegans (animal)

Fig. 1. Predicted secondary structures of miRNA precursors

However, the cloning method cannot identify miRNAs that express at low level or in highly constrained tissue- and time-specific expression patterns or response under conditional stimuli. Consequently, bioinformatics prediction becomes a useful complement.

MiRNA precursors can fold and form low free energy stem-loop structures [4], as predicted by Mfold or other softwares [29], which is one of the major characteristics to be identified by computational approaches. Unfortunately, a genome may contain large amounts of similar stem-loop structures other than miRNAs. For example, about 11 million hairpins can be folded in the human genome [2], and some 312,236 hairpin candidates were extracted in the Arabidopsis genome intergenic regions [25]. Nowadays, several computational approaches have been developed to detect miRNA genes (please refer to related work in Sect. 2.2). However, all these studies have been focused on searching miRNAs in animal rather than in plant genomes. It is necessary to develop a plant aimed miR-NAs detector. First, the length of known plant pre-miRNAs vary from 47 nt to 689 nt, while the length of animals' vary only from 44 nt to 180 nt (miRBase 16); Second, the plant miRNAs precursors have more complex RNA secondary structures rather than a simple stem-loop structure (Fig. 1). These characteristics make it inappropriate for those animal data based approaches to be applied in plants, and make correctly classifying real miRNA precursors from pseudo's facing a big challenge.

In this paper, we propose a novel SVM-based approach (MiR-PD) for genome-wide plant miRNAs prediction in a single genome. Twelve features in total are selected to construct an effective classifier. Experimental results upon real data sets show that our MiR-PD method achieves high sensitivity and specificity and is capable of detecting miRNAs from the pool of small RNA sequences.

The rest of paper is organized as below. We introduce preliminary knowledge in Sect. 2, including SVM technique and some related work. We then describe our detector in detail in Sect. 3 and report experimental studies in Sect. 4. Finally, we conclude the paper briefly in the last section.

2   Preliminary Knowledge

2.1   Support Vector Machine

Support vector machine (SVM) is an effective classification method. The principle of SVM is to find the largest separating margin in the feature space. Samples are separated in the higher Hilbert feature space that is implicitly defined by the kernel function $K(x, y) = \langle \phi(x) \cdot \phi(y) \rangle$, where $\phi(\cdot)$ is a map from the original feature space to high-dimensional space. SVM allows the slack variable ξ to tolerate noise and outliers in the sample set. Besides, the formulation which supports to analyze the unbalanced data is used, with different penalty parameters for different classes [20]. Then, for C-SVM, the optimal problem is

$$\text{MINIMIZE} \ \frac{1}{2} \sum_{i,j} \alpha_i K(x_i, x_j) \alpha_j - \sum_i \alpha_i \tag{1}$$

$$\text{SUBJECT TO} \ 0 \leq \alpha_i \leq C \cdot w_1, \ y_i = 1$$

$$0 \leq \alpha_i \leq C \cdot w_{-1}, y_i = -1$$

C is the penalty parameter of the error term which represents the tolerance of misclassification. w_1 and w_{-1} are the penalty weight of positive (plant pre-miRNAs) and negative (pseudo pre-miRNAs) classes, respectively. Increasing the weight of real or pseudo pre-miRNAs helps to increase the sensitivity or specificity, respectively. Considering the unbalanced number of these two-class points, the weight should be tuned. In this work, the LibSVM package version 2.8, which supplies an implement of SVM [6], is employed and 5-fold cross-validation is used to estimate the accuracy of the classifier and guide the selection of parameters.

2.2   Related Work

A few literatures have studied how to predict miRNA genes, including miRscan [18,19], miRseeker [16], miRAlign [24], PalGrade [2] for animals, and miRcheck [14], miRFinder [3] for plants. Generally, these approaches rely on the sequence conservation in related organisms and the conservation patterns of the 3' and 5' arms of pre-miRNA as well as the stem-loop hairpin structure. Consequently, such methods can not detect miRNAs without close homologues either due to the limited sequence data or the rapid evolution of some miRNA genes. Pathogenic viral-encoded pre-miRNAs represent a good example that it is necessary to explore new methods to predict miRNAs based on a single genome [8].

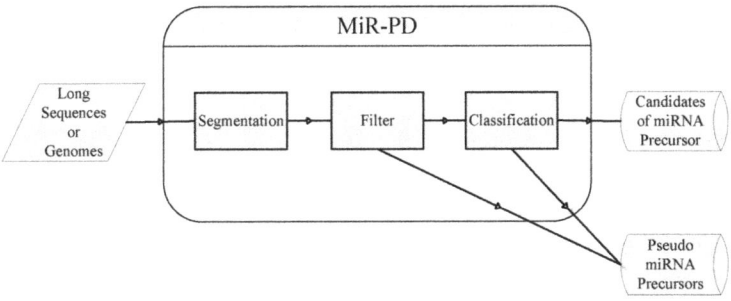

Fig. 2. Architecture of MiR-PD

More recently, several SVM-based methods have been developed. For example, Sewer et al. have built an SVM classifier model based on 40 sequence and structure features to search clustered miRNAs in the genomes of human, mouse and rat, respectively [23]. Triplet elements representing the local structure-sequence features were selected by Xue et al. to train SVM model and achieved performance ∼ 90% accuracy for human dataset [27]. RNAmicro, as a sub-screen for large-scale non-coding RNA surveys with RNAz [26] or Evofold [21], trained another SVM model and achieved 91.16% sensitivity and 99.47% specificity in animals [11]. MiPred is developed to identify pre-miRNAs with 84.55% sensitivity and 97.75% specificity in human dataset [15].More recently, miR-KDE (the kernel density estimator), G2DE (the generalized Gaussian density estimator) based classifier and microPred were developed for prediction of human pre-miRNAs [1,7,13]. All these existing studies focus on animal datasets, while our focus is plant datasets.

3   MiR-PD Approach

In this section, we begin to depict our SVM based plant miRNA detector, named miR-PD (Fig. 2). MiR-PD contains three phases:

Phase 1: Segmentation – cutting and parsing sequences with stem-loop structures from long sequences or genomes;

Phase 2: Filter – roughly selecting pre-miRNA candidates from the results of phase 1;

Phase 3: Classification – SVM based classifier, which is contributed by a number of training samples, takes responsibility for detecting the real pre-miRNAs from candidates.

3.1   Segmentation

The goal of Segmentation is to slide the given sequences (long sequences or genomes) to parse all possible candidates that contain stem-loop structures by using Vienna RNA Package [12]. The outputs of Segmentation are sequences with RNA secondary structures.

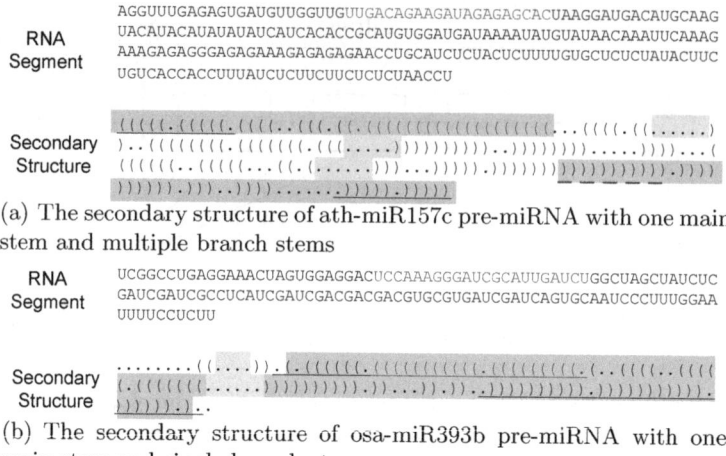

(a) The secondary structure of ath-miR157c pre-miRNA with one main stem and multiple branch stems

(b) The secondary structure of osa-miR393b pre-miRNA with one main stem and single branch stem

Fig. 3. Secondary structures and features

Fig. 3 (a) illustrates the miRNA precursor and its secondary structure. Each nucleotide is at one of the two states, *paired* or *unpaired*, described by brackets ("(" and ")") and dot ("."), respectively. Mature miRNA sequence is labeled with red. "(" and ")" means a pair of nucleotides, AU pair, GC pair or GU wobble pair. Otherwise, unpaired nucleotides are denoted as dots.

3.2   Filter

In general, the secondary structures of pre-miRNAs of plants are more complex than simple stem-loop structures. For convenient discussion, the stem containing mature miRNA is called *main stem* and the rest are called *branch stems* in stem-loop structure.

Besides the stem-loop structure of pre-miRNAs, currently known miRNAs also have some computable characteristics, as summarized below. It is convenient to filter parts of input sequences with such information.

Rule 1. *Each candidate contains at least 57 nucleotides.*

Rule 2. *The free energy of secondary structure of each candidate should be no more than -9 kcal/mol.*

Rule 3. *The main stem has at least 16 base pairs (including the GU wobble pairs).*

At last, redundant candidates and sequences with non-AGCU nucleotides are discarded. These criteria ensure that sequence segments parsed here have similar characteristics with the real pre-miRNAs as we've studied according known plant pre-miRNAs deposited on miRBase [10].

3.3 Classification

The capability of distinguishing real and pseudo pre-miRNAs is a basic principle to select features. Plant miRNA precursors are known to have higher negative minimal free energies (-MFE), higher minimal free energy index (MFEI) and more U nucleotides ($\sum U$) than other RNAs [28]. The loop is also a necessary component of the RNA secondary structure, so we treat the number of nucleotides involved in loop(s) as one feature (SUMloop, the highlight region shown in Fig. 3). The lengths of plant pre-miRNAs vary, and lengths of pseudo pre-miRNAs we extracted also differ from one another. Thus, the length of pre-miRNA (Lp) is adopted as another global feature to normalize other features.

Since the main stem is a critical part of the miRNA, we also try to select several features from this sub-structure, inclusive of the length of main stem (Ls), the number of A-U and G-C base pairs (AUpairs, GCpairs), the maximum continuous base pairs (MAXpairs, marked as dashed line in Fig. 3), the maximum length of symmetrical structure (MAXsym, shown as the underline in Fig. 3) and local contiguous structure-sequence features (KL1, KL2 are the compressed features of 32 triplet elements features, detailed data not shown). Finally, five global features and seven sub-structure features (obtained from the main stem) are selected in our model, as shown in Table 1.

Table 1. Features used for SVM classification model

Index	Feature description	descriptors
	global features	
1	length of pre-miRNA	Lp
2	negative minimal free energies	-MFE
3	minimal free energy index	MFEI
4	number of U nucletides	$\sum U$
5	number of nucletides involved in loops	SUMloop
	sub-structure features	
6	length of main stem	Ls
7/8	number of A-U,G-C base pairs	AUpairs, GCpairs
9	maximum continuous base pairs	MAXpairs
10	maximum length of symmetrical structure	MAXsym
11/12	local contiguous structure-sequence	KL1, KL2

4 Experimental Study

In this section, we present an experimental study upon real data. All the algorithms are implemented in Perl and C++ and the experiments are performed on a Linux system with Intel Core 2 CPU (2.4GHz) and 4G memory.

4.1 Data Source

We use following data sets in our study.

Testing Sets for Filter. 131 Arabidopsis and 230 rice pre-miRNAs that are described in Genome coordinates on miRBase were cut from their genome region with randomly 100 ∼ 300 nucleotides extending at both ends. These sequences, termed TF-a for Arabidopsis and TF-r for rice, are used for testing the efficiency of the Filter.

SVM Training Sets. 790 Plant pre-miRNAs and 7,900 sequence segments with the similar stem-loop structures (we call them pseudo pre-miRNAs) are collected as training sets.

Plant Pre-miRNAs Set. Plant miRNAs and their precursor sequences were downloaded from the miRBase9.0. A total of 790 miRNAs from 8 plant species were obtained and selected as positive training set (TR-P).

Pseudo Pre-miRNAs Set. Pseudo pre-miRNAs set comprises five independent subsets: protein coding sequences (CDSs), intergenic and intronic sequences, genome sequences and tRNAs/rRNAs. CDSs on Arabidopsis chromosome 1, intergenic and intronic sequences on Arabidopsis chromosome 2 were obtained from TAIR6 (ftp://ftp.arabidopsis.org). Genome sequence was extracted from the genome region of position 5,000,000 to 6,000,000 on rice chromosome 10 downloaded from TIGR version 4.0 (ftp://ftp.tigr.org). The tRNA/rRNA sequences were obtained from Rfam [9]. Firstly, these sequences are parsed by Filter. Then, redundant sequences are excluded according to the similarity calculated by BLAST-CLUST (S=10, L=0.5, W=16). We randomly pick out 2,400 CDSs, 2,500 intergenic sequences, 500 intronic sequences, 2,000 genome sequences and 500 tRNAs/rRNAs from those five subsets as negative training set (TR-N).

SVM Testing Sets. Plant data and cross species data are collected as testing sets.

Updated Plant miRNAs Set. Latest updated 441 plant miRNAs from miRBase 16.0 were used as positive testing data (TE-P).

Pseudo Pre-miRNAs testing set Except for the 2,400 CDS subset in the training set, the rest 62,883 CDS pseudo pre-miRNAs are chosen as negative testing set (TE-N).

Non-plant Pre-miRNAs Set. With the same criteria of selecting TR-P, 3,447 non-plant pre-miRNAs are obtained from miRBase as cross-species testing set (TE-C).

4.2   The Filter Efficiency

Filter is designed to extract stem-loop candidates from long sequences. Datasets TF-a and TF-r are used to evaluate the efficiency of Filter. There are 131 and 230 sequence segments in the TF-a and TF-r, respectively. After parsed by Filter, it produces 507 candidates from TF-a and 879 candidates from TF-r. Of these candidates, 122 from TF-a and 222 from TF-r cover more than 80% region of the corresponding pre-miRNAs. Notably, 35 TF-a and 53 TF-r candidates are almost identical with known pre-miRNAs. Only 1 out of 131 Arabidopsis miRNAs and 2 out of 230 rice miRNAs are missed in these candidates. It shows that Filter achieves fairly satisfactory sensitivity (99.17%) while parsing candidates within a reasonable number.

4.3   Performance of SVM Features

In this SVM classification model, 12 global and sub-structure features of plant pre-miRNAs are employed. The Parzen window density estimation method is used to estimate the probability distribution of each feature using the datasets TR-P and TR-N (Fig. 4), which can visually show the performance of selected features. Different distribution between real and pseudo pre-miRNAs means the large probabilistic distance. Probabilistic distance, such as divergence, measures the separability–the larger distance, the better separability. The probability distribution of SUMloop and seven sub-structure features are shown in Fig. 4. Since other features like -MFE, MFEI and $\sum U$ have been evaluated in other literatures, we won't list the corresponding analysis in this paper.

It's worth to mention that increments of the dimension of feature space can improve the separability, but it comes with the problem of overfitting. Reducing the number of features can decrease induction time and VC dimension. Furthermore, irrelevant or redundant features may ruin the accuracy of the classification model. Local contiguous structure-sequence features, as described in Xue et al.(2005)'s work [27], contain 32 possible triplet elements. It provides too many features concerning the same subject, which may bias the classification model in this case. Therefore, we compress 32 features into 2, through Karhunen-Loeve transformation that incorporate class information, which achieves good performance (Fig. 4 (e, f)).

4.4   SVM Training

Support vector machine (SVM) is used to classify real and pseudo pre-miRNAs with 12 selected features. Training sets TR-P and TR-N are used to calibrate the SVM classifier, and then the classifier is applied to the testing datasets. The model and parameter is settled based on five-fold cross-validation accuracy, which is C-SVM using RBF kernel

$$K(x_i, x_j) = \exp(-\mu|x_i - x_j|^2) \qquad (\mu = 1/12), \qquad (2)$$

and penalty parameter $C = 32$ with weights $w_1 = 10$, $w_{-1} = 1$. It achieves five-fold cross-validation accuracy rate 96.43%, which basically reflects the ability of discrimination and shows that the SVM classification model performs well in classifying real and pseudo pre-miRNAs. A score with range [0.0, 1.0] is assigned to each sample by the classification model. The sample will be classified as positive when its score is beyond a specified threshold (default is 0.5, Fig. 5).

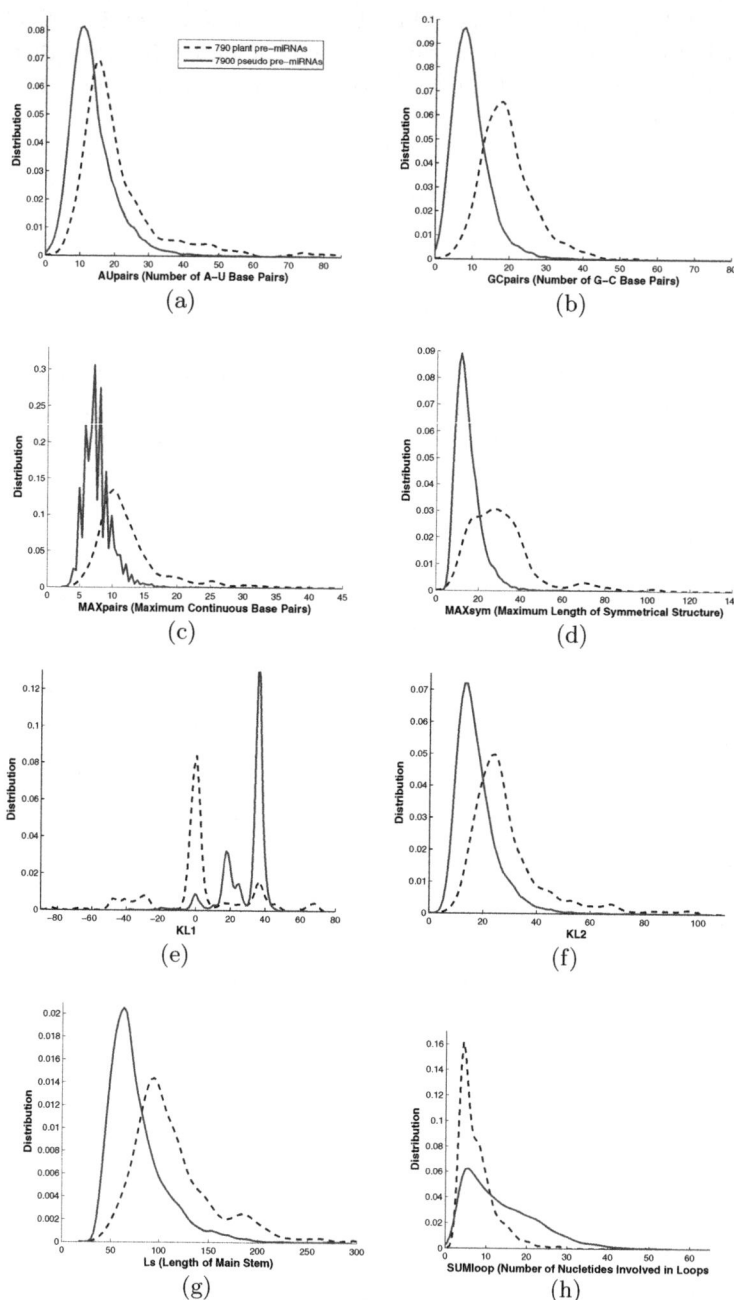

Fig. 4. Based on the TR-P and TR-N datasets, the conditional density distribution of each feature is estimated using the Parzen window density estimation method. Except for features that were studied by Zhang, et al. (2006), the other features AUpairs (A), GCpairs (B), SUMloop (C), Ls (D), MAXpairs (E), MAXsym (F), KL1 (G), KL2 (H) are estimated.

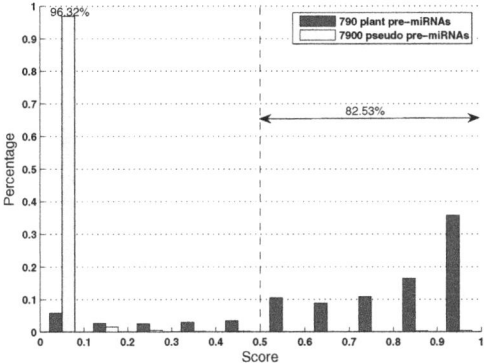

Fig. 5. SVM training set score distribution. Each sample is assigned a score by the classification model, which represents the reliability of being positive by probability estimation. The default cutoff of our classifier is 0.5, and the threshold could be easily adjusted to meet the balance between sensitivity and specificity.

4.5   Classification Testing

When applying the classifier to the sets TE-P and TE-N, 342 out of 441 updated plant pre-miRNAs are recognized as positive and 62,799 out of 62,883 pseudo pre-miRNAs are correctly classified as negative, which gives the sensitivity 77.55% and specificity 99.87%. The classifier achieves a bit lower sensitivity, and some pre-miRNAs are classified as negative with very low classification scores, such as ath-MIR855 (0.0577), ath-MIR824 (0.0041), ath-MIR854a (0.0031), ath-MIR854b (0.0036), ath-MIR854c (0.0027), ath-MIR854d (0.0036). With no exception, the predicted secondary structures of these newly discovered pre-miRNAs are very complex with at least six branch stems, which may affect the accuracy of our classification model.

4.6   Cross Species Testing

Our classification model is designed based on features of plant pre-miRNAs. It is interesting to see whether it works on non-plant pre-miRNAs prediction or not. We then test the model on 3,447 non-plant pre-miRNAs set and it only achieves 46.53% sensitivity (human 56.18%, animals 45.25%, viruses 37.71%). It suggests that some features of pre-miRNAs are specific in plants.

5   Conclusion

As the RNA secondary structure of plant pre-miRNA is more complex than a simple hairpin, it is difficult and insufficient to predict plant miRNAs using current computational approaches in a single genome. In this study, a novel SVM-based detector, MiR-PD, is proposed to distinguish real pre-miRNAs from pseudo's in plants using 12 global and sub-structure features. The detector achieves 96.43% five-fold cross-validation accuracy with the training dataset and reports the sensitivity 77.55% and specificity 99.87% after applying to the testing dataset. Since there are large amounts of possible candidate hairpins in a genome, reducing the false positive rate is thus necessary for its genome wide application. One of the strategies is raising the threshold to increase the specificity, with the cost of reducing the sensitivity. In our case, 62,856 out of 62,883 (99.96%) samples of TE-N were correctly classified as negative when the threshold is adjusted to 0.7, at the cost of reducing the sensitivity to 71.70%. Due to its good specificity, a genome-wide miRNAs prediction approach is also proposed without relying on phylogenetical conservation, which would be important to detect those species-specific miRNAs.

At present, large amounts of sequence data have been obtained by sequencing small RNA libraries using high-thoughput sequencing approaches. To dig out miRNAs from those huge databases remains a hard work without an efficient classification model. Thus, our study also offers a useful tool to detect miRNAs from the pool of small RNA sequences.

References

1. Batuwita, R., Palade, V.: microPred: effective classification of pre-miRNAs for human miRNA gene prediction. Bioinformatics 25(8), 989 (2009)
2. Bentwich, I., Avniel, A., Karov, Y., Aharonov, R., Gilad, S., Barad, O., Barzilai, A., Einat, P., Einav, U., Meiri, E., et al.: Identification of hundreds of conserved and nonconserved human microRNAs. Nature Genetics 37(7), 766–770 (2005)
3. Bonnet, E., Wuyts, J., Rouzé, P., Van de Peer, Y.: Detection of 91 potential conserved plant microRNAs in Arabidopsis thaliana and Oryza sativa identifies important target genes. PNAS 101(31), 11511 (2004)
4. Bonnet, E., Wuyts, J., Rouzé, P., Van de Peer, Y.: Evidence that microRNA precursors, unlike other non-coding RNAs, have lower folding free energies than random sequences. Bioinformatics (2004)
5. Carrington, J.C., Ambros, V.: Role of microRNAs in plant and animal development. Science 301(5631), 336 (2003)
6. Chang, C., Lin, C.: LIBSVM: a library for support vector machines (2001)
7. Chang, D., Wang, C., Chen, J.: Using a kernel density estimation based classifier to predict species-specific microRNA precursors. BMC Bioinformatics 9(suppl.12), 2 (2008)
8. Cullen, B.: Viruses and microRNAs. Nature Genetics 38, S25–S30 (2006)
9. Griffiths-Jones, S., Moxon, S., Marshall, M., Khanna, A., Eddy, S., Bateman, A.: Rfam: annotating non-coding RNAs in complete genomes. Nucleic Acids Research 33(Database Issue), D121 (2005)
10. Griffiths-Jones, S., Saini, H., Dongen, S., Enright, A.: miRBase: tools for microRNA genomics. Nucleic Acids Research (2007)
11. Hertel, J., Stadler, P.: Hairpins in a Haystack: recognizing microRNA precursors in comparative genomics data. Bioinformatics 22(14), e197 (2006)
12. Hofacker, I., Fekete, M., Stadler, P.: Secondary structure prediction for aligned RNA sequences. Journal of Molecular Biology 319(5), 1059–1066 (2002)
13. Hsieh, C., Chang, D., Hsueh, C., Wu, C., Oyang, Y.: Predicting microRNA precursors with a generalized Gaussian components based density estimation algorithm. BMC Bioinformatics 11(suppl.1), 52 (2010)

14. Jones-Rhoades, M., Bartel, D.: Computational identification of plant microRNAs and their targets, including a stress-induced miRNA. Molecular Cell 14(6), 787–799 (2004)
15. Kwang Loong, S., Mishra, S.: De novo SVM classification of precursor microR-NAs from genomic pseudo hairpins using global and intrinsic folding measures. Bioinformatics (2007)
16. Lai, E., Tomancak, P., Williams, R., Rubin, G.: Computational identification of Drosophila microRNA genes. Genome Biol. 4(7), R42 (2003)
17. Lee, R., Feinbaum, R., Ambros, V.: The C. elegans heterochronic gene lin-4 encodes small RNAs with antisense complementarity to lin-14. Cell 75(5), 843–854 (1993)
18. Lim, L., Glasner, M., Yekta, S., Burge, C., Bartel, D.: Vertebrate microRNA genes. Science 299(5612), 1540 (2003)
19. Lim, L., Lau, N., Weinstein, E., Abdelhakim, A., Yekta, S., Rhoades, M., Burge, C., Bartel, D.: The microRNAs of Caenorhabditis elegans. Genes & Development 17(8), 991 (2003)
20. Osuna, E., Freund, R., Girosi, F.: Support vector machines: Training and applications. CBCL-144 (1997)
21. Pedersen, J., Bejerano, G., Siepel, A., Rosenbloom, K., Lindblad-Toh, K., Lander, E., Kent, J., Miller, W., Haussler, D.: Identification and classification of conserved RNA secondary structures in the human genome. PLoS Comput. Biol. 2(4), e33 (2006)
22. Reinhart, B., Slack, F., Basson, M., Pasquinelli, A., Bettinger, J., Rougvie, A., Horvitz, H., Ruvkun, G.: The 21-nucleotide let-7 RNA regulates developmental timing in Caenorhabditis elegans. Nature 403(6772), 901–906 (2000)
23. Sewer, A., Paul, N., Landgraf, P., Aravin, A., Pfeffer, S., Brownstein, M., Tuschl, T., Van Nimwegen, E., Zavolan, M.: Identification of clustered microRNAs using an ab initio prediction method. BMC Bioinformatics 6(1), 267 (2005)
24. Wang, X., Zhang, J., Li, F., Gu, J., He, T., Zhang, X., Li, Y.: MicroRNA identification based on sequence and structure alignment. Bioinformatics 21(18), 3610 (2005)
25. Wang, X., Reyes, J., Chua, N., Gaasterland, T.: Prediction and identification of Arabidopsis thaliana microRNAs and their mRNA targets. Genome Biology 5(9), R65 (2004)
26. Washietl, S., Hofacker, I., Stadler, P.: Fast and reliable prediction of noncoding RNAs. Proceedings of the National Academy of Sciences 102(7), 2454 (2005)
27. Xue, C., Li, F., He, T., Liu, G., Li, Y., Zhang, X.: Classification of real and pseudo microRNA precursors using local structure-sequence features and support vector machine. BMC Bioinformatics 6(1), 310 (2005)
28. Zhang, B., Pan, X., Cox, S., Cobb, G., Anderson, T.: Evidence that miRNAs are different from other RNAs. Cellular and Molecular Life Sciences 63(2), 246–254 (2006)
29. Zuker, M.: Mfold web server for nucleic acid folding and hybridization prediction. Nucleic Acids Research 31(13), 3406 (2003)

Towards Recommender System Using Particle Swarm Optimization Based Web Usage Clustering

Shafiq Alam, Gillian Dobbie, and Patricia Riddle

Department of Computer Science, University of Auckland, New Zealand
sala038@aucklanduni.ac.nz, {gill,pat}@cs.auckland.ac.nz

Abstract. Efficiency and quality of the product of data mining process is a challenging question for the researchers. Different methods have been proposed in the literature to tackle these problems. Optimization based methods are a way to address this issue. We addressed the problem of data clustering by implementing swarm intelligence based optimization technique called Particle Swarm Optimization (PSO). We scaled the approach to implement it in a hierarchical way using Hierarchical Particle Swarm (HPSO) clustering. The paper also aims to outline our novel outlier detection technique. The research will lead us to provide a benchmark for web usage mining and propose a collective intelligence based recommender system for the usage of Java API documentation.

Keywords: Swarm intelligence, clustering, recommender system, outlier detection.

1 Introduction

Data mining or Knowledge discovery in databases (KDD) is an automated process of extracting useful, comprehendible, interesting and previously unknown patterns from large amount of data using techniques from machine learning, statistics and computational intelligence [11] [8]. The contribution of data mining methods for information comprehension is incredible however the efficiency of the mining process and the quality of the resultant patterns is still a threatening question to the data mining researchers. To tackle the efficiency problem various optimization based methods have been proposed, which are either implemented as an independent system to perform the data mining task or hybridized with the existing data mining methods. The use of intelligent optimization techniques is effective in enhancing the performance of complex, real time, and costly data mining processes. One such optimization technique is swarm intelligence which took inspiration from social and cognitive learning properties of vertebrates and insects, and implements it through multi agent systems in the form of software components communicating with each other in a highly decentralized environment. The cooperative behavior [4] ensures these agents converge on some optimum solution. The two basic algorithms, ant colony optimization (ACO) and

L. Cao et al. (Eds.): PAKDD 2011 Workshops, LNAI 7104, pp. 316–326, 2012.

particle swarm optimization (PSO), have been found to be efficient in various areas of data mining.

The application of data mining techniques are widely spread but most extensive use of the data mining methods is reported in the last decade in the field of exponentially growing World Wide Web (WWW). Web mining, a sub domain of data mining is the implementation of data mining algorithms and standard KDD techniques that performs pre-processing, transformation, pattern extraction, post processing and pattern analysis on web data. It uses clustering, classification, association mining and prediction analysis to reveal the hidden patterns in structure, contents and usage of WWW. Web mining provides the foundation of web usage based web recommendation system to help the user to be directed to their required information based on the pattern extracted for the web user's browsing information.

The scope of this paper is the study and application of swarm intelligence in data clustering and to provide an intelligent clustering approach to develop usage mining based recommender systems. We will also discuss our novel agglomerative hierarchical clustering based on particle swarm optimization (PSO), our outlier detection mechanism and a case study of web log of Java API usage.

2   Related Work

The credit of starting a research initiative towards PSO based data clustering goes to Van der Merwe and Engelbrecht [15], who presented the idea of using PSO with k-means clustering for refining the k-means clustering technique. The work presented by Xiao et al. [18] is another example where PSO is used alongside other clustering technique. In their proposed approach PSO was hybridized with self-organizing maps (SOM) to improve the efficiency of clustering process and named the approach as SOM/PSO. Chen and Ye [14] also employed a representation in which each particle represent the centroids of the clusters but as opposed to the previous work there is no hybridization with any other algorithm, so the execution time which is not reported must be higher. Another hybrid work is done by Omran et al. [16] who used a Dynamic Clustering algorithm based on PSO(DCPSO). The proposed approach is a hybridization of PSO and K-means where PSO performs clustering and K-means does refinement of the clusters. Cohen et. al. [6] has a novel approach to use PSO as an independent clustering technique where the centroids of the clusters were guided by the help of social and cognitive learning of the particles. As opposed to earlier versions of PSO based clustering techniques, in this approach each particle represents a portion of the solution instead of representing the whole solution as a single particle. In [1] the authors proposed Evolutionary PSO Clustering (EPSO-clustering) technique which uses the concept of merging different particles to form strong particles where particles represent cluster centroids which provide a hierarchical solution for clustering problem. For PSO based web usage clustering we could only trace a single work [5] where the authors reported web sessions clustering using PSO, with the help of improved velocity PSO and k-means. In [2] we extended our clustering approach and used it for clustering web usage session based on time attributes of the session. For web usage clustering the related work include [12], [17] and [3] who used data clustering as a part of their usage mining model.

Regardless of how the technique works, and what are their strength and weakness, the use of PSO in data mining and particularly in data clustering is increasing. A number of applications of these techniques have been reported in the literatures which verify the applicability and suitability of PSO for data mining applications.

3 Swarm Intelligence

Swarm intelligence is a collective intelligence based optimization technique originally inspired by the behavior of animal's group activities. The technique implements the communication and cooperation nature of such animals, by mimicking through autonomous agents in a multi agent environment. Ant Colony Optimization (ACO) and Particle Swarm Optimization (PSO) are two main swarm intelligence based optimization techniques widely used for data mining, distributed systems, power systems, hybrid systems, and complex systems. PSO, originally proposed by Kennedy and Eberhart [13] is inspired by the swarm behavior of birds, fish and bees while searching for food and communicating with each other. The approach is highly decentralized and is based upon cooperation among the agents or particles [9] [7].

4 Particle Swarm Based Clustering

The approaches so far proposed for data clustering based on particle swarm optimization exploits the distance between the data points as similarity measure. The problem formulation is either done to select a single particle for the complete solution or the entire swarm represents the solution of the problem. Some researchers have hybridized PSO with a traditional clustering algorithm to optimize the selection of centroids or other input parameters [9]. Some approaches use fitness function to measure the quality of the solution while other approaches use traditional termination criteria to find the final solution.

4.1 EPSO-Clustering

We proposed Evolutionary particle swarm optimization based clustering EPSO-Clustering [1] based on generation based evolution of the swarm. The swarming process of the EPSO starts with the initial representation of clustering centroids using particles. During generation evolution each weaker particle (particle with less data) is consumed by the stronger particle and the process is iteratively performed until the required number of particles is remained or some distance minimization or maximization function is achieved. Minimum number of elements

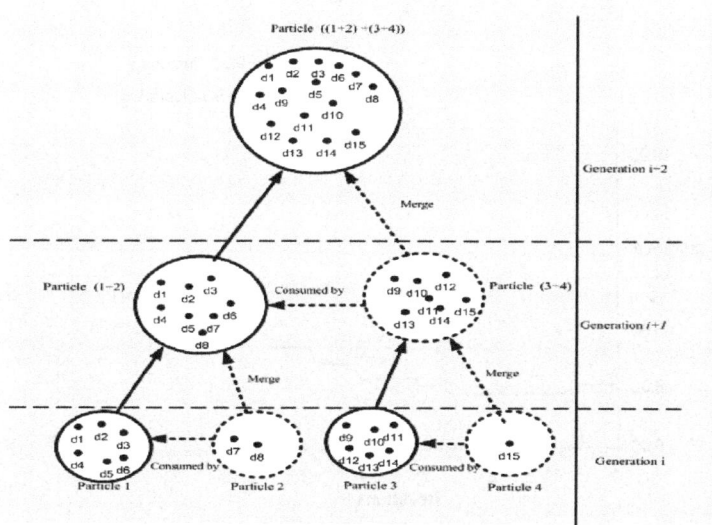

Fig. 1. Particle merging and evolution

per cluster is another stopping criterion which can be enforced. Three main component of the EPSO clustering are the cognitive component i.e. learning from experience, the social component which shows learning from the environment and the self-organizing component which take into account the density of a cluster. Figure 1 shows an example of how the particles evolve during different generations. During *generation i* + 1 two particles particle 2 and particle 4 are consumed by particle 1 and particle 3 respectively. The resultant configuration has stronger particles thus formed from acquiring the food (data elements) of the weaker particles. The performance of the EPSO against the PSO clustering is given in Figure 2.

Apart from the advantages of EPSO-clustering there were some deficiencies in the approach such as selection criteria for merging of particles and numbers of particles to be merged per generation, and generalized values of different parameters. We tried to address these deficiencies by proposing Hierarchical Particle Swarm based clustering (HPSO-clustering) technique. Next section explains the approach.

4.2   HPSO-Clustering

To tackle the problem of degeneracy and background knowledge for data clustering, and deficiencies in EPSO-clustering, we proposed the idea of hierarchical clustering and provide a framework based on particle swarm based hierarchical agglomerative clustering. HPSO-clustering use the same mechanism as EPSO but it split the data into partitions before it generates the hierarchy of clusters. It contains the advantages of both partition based clustering as well as hierarchical clustering. In HPSO-clustering after each generation the candidate

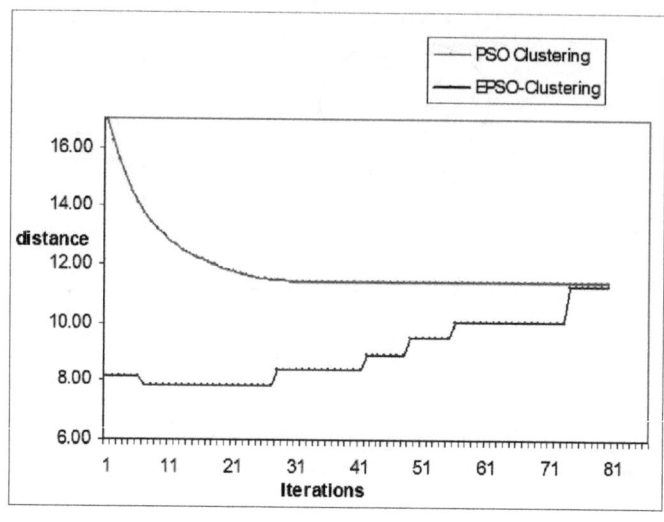

Fig. 2. PSO clustering v/s EPSO-clustering

clusters (only two) are merged to move to the next generation. Merging of two clusters give more populated clusters and proper direction to the particle to move into better centroid position. This provides a solution to the clustering problem based upon different level of compactness. Figure 3 shows the consistency of the approach against different parameters. Number of observations to calculate these parameters were taken as 1000 with two dimension each or as stated otherwise. In Figure 4 the execution time of HPSO was compared with DBSCAN [10] clustering approach. The proposed approach was also compared with other clustering approaches on benchmark data.

4.3   PSO Bases Web Usage Clustering

We used PSO-clustering and EPSO-clustering in the web usage domain to verify its applicability on real world data. We chose benchmark web usage data, i.e. NASA web log file, which contains 1891715 HTTP requests to NASA Kennedy Space Centre's web server, for verification of our proposed approach. Our clustering problem formulation includes the encoding of web usage sessions as data points for the PSO-clustering algorithm with each data point possessing time dimension information of the web users activity i.e. session length, the amount of data downloaded and number of visits during a particular session. The data was passed through a sophisticated pre-processing stage. The results reveal the fact that more than 60% of the web requests recorded in the web logs are image requests and are useless in the context of web session clustering. The importance of the pre-processing stage is verified by the ratio of successfully selected requests to total requests i.e. 1/3 of the total request are selected for analysis

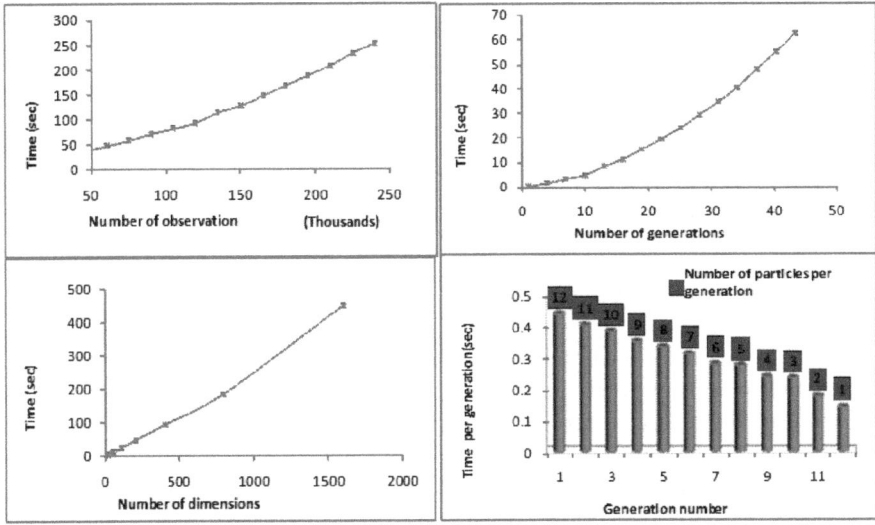

Fig. 3. HPSO clustering execution time

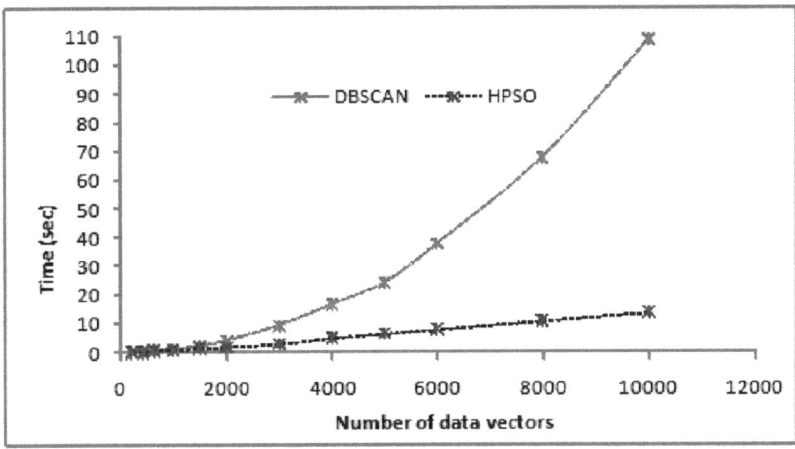

Fig. 4. HPSO clustering execution time

to find the real sessions, which can contribute in finding the hidden patterns in the activity data. After fulfilling standard constraints on selecting sessions for clustering the remaining 400 sessions were divided into four logs of 100 sessions each. We compared the approaches and found that PSO has better results than k-means clustering. Some preliminary results are shown in Table 1.

Table 1. K means and EPSO clustering

Log	K means		EPSO	
	Mean Intra-Clus Dist.	**Fitness**	**Mean Intra-Clus Dist.**	**Fitness**
1	81.82114	245.463	81.48635	**244.45905**
2	35.0334	105.10021	34.85	**104.55**
3	27.7769	55.55389	25.836569	**51.673138**
4	59.1294	118.25885	59.136744	118.27349

4.4   HPSO Based Outlier Detection

Our novel idea for outlier detection exploits HPSO-clustering. The first generation particles are mapped to input data vectors and become the representative
of the initial clustering centroids. The number of the particles is kept large and
uniform so that they can cover the entire input data space. During iteration every particle calculates its Euclidean distance from the input data vectors picks
up the nearest data vector and updates its velocity and position. The particle
with the minimum number of data vector and falling at a considerable distance
from other clusters will be treated as an outlier and is not consumed in any
generation. An outlier particle may be a single outlier or a representative of an
outlier cluster.

In Figure 5 cluster 3 and cluster 4 are considered to be outliers on the basis
of the distances from their nearest particles i.e. the particles of cluster 2 and
cluster 1 respectively. The cluster may contain one or more data vectors decided
by the consumption and merge operation. In Figure 5 bold arrows show the
distance to the nearest particle. The threshold distance is compared only with
the nearest particle. The doted arrows represent the distance between potential
outliers and other clusters. This distance will not be compared with the outlier
threshold distance. The threshold distance D_t is compared with the average of
intra-cluster distance of the nearest cluster i.e. $\|X_i - X_{(nearest)}\|$ where D_t is
defined as:

$$OD(X_i) = \frac{D_t}{k} \times \sqrt{\sum_{j=1}^{k}(Y_j - X_i)^2} \tag{1}$$

where Y_i represents data vectors and k is the number of total data associated
with a particular particle X_j. The outlier distance evolves through different generations. Initial generations have a smaller outlier distance compared to the later
generations where the intra cluster distance of the individual clusters increases
due to merging of clusters and the population of clusters increases. During each
generation of the swarm, the particle with the less number of data vectors won
is targeted as proposed outlier. The nearest neighbor of the proposed outlier
cluster helps in deciding the outlier cluster to be confirmed as outlier. The intra-
cluster distance of the nearest neighbor is calculated and is multiplied by the
density factor to find the threshold distance. This approach is scalable to cluster
the data as well as detect the outliers in the data.

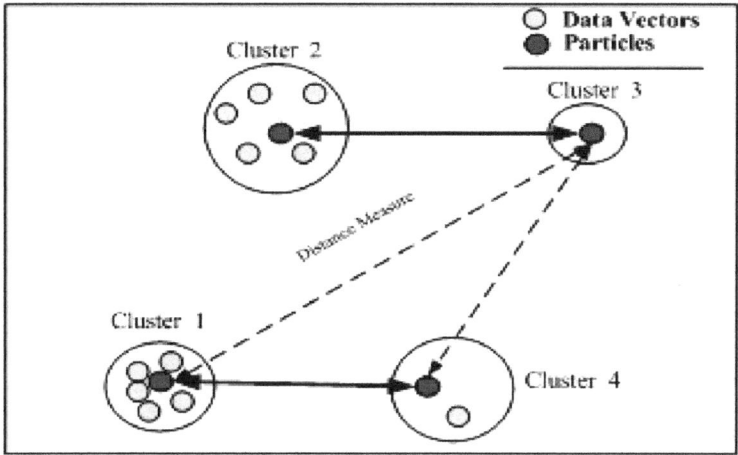

Fig. 5. Individual and cluster of outliers

5   JAVA API Usage Log and Recommender System

The Java API (Application Programming Interface) is a description of all the packages, classes, and interfaces that a standard Java language provides. The API possesses the details of each method of the classes along with the information about parameters, fields, constructors and usage examples. The purpose of the documentation is to provide a convenient way to help the developer find details of a desired package, interface, class or method. The total number of classes and interfaces has grown to more than 4000. The more the API grows the more difficult it becomes to use, as the programmer has to either remember the desired class or find the class from the growing number of classes using a key word search. Memorizing the details of classes, methods and interfaces is difficult and to use a key word search the developer needs to have prior knowledge of at least the name of the particular class. Table 2 shows a snap shot of the data which is collected from the usage log of Java API documentation after performing some initial pre-processing and cleaning of the log file.

We aim to develop a usage clustering based recommender system for the usage of Java API documentation. As a case study we have chosen the usage data of Java API documentation collected in the Department of Computer Science's web usage logs. We aim to have sufficient user requests which could give us a realistic sequence of individual sequences as well as API usage sequence. The logs contain requests from 2006 to 2010, each containing raw data including Java API usage requests. Table 3 shows some of the initial statistics about the data. We will pass this data through a sophisticated pre-processing stage to extract usage sessions as shown in previous sections.

The data is based for our proposed PSO powered clustering based intelligent recommender system. The outline of the proposed system is shown in Figure 6 follows. A Java web user's session is tracked and searched for the corresponding

Table 2. Java API navigation requests

Java API Requests
/references/java/java1.5/api/java/nio/package-frame.html
/references/java/java1.5/api/java/awt/font/package-frame.html
/references/java/java1.5/api/java/rmi/server/package-frame.html
/references/java/java1.5/api/javax/crypto/package-frame.html
/references/java/java1.5/api/javax/security/sasl/class-use/AuthenticationException.html
/references/java/java1.5/api/index.html?java/awt/image/renderable/class-useRenderableImageProducer.html
/references/java/java1.5/api/index.html?java/util/PriorityQueue.html
/references/java/java1.5/api/index.html?java/awt/image/renderable/class-useRenderableImageProducer.html
/references/java/java1.5/api/index.html?java/nio/channels/class-useCancelledKeyException.html
/references/java/java1.5/api/java/awt/event/HierarchyEvent.html
/references/java/java1.5/api/javax/print/attribute/standard/PrinterState.html

Table 3. Statistics about the data

log statistics	
Start Date:	24/12/2006
End date:	31/12/2009
Total requests:	38387774
Java API Request:	5206947
Images requests:	1537978
CSS requests:	44569
Distinct IPs:	74211
Distinct Pages:	60140

cluster of users based on some similarity measure. The user session is treated as data point and the corresponding cluster is represented by an agent/ particle. The agent then search recommendations for the input user session within its given scope and rank them on the basis of its similarity to the input web user session. The corresponding recommendation enables the user to be directed to its desired page. This approach help in reducing the input search space for recommendation and ranking. To test the performance of the proposed approach we aim to provide a standard benchmark for WUM research. The benchmark we propose will be a representative of a similar class of people with different session at different time. The time span of this benchmark is proposed to be sufficiently long to demonstrate real trends in the usage data.

6   Conclusion and Future Work

In this paper we highlighted the use of particle swarm optimization for data clustering, outlier detection and usage based recommender systems. We used

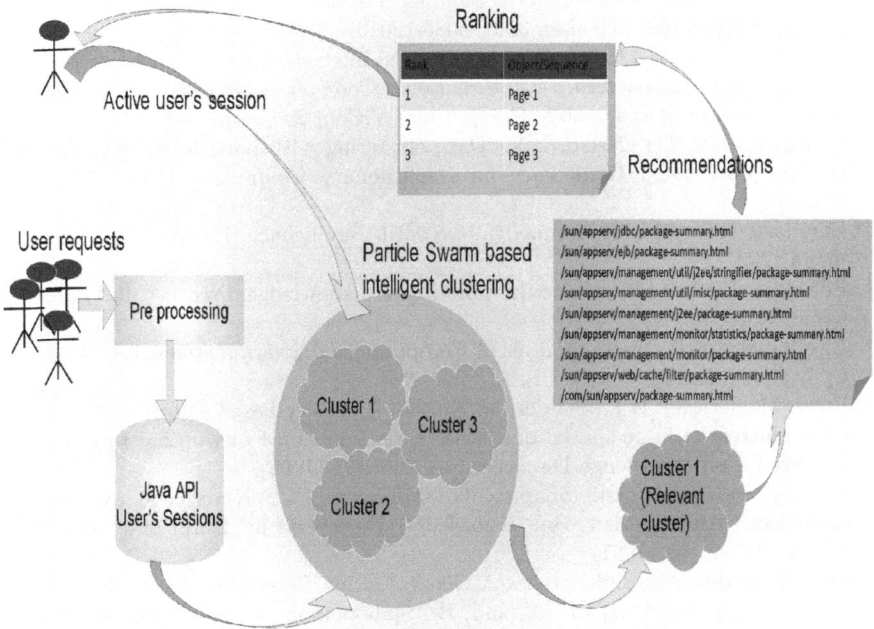

Fig. 6. The proposed recommender system

particle swarm based clustering in different scenarios; particle swarm clustering in evolutionary manner, and particle swarm clustering in hierarchical evolutionary manner. The approaches were tested against benchmark data clustering data and benchmark web session data. The comparison of the results verifies the validity of the approach. We have also provided a framework for detection of density and cluster based outliers in data. Our future work includes automation of different parameters of our technique, and development of swarm intelligent web usage recommender system. The last but not least goal of this project is to provide benchmark data for web usage mining. We aim to provide web session data of real web usage requests for a particular domain.

References

1. Alam, S., Dobbie, G., Riddle, P.: An evolutionary particle swarm optimization algorithm for data clustering. In: Swarm Intelligence Symposium, SIS 2008, pp. 1–6. IEEE (2008)
2. Alam, S., Dobbie, G., Riddle, P.: Particle swarm optimization based clustering of web usage data. In: IEEE/WIC/ACM International Conference on Web Intelligence and Intelligent Agent Technology, WI-IAT 2008, vol. 3, pp. 451–454 (2008)
3. Banerjee, A., Ghosh, J.: Clickstream clustering using weighted longest common subsequences. In: Proceedings of the Web Mining Workshop at the 1st SIAM Conference on Data Mining, pp. 33–40 (2001)

4. Cao, L.: In-depth behavior understanding and use: the behavior informatics approach. Information Sciences, 3067–3085 (2010)
5. Chen, J., Zhang, H.: Research on application of clustering algorithm based on pso for the web usage pattern. In: International Conference on Wireless Communications, Networking and Mobile Computing, WiCom 2007, pp. 3705–3708 (2007)
6. Cohen, S.C.M., De Castro, L.N.: Data clustering with particle swarms. In: 2006 IEEE International Conference on Evolutionary Computation, pp. 1792–1798 (2006)
7. Eberhart, R.C., Shi, Y., Kennedy, J.: Swarm Intelligence, 1st edn. Morgan Kaufmann (2001)
8. Edelstein, H.: Introduction to data mining and knowledge discovery, 3rd edn. Two Crows Corp. (1999)
9. Engelbrecht, A.P.: Fundamentals of Computational Swarm Intelligence. John Wiley & Sons (2006)
10. Ester, M., Kriegel, H.P., Sander, J., Xu, X.: A density-based algorithm for discovering clusters in large spatial databases with noise. In: Proc. of 2nd International Conference on Knowledge Discovery, pp. 226–231 (1996)
11. Frawley, W.J., Piatetsky-Shapiro, G., Matheus, C.J.: Knowledge discovery in databases: An overview. In: Knowledge Discovery in Databases, pp. 1–30. AAAI/MIT Press (1991)
12. Fu, Y., Sandhu, K., Shih, M.-Y.: A Generalization-Based Approach to Clustering of Web Usage Sessions. In: Masand, B., Spiliopoulou, M. (eds.) WebKDD 1999. LNCS (LNAI), vol. 1836, pp. 21–38. Springer, Heidelberg (2000)
13. Kennedy, J., Eberhart, R.C.: Particle swarm optimization. In: Proc. of International Conference on Neural Networks IV, pp. 1942–1948 (1995)
14. Kuo, R., Wang, M., Huang, T.: An application of particle swarm optimization algorithm to clustering analysis. Soft Computing - A Fusion of Foundations, Methodologies and Applications 15, 533–542 (2011)
15. van der Merwe, D., Engelbrecht, A.: Data clustering using particle swarm optimization. In: The 2003 Congress on Evolutionary Computation, CEC 2003, vol. 1, pp. 215–220 (2003)
16. Omran, M., Salman, A., Engelbrecht, A.: Dynamic clustering using particle swarm optimization with application in image segmentation. Pattern Analysis and Applications 8, 332–344 (2006)
17. Shahabi, C., Zarkesh, A., Adibi, J., Shah, V.: Knowledge discovery from users webpage navigation. In: Proceedings of Seventh International Workshop on Research Issues in Data Engineering, pp. 20–29 (April 1997)
18. Xiao, X., Dow, E.R., Eberhart, R., Miled, Z.B., Oppelt, R.J.: Gene Clustering using Self-Organizing Maps and Particle Swarm Optimization. In: Guo, M. (ed.) ISPA 2003. LNCS, vol. 2745, pp. 154–160. Springer, Heidelberg (2003)

Weighted Association Rule Mining Using Particle Swarm Optimization

Russel Pears[1] and Yun Sing Koh[2]

[1] School of Computing and Mathematical Sciences, Auckland University of Technology
`rpears@aut.ac.nz`
[2] Dept of Computer Science, University of Auckland
`ykoh@cs.auckland.ac.nz`

Abstract. Association rule mining is an important data mining task that discovers relationships among items in a transaction database. Most approaches to association rule mining assume that the items within the dataset have a uniform distribution. Therefore, weighted association rule mining (WARM) was introduced to provide a notion of importance to individual items. In previous work most of these approaches require users to assign weights for each item. This is infeasible when we have millions of items in a dataset. In this paper we propose a novel method, *Weighted Association Rule Mining using Particle Swarm Optimization (WARM SWARM)*, which uses particle swarm optimization to assign meaningful item weights for association rule mining.

Keywords: Weighted Items, Particle Swarm Optimization, Association Rule Mining.

1 Introduction

Association rule mining was introduced by [1] and is widely used to derive meaningful rules containing items that are statistically related to each other. It aims to extract interesting correlations, frequent patterns, associations or casual structures among sets of items in transaction databases. The relationships are not based on the inherent properties of the data themselves but rather based on the co-occurrence of the items within the database. The original motivation for seeking association rules came from the need to analyze supermarket transaction data also known as market basket analysis. An example of a common association rule is {*bread*} → {*butter*}. This indicates that a customer buying bread would tend to buy butter. With traditional rule mining techniques even a modest sized dataset can produce thousands of rules, and as datasets get larger, the number of rules produced becomes unmanageable. This highlights a key problem in association rule mining; keeping the number of generated itemsets and rules in check, whilst identifying interesting rules amongst the plethora generated.

In the classical model of association rule mining, the items are treated as being uniformly distributed. In reality, most datasets are skewed with unbalanced data. By applying the classical model to these datasets, important but critical rules which occur infrequently may be missed. For example consider the rule: {*stiff neck, fever, aversion to light*} → {*meningitis*}. Meningitis occurs relatively infrequently in a medical

L. Cao et al. (Eds.): PAKDD 2011 Workshops, LNAI 7104, pp. 327–338, 2012.

dataset, however if it is not detected early the consequences can be fatal. Recent techniques have tackled this problem by weighting individual items differently to reflect their importance [5,14,17,18]. For example, items in a market basket dataset may be weighted based on the profit they generate. However, most datasets do not come with preassigned weights and so the weights must be manually assigned in a time consuming and error-prone fashion.

In this paper, we propose to use particle swarm optimization (PSO) to assign item weights. PSO searches the vector space of possible solutions of a problem in an effort to find the optimal solution. Conceptually a swarm of particles traverse the multidimensional solution space, with each particle drawn stochastically toward its best previous position (or solution) and the position of the best particle in the swarm. As the particles swarm across the solution space, they test many possible solutions, and quickly and effectively converge on low error solutions [6].

Our approach to itemset generation is similar to traditional Apriori, except that during candidate itemset generation, the support of an itemset is multiplied by the sum of the weights of its constituent items. This weighted support is then compared to a weighted minimum support threshold, and itemsets whose weighted support is below the weighted minimum support are not considered for further itemset generation, in the same way as traditional Apriori. The item weights are fit using particle swarm optimization with a fitness function which encourages items which co-occur with other items with high confidence to have a larger weight.

The rest of this paper is organized as follows. In the next section, we look at some of the related work. Section 3 presents the related concepts used in our approach namely, association rule mining and particle swarm optimization. Section 4 describes our proposed algorithm and Section 5 presents the performance and the analysis of the results produced. Finally we summarize our contributions in Section 6 and outline directions for future work.

2 Related Work

The concept of an association rule was first introduced by [1]. Given a set of items, $I = \{i_i, i_2, \ldots, i_n\}$, a transaction may be defined as a subset of I and a dataset may therefore be defined as a set D of transactions. A set X of items is called an itemset. The support of X, $\sup(X)$, is the proportion of transactions containing X in the dataset. An *association rule* is an implication of the form $X \to Y$, where $X \subset I, Y \subset I$, and $X \cap Y = \emptyset$. The rule $X \to Y$ has *support* of s in the transaction set D, if $s = \sup(XY)$. The rule $X \to Y$ holds in the transaction set D with *confidence* c where $c\mathrm{conf}(X \to Y) = \frac{\sup(XY)}{\sup(X)}$.

Given a transaction database D, a support threshold *minsup* and a confidence threshold *minconf*, the task of association rule mining is to generate all association rules that have support and confidence above the user-specified thresholds. Using the classical approach, the algorithms are open to the rare items problem [9]. Items which are rare but have high confidence levels are pruned out. Rare items have low support which is unlikely to reach the minimum support threshold. For example, [7] noted that in market basket analysis rules such as $\{caviar\} \to \{vodka\}$ will not be generated using normal frequent itemset mining algorithms. This is because both caviar and vodka are expensive items and are not purchased frequently, and thus are unlikely to have support above the minimum support threshold.

Numerous algorithms have been proposed to overcome this problem. Many of these algorithms follow the classical framework but substitute an item's support with a weighted support. Each item is assigned a weight to represent the importance of individual items, with items that are considered interesting having a larger weight. This approach is called *weighted association rule mining* (WARM) [10,5,15,17,18].

In weighted association rule mining, we assign a weight w_i for each item i, where $-1 \leq w_i \leq 1$, to show the importance of the item. The weighted support of an item i is $w_i \text{sup}(i)$. Similar to the traditional association rule mining, a weighted support threshold and a confidence threshold will be assigned to measure the strength of the association rules. Here a k-itemset, X, is considered a large itemset if the weighted support of this itemset is greater than the user-defined minimum weighted support (wminsup) threshold.

$$\left(\sum_{i \in X} w_i \right) \text{sup}(X) \geq wminsup \tag{1}$$

The weighted support of a rule $X \rightarrow Y$ is:

$$\left(\sum_{i \in X \cup Y} w_i \right) \text{sup}(XY) \tag{2}$$

An association rule $X \rightarrow Y$ is called an interesting rule if $X \cup Y$ is a large itemset and the confidence of the rule is greater than or equal to a minimum confidence threshold.

Sanjay et al. [10] introduced weighted support to association rule mining by assigning weights to both items and transactions. In their approach rules whose weighted support is larger than a given threshold are kept for candidate generation, much like in traditional Apriori. A similar approach was adapted by [5], but they did not apply weights to transactions, only items. They also proposed two different ways to calculate the weight of an itemset, either as the sum of all the constituent items' weights or as the average of the weights. However, both of these approaches violated the downward closure property. The downward closure property of the support measure in classical association rule mining cannot be applied to weighted items. As a result the mining algorithm proposed became more complex.

Tao et al. [15] later proposed a "weighted downward closure property" as the adjusted support values violate the original downward closure property. In their approach, two types of weights were assigned, item weight and itemset weight. The goal of using weighted support is to make use of the weight in the mining process and prioritize the selection of targeted itemsets according to their significance in the dataset, rather than by their frequency alone.

Yan and Li [18] proposed to incorporate pageview weights into association rule mining. The earlier approaches [10,5,15,17] assumed a fixed weight for each item. Yan and Li allowed their weights to vary. They used the duration of a pageview as the weights. The rules generated in these approaches rely heavily on the weights used. Thus to ensure the results generated are useful, we must determine a way to assign the item weights effectively.

2.1   Related Concepts: Particle Swarm Optimization (PSO)

Particle swarm optimization (PSO) was first introduced by Kennedy and Eberhart [8] as a population-based heuristic method to find optimal solutions to non linear numerical problems. PSO has recently been used in the areas of classification [12,16] and clustering [13]. However, research combining both PSO and association rule mining [3] has been not been prolific.

The swarm consist of a fixed number of particles, with each particle having the following properties: a current position vector x_i in the search space, current velocity vector v_i, a personal best position vector p_i, and the global best position p_{gb} common to all the p_i. In every iteration each particle in the swarm is updated using the following equation:

$$v_i = v_i + r_1(p_i - x_i) + r_2(p_{gb} - x_i) \tag{3}$$

$$x_i = x_i + v_i \tag{4}$$

Here r_1 and r_2 are random numbers uniformly distributed within [0,1]. The PSO algorithm performs the update operation using Equation 3 and 4 repeatedly. The quality of each particle is measured using a fitness function which reflects the fitness of the particle's solution, and particles update their position until the fittest particle has not improved its fitness significantly for a number of iterations.

PSO has been applied to the learning problems of neural networks and function optimization. The main advantages of PSO are the robustness to control parameters and its computational efficiency when compared with mathematical algorithms and other heuristic optimization techniques.

In the next section, we combine both weighted association rule mining and particle swarm optimization to discover interesting itemsets. We use particle swarm optimization to fit the weights for each item in a dataset.

3   Weighted Association Rule Mining Using Particle Swarm Optimization (WARM SWARM)

In classical association rule mining, itemsets are evaluated by their support, a measure based solely on an itemset's occurrence in a dataset. In previous research, items were assigned weights in relation to the importance of the particular item. However, knowing the relative importance of different items in a dataset requires expert knowledge, and manually weighting all the items in a large dataset is error prone and time consuming. In reality a dataset may contain millions of items; manually assigning a weight to every item is infeasible. WARM SWARM solves this problem, by quickly fitting the item weights using PSO before doing weighted rule mining. The item weights are fit using PSO with a fitness function which encourages items which co-occur with other items with high confidence to have a larger weight.

3.1   Weighting Function

In this section, we first describe the weighting function which we use to fit the weights for the items within the dataset. Intuitively an item should be weighted based on the strength of its connections to other items, and the weights of the other items connected to it. We say that two items are connected if they have occurred together within the same transaction. Items that appear often with similarly connected items should be considered good items, and should be highly weighted.

Table 1. Transaction Database

(a) Transaction database

tid	Itemset
100	A, B, C
200	A, C
300	A, B, D
400	A, E
500	E, F
600	E, F

(b) Unique Items

Item	Related Unique Items
A	B, C, D, E
B	A, C, D
C	A, B
D	A, B
E	A, F
F	E

(c) Example graph derived from Table 1

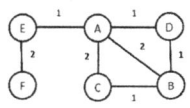

Table 1 shows an example transaction dataset. In Table 1(a) the left column represents the list of transaction IDs and the right column represents the items belonging to each particular transaction. In Table 1(b) the left column represents the list of items belonging to each particular transaction and the right column represents the unique items which appear together with the item in the left column. Table 1(b) can be represented as a graph whereby the nodes represent an item and the edges represent the support of the two items as an itemset. For example, the edge between node A and node B has a strength of 2, meaning that A and B occur together twice in the dataset.

The weighting technique we developed for our approach is inspired by Inverse Distance Weighting by Shepard [11]. Inverse distance weighting is an interpolation technique which generates values for unknown points as a function of the values of a set of known points scattered throughout the dataset.

The contribution to an item's weight relies on the weights of its neighboring items. Given the item k which is connected to n items and given that the weight of an item is w_i, we can determine the contribution of item k's connected weights, c_k as:

$$c_k = \sum_{i}^{n} \alpha_{ik} w_i \tag{5}$$

The α_{ik} value measures the strength of the connection between item k and item i as the co-occurrence count of item k and i, divided by the occurrence count of item k. The α_{ik} value is defined below.

$$\alpha_{ik} = \frac{count(ik)}{count(k)} \tag{6}$$

For example, in Table 1 item A appears together with item B, C, D, and E. Here we are able to find the connected weight for A as:

$$c_A = \sum_{i \in \{B,C,D,E\}}^{n} \alpha_{iA} w_i$$

During particle swarm optimization, every particle has a weight for each item in the dataset. The weights are randomly assigned a value, and are then updated to move towards the fittest particle in every iteration. The fitness of a particle is calculated as the absolute value of the sum of the difference between each item's weight w_k and that item's connections c_k. Values closer to zero show a closer fit, and are considered fitter.

3.2 Weighted Association Rule Mining Using Particle Swarm Optimization (WARM SWARM)

In particle swarm optimization each particle or solution contains a position and velocity. As particles move around the solution space, they sample different locations. Each location has a fitness value according to how good it is at satisfying the objective or fitness function. Particles will eventually swarm around the area in the space where the fittest solutions are.

The WARM SWARM algorithm is used to find the appropriate weights for each item within a transaction dataset. Each item weight can be represented as a dot in the problem space. One particle in the swarm represents one possible set of item weights $W = \{w_{i_1}, w_{i_2}, \ldots, w_{i_n}\}$, where, w_{i_1} is the weight for item i_1 and n is the number of items in the dataset. According to its own experience and those of its neighbors, the particle adjusts its weights position, in the vector space.

4   Experimental Results

In this section, we discuss the results from the experiments carried out to evaluate the performance of our algorithm. We tested our algorithm using both synthetic and real data. All the experiments were performed on a 2GB Intel Core 2 machine with 2GB of main memory running Windows Vista. All the programs are written in C++. All experiments were run 10 times for every dataset.

4.1   Synthetic Datasets

To assess the performance of our algorithm, we used a modified version of the data generator proposed by [2]. These experiments were carried out to ensure that all the weights fitted by PSO were able to find the itemsets that were deliberately injected. Table 2 summarizes the characteristics of several of the datasets generated during our tests. To create a dataset D, our synthetic data generation program takes the following parameters: number of transactions $|D|$, average size of transactions $|T|$, number of unique items $|I|$, and number of large itemsets $|L|$.

To generate the synthetic dataset, we first determine the size of the next transaction. This is generated using a Poisson distribution whose mean is the average size of the transaction. We then fill the transactions with items. Each transaction T is assigned a series of potential frequent itemsets. Each itemset in T has a weight associated with it, which corresponds to the probability that this itemset will be picked. Random items are also injected into the transactions. These items are treated as noise and should be pruned out by the weighted association rule mining algorithm.

Table 2. Parameter settings

Name	$|T|$	$|L|$	$|I|$	$|D|$	Size (MB)
T5L10I10KD1K	5	10	10K	1K	0.03
T5L10I10KD10K	5	10	10K	10K	0.30
T5L10I10KD100K	5	10	10K	100K	3.49

In this section, we measure the quality of the results produced by WARM SWARM and Apriori. In the experiments, we varied (1) the number of transactions, and (2) number of large itemsets in the dataset. We ran WARM SWARM on each dataset 10 times. This was to ensure that the results generated by WARM SWARM were consistent. We compared the itemsets that were found by WARM SWARM against the known itemsets from the synthetic datasets. We then ran Apriori on each dataset. For all the experiments using Apriori we lowered the minimum support threshold to 0.05. This was a conscious attempt to allow Apriori a chance to find all the synthetic itemsets that were deliberately injected. In WARM SWARM we set the weighted minimum support threshold to 0.10 for all the experiments.

$$\text{Precision} = \frac{|\text{relevant itemsets found}|}{|\text{itemsets found}|} \tag{7}$$

$$\text{Recall} = \frac{|\text{relevant itemsets found}|}{|\text{relevant itemsets injected}|} \tag{8}$$

We measured the quality of the results based on accuracy by precision and recall. Precision is defined as the number of relevant itemsets found divided by the total number of itemsets generated. In our case relevant itemsets produced are itemsets that were deliberately injected into the synthetic dataset. Recall is defined as the number of relevant itemsets found divided by the total number of existing relevant itemsets that are deliberately injected into the synthetic dataset.

Varied Number of Transactions. In the first experiment, we varied the number of transaction from 10^3 to 10^6 over the T5L10I10K dataset. The average size of a large itemset was 3. This was to ensure that WARM SWARM was still able to find the relevant large itemsets injected when the number of transactions changed. The results generated are very promising.

Table 3 displays the recall and precision values generated for the synthetic datasets using WARM SWARM and Apriori when we varied the number of transactions.

Table 3. Recall and Precision Values (T5L10I10K)

Dataset	WARM SWARM		Apriori	
	Avg Recall	Avg Precision	Recall	Precision
T5L10I10KD1K	1.00	0.76	1.00	0.11
T5L10I10KD5K	1.00	1.00	1.00	0.35
T5L10I10KD10K	1.00	0.96	1.00	0.11
T5L10I10KD50K	1.00	0.93	1.00	0.22
T5L10I10KD100K	1.00	0.97	1.00	0.38

For WARM SWARM, we take the average recall and precision value. Overall the recall values for both WARM SWARM and Apriori were very notable; on average both these techniques had recall values of 1.00. On the other hand, there was a significant difference between the precision value produced by WARM SWARM and Apriori. On average the precision value produced by WARM SWARM was 0.92, whereas, the average precision value produced by Apriori was 0.23. From these results, we note that WARM SWARM was finding only 8% irrelevant itemsets and Apriori was finding 77% irrelevant itemsets. By comparison Apriori was producing a vast number of itemsets that were not relevant when compared to WARM SWARM. These itemsets may have appeared together due to noise and should have been pruned out. Figure 1 and 2 compares the recall and precision values for results based on WARM SWARM and Apriori.

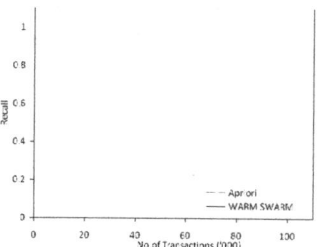

Fig. 1. Recall (T5L10I10K)

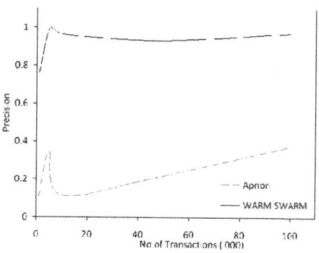

Fig. 2. Precision (T5L10I10K)

Figure 3 shows the recall value generated in the ten different runs using WARM SWARM. Overall average recall is 1.0. Having a recall value of 1.0 means that we have managed to find all the itemsets that were deliberately injected into the dataset.

Figure 4 shows the precision value generated by the ten different runs using WARM SWARM. From the figure we notice there was fluctuation in the precision value caused by the number of additional itemsets found by WARM SWARM that were not deliberately injected. The average precision value of 0.92 is still at a significant level. Our algorithm is not overly greedy and it was still able to find a compact set of itemsets which were deliberately injected into the dataset. Notice that the average precision shows signs of increasing as the number of the transactions in the dataset increases from 10^3 to 10^6 transactions.

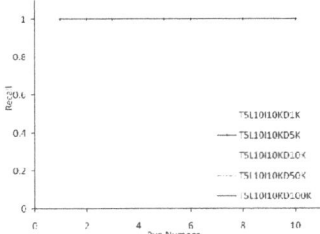

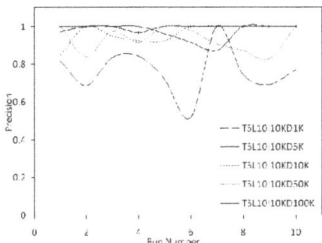

Fig. 3. Recall Value based on WARM SWARM

Fig. 4. Precision Value based on WARM SWARM

Varied Number of Large Itemsets. In the second experiment, we varied the number of large itemsets from 5 to 30 over the T5I10KD10K dataset with an interval of size 5. The average size of the large itemsets was 3. This experiment was conducted to ensure that WARM SWARM was still able to find the relevant large itemsets injected when the number of itemsets injected changed. Table 4 displays the recall and precision values generated for the synthetic datasets using WARM SWARM and Apriori when we varied the number of itemsets. As in the previous experiment we take the average recall and precision value for WARM SWARM. Overall both WARM SWARM and Apriori had recall values of 1.0. WARM SWARM had an average precision value of 0.95, whereas Apriori had an average precision value of 0.61. From these results, WARM SWARM was finding 5% irrelevant itemsets and Apriori was finding 39% irrelevant itemsets.

Table 4. Recall and Precision Values (T5I10KD10K)

Dataset	WARM SWARM		Apriori	
	Avg Recall	Avg Precision	Recall	Precision
T5L5I10KD10K	1.00	0.71	1.00	0.07
T5L10I10KD10K	1.00	1.00	1.00	0.62
T5L15I10KD10K	0.99	1.00	1.00	0.52
T5L20I10KD10K	0.99	1.00	1.00	0.46
T5L25I10KD10K	1.00	0.98	1.00	1.00
T5L30I10KD10K	1.00	0.98	1.00	1.00

Figure 5 and 6 compares the recall and precision values for results in Table 4 based on WARM SWARM and Apriori. Apriori did not perform well when the number of large items fell below 20, and the datasets were unbalanced. In those datasets, Apriori was finding a significant amount of irrelevant large itemsets. Clearly, Apriori was not able to handle unbalanced datasets as well as WARM SWARM.

Figure 7 shows the recall value generated by the ten different runs using WARM SWARM. Overall average recall is close to 1.0. All the datasets had an average recall value of 1.0 except for T5L15I10KD10K and T5L20I10KD10K. Taken as a whole, these two cases reduced the overall average recall values for their datasets.

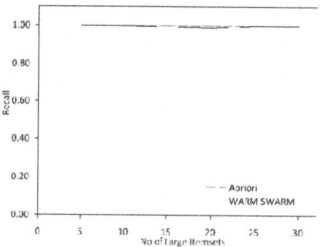

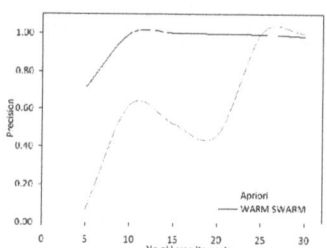

Fig. 5. Recall (T5I10KD10K)   **Fig. 6.** Precision (T5I10KD10K)

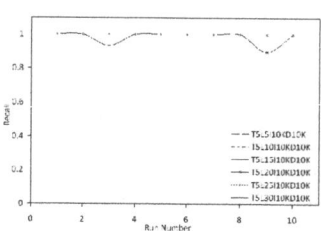

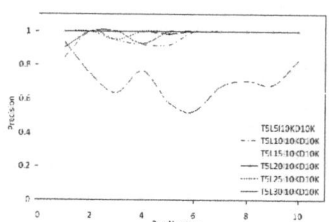

Fig. 7. Recall Value   **Fig. 8.** Precision Value

Figure 8 shows the precision value generated by the ten different runs using WARM SWARM. The fluctuation of the precision values for the datasets is relatively minimal except for the T5L5I10KD10K dataset. It was slightly harder for WARM SWARM to fit the weights for the T5L5I10KD10K dataset. Nonetheless WARM SWARM was still able to produces results with recall value of 1.00.

In this section we showed that WARM SWARM can effectively generate all the rules that were injected into synthetic datasets as measured by precision and recall. In the next section, we investigate the results produced by WARM SWARM on real datasets.

4.2 Real-World Datasets

We compared the performance of the WARM SWARM algorithm and the standard Apriori algorithm on six different datasets from the UCI Machine Learning Repository [4]. Table 5 displays the results produced by WARM SWARM and Apriori. Each row of the Table represents an attempt to find rules with minimum confidence of 0.90 and lift greater than 1.0 from the dataset in the left most column. In column *Rules(sup > 0.1)* we show the number of rules found by WARM SWARM which have support less than 0.10.

In the WARM SWARM algorithm, we allowed items that have count greater than one to be considered. We ran WARM SWARM on each of the datasets 10 times and chose the result with the best fitness. Due to the nature of Apriori, it can become unreasonably slow even for small datasets. If we set the minimum support to 0.0, it was not clear that many datasets would finish in reasonable time. Thus we raised the minimum support value until the runtime was no longer prohibitively expensive.

Table 5. Comparison of results of WARM SWARM with Apriori

Dataset	WARM SWARM						Apriori			
	wminsup	Rules sup<0.10	Rules	Itemsets	Fitness Value	Time(s)	*minsup*	Rules	Itemsets	Time(s)
TeachingEval.	0.1	2	3	13	0.00004	37	0.01	5514	1856	1
Zoo	0.1	1465	2591	633	0.00013	71	0.20	6535229	62771	1283
Soybean-Large	0.4	3459	22903	5079	0.00007	224	0.45	5098877	94789	2140
Dermatology	0.3	321	2254	1466	0.00050	343	0.40	9695623	105689	2823
Mushroom	0.1	3308	3892	2232	0.00007	231	0.20	8824926	45349	7480
Adult	0.1	105	171	216	0.00002	139	0.03	58636	21296	2147

Using WARM SWARM we were able to generate some interesting rules with support less than 0.10. This was important as the percentage of rules found with support less than 0.10 for the UCI datasets ranged between 14% and 85%. Using Apriori we were not able to push the minimum support less than 0.10 in four of the datasets without extreme runtime. In Teaching Evaluation and Adult dataset, we were able to push the support below the 0.10 level. However the set of rules generated by these two datasets using Apriori were drastically larger than that of WARM SWARM.

5   Conclusions and Future Work

In this paper we present a novel approach, WARM SWARM, which combines particle swarm optimization with weighted association rule mining. Fitting the weights using particle swarm optimization eliminates the error-prone manual assignment of weights required for traditional weighted association rule mining. In this paper, we fit the weights of each item based on the connection strength of items with their neighbors. In the experiments, we show that even with the overhead of weight fitting using particle swarm optimization, WARM SWARM was significantly faster than standard Apriori, but still managed to produce interesting and meaningful rules. This was especially evident when the size of the datasets increased.

The drawback WARM SWARM is that particle swarm optimization is non-deterministic and can produce multiple solutions for any one problem. WARM SWARM requires several runs in order to get adequate coverage over the output rules generated. Despite that, due to its fast speed, performing multiple runs is not a problem, and as future work, we can consider various modifications of particle swarm optimization to improve the overall consistency of our results.

References

1. Agrawal, R., Imielinski, T., Swami, A.N.: Mining association rules between sets of items in large databases. In: Buneman, P., Jajodia, S. (eds.) Proceedings of the 1993 ACM SIGMOD International Conference on Management of Data, pp. 207–216 (1993)
2. Agrawal, R., Srikant, R.: Fast algorithms for mining association rules in large databases. In: Bocca, J.B., Jarke, M., Zaniolo, C. (eds.) Proceedings of the 20th International Conference on Very Large Data Bases, VLDB, Santiago, Chile, pp. 487–499 (1994)

3. Alatas, B., Akin, E.: Rough particle swarm optimization and its applications in data mining. In: Soft Computing - A Fusion of Foundations, Methodologies and Applications. Springer, Heidelberg (2008)
4. Asuncion, A., Newman, D.: UCI machine learning repository (2007), http://www.ics.uci.edu/~mlearn/MLRepository.html
5. Cai, C.H., Fu, A.W.C., Cheng, C.H., Kwong, W.W.: Mining association rules with weighted items. In: IDEAS 1998: Proceedings of the 1998 International Symposium on Database Engineering & Applications, p. 68. IEEE Computer Society, Washington, DC, USA (1998)
6. Clerc, M., Kennedy, J.: The particle swarm-explosion, stability, and convergence in a multi-dimensional complex space. IEEE Trans. Evolutionary Computation 6(1), 58–73 (2002)
7. Cohen, E., Datar, M., Fujiwara, S., Gionis, A., Indyk, P., Motwani, R., Ullman, J.D., Yang, C.: Finding interesting association rules without support pruning. IEEE Transactions on Knowledge and Data Engineering 13, 64–78 (2001)
8. Kennedy, J., Eberhart, R.C.: Particle swarm optimization. In: Proceedings of the 1995 IEEE International Conference on Neural Networks, pp. 1942–1948. IEEE Service Center, Piscataway (1995)
9. Liu, B., Hsu, W., Ma, Y.: Mining association rules with multiple minimum supports. In: Proceedings of the 5th ACM SIGKDD International Conference on Knowledge Discovery and Data Mining, pp. 337–341 (1999)
10. Sanjay, R., Ranka, S., Tsur, S.: Weighted association rules: Model and algorithm (1997), http://citeseer.ist.psu.edu/185924.html
11. Shepard, D.: A two-dimensional interpolation function for irregularly-spaced data. In: Proceedings of the 1968 23rd ACM National Conference, pp. 517–524. ACM (1968)
12. Sousa, T., Silva, A., Neves, A.: A Particle Swarm Data Miner. In: Pires, F.M., Abreu, S.P. (eds.) EPIA 2003. LNCS (LNAI), vol. 2902, pp. 43–53. Springer, Heidelberg (2003)
13. Srinoy, S., Kurutach, W.: Combination artificial ant clustering and K-PSO clustering approach to network security model. In: ICHIT 2006: Proceedings of the 2006 International Conference on Hybrid Information Technology, pp. 128–134. IEEE Computer Society, Washington, DC, USA (2006)
14. Sun, K., Bai, F.: Mining weighted association rules without preassigned weights. IEEE Trans. on Knowl. and Data Eng. 20(4), 489–495 (2008)
15. Tao, F., Murtagh, F., Farid, M.: Weighted association rule mining using weighted support and significance framework. In: KDD 2003: Proceedings of the Ninth ACM SIGKDD, pp. 661–666. ACM, New York (2003)
16. Veenhuis, C., Köppen, M.: Data swarm clustering. In: Abraham, A., Grosan, C., Ramos, V. (eds.) Swarm Intelligence in Data Mining. SCI, vol. 34, pp. 221–241. Springer, Heidelberg (2006)
17. Wang, W., Yang, J., Yu, P.S.: Efficient mining of weighted association rules (WAR). In: KDD 2000: Proceedings of the Sixth ACM SIGKDD International Conference on Knowledge Discovery and Data Mining, pp. 270–274. ACM, New York (2000)
18. Yan, L., Li, C.: Incorporating Pageview Weight into an Association-Rule-Based Web Recommendation System. In: Sattar, A., Kang, B.H. (eds.) AI 2006. LNCS (LNAI), vol. 4304, pp. 577–586. Springer, Heidelberg (2006)

An Unsupervised Feature Selection Framework Based on Clustering

Sheng-yi Jiang and Lian-xi Wang

School of Informatics, Guangdong University of Foreign Studies,
Guangzhou, China
jiangshengyi@163.com, wanglianxi20082008@126.com

Abstract. Feature selection plays an important part in improving the quality of learning algorithms in machine learning and data mining. It has been widely studied in supervised learning, whereas it is still relatively rare researched in unsupervised learning. In this work, a clustering-based framework formed by an unsupervised feature selection algorithm is proposed. The proposed framework is mainly concerned with the problem of determining and choosing important features, which are selected by ranking the features according to the importance measure scores, from the original feature set without class information. Theory analyzed indicates that the time complexity of each algorithm is nearly linear with the size and the number of features of dataset. Experimental results on UCI datasets show that algorithm with different scores in the framework are able to identify the important features with clustering, and the proposed algorithm have obtained competitive results in terms of classification error rate and the degree of dimensionality reduction when compared with the state-of-the-art supervised and unsupervised feature selection approaches.

Keywords: Feature Selection, Unsupervised Learning, Feature Importance Measure Score, Clustering.

1 Introduction

Feature selection plays an important part in improving the quality of classification and clustering in machine learning and data mining. With the rapid accumulation of high dimensional data, it poses a serious challenge to feature selection algorithms. In theory, more features should provide more discriminating power, but in practice, with a limited amount of training data, redundant and uninformative features not only significantly slow down the learning process, but also cause the classifier to over-fit the training. Therefore, feature selection becomes very necessary for machine learning tasks when facing data with enormity on both size and dimensionality.

Researchers have studied feature selection in various aspects. A comprehensive survey of feature selection algorithms is presented in [13]. According to whether the class information is available or not, feature selection can be separated into two fundamental groups, supervised feature selection [2,6,10,12,18,21,22] and unsupervised feature selection [3,4,5,7,16,17,19,23]. Generally, feature selection in

L. Cao et al. (Eds.): PAKDD 2011 Workshops, LNAI 7104, pp. 339–350, 2012.

supervised learning has been well studied. However, for unsupervised learning, the research is relatively rare. When the class information is sufficient, supervised feature selection methods usually outperform unsupervised feature selection ones [3]. In this paper, we are particularly interested in researching on feature selection methods in unsupervised learning.

On the basis of single-pass clustering algorithms [8], a series of feature importance measure scores formed a framework, which are concerned with tree creation phase, are proposed in this work. The methods can distinguish the important and noisy features effectively from the original data. Empirical results illustrate the performance of the developed algorithms, which in general have obtained competitive results in terms of classification error rate when compared with other classic feature selection methods, and the approaches in the framework get approximate results.

The rest of this paper is organized as follows. Section 2 presents previous related work. In section 3, some preliminaries for feature selection approaches of a framework are introduced. In section 4, several feature importance measure scores and an unsupervised feature selection algorithm, which rely on the measure feature discriminating for "goodness of split" in the framework, are elaborated. Experimental results are developed in Section 5. Finally, we give conclusions and discuss the future work in section 6.

2   Related Work

In fact, feature selection has been paid more and more attention for researchers to study in unsupervised learning recently. In brief, there are two fundamentally different approaches for unsupervised feature selection based on clustering: feature (attribute) clustering and object clustering. The former approach removes redundant features by partitioning the original feature set into some distinct clusters, which are formed by similar features. Generally speaking, after grouping features into clusters, such filters select a subset of features from each cluster, discarding some of less discriminative features. Mitra [16] and Covões [4] used some new indexes to measure feature similarities so that feature redundancy is detected. Although promising results obtained, they required data with continuous types only. Wang [19] proposed a feature-weight learning approach based on a defined similarity measure and an evaluation function to improve the performance of FCM. However, the defined similarity measure and evaluation function are complicated and difficult to interpret. Moreover, a global feature selection approach which selects feature by clustering (for k-means) and evaluates the quality of clusters using normalized cluster separability [25]. It can only find one relevant feature subset for all clusters. However, such a global approach cannot identify individual clusters that exist in different feature subspaces.

The latter approach firstly partitioned a set of objects into clusters by a certain distance measure, such objects in the same cluster are more similar to each other than objects in different clusters, and then selected discriminate features based on the importance of the features according to some certain defined criteria.

Modha [17] published a method for feature weighting in k-means clustering based on within-cluster and between cluster matrices. But this method of finding optimal weights from a predefined set of variable weights may not guarantee that the predefined set of weights would contain the optimal weights. Huang [7] have presented W-k-means algorithm that can calculate feature weights automatically. Based on the current partition in the iterative k-means clustering process, the algorithm calculates a new weight for each feature based on the variance of the within cluster distances. The new weights are used in deciding the cluster memberships of objects in the next iteration. The optimal weights are found when the algorithm converges. However, this approach is too dependant on the selection of k value. Very recently, Zeng [23] presented a new feature selection approach for Gaussian mixture clustering. In the algorithm, a new feature relevance measurement index was introduced to identify the most relevant features. Furthermore, many developments show that feature importance ranking methods are well adopted to feature selection for clustering with unlabeled data [5]. From the previous works, we conclude that the effectiveness of unsupervised feature selection methods may be affected by the performance of the clustering algorithms. In this work, we explore a framework formed by several unsupervised feature selection methods based on the partitioning the objects into clusters of similar objects.

3   Basic Concept on Clustering

In this section, a robust clustering algorithm is introduced. Section 3.1 gives some preliminaries of single-pass clustering algorithm. The description of single-pass clustering algorithm is presented in section 3.2.

3.1   Preliminaries

We suppose that dataset D consists of N instances, and each instance with m categorical features, where $D_i (1 \leq i \leq m)$ denotes the i-th feature.

Definition 1. For a cluster C and an feature value $a_i \in D_i$, the frequency of a in C with respect to D_i is defined as: $Freq_{C|D_i}(a_i) = |\{object \mid object \in C, object.D_i = a_i\}|$.

Definition 2. For a cluster C, the cluster summary information (CSI) is defined as: $CSI = \{n, Summary\}$, where n is the size of the cluster $C(n=|C|)$, $Summary$ is given as the frequency information for categorical feature values: $Summary = \{< Stat_i > \mid Stat_i = \{(a, Freq_{C|D_i}(a)) \mid a \in D_i\}, 1 \leq i \leq m\}$.

3.2   Single-Pass Clustering Algorithm

Before giving the feature selection scheme, we firstly introduce the single-pass clustering algorithm. The single-pass clustering algorithm employs the least distance principle to divide the dataset. The clustering algorithm is described as follows:

(1) Initialize the set of clusters S, as the empty set, and read a new object p.

(2) Create a cluster with the object p.

(3) If no objects are left in the database, go to (6), otherwise read a new object p, and find the cluster C_1 in S which is closest to the object p. Namely, find a cluster C_1 in S, such that $d(p,C_1) \le d(p,\overline{C})$ for all $\overline{C}$ in S.

(4) If $d(P,C_1) > r$, go to (2).

(5) Merge object p into cluster C_1 and modify the *CSI* of cluster C_1, go to (3).

(6) Stop.

4   Unsupervised Feature Selection Framework

In this section, we introduce some measure scores for determining the relatively important features according to decision tree formation process. These measure scores, which formed a framework, are applicable to nominal data and needn't class information to evaluate the features. Therefore, when facing continuous features, they can be discretized properly in advance.

4.1   Feature Importance Measure Scores

Now we discuss how to evaluate the goodness of features for unsupervised data. A feature is discriminative if it can separate the clusters well. That is to say, the goodness of split measure calculates the extent to which the particular feature values split the data into separate clusters. A feature value of good features should occur in a certain cluster only or more frequently, while in the remaining clusters appear less or even do not appear. A perfect feature would have each feature value associated with only one cluster. In this sense, the more pure clusters of the corresponding feature are, the more important the feature is. In one word, the problem of feature selection in unsupervised learning can be solved by the purity in different clusters on features.

Consider an instance in a data, which is described by a set of features. If the data has distinct clusters, this instance should belong to some clusters, and the instance must be well separated from some instances and very close to others. Based on theory of the decision tree formation [15], we develop a series of scores for ranking features in terms of importance.

Definition 3. Suppose the dataset D with N instances, each instance is described by m features. The dataset D is partitioned into k clusters after clustering, let $C = \{C_1, C_2, \cdots, C_k\}$ be the results of clustering.

(a) the importance score of cluster C_j on feature D_i is computed as: $f(D_i | C_j)$.

(b) the importance score of feature D_i is defined as: $F(D_i) = \sum_{j=1}^{k} \frac{|C_j|}{N} f(D_i | C_j)$

where $|C_j|$ is the size of the cluster C_j.

Some specific approaches for calculating the importance score of cluster C_j on feature D_i with decision tree induction are given as follows:

*1)   Gini Index Measure(**GINI**):*

$$f_1(D_i \mid C_j) = 1 - \sum_{t=1}^{n_i^{(j)}} p(c_{it}^{(j)})^2 \tag{1}$$

*2)   Gini-ratio Measure(**GIR**):*

$$f_2(D_i \mid C_j) = (1 - \sum_{t=1}^{n_i^{(j)}} p(c_{it}^{(j)})^2) / GV(D_i) \tag{2}$$

*3)   Entropy Measure (**EM**):*

$$f_3(D_i \mid C_j) = -\sum_{t=1}^{n_i^{(j)}} p(c_{it}^{(j)}) \log p(c_{it}^{(j)}) \tag{3}$$

*4)   Expect Cross Entropy Measure(**ECE**):*

$$f_4(D_i \mid C_j) = -\sum_{t=1}^{n_i^{(j)}} [p(c_{it}^{(j)}) \log p(c_{it}^{(j)}) + (1 - p(c_{it}^{(j)})) \log(1 - p(c_{it}^{(j)}))] \tag{4}$$

*5)   Gain-ratio Measure(**GAR**):*

$$f_5(D_i \mid C_j) = -\sum_{t=1}^{n_i^{(j)}} p(c_{it}^{(j)}) \log p(c_{it}^{(j)}) / IV(D_i) \tag{5}$$

In the above formulas, the symbol $n_i^{(j)}$ and $p(c_{it}^{(j)})$ is the number of different values of the cluster C_j on feature D_i and the relative probability of the *t-th* value of the cluster C_j on feature D_i, respectively. Furthermore, the whole entropy and Gini value of feature D_i are denoted as $IV(D_i)$ and $GV(D_i)$ respectively.

According to the definitions described above, we conclude that all scores are similar to some extent when handle the same data. The smaller score is, the more important the feature is.

4.2   Feature Selection Framework

Based on the previous work proposed before, we develop a framework formed by an unsupervised feature selection method using five measure scores.

The whole procedures of the proposed algorithm are summarized as follows:

Step 1. Discretization and initialization: use the approximate equal frequency discretization [9], called AEFD, to discretize the continuous features, and set the initialized value $Sum(D_i)$ of the importance of feature D_i. Let $Sum(D_i) = 0$;

Step 2. Calculate the importance of each feature: run the following steps s times until the results are stable.

Step 2.1. Clustering: clustered the dataset D by single-pass clustering algorithm according to the selected threshold r presented in section 3.2.

Step 2.2. Calculate the importance score of $F(D_i)$ of feature D_i after clustering according to definition 3.

Step 2.3. Modify the importance score of feature D_i : $Sum(D_i) = Sum(D_i) + F(D_i)$.

Step 3. Rank the features in terms of importance score: rank the features $D_1 \sim D_m$ in descending order according to the value of $Sum(D_i)$ for each feature. The newly obtained feature order denotes as $D_i^*(i = 1, \cdots, m)$.

Step 4. Select the feature subset in terms of feature importance score: find the rapid changing point or inflection point i_0 in $D_i^*(i = 1, \cdots, m)$.

Step 5. Remove the features whose order number are less than i_0.

In step 2, iterate the clustering algorithm for many times is to decrease the influence of clustering by the thresholds. In the implementation, we observe from our experiments that the proposed algorithms set s in less than 10 iterations, and typically within 4 iterations.

5   Empirical Results

The empirical results in this section compare the technique presented here with other methods. We present the datasets, methods and classifiers used in the validation process, and the comparison results in the following sections.

5.1   Experimental Setup

In this section, we choose two state-of-the-art supervised and unsupervised feature selection algorithms in comparison with the introduced methods in the framework.

Table 1. Summary of bench-mark data sets

Dataset	Cat./Con.	Instances	Classes	Dataset	Cat./Con.	Instances	Classes
Australian	8/6	690	2	Mushroom	22/0	8124	2
Breast	0/9	699	2	Pima	0/8	768	2
Credit	9/6	690	2	Promoters	57/0	106	7
German	13/7	1000	2	Segmentation	0/19	2310	2
Glass	0/9	214	7	Sonar	0/60	208	2
Heart	7/6	270	2	Ticdata2000	85/0	5822	2
Iris	0/4	150	3	Vote	16/0	435	2
Liver	0/6	345	2	Wine	0/13	178	3

In our experiments, we evaluate the performance of our proposed unsupervised feature selection algorithms on sixteen datasets from UCI machine learning repository [1]. Table 1 describes the summaries of each dataset, including the number of instances (Instances), continuous features (Con.), categorical features (Cat.) and classes (Class). These datasets contain various number of instances and dimensionalities of the feature space, as well as classes. The size of the datasets varies between 106 and 8 124 instances and the total number of original features of the datasets is up to 85.

5.2   Results on UCI Datasets

In this work, we compare the performance of our proposed algorithms with Variance method [24], GFS-k-means method [25], CFS method [26], FCBF algorithm [21], FarVPKNN algorithm [27] and NDEM algorithm [28] on several UCI datasets from different aspects with dimensionality reduction and classification error rates of C4.5 and Naïve Bayes learning algorithms. Among them, the first two algorithms are unsupervised, while the last four are supervised. In supervised methods, FarVPKNN and NDEM can deal data with mixed types, while CFS and FCBF require every continuous feature to be discretized. In the experiments, we keep special search strategy for each algorithm.

To increase the statistical significance of the results when using datasets with a limited number of samples, we choose the first 2/3 of samples from each class as the training data, and the remaining for testing of each dataset.

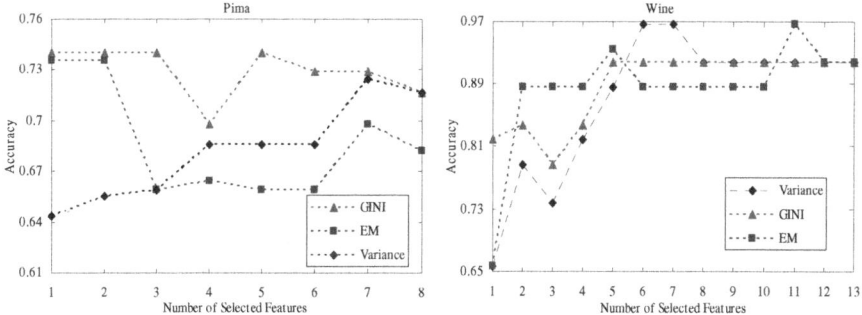

Fig. 1. Classification accuracies vs. number of selected features on Pima and Wine data sets

Table 2. Comparison of our algorithms and GFS-k-means on three UCI datasets

Dataset	Feature Subset			Error Rate (%)		
	EM	GINI	GFS	EM	GINI	GFS
Glass	3,6,8	1,3,5,6	2,3,5,6,7,8,9	43.84	41.09	34.25
Wine	6,7,10,12,13	6,7,10,13	1,2,3,4,5,6,8,9, 10,11,12,13	6.56	3.28	8.20
Sonar	11,15,18,19,20, 21,22,36,37,39	1,2,3,5,10,11, 21,52,59,60	35,36,37,38,41,42, 44,46,47,51,55,56, 57,58,59,60	34.38	30.96	30.99

Firstly, we choose Gini and Entropy as the art measure scores for comparing our algorithm with variance score to evaluate the importance of features on Pima and Wine data. Fig.1 shows the plots for accuracy vs. numbers of selected features.

Fig.1 indicates that, the feature importance measures of GINI and Entropy are comparable to Variance and the performance of Gini significantly better than that of Entropy. This verifies that our scores are effective to evaluate the feature importance.

We also evaluate our algorithms and compare the results with GFS-k-means, which need not to discrete continuous features, on three UCI datasets (Glass, Sonar and Wine). Table 2 shows the experiment results. For Glass dataset, GFS-k-means keeps 7 out 9 features with the error rate of 34.25%. Compared to GFS-k-means, our proposed algorithms obtain less features and lower error rates.

Experiments on Wine and Sonar datasets give similar results. In summary, the performance of our algorithms outperforms GFS-k-means. Then, we verify the performance of our algorithms with FarVPKNN and NDEM on Credit and Heart datasets, which contain the mixture of continuous and discrete features.

The selected feature with different feature selection algorithms, such as NDEM, FarVPKNN and our algorithms with feature importance measure of Entropy and GINI, are presented on the order of selecting in Table 3. From the table, we can find that the selected features are distinct when applying different algorithms and our method selects fewer features than other two supervised approaches except FarVPKNN on Credit data.

Table 3. Features that were sequentially selected when using different algorithms

Data	EM	GINI	FarVPKNN	NDEM
Credit	13,4,5,10,9	1,9,12,10	3	9,13,6,10,7,1,2,12,3,4,8
Heart	6,9,2,7,13,11,12,3	2,3,7,13, 9,12	13,5,10,12,3,1,4	13,12,3,11,7,4,1,8,2,5,6

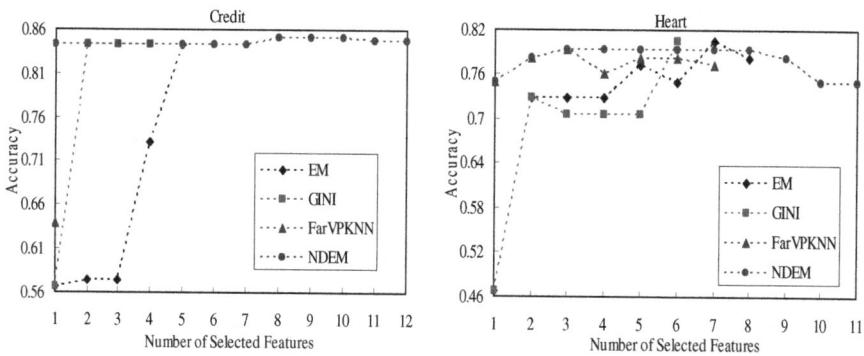

Fig. 2. Accuracies vs. Different number of selected features on Credit and Heart data sets

Fig.2 visualizes a change of classification accuracy with respect to the number of selected features for four data sets. As shown in the figure, in most cases, the performance of our method is similar to those of FarVPKNN and NDEM. Generally, increasing the number of features greatly improves the accuracy on the two datasets by the proposed algorithms. Except for our method, the other algorithms climb slowly when the features increasing of the selection process, and sometimes even drop. This phenomenon occurs because there are many redundant or noisy features in the feature sets. The previous experimental results confirm that our algorithms are not influenced by the discreting methods.

In the following, we test the proposed algorithms with all the feature importance scores. Table 4-6 summarize the average results obtained by each feature selection algorithm. The acronyms GINI, EM, ECE, GAR, GIR, MVM and Full refer to GINI, Entropy Measure, Expect Cross Entropy, Gain-ratio measure, Gini-ratio measure, Majority Voting Mechanism in the unsupervised framework, and finally to the results found by using all features of the dataset. The selected feature subsets of MVM method are formed by choosing the features appeared more than half of the approaches in the framework.

Table 4. Number of features selected by different feature selection algorithms

Dataset	Full	CFS	FCBF	GINI	GIR	EM	ECE	GAR	MV
Australian	14	8	7	5	5	4	5	4	5
Breast	9	4	8	4	4	7	4	4	3
Credit	15	8	6	4	3	5	7	5	5
German	34	4	5	8	6	10	10	10	5
Glass	9	7	5	4	4	3	3	3	3
Heart	13	7	7	6	5	8	8	8	7
Iris	4	1	2	1	2	3	2	3	2
Liver	6	3	1	2	4	3	4	4	4
Mushroom	22	11	4	12	10	10	10	10	10
Pima	8	3	3	3	3	4	4	2	4
Promoters	57	21	6	8	11	9	14	12	9
Segmentatio	19	4	6	7	9	9	10	6	9
Sonar	60	3	10	10	14	10	11	10	9
Ticdata2000	85	2	6	8	9	8	8	8	8
Vote	16	4	3	3	3	3	3	3	3
Wine	13	5	10	4	4	5	5	4	5
Average	24	5.94	5.56	5.56	6	6.31	6.75	6	5.69
Average dimensiona-lity reduction (%)		75.26	76.82	76.82	75	73.69	71.88	75	76.3

Table 4 records the number of newly selected features for each feature selection algorithm. The eighteenth row of the Table 4 shows the average dimensionality reduction of all the datasets due to each one of the feature selection methods used. All the approaches tend to remove more than 1/4 of the original features. The last row of Table 4 indicates that the unsupervised feature selection algorithms in the framework give the similar degree of dimensionality reduction.

Table 5. The averaged classification error rates (%) of C4.5 on UCI datasets

Dataset	Full	CFS	FCBF	GINI	GIR	EM	ECE	GAR	MVM
Australian	16.09	16.68	16.51	15.11	14.38	14.60	16.17	16	16.04
Breast	5.13	5.13	4.62	4.16	5.04	5.17	5.80	6.00	5.50
Credit	14.17	14.72	15.23	14.68	14.55	14.72	14.77	14.72	14.64
German	29.74	29.53	28.68	29.76	30.21	31.12	31.12	31.12	29.44
Glass	31.64	32.88	31.91	35.07	31.91	40.00	40.00	40.00	40.00
Heart	25.00	23.91	22.58	20.11	24.15	23.04	23.04	23.04	22.17
Iris	4.90	4.79	4.79	4.79	4.79	4.90	4.71	4.90	4.71
Liver	38.73	40.49	40.49	38.73	40.50	33.98	35.34	35.34	35.34
Mushroom	0	0.95	0.97	0.89	3.84	3.84	3.84	3.84	3.84
Pima	27.52	26.28	33.13	25.95	32.91	33.63	33.09	37.19	33.09
Promoters	25.68	20.56	22.43	19.73	20.28	23.56	22.16	21.94	22.16
Segmentation	4.05	4.08	4.27	5.11	5.11	11.48	11.48	15.54	14.52
Sonar	29.58	30.42	30.42	31.83	22.82	34.08	34.22	34.08	34.79
Ticdata2000	5.92	5.86	5.19	5.85	5.91	5.98	5.98	5.98	5.98
Vote	4.05	4.05	4.66	4.26	4.26	4.26	4.26	4.26	4.26
Wine	9.84	7.54	9.67	9.84	8.03	7.21	11.15	8.03	7.21
Average	17.00	16.74	17.22	16.53	16.79	18.22	18.57	18.87	18.36

Table 6. The averaged classification error rates (%) of Naive Bayes on UCI datasets

Dataset	Full	CFS	FCBF	GINI	GIR	EM	ECE	GAR	MVM
Australian	23.91	24.68	26.43	14.60	14.38	32.04	31.98	23.87	28.30
Breast	3.28	3.32	3.24	3.36	4.83	4.29	4.50	4.50	4.50
Credit	21.87	25.19	24.77	14.55	14.55	13.74	13.70	13.70	13.62
German	17.50	26.06	14.83	28.91	29.41	26.64	26.64	26.64	29.41
Glass	51.78	52.05	53.84	58.08	52.05	50.27	50.27	50.27	50.27
Heart	15.33	16.30	15.76	17.61	17.68	17.39	17.39	17.39	17.39
Iris	3.33	3.33	3.33	3.33	3.33	3.33	3.33	3.33	3.33
Liver	45.76	44.41	44.15	45.08	43.81	47.71	47.71	47.71	47.71
Mushroom	4.26	1.43	1.44	1.08	9.92	9.92	9.92	9.92	9.92
Pima	25.69	23.98	34.66	26.11	34.94	34.73	35.61	35.61	35.61
Promoters	8.11	7.22	7.37	10.56	15.83	13.78	10.56	10.56	13.77
Segmentation	17.45	13.04	12.56	14.78	16.25	12.13	13.65	16.74	13.91
Sonar	29.72	29.30	26.34	25.77	25.77	38.03	37.89	39.15	37.18
Ticdata2000	18.23	7.17	5.91	6.18	6.18	6.39	6.39	6.39	6.39
Vote	9.46	4.46	3.98	5.47	5.47	5.47	5.47	5.47	5.47
Wine	2.30	2.30	1.80	5.24	3.11	6.56	3.11	3.11	6.56
Average	18.62	17.77	17.53	17.54	18.59	20.15	19.88	19.65	20.21

Table 5 and 6 show the learning error rates of C4.5 and Naïve Bayes respectively on different feature sets. The results of the last row of Table 5 and 6 indicate that almost all the unsupervised feature selection algorithms classified on the newly selected features can obtain similar error rates except with Gini score and it also can illustrate that the proposed algorithm with different scores have approximate performance and the error rates are nearly close to the other two classic supervised feature selection approaches, as well as the full feature sets.

From the previous empirical study, we can conclude that the unsupervised feature selection algorithms in the framework can efficiently achieve high degree of dimensionality reduction and enhance or maintain predictive accuracy (or decrease the classification error rates) with selected features. In this framework, the performance of Gini score outperforms the other scores, and it is very close to the supervised approaches.

6 Conclusions and Future Work

This paper has introduced both several scores for computing the feature importance and a clustering-based framework of unsupervised feature selection. A series of unsupervised feature selection algorithms, which can identify the important features, are presented according to the selection measures for decision-tree induction. Theory analyzed in this paper indicates that the time complexities for the algorithms are nearly linear with the size of dataset, the number of features. The experimental results on UCI datasets demonstrate that the performance of the introduced clustering-based unsupervised feature selection approaches of the framework outperforms or similar with the existing analogous methods with respect to classification error rate and the degree of dimensionality reduction.

Future work is directed to the following issues: (1) investigate new effective definitions to measure the importance of features more accurately; (2) extend our approaches to work on data with higher dimensionality; (3) apply our methods in explorative analysis; (4) experiment our approaches to any particular clustering algorithms.

Acknowledgments. The author would like to thank Anonymous reviewers. This work is supported by the National Natural Science Foundation of China (No.61070061).

References

1. Asuncion, A., Newman, D. J.: UCI Machine Learning Repository (2007), http://www.ics.uci.edu/~mlearn/MLRepository.html
2. Au, W., Chan, K.C.C., Wong, A.K.C.: Attribute Clustering for Grouping, Selection, and Classification of Gene Expression Data. IEEE/ACM Transactions on Computational Biology and Bioinformatics 2, 83–101 (2005)
3. Bishop, C.M.: Neural Networks for Pattern Recognition. Oxford University Press (1995)
4. Covões, T.F., Hruschka, E.R., de Castro, L.N., Santos, Á.M.: A Cluster-Based Feature Selection Approach. In: Corchado, E., Wu, X., Oja, E., Herrero, Á., Baruque, B. (eds.) HAIS 2009. LNCS, vol. 5572, pp. 169–176. Springer, Heidelberg (2009)
5. Dash, M., Liu, H., Yao, J.: Dimensionality Reduction of Unsupervised Data. Newport Beach. In: Proc 9th IEEE Int'l Conf. Tools with Artificial Intelligence, pp. 532–539 (1997)

6. Guyon, I., Elisseeff, A.: An Introduction to Variable and Feature Selection. Journal of Machine Learning Research 3, 1157–1182 (2003)
7. Huang, J.Z., Ng, M.K., Rong, H.Q.: Automated Variable Weighting in k-Means Type Clustering. IEEE Transactions on Pattern Analysis and Machine Intelligence 27, 657–668 (2005)
8. Jiang, S.Y., Song, X.Y.: A Clustering-based Method for Unsupervised Intrusion Detections. Pattern Recognition Letters 5, 802–810 (2006)
9. Jiang, S.Y., Li, X., Zheng, Q., et al.: Approximate Equal Frequency Discretization Method. In: GCIS, vol. 5, pp. 514–518 (2009)
10. Sotoca, J., Pla, F.: Supervised Feature Selection by Clustering Using Conditional Mutual Information-based Distances. Pattern Recognition 43, 2068–2081 (2010)
11. Kira, K., Rendell, L.: The Feature Selection Problem: Traditional Methods and a New Algorithm. In: Proceedings of AAAI 1992, San Jose, CA, pp. 129–134 (1992)
12. Last, M., Kandel, A., Maimon, O.: Information-theoretic Algorithm for Feature Selection. Pattern Recognition Letters 22, 799–811 (2001)
13. Liu, H., Yu, L.: Toward Integrating Feature Selection Algorithms for Classification and Clustering. IEEE Transactions on Knowledge and Data Engineering 17, 1–12 (2005)
14. Liu, H., Motoda, H.: Feature Selection for Knowledge Discovery and Data Mining, vol. 454, pp. 121–135. kluwer Academic Publishers, Boston (1998)
15. Mingers, J.: An Empirical Comparison of Selection Measures for Decision-Tree Induction. Machine Learning 3, 19–342 (1989)
16. Mitra, P., Murthy, C.A.: Unsupervised Feature Selection Using Feature Similarity. IEEE Transactions on Pattern Analysis and Machine Intelligence 24, 301–312 (2002)
17. Modha, D.S., Spangler, W.S.: Feature Weighting in k-means Clustering. Machine Learning 52, 217–237 (2003)
18. Singh, S., Murthy, H., Gonsalves, T.: Feature Selection for Text Classification Based on Gini Coefficient of Inequality. In: 4th Workshop on Feature Selection in Data Mining, pp. 76–85 (2010)
19. Wang, X.Z., Wang, Y.D.: Improving Fuzzy C-means Clustering Based on Feature-weight Learning. Pattern Recognition Letters 25, 1123–1132 (2004)
20. Witten, I.H., Frank, E.: Data Mining: Practical Machine Learning Tools and Techniques, 2nd edn. Morgan Kaufmann, San Francisco (2005), http://www.cs.waikato.ac.nz/ml/weak/
21. Yu, L., Liu, H.: Efficient Feature Selection via Analysis of Relevance and Redundancy. Journal of Machine Learning Research 5, 1205–1224 (2004)
22. Zhang, D., Chen, S., Zhou, Z.: Constraint score: A New Filter Method for Feature Selection with Pair-wise Constraints. Pattern Recognition 41, 1440–1451 (2008)
23. Zeng, H., Cheung, Y.: A New Feature Selection Method for Gaussian Mixture Clustering. Pattern Recognition 42, 243–250 (2009)
24. Bishop, C.M.: Neural Networks for Pattern Recognition. Oxford University Press (1995)
25. Dy, J.G., Brodley, C.E.: Feature Selection for Unsupervised Learning. Journal of Machine Learning Research 5, 845–889 (2004)
26. Hall, M.A.: Correlation-based Feature Subset Selection for Machine Learning, Hamilton, New Zealand (1998)
27. Hu, Q., Liu, J., Yu, D.: Mixed Feature Selection Based on Granulation and Approximation. Knowledge based Systems 21, 294–304 (2008)
28. Hu, Q., Pedrycz, W., Yu, D.: Selecting Categorical and Continuous Features Based on Neighborhood Decision Error Minimization. IEEE Trans. on Systems, Man, and Cybernetics-Part B: Cybernetics 40, 137–150 (2010)

Workshop
on Advances and Issues
in Traditional Chinese Medicine
Clinical Data Mining
(AI-TCM 2011)

Discovery of Regularities in the Use of Herbs in Traditional Chinese Medicine Prescriptions

Nevin L. Zhang[1], Runsun Zhang[2], and Tao Chen[3]

[1] Department of Computer Science & Engineering,
The Hong Kong University of Science & Technology,
Clear Water Bay, Kowloon, Hong Kong
lzhang@cse.ust.hk
[2] Guanganmen Hospital,
Chinese Academy of Chinese Medical Sciences,
Beijing, China
[3] EMC Labs China
Beijing, China
tao.chen2@emc.com

Abstract. Traditional Chinese medicine (TCM) is a discipline with its own distinct methodologies and philosophical principles. The main method of treatment in TCM is to use herb prescriptions. Typically, a number of herbs are combined to form a formula and different formulae are prescribed for different patients. Regularities on the mixture of herbs in the prescriptions are important for both clinical treatment and novel patent medicine development. In this study, we analyze TCM formula data using latent tree (LT) models. Interesting regularities are discovered. Those regularities are of interest to students of TCM as well as pharmaceutical companies that manufacture medicine using Chinese herbs.

Keywords: Herb regularities, latent tree model, traditional Chinese medicine prescription.

1 Introduction

Traditional Chinese medicine (TCM) is a discipline with its own distinct methodologies and philosophical principles [1]. It has successfully prevented the Chinese and East Asia people from serious diseases for thousands of years. As one of the oldest healing systems, TCM includes the therapies like herbal medicine, acupuncture, moxibustion, massage, food therapy, and physical exercise [2]. They can be practically used for various diseases treatment [3]. The herbal medicine, generally called formula, is one of the most important TCM therapies. It tries to acquire maximal therapeutic efficacy with minimal adverse effects. Typically, a formula consists of several medicinal herbs or minerals (we will use the word "herb" to refer to medicinal materials in formula). Different components in a formula have different 'roles' for disease treatment [4]. Clinical herb prescription is

L. Cao et al. (Eds.): PAKDD 2011 Workshops, LNAI 7104, pp. 353–360, 2012.

a complicated and flexible procedure that integrates the knowledge of syndrome differentiation (i.e., TCM diagnosis), TCM herb and formula theories, treatment principles, and empirical herb prescription knowledge inherited from the ancient literatures and acquired through individual experiences. In contrast to the modern drug therapies that often adhere to the common and operational clinical guidelines, TCM physicians emphasize more on individuality when prescribing formulae in TCM clinical practices. The formulae prescribed for different patients are almost never the same. A large amount of formula data, along with other clinical information, has been accumulated over the years. To manage all the data, Zhou et al. [5] have developed a clinical data warehouse.

In this study, we are interested in discovering the regularities on herb combination from large-scale clinical herb prescription databases. Several data mining methods have been used for the purpose before [6]. The results are not satisfying. Take association rules as an example. It is the most commonly used method. A key drawback is that it produces a large number of rules, often in the thousands. Clinical researchers have to painstakingly go through all the rules to get the final discoveries. This takes a lot of time and efforts. Moreover, association rules are concerned with only co-occurrence patterns, while it is far more interesting to analyze the complicate interactions, e.g. synergy, mutual detoxification and mutual inhibition, among the herbs in the clinical formulae. Furthermore, the co-occurrence frequency-based methods can not detect the negative dependence between herbs, which is important for clinical practices.

Zhang et al. [7,8] have studied the discovery of TCM diagnosis knowledge from the clinical data using a new class of statistical methods called latent tree (LT) models. The models enable one to discover the latent structures based on local dependences between the manifestation variables [9]. Technically, an LT model is a tree-structured Bayesian network where variables at leaf nodes are observed and are hence called "manifest variables", whereas variables at internal nodes are hidden and hence are called "latent variables". All variables are assumed to be discrete. Arrows represent direct probabilistic dependence.

In this study, we use LT models to analyze TCM formula data and thereby reveal the underlying latent structures. In particular, we analyzed the herb prescription data for patients in a condition known in TCM as "disharmony between liver and spleen (DBLS) syndrome". The data were extracted from a data warehouse [5], and the prescriptions were made by senior and well known TCM experts. The analysis has revealed some clinically useful regularities. Common herb combinations and their modifications for the treatment of DBLS were discovered. The results are useful for the students of TCM as well as pharmaceutical companies that manufacture medicine using Chinese herbs.

The rest of this paper is organized as follows. We introduce the latent tree models and the clinical data in turn in section 2 and 3. The results are presented in section 4. Finally, we discuss the clinical significance of the study and the future work in section 5.

2   Latent Tree Models

A *latent tree (LT) model* is a Bayesian network where (1) the network structure is a rooted tree; (2) the internal nodes represent latent variables and the leaf nodes represent manifest variables; and (3) all the variables are categorical. *Latent class (LC) models* are LT models with a single latent node. The terms variable and node are interchangeable throughout this paper.

Figure 1 (a) shows the structure of an LT model. In the model, there is an arrow from variable Y_1 to variable Y_2. This means that Y_2 depends on Y_1 directly. The dependence is quantified by a conditional distribution $P(Y_2|Y_1)$, which gives a distribution for Y_2 for each value of Y_1. All these distributions forms the *parameters* of a latent tree model. We write an LT model as a pair $M = (m, \theta)$, where θ is the collection of parameters. The first component m consists of the variables, the cardinalities of the variables, and the model structure. We sometimes refer to m also as an LT model.

Assume that there is a collection $\mathcal{D}$ of i.i.d. samples on manifest variables generated by an unknown LT model. By *LT model learning* we mean the effort to reconstruct the generative model from the data. The search-based approach aims at maximizing a scoring function. The BIC score [10] of a model m is:

$$BIC(m|\mathcal{D}) = \max_\theta \log P(\mathcal{D}|m, \theta) - \frac{d(m)}{2} \log N,$$

where $d(m)$ is *dimension*, i.e., the number of independent parameters of the model, and N is the sample size. The first term on the right hand side is known as the *maximized loglikelihood of m*. It measures how well model m fits the data $\mathcal{D}$. The second term is a penalty term for model complexity.

Consider two LT models m and m' that share the same manifest variables $X_1, X_2, \ldots, X_n$. We say that m *includes* m' if for any parameter value θ' of m', there exists parameter value θ of m such that

$$P(X_1, \ldots, X_n|m, \theta) = P(X_1, \ldots, X_n|m', \theta').$$

When this is the case, m can represent any distributions over the manifest variables that m' can. If m includes m' and vice versa, we say that m and m' are *marginally equivalent*. Marginally equivalent models are *equivalent* if they have the same number of independent parameters. It is impossible to distinguish between equivalent models based on data if penalized likelihood score is used for model selection.

Let Y_1 be the root of a latent tree model m. Suppose Y_2 is a child of Y_1 and it is also a latent node. Define another latent tree model m' by reversing the arrow $Y_1 \rightarrow Y_2$. Variable Y_2 becomes the root in the new model. The operation is called *root walking*; the root has walked from Y_1 to Y_2. The model m' in Figure 1 (b) is the model obtained by walking the root from Y_1 to Y_2 in model m.

It has been shown that root walking leads to equivalent models [11]. Therefore, the root and edge orientations of an LT model cannot be determined from data. We can only learn *unrooted LT models*, which are LT models with all directions

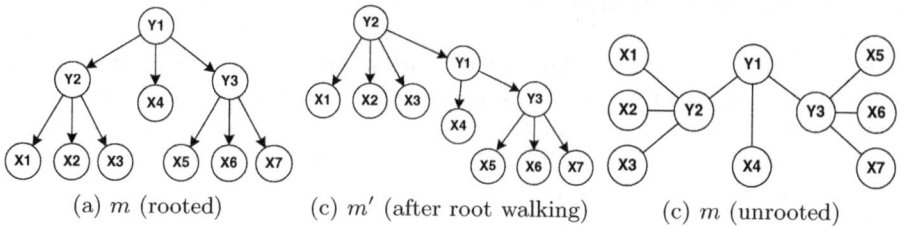

(a) m (rooted)   (c) m' (after root walking)   (c) m (unrooted)

Fig. 1. Rooted latent tree models, latent tree model obtained by root walking, and unrooted latent tree model. The Xs are manifest variables and the Ys are latent variables.

on the edges dropped. An example of an unrooted LT model is given in Figure 1 (c).

An unrooted LT model represents an equivalent class of LT models. Members of the class are obtained by rooting the model at various nodes. Semantically it is a Markov random field over an undirected tree. The leaf nodes are observed while the interior nodes are latent. Marginal equivalence and equivalence can be defined for unrooted LT models in the same way as for rooted models. Henceforth LT models always mean unrooted LT models in this paper unless it is explicitly stated otherwise.

3   The Clinical Data

TCM differentiates between different patients based mainly on their symptoms. The classification is known as *syndrome*, or *pattern*. Patients with different diseases might have the same syndrome manifestations, and the same disease might have different syndrome manifestation at different stages. TCM treatment is mainly targeted at the syndromes rather than the diseases.

DBLS is a general syndrome that can manifest in many chronic diseases, such as chronic gastritis, fatty liver, infertility and liver cirrhosis. The main symptoms of DBLS are *"irritability, mental-emotional depression, chest, rib-side and abdominal distention or pain, premenstrual breast distention and pain, painful menstruation, fatigue, reduced food intake, stomach and epigastric distention and fullness after eating"*, etc [3].

The DBLS syndrome has complicated symptoms and hierarchical pathological structures. There are a large variety of herb prescriptions [12]. The data set we analyzed consists of 1,287 clinical formulae for the DBLS syndrome. Most of them are prescribed by famous senior TCM physicians from the best TCM hospitals in Beijing. The formula data contain totally 367 distinct herbs. Each prescription contains 15 herbs on average. The top 5 most frequently used herbs are *indian bread* (1,107), *stir-frying largehead atractylodes rhizome* (1,011), *Chinese thorowax root* (974), *white peony root* (880) and *Chinese angelica* (719). This means that most formulae for DBLS syndrome have the above five herbs as

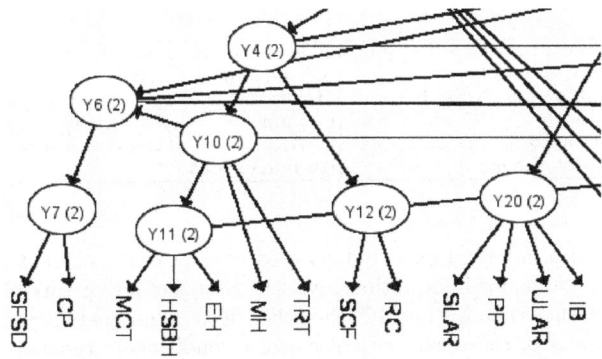

Fig. 2. Part of the LT model for the DBLS herb prescription data. The herb variables are at the leaf nodes and the latent variables are at the internal nodes. The numbers in parentheses are the numbers of states of the latent variables. The full names of the herbs are provided in Table 1.

ingredients. These 5 herbs are the core ingredients of a famous ancient formula known as *Xiao Yao San*. Most of the other herbs are in relatively low frequency.

Due to the intrinsic complexity in learning LT models, we reduced the number of variables and only the top 102 frequent herbs were included in the analysis. Each of those herbs appeared in at least 30 formulae. The herb variables were converted into binary variables, taking value '0' or '1'. The final data set is represented as a table with 102 attributes and 1,287 records.

4   The Results

We conducted LT analysis on the data set. The analysis was run on a 20-node cluster, where every node is of hard configuration: 2 x dual core AMD Opteron 2216 (2.4GHz) processors. The process took 156.9 hours to finish. The BIC score of the resultant model is -32,739. The resulting LT model has 38 latent variables. A portion of the model structure is shown in Fig. 2. This means that our analysis has identified 38 latent factors from the DBLS data set.

Each of the latent variables has a number of states. For example, the variable $Y1$ has 2 states. This means that we have grouped the data set into 2 clusters according to the latent factor $Y1$. The variable $Y11$ also has 2 states. This means that we have grouped the data set in another way into 2 clusters. Thus, we have simultaneously clustered the data set in multiple ways.

In the following, we examine several latent variables and their states to demonstrate the interestingness of the results.

The latent variable $Y7$ is connected to herb variables SFSD and CP, and latent variable $Y6$. $Y7$ has two states $S0$ and $S1$, which indicates that it represents a partition of the prescriptions into two classes. In the following, we discuss the meanings of this partition and hence the variable $Y7$ itself.

Table 1. The full name of some of the herbs

SFSD: stir-fry flying squirrel's droppings	SCF: szechwan chinaberry fruit
MCT: medicinal cyathula toot	RC: rhizoma corydalis
HSBH: hirsute shiny bugleweed herb	SLAR: stir-fry largehead atractylodes rhizome
EH: epimedium herb	PP: purified pinellia [tuber]
MH: motherwort herb	ULAR: uncooked largehead atractylodes rhizome
TRT: turmeric root tuber	IB: indian bread

Table 2. Numerical information about selected latent variables. MI stands for mutual information, IC stands for information coverage. States of latent variables are denoted by $S0$ and $S1$. Other than MI and IC, the other decimal numbers are marginal probabilities of the states of the latent variables or the conditional probability distributions of the herb variables. See the discussion of $Y7$ to get the precise meanings of the terms and the numbers.

Y7	MI	IC	$S0=.04$	$S1=.96$
SFSD	.85	.85	.81	0
CP	.60	.98	.66	0

Y12	MI	IC	$S0=.013$	$S1=.87$
RC	1	1	.1	0
SCF	.16	1	.34	.02

Y11	MI	IC	$S0=.95$	$S1=.06$
MH	.71	.71	.05	1
HSBH	.70	.93	.02	.9
MCT	.46	.98	0	.56

Y20	MI	IC	$S0=.79$	$S1=.21$
SLAR	1	1	1	0
ULAR	.08	1	0	.12

Various numerical information about Y7 is given in Table 2. In terms of mutual information (MI), Y7 is the closest to SFSD and CP. The ratio of the MI between Y7 and SFSD over the MI between Y7 and all manifest variables is called the information coverage (IC) of SFSD with respect Y7. It is .85. Similarly, the information of the two variables SFSD and CP with respect Y7 is .98. Because of this, we can say that Y7 represents a partition of the prescriptions almost totally based on whether they use the two herbs SFSD and CP.

The partition consists of two classes Y7=S0 and Y7=S1. The former class has marginal probability 0.04. This indicates the class Y7=S0 consists of 4% the prescriptions. This class of prescriptions has high probabilities, 0.81 and 0.66 respectively, to prescribe the herbs SFSD and CP. On the other hand, the rest of the prescriptions do not use those two herbs at all.

These findings turn out to be very interesting. As a matter of fact, the combination of the two herbs SFSD and CP is a classical formula, called *Shi Xiao San*. It is mainly used to *promote blood circulation* and *remove blood stasis*. Therefore, the two herbs are often used for the treatment of DBLS patients accompanied with *blood stasis* syndrome.

Similar inspection reveals that states of other latent variables represent other interesting combinations of herbs: $Y11=S1$ represents the combination of *motherwort herb* and *medicinal cyathula toot*, which is for the DBLS patients with diseases, such as polycystic ovary syndrome and primary infertility; $Y12=S0$

represents the combination of *rhizoma corydalis and szechwan chinaberry fruit*, which is for the DBLS patients with symptoms of all kinds of pains; and so on.

Negative associations have also been uncovered. Prescriptions in the class $Y20=S0$ all use *stir-fry largehead atractylodes rhizome*, but never *uncooked largehead atractylodes rhizome*, while prescriptions in the class $Y20=S1$ have some probabilities of using the latter, but never former. This is consistent with reality. In practice, the two herbs are never used together. Other data mining methods such as association rules cannot find this kind of negative dependence.

5   Concluding Remarks

Prescription is a skill that TCM students take years to learn and master. By revealing patterns of herb combinations in the prescriptions by seasoned experts, our study can help the students to acquire the skill faster and better. They can also help pharmaceutical companies to decide what combination of Chinese herbs to test.

One future direction is to correlate the patterns that we found with symptoms. The results would be even more interesting to TCM researchers, practitioners, and students.

Acknowledgements. This work is partially supported by Program of Beijing Municipal S&T Commission, China (D08050703020803, D08050703020804), China NSFC project (90709006), National Key Technology R&D Program (2007BA110B06), and China 973 project (2011CB505101).

References

1. Anonymous: The Inner Canon of Emperor Huang. Chinese Medical Ancient Books Publishing House, Beijing (2003)
2. Tang, J.L., Liu, B.Y., Ma, K.W.: Traditional chinese medicine. Lancet 372 (2008)
3. Flaws, B., Sionneau, P.: The treatment of modern western medical diseases with Chinese medicine: a textbook and clinical manual, 2nd edn. Blue Poppy Press (2005)
4. Wang, L., Zhou, G.B., Liu, P., Song, J.H., Liang, Y., Yan, X.J., Xu, F., Wang, B.S., Mao, J.H., Shen, Z.X., Chen, S.J., Chen, Z.: Dissection of mechanisms of chinese medicinal formula realgar-indigo naturalis as an effective treatment for promyelocytic leukemia. PNAS 105 (2008)
5. Zhou, X., Chen, S., Liu, B., Zhang, R., Wang, Y., Li, P., Guo, Y., Zhang, H., Gao, Z., Yan, X.: Development of traditional chinese medicine clinical data warehouse for medical knowledge discovery and decision support. Artif. Intell. Med. 48(2-3) (2009)
6. Feng, Y., Wu, Z., Zhou, X., Zhou, Z., Fan, W.: Knowledge discovery in traditional chinese medicine: State of the art and perspectives. Artif. Intell. Med. 38(3) (2006)
7. Zhang, N.L., Yuan, S., Chen, T., Wang, Y.: Latent tree models and diagnosis in traditional chinese medicine. Artif. Intell. Med. 42 (2008)
8. Zhang, N.L., Yuan, S., Chen, T., Wang, Y.: Statistical validation of traditional chinese medicine theories. J. Altern. Complement Med. 14(5) (2008)

9. Jakulin, A., Bratko, I.: Testing the significance of attribute interactions. In: ICML 2004 (2004)
10. Schwarz, G.: Estimating the dimension of a model. Annals of Statistics 6, 461–464 (1978)
11. Zhang, N.L.: Hierarchical latent class models for cluster analysis. Journal of Machine Learning Research 5, 697–723 (2004)
12. Zhang, R.: Clinical research on the syndrome structure and syndrome hierarchical differentiation of disharmony of liver and spleen syndrome. PhD dissertation, Guanganmen hospital, China academy of Chinese medical sciences (2008) (in chinese)

COW: A Co-evolving Memetic Wrapper for Herb-Herb Interaction Analysis in TCM Informatics

Dion Detterer and Paul Kwan

School of Science and Technology, University of New England,
Armidale NSW 2351, Australia
{ddettere,kwan}@turing.une.edu.au

Abstract. Traditional Chinese Medicine (TCM) relies heavily on interactions between herbs within prescribed formulae. However, given the combinatorial explosion due to the vast number of herbs available for treatment, the study of herb-herb interactions by pure human analysis is impractical, with computer-aided analysis computationally expensive. Thus feature selection is crucial as a pre-processing step prior to herb-herb interaction analysis. In accord with this goal, a new feature selection algorithm known as a Co-evolving Memetic Wrapper (COW) is proposed: COW takes advantage of recent developments in genetic algorithms (GAs) and memetic algorithms (MAs), evolving appropriate feature subsets for a given domain. As part of preliminary research, COW is demonstrated to be effective in selecting herbs in the TCM insomnia datatset. Finally, possible future applications of COW are examined, both within TCM research and in broader data mining contexts.

Keywords: Traditional Chinese Medicine, memetic algorithm, wrapper, feature selection, data mining.

1   Introduction

Theory behind Traditional Chinese Medicine (TCM) relies heavily on the idea of herb-herb interaction when formulating treatment [6]. In contrast to Western medical approaches which favor the use of components in isolation, TCM places an emphasis on holistic treatment by combining elements to form a customized formulation on a case-by-case basis.

It is thus of particular importance that any careful analysis of the TCM approach to treatment should not simply take note of the roles played by individual herbs but should also account for any potential complex, non-linear interactions between them.

In this paper, a new algorithm for selecting subsets of herbs with a high potential for non-linear interaction is given a preliminary investigation, as applied to the TCM insomnia dataset. By utilizing approaches to co-evolving memetic algorithms and applying them in the direction of feature selection, especially with regard to TCM data mining, it is hoped that significant progress can be made in the study of interacting herbs.

L. Cao et al. (Eds.): PAKDD 2011 Workshops, LNAI 7104, pp. 361–371, 2012.

The details of the TCM insomnia dataset are given in Section 2, while prior work relevant to the present work (including work both in and outside the area of TCM data mining) is given in Section 3; the COW algorithm is detailed in Section 4, with Section 5 presenting analysis of the TCM insomnia dataset using COW and multifactor dimensionality reduction (MDR). Discussion of the results is provided in Section 6, and finally, suggestions for further research are proposed in Section 7.

2   The TCM Insomnia Dataset

Insomnia is a rather pernicious and troubling condition, with clinical data showing that it affects 1 in 5 patients who are attended to by general practitioners [6]. It is thus vital that we apply existing knowledge discovery and data mining techniques to TCM data in order to uncover the "hidden gems" within the ancient formulations in TCM. New insomnia treatments, based on rigorous, modern clinical trials, could then be formulated, based (at least initially) on herbs and herb combinations discovered through modern computational analysis.

The TCM insomnia dataset is extracted from a clinical data warehouse containing both inpatient and outpatient encounters [7]. It includes 460 instances (cases) with 111 features (herbs), where each instance gives a prescribed formula that includes some combination of herbs. Only those herbs which appeared in at least 10 prescriptions are included: the 111 herbs given are taken from a larger set of 261 herbs. The class label then gives the outcome, such that it is either positive or negative: 392 cases are labeled as having a positive outcome, with only 68 labeled as negative.

Each herb is listed with a generic variable name; however, a separate table then maps variables to herbs, with each herb typically listing both the Chinese name for the herb and its English translation. For example, the 27th herb in the dataset is named "VAR46" (i.e. herb 46 of the original 261 herbs), which then maps to 炒决明子 (stir-frying cassia seed).

3   Prior Work

3.1   Multifactor Dimensionality Reduction and Hierarchical Core Sub-networks

Of particular interest in regard to TCM analysis is the development of multifactor dimensionality reduction (MDR). MDR was developed with the goal of analyzing gene-gene interaction (i.e. epistasis) in bioinformatics [4]. By transforming an n-dimensional space into a 1-dimensional space (i.e. transforming a set of selected features to a single feature), a measurement of the new feature's ability to predict a class label can then be derived—this measurement suggests how closely the selected features interact to influence the outcome.

For example, Fig. 1 depicts MDR performed on the herbs represented by features VAR40 and VAR196 in the TCM insomnia dataset. First, all combinations of the two herbs are considered, and within each combination, counts for each class are calculated: for example, where neither VAR40 nor VAR196 occur, there are 165 positive outcomes and 53 negative outcomes, while where both VAR40 and VAR146 occur, there are 19 positive outcomes and 1 negative outcome. Next, each combination is determined to have either a high or low probability for a positive outcome by testing the ratio of positive to negative counts against a given threshold T: the ratio of positive to negative counts must be greater than T for the combination to be determined to have a high probability of a positive outcome: in this example, if either VAR40 or VAR196 (or both) are present, there is a high probability of a positive outcome; if neither VAR40 nor VAR196 are present, there is a low probability of a positive outcome. We now have a new constructed feature, VAR40_VAR196, which maps low probability positive outcomes to its value '0' and high probability positive outcomes to its value '1'. Accuracy is then determined by assessing how well this constructed feature matches the actual outcome for each instance.

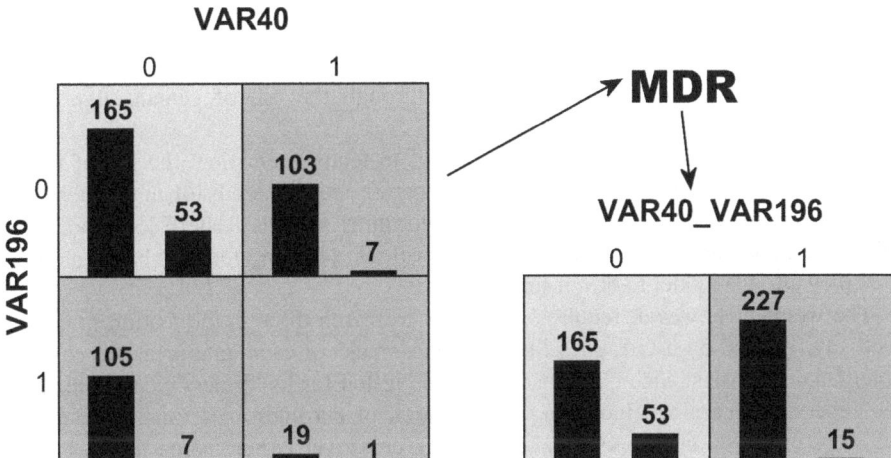

Fig. 1. MDR performed on two herbs from the TCM insomnia dataset. Dark-shaded cells have a high probability of a positive outcome; light-shaded cells have a low probability of a positive outcome.

MDR typically performs an exhaustive search through feature-space, such that the best combination of 1, 2, 3,..., n features is returned (where n is the total number of features in the dataset). Once the number of features in the dataset grows above 10, however, the task of searching for all combinations becomes increasingly infeasible.

Both the application of MDR to the insomnia dataset—and the problems one faces in doing so—should be obvious. While the insomnia dataset is nowhere near the order of magnitude of typical microarray data, MDR still suffers a computationally-intensive task when attempting to analyze all potential herb subsets within the 111 herbs given.

This problem, however, can be alleviated through careful feature selection. In the case of TCM herb-herb analysis, this simply means that we should significantly reduce the number of herbs to consider. In addition, if the purpose is to discover herb combinations that suggest further investigation, ideally, there should be less than 111 herb combinations suggested (as would be ordinarily returned by an MDR analysis of the insomnia dataset): this is simply a matter of practicality when finding a starting point for rigorous (and expensive) clinical trials.

One approach in initial herb selection prior to MDR analysis is in the use of hierarchical core sub-networks [8]. By extracting herbs that form "hubs" within the larger herb network implied by the data, a herb subset with a high potential for herb-herb interactions can then be subjected to MDR.

3.2   Feature Selection via Genetic Algorithms

A more general approach (i.e. beyond the application to datasets exhibiting complex network properties) to feature selection is through the use of genetic algorithms (GAs).

Classically, there are two broad approaches to feature selection: the use of filters and the use of wrappers [2]. While filters operate independently of any supervised learning algorithm (selecting features via algorithms such as ReliefF [3]), wrappers rely directly on such supervised learning algorithms, selecting features based on how well they allow a model to be constructed.

The wrapper, however, requires some way to search through the feature space effectively. Guided by a GA, so-called "chromosomes" or candidate solutions can encode feature subsets; each chromosome is then tested for its "fitness", determined by the supervised learning algorithm. A population of chromosomes finally undergoes the steps of selection (proportional to fitness), crossover (where some features from one parent are combined with some features from another parent) and mutation (where individual features are selected or deselected at random within each chromosome) in order to simulate the process of natural selection. After several generations, the population should then exhibit an average fitness that converges towards an optimal fitness, suggesting that close to the best feature subsets (with regard to classification) have been selected. Fig. 2 gives a general algorithm for a GA-based wrapper.

Recently, the idea of a wrapper-filter hybrid was proposed, using a GA to guide the wrapper's search in feature-space while a filter is then used as a local search (LS) optimization [9]. While the population of candidate solutions evolves via evolutionary operators, the local search relies solely on static data gathered from feature-ranking filters. In the case of TCM herb-herb analysis, this is particularly problematic, given that we specifically wish to avoid the consideration of herbs in isolation.

In this paper, then, a new feature selection algorithm is proposed. Entitled COW or Co-evolving Memetic Wrapper, this algorithm not only includes a population of candidate solutions (i.e. a set of candidate feature subsets) but also a population of co-evolving memes for local search: sets of likely key features and possibly irrelevant or redundant features. Both populations undergo evolutionary operations, with the meme population co-evolving based on the individual memes' ability to improve the candidate solutions.

---

1. Generate a population of candidate solutions from randomly selected feature subsets
2. Do while stopping condition is not met:
   a. Evaluate the fitness (i.e. classification accuracy) of each candidate solution
   b. Generate new population of candidate solutions by performing crossover and mutation on selected individuals in the prior population

---

Fig. 2. General algorithm for GA as feature selection wrapper

4   A Closer Look at COW

The ideas behind COW are relatively straightforward, though they result in a fairly sophisticated algorithm.

With the recent publication of papers such as [5], the notion of co-evolving memetic algorithms (COMAs) is bringing about a new generation of memetic algorithms. However, this author is unaware of any other attempts to employ a form of COMA in pursuit of feature selection.

The overall algorithm in COW (given in Fig. 3) is similar to most GAs, in that a population of candidate solutions undergoes evolutionary operations in order to evolve over several generations.

---

1. Initialize populations of candidate solutions and memes
2. Evaluate initial fitness of candidate solutions
3. Apply memes to selected elite candidate solutions
4. Do while stopping condition is not met:
   a. Generate the next generation of candidate solutions and memes
   b. Evaluate fitness of new candidate solutions
   c. Apply memes to selected candidate solutions, modifying the fitness of each applied meme accordingly
5. Return selected candidate solutions

Fig. 3. The general algorithm employed by COW

The differences between COW and standard GAs, however, are twofold: firstly, as with other MAs, memes are used as local search optimizations on individual candidates; secondly, as with COMAs, the meme population evolves alongside the population of candidate solutions. The advantage to a co-evolving population of memes in the case of COW is that local search optimizations themselves improve via evolutionary operations; thus it is expected that the improvement in candidate solutions' fitness will accelerate over the course of a run of COW. Whereas employing pure filter-based memes means WFFSA as proposed in [9] relies solely on individually-ranked features for memetic improvement, COW evolves complete feature subset optimizations that adapt to meet the needs of the best candidate solutions.

The initial population of candidate solutions is generated randomly and represented using bit-strings, where '1' means a feature is selected and '0' means it is deselected. Unlike in regular GAs, however, the candidate solutions of COW are restricted so that if there are n total features in the dataset (not including the class feature), $m \leq n$ features are selected in any one candidate solution. This restriction is incorporated into all evolutionary operations performed on the candidates (i.e. crossover and mutation); mutation uses the algorithm given in [9], but crossover uses a variant developed for COW:

Let c_1 and c_2 be two candidate solutions.

1. Find all indices where c_1 is '0' and c_2 is '1', storing indices in list l_1
2. Find all indices where c_1 is '1' and c_2 is '0', storing indices in list l_2
3. Shuffle within both l_1 and l_2
4. Let $k = \min(\text{size}(l_1), \text{size}(l_2))$
5. For $i = 1$ to k

     a. Generate random number rnd in interval $[0, 1)$

     b. If $rnd < 0.5$ then

          i. Set c_1 to '1' at index $l_1[i]$ and to '0' at index $l_2[i]$

          ii. Set c_2 to '0' at index $l_1[i]$ and to '1' at index $l_2[i]$

Fig. 4. Restricted crossover as used in COW

Fitness is calculated based on the reported accuracy of a model generated by an induction algorithm, given a copy of the dataset that includes only those features specified per candidate. However, as with WFFSA as specified in [9], candidates whose difference in fitness is less than a small value (ε) are given higher ranking and thus a greater chance of selection.

The crux of COW, however, and the property that sets it apart from any other GA-based wrapper, is the purpose and behavior of its memes in service of local search optimization.

4.1 Local Search in COW

The memes used in COW facilitate the local search by both improving candidate solutions and co-evolving in accord with how well they perform in terms of candidate improvement.

Each individual meme is composed of two halves. Firstly, there is the so-called "delete" chromosome. Memetic chromosomes in COW, like their candidate solution counterparts, are encoded as bit-strings: for a delete chromosome, a '1' represents a feature proposed as being irrelevant or redundant. By inverting the chromosome (using NOT), a simple AND operation can then be used to apply the delete portion of a meme to a candidate solution. As the name implies, the purpose is to delete features that do not significantly contribute to the classification task.

In addition to a delete chromosome, there is also an "add" chromosome in each meme as well. Here a '1' indicates a feature proposed as being worth adding to candidate solutions.

Add and delete chromosomes are strictly paired individually in the same meme so that both must always be applied together. The pair thus functions as a hypothesis for the features most relevant and least relevant to the dataset. The delete operation always precedes the add operation so as to preserve the restricted nature of the candidate solution. Whereas all features specified in the delete chromosome are deselected in the candidate solution to which it's applied, features specified in the add chromosome are selected individually by random in the candidate solution until either the candidate has its maximum number of features selected (per restriction) or there are no more features from the add chromosome to select in the candidate.

Local search on a given candidate proceeds as follows:

---

1. Select a meme based on each meme's prior fitness (inferred from the average of the meme's parents' fitnesses, if this meme has not yet been applied previously)
2. Apply delete and then add chromosomes from the meme to the candidate solution
3. Re-evaluate the candidate's fitness
4. If the difference (Δ) between the new fitness and prior fitness is greater than or equal to a small value (ε), or less than ε but with fewer features selected, retain the improved candidate, else revert the candidate
5. Add Δ to the meme's current improvement tally and increment the number of times this meme has been tested
6. Set the meme's fitness to the value of Δ averaged over the number of tests for this meme

---

Fig. 5. COW local search algorithm

The meme's own fitness can then be used in its selection for crossover within its own population, as well as when selecting memes to apply to a given candidate solution.

5   Experimental Evaluation

For this study, the TCM insomnia dataset was used to test the effectiveness of COW on TCM data. COW was initialized using 30 candidate solutions and 30 memes. The memes were themselves initialized by randomly selecting within the 37 top-ranked herbs (for add chromosomes) and the 37 lowest-ranked herbs (for delete chromosomes) as per ReliefF rankings.

Candidate solutions were restricted to at most selecting 10 herbs each, with $\varepsilon = 0.001$ (as reducing further than 10 herbs was not essential), giving preference to candidates with a smaller herb subset where the difference in fitness is less than ε.

Up to 10 memes were applied to a selected candidate, on an improvement-first [9] basis. The top 5 elite candidates in any given generation were selected for memetic improvement.

Crossover for both memes and candidate solutions was performed in all tests with a probability of 0.6, with a further mutation factor of 0.1. Linear ranking selection [1] was used for all selection tasks, with a selection pressure of 1.5.

With each new generation, the top two elite candidate solutions from the prior generation were carried over to ensure that the best solutions were never lost. (They are still subject to mutation, as were all other candidate solutions.)

For the candidate solution fitness function, an implementation of MDR was used without cross-validation, in order to provide relatively fast fitness computations that were directly relevant to the problem of TCM herb-herb interactions. The MDR threshold was set to $T \cong 5.76$, the overall ratio of positive to negative outcomes in the TCM insomnia dataset, in order to correct for the imbalance in outcomes. 1NN with LOOCV was initially tested but proved computationally infeasible.

Using this MDR-based fitness function in COW, with further parameters as detailed above, gave results where rapid increase in overall fitness in the population was observed in Fig. 6, as expected.

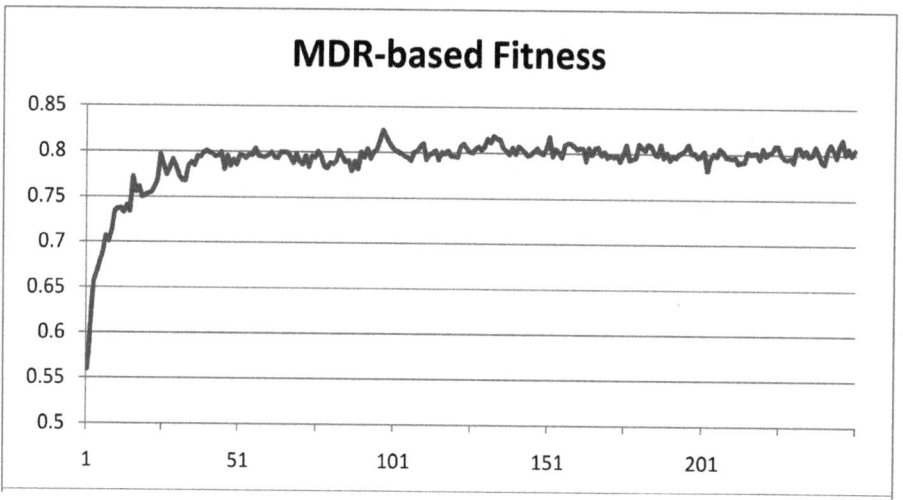

Fig. 6. Average population fitness over 250 generations, averaged over 10 independent runs using COW with MDR-based fitness measure on TCM insomnia dataset

After 20,000 fitness calls (which resulted in between 266 and 273 generations over the 10 independent runs) the herbs given in Table 1 were selected by COW. Running each of these herb sets through complete MDR with 10-fold cross-validation, then selecting the subset containing between 2 and 5 herbs with the greatest testing accuracy, the results given in Table 2 were returned.

Table 1. Herbs selected by COW over 10 independent runs after 20,000 fitness calls

Run	Herbs selected	Candidate fitness
1	VAR1, VAR21, VAR35, VAR40, VAR89, VAR92, VAR176, VAR196, VAR215, VAR246	0.917
2	VAR19, VAR34, VAR43, VAR45, VAR117, VAR125, VAR126, VAR176, VAR215, VAR254	0.913
3	VAR15, VAR35, VAR36, VAR45, VAR47, VAR90, VAR147, VAR176, VAR211, VAR242	0.900
4	VAR6, VAR19, VAR21, VAR45, VAR126, VAR172, VAR175, VAR176, VAR215, VAR237	0.917
5	VAR6, VAR35, VAR41, VAR90, VAR176, VAR178, VAR215, VAR217, VAR238, VAR254	0.915
6	VAR2, VAR5, VAR19, VAR33, VAR34, VAR142, VAR155, VAR176, VAR239, VAR254	0.902
7	VAR3, VAR39, VAR41, VAR89, VAR90, VAR92, VAR124, VAR210, VAR215, VAR246	0.911
8	VAR2, VAR41, VAR43, VAR45, VAR90, VAR120, VAR137, VAR176, VAR236, VAR239	0.911
9	VAR1, VAR2, VAR3, VAR19, VAR35, VAR41, VAR178, VAR201, VAR215, VAR238	0.909
10	VAR1, VAR6, VAR19, VAR36, VAR90, VAR126, VAR176, VAR179, VAR202, VAR254	0.913

Table 2. Herb subsets with highest testing accuracy returned by MDR with 10-fold cross-validation

Run	Herb subset	Accuracy
1	VAR21, VAR89, VAR176, VAR215, VAR246	0.7105
2	VAR19, VAR126, VAR176, VAR215, VAR254	0.6958
3	VAR90, VAR176	0.6161
4	VAR6, VAR126, VAR215, VAR237	0.6867
5	VAR41, VAR90, VAR176, VAR215	0.6318
6	VAR2, VAR155, VAR176, VAR254	0.6056
7	VAR41, VAR89, VAR90, VAR215, VAR246	0.6507
8	VAR2, VAR41, VAR90, VAR176	0.5989
9	VAR2, VAR19, VAR41, VAR215	0.6615
10	VAR19, VAR90, VAR126, VAR176	0.6491

6   Discussion

As is apparent in Fig. 6, convergence towards optimal solutions occurs approximately within the first 50 generations, at which point average fitness oscillates around 0.8.

Without cross-validation, this means that the herbs selected by this implementation of COW after 50 generations have a relatively high likelihood to exhibit herb-herb interactions, which in turn suggests that rapid convergence is taking place.

Given the results obtained in the previous section, the herb subset returned by MDR in Run 1 has the highest likelihood for herb-herb interactions in the TCM insomnia dataset. Shown in Table 3, these herbs are as follows:

Table 3. Herbs with high likelihood of interaction in TCM insomnia dataset

Variable name	Chinese name	English name
VAR21	麦芽	Germinated barley
VAR89	玄参	Figwort root
VAR176	白茅根	Lalang grass rhizome
VAR215	青皮	Green tangerine peel
VAR246	胆南星	Bile arisaema

In particular, given their frequent appearance in all analysis in this paper (including four pairings in the initial data returned by COW), lalang grass rhizome and green tangerine peel seem like excellent candidates for further research into their combined effects in the treatment of insomnia.

7   Conclusion and Future Work

The co-evolving memetic wrapper for feature selection (COW) differs from other proposed memetic algorithm-based wrappers in that it evolves local search optimizations that are best suited to a given domain.

By applying COW to the TCM insomnia dataset, followed by further analysis via MDR, five out of 111 potential herbs have been suggested for the treatment of insomnia, due to their high likelihood of herb-herb interaction. This will be subjected to further validation by consulting TCM doctors at the China Academy of Chinese Medical Sciences, Beijing, China from which the dataset was originally obtained.

Further avenues of investigation for COW include the questions of how best to initialize the meme population, what other means of memetic evolution may be developed and how best to integrate herb-herb interaction detection into COW. In addition, we will test the proposed algorithm on other TCM clinical datasets to verify its effectiveness.

In particular, the use of MDR-based approaches requires further consideration. On a more general note, the use of COW in comparison to other feature selections approaches will be investigated, using publicly available datasets in order to benchmark performance.

However, COW's viability has been demonstrated with regard to the TCM insomnia dataset, with the possibility for future clinical trials, given the suggested herbs in the previous section.

Acknowledgements. Special thanks are given to Professor Baoyan Liu and Dr Xuezhong Zhou and their colleagues at the China Academy of Chinese Medical Sciences, Beijing, China for permission to use the insomnia dataset in our experiments.

References

1. Baker, J.E.: Adaptive Selection Methods for Genetic Algorithms. In: Proc. Int'l Conf. Genetic Algorithm and Their Applications, pp. 101–111. Lawrence Erlbaum Associates, New Jersey (1985)
2. Guyon, I., Elisseeff, A.: An introduction to variable and feature selection. The J. of Machine Learning Research, 1157–1182 (2003)
3. Kononenko, I., Šimec, E., Robnik-Šikonja, M.: Overcoming the myopia of inductive learning algorithms with RELIEFF. Applied Intelligence, 39–55 (1997)
4. Ritchie, M.D., Hahn, L.W., Roodi, N., Bailey, L.R., Dupont, W.D., Parl, F.F., Moore, J.H.: Multifactor-dimensionality reduction reveals high-order interactions among estrogen-metabolism genes in sporadic breast cancer. American J. of Human Genetics, 138–147 (2001)
5. Smith, J.E.: Coevolving memetic algorithms: a review and progress report. IEEE Transactions on Systems, Man, and Cybernetics, Part B, 6–17 (2007)
6. Wing, Y.K.: Herbal treatment of insomnia. Hong Kong Med. J., 392–402 (2001)
7. Zhou, X., Chen, S., Liu, B., et al.: Development of Traditional Chinese Medicine Clinical Data Warehouse for Medical Knowledge Discovery and Decision Support. Artificial Intelligence in Medicine 48(2-3), 139–152 (2010)
8. Zhou, X., Poon, J., Kwan, P., Zhang, R., Wang, Y., Poon, S., Liu, B., Sze, D.: Novel Two-Stage Analytic Approach in Extraction of Strong Herb-Herb Interactions in TCM Clinical Treatment of Insomnia. In: Zhang, D. (ed.) ICMB 2010. LNCS, vol. 6165, pp. 258–267. Springer, Heidelberg (2010)
9. Zhu, Z., Ong, Y.S., Dash, M.: Wrapper–filter feature selection algorithm using a memetic framework. IEEE Transactions on Systems, Man, and Cybernetics, Part B, 70–76 (2007)

Selecting an Appropriate Interestingness Measure to Evaluate the Correlation between Syndrome Elements and Symptoms

Lei Zhang[1], Qi-ming Zhang[2], Yi-guo Wang[2], and Dong-lin Yu[1]

[1] Shandong University of Chinese Medicine, Jinan (250355), China
{tcmxpzl,ydlin03}@126.com
[2] Institute of Basic Research in Clinical Medicine,
China Academy of Chinese Medical Sciences,
Beijing (100700), China
zhang_917@126.com, ygw541@yahoo.com.cn

Abstract. In order to select the best interestingness measure appropriate for evaluating the correlation between syndrome elements and symptoms, 60 objective interestingness measures were selected from different subjects. Firstly, a hypothesis for a good measure was proposed. Based on the hypothesis, an experiment was designed to evaluate the measures. The experiment was based on the clinical record database of past dynasties including 51,186 clinical cases. The selected dataset in this study had 44,600 records. Han and Re were selected as the experimental syndrome elements. Three indicators calculated according to the distances between two syndrome elements were obtained in the experiment and were combined into one indicator. The *Z score*, *ϕ-coefficient* and *Kappa* were selected from 60 measures after the experiment. The *Z score* and *ϕ-coefficient* were selected according to subjective interestingness. Finally, the *ϕ-coefficient* was selected as the best measure for its low computational complexity. The method introduced in this paper may be used in other similar territories. Further research of traditional Chinese medicine can be made based on the conclusion made in this paper.

Keywords: Interestingness measure, syndrome element, symptom, traditional Chinese medicine.

1 Introduction

In traditional Chinese medicine, syndrome is an important foundation in the treatment of diseases. There are huge numbers of syndromes, making them difficult to study. Therefore, syndrome element, i.e. the components of a syndrome, was proposed in order to simplify the study of syndromes, and the number of syndrome elements is limited [1, 2]. Syndrome differentiation is mainly based on symptoms, the same as for syndrome elements. For example, the diagnosis of syndrome element Re is based on symptoms such as a bitter mouth, yellow urine, a red tongue, and so on. In other words, these symptoms closely correlate with syndrome element Re, whereas other

L. Cao et al. (Eds.): PAKDD 2011 Workshops, LNAI 7104, pp. 372–383, 2012.

symptoms do not. So it is important to evaluate the correlation between a syndrome element and different symptoms. Many authors have tried to study the correlation between syndrome elements and symptoms using many different methods [3-5]. However, different methods always correspond to different results. Association rule mining, a method of data mining, was used in this study to determine the correlation between syndrome elements and symptoms.

Association rule mining was first introduced by Agrawal, Imielinski and Swami in 1993[6]. It is widely used to find the correlation between items in a dataset over the past two decades. In the classical framework, an association rule is regarded as a strong association rule if it satisfies the minimum threshold of support and confidence. However, the correlation between items cannot be exactly reflected by this framework [7-9]. Interestingness was proposed for evaluating the correlation between items. Interestingness is generally divided into two categories: objective measures based on the statistical strengths or properties of the patterns discovered and subjective measures that are derived from the user's beliefs or expectations of their particular problem domain [10]. Over recent years, many objective interestingness measures have been proposed and each measure has both advantages and disadvantages. There is no single measure that is appropriate for all conditions. Therefore, a measure appropriate for finding the correlation between syndrome elements and symptoms needs to be chosen.

An object interestingness measure that is appropriate for evaluating the correlation between syndrome elements and symptoms was selected in this paper. The rest of the article is organized as follows. Section 2 introduces the objective interestingness measures involved in this study. A hypothesis for choosing an objective interestingness measure is introduced in section 3, and several measures are selected on the basis of this hypothesis in section 4. Section 5 describes how the best measure is selected based on subjective interestingness and computational complexity. Finally, section 6 comprises the conclusion.

2   Related Objective Interestingness Measures

Objective interestingness measures always come from different subjects. The measures in this paper were selected from data mining, machine learning, statistics and epidemiology. Table 1 shows all 60 measures involved in this paper. They were compiled from different sources: numbers 1-21 [11], number 22 [12], number 23 [13], number 24 [14], numbers 25-32 [15], numbers 33-59 [16], number 60 [17].

Table 1. Objective interestingness measures

Name	Formula	Name	Formula
1. $\phi - coefficient$	$\dfrac{P(A,B) - P(A)P(B)}{\sqrt{P(A)P(B)(1 - P(A))(1 - P(B))}}$	31. *Positive Likelihood Ratio*	$\dfrac{a(c+d)}{c(a+b)}$
2. *Goodman − Kruskal's*	$\dfrac{(\sum_j \max_k P(A_j, B_k) + \sum_k \max_j P(A_j, B_k) - \max_j P(A_j) - \max_k P(B_k))}{/(2 - \max_j P(A_j) - \max_k P(B_k))}$	32. *Negative Likelihood Ratio*	$\dfrac{b(c+d)}{d(a+b)}$

Table 1. (*Continued*)

3. *Odds Ratio*	$\dfrac{P(A,B)P(\overline{A},\overline{B})}{P(A,\overline{B})P(\overline{A},B)}$	33.	*Pointwise Mutual Information*	$\log\dfrac{P(xy)}{P(x^*)P(^*y)}$
4. *Yule's Q*	$\dfrac{P(A,B)P(\overline{A},\overline{B}) - P(A,\overline{B})P(\overline{A},B)}{P(A,B)P(\overline{A},\overline{B}) + P(A,\overline{B})P(\overline{A},B)} = \dfrac{\alpha - 1}{\alpha + 1}$	34.	*Mutual Dependency*	$\log\dfrac{P(xy)^2}{P(x^*)P(^*y)}$
5. *Yule's Y*	$\dfrac{\sqrt{P(A,B)P(\overline{A},\overline{B})} - \sqrt{P(A,\overline{B})P(\overline{A},B)}}{\sqrt{P(A,B)P(\overline{A},\overline{B})} + \sqrt{P(A,\overline{B})P(\overline{A},B)}} = \dfrac{\sqrt{\alpha}-1}{\sqrt{\alpha}+1}$	35.	*Log Frequency Biased MD*	$\log\dfrac{P(xy)^2}{P(x^*)P(^*y)} + \log P(xy)$
6. *Kappa*	$\dfrac{P(A,B) + P(\overline{A},\overline{B}) - P(A)P(B) - P(\overline{A})P(\overline{B})}{1 - P(A)P(B) - P(\overline{A})P(\overline{B})}$	36.	*Normalized Expectation*	$\dfrac{2f(xy)}{f(x^*) + f(^*y)}$
7. *Mutual Information*	$\dfrac{\sum_i \sum_j P(A_i,B_j)\log\dfrac{P(A_i,B_j)}{P(A_i)P(B_j)}}{\min(-\sum_i P(A_i)\log P(A_i), -\sum_j P(B_j)\log P(B_j))}$	37.	*Mutual Expectation*	$\dfrac{2f(xy)}{f(x^*) + f(^*y)} * p(xy)$
8. *J − Measure*	$\begin{array}{l}\max(P(A,B)\log(\dfrac{P(B\mid A)}{P(B)}) + P(A\overline{B})\log(\dfrac{P(\overline{B}\mid A)}{P(\overline{B})}),\\ P(A,B)\log(\dfrac{P(A\mid B)}{P(A)}) + P(\overline{A}B)\log(\dfrac{P(\overline{A}\mid B)}{P(A)}))\end{array}$	38.	*Salience*	$\log\dfrac{P(xy)^2}{P(x^*)P(^*y)} * \log f(xy)$
9. *Gini Index*	$\begin{array}{l}\max(P(A)[P(B\mid A)^2 + P(\overline{B}\mid A)^2]\\ + P(\overline{A})[P(B\mid \overline{A})^2 + P(\overline{B}\mid \overline{A})^2]\\ -P(B)^2 - P(\overline{B})^2, P(B)[P(A\mid B)^2 + P(\overline{A}\mid B)^2]\\ + P(\overline{B})[P(A\mid \overline{B})^2 + P(\overline{A}\mid \overline{B})^2] - P(A)^2 - P(\overline{A})^2)\end{array}$	39.	*Pearson's χ^2 test*	$\sum_{ij}\dfrac{(f_{ij} - \hat{f}_{ij})^2}{\hat{f}_{ij}}$
10. *Support*	$P(A,B)$	40.	*T test*	$\dfrac{f(xy) - \hat{f}(xy)}{\sqrt{f(xy)(1 - (f(xy)/N))}}$
11. *Confidence*	$\max(P(B\mid A), P(A\mid B))$	41.	*Z score*	$\dfrac{f(xy) - \hat{f}(xy)}{\sqrt{\hat{f}(xy)(1 - (\hat{f}(xy)/N))}}$
12. *Laplace*	$\max(\dfrac{NP(A,B) + 1}{NP(A) + 2}, \dfrac{NP(A,B) + 1}{NP(B) + 2})$	42.	*Log Likelihood Ratio*	$-2\sum_{ij} f_{ij}\log\dfrac{f_{ij}}{\hat{f}_{ij}}$
13. *Conviction*	$\max(\dfrac{P(A)P(\overline{B})}{P(A\overline{B})}, \dfrac{P(B)P(\overline{A})}{P(B\overline{A})})$	43.	*Squared Log Likelihood Ratio*	$-2\sum_{ij}\dfrac{\log f_{ij}^2}{\hat{f}_{ij}}$
14. *Interest*	$\dfrac{P(A,B)}{P(A)P(B)}$	44.	*Rogers − Tanimoto*	$\dfrac{a + d}{a + 2b + 2c + d}$
15. *Consine*	$\dfrac{P(A,B)}{\sqrt{P(A)P(B)}}$	45.	*Hamann*	$\dfrac{(a + d) - (b + c)}{a + b + c + d}$
16 *Piatetsky − Skapiro's*	$P(A,B) - P(A)P(B)$	46.	*Third Sokal − Sneath*	$\dfrac{b + c}{a + d}$
17. *Certainty Fator*	$\max(\dfrac{P(B\mid A) - P(B)}{1 - P(B)}, \dfrac{P(A\mid B) - P(A)}{1 - P(A)})$	47.	*First Kulczynsky*	$\dfrac{a}{b + c}$
18. *Added Value*	$\max(P(B\mid A) - P(B), P(A\mid B) - P(A))$	48.	*Second Sokal − Sneath*	$\dfrac{a}{a + 2(b + c)}$
19. *Collective Strength*	$\begin{array}{l}\dfrac{P(A,B) + P(\overline{AB})}{P(A)P(B) + P(\overline{A})P(\overline{B})}\\ \times \dfrac{1 - P(A)P(B) - P(\overline{A})P(\overline{B})}{1 - P(A,B) - P(\overline{AB})}\end{array}$	49.	*Second Kulczynski*	$\dfrac{1}{2}(\dfrac{a}{a + b} + \dfrac{a}{a + c})$
20. *Jaccard*	$\dfrac{P(A,B)}{P(A) + P(B) - P(A,B)}$	50.	*Fifth Sokal − Sneath*	$\dfrac{ad}{\sqrt{(a + b)(a + c)(d + b)(d + c)}}$
21. *Klosgen*	$\begin{array}{l}\sqrt{P(A,B)}\max(P(B\mid A) - P(B),\\ P(A\mid B) - P(A))\end{array}$	51.	*Baroni − Urbani*	$\dfrac{a + \sqrt{ad}}{a + b + c + \sqrt{ad}}$

Table 1. (*Continued*)

22. Validity($X \Rightarrow Y$)	$P(XY) - P(\overline{X}Y)$	52. *Michael*	$\dfrac{4(ad-bc)}{(a+d)^2+(b+c)^2}$
23. *Match*($A \Rightarrow B$)	$\dfrac{P(AB)}{P(A)} - \dfrac{P(\overline{A}B)}{P(\overline{A})} = \dfrac{P(AB)-P(A)\times P(B)}{P(A)\times(1-P(A))}$	53. *Mountford*	$\dfrac{2a}{2bc+ab+ac}$
24. *Influence* ($X \Rightarrow Y$)	$\dfrac{P(Y\mid X)-P(Y)}{\sigma_p}$, $\sigma_p=\sqrt{\dfrac{P(Y)(1-P(Y))}{n}}$	54. *Fager*	$\dfrac{a}{\sqrt{(a+b)(a+c)}} - \dfrac{1}{2}\max(b,c)$
25. *Sensitivity*	$\dfrac{a}{a+b}$	55. *Unigram Subtuples*	$\log\dfrac{ad}{bc} - 3.29\sqrt{\dfrac{1}{a}+\dfrac{1}{b}+\dfrac{1}{c}+\dfrac{1}{d}}$
26. *Specificity*	$\dfrac{d}{c+d}$	56. *U Cost*	$\log(1+\dfrac{\min(b,c)+a}{\max(b,c)+a})$
27. *Youden*	$\dfrac{ad-bc}{(a+b)(c+d)}$	57. *S Cost*	$\log(1+\dfrac{\min(b,c)}{a+1})^{-\frac{1}{2}}$
28. Crude Agreement	$\dfrac{a+d}{a+b+c+d}$	58. *R Cost*	$\log(1+\dfrac{a}{a+b})*\log(1+\dfrac{a}{a+c})$
29. Adjust Agreement	$\dfrac{1}{4}(\dfrac{a}{a+b}+\dfrac{a}{a+c}+\dfrac{d}{c+d}+\dfrac{d}{b+d})$	59. *T Combined Cost*	$\sqrt{U\times S\times R}$
30. Percent Positive Agreement(PPA)	$\dfrac{2a}{2a+b+c}$	60. Simplified Pointwise Mutual Information	$\dfrac{f(x,y)^2}{f(x)f(y)}$

	y	$\overline{y}$	Σ
x	$f(xy)=a$	$f(x\overline{y})=b$	$f(x*)$
$\overline{x}$	$f(\overline{x}y)=c$	$f(\overline{x}\overline{y})=d$	$f(\overline{x}*)$
Σ	$f(*y)$	$f(*\overline{y})$	N

A 2×2 contingency table of the frequency of x and y is shown in the above table. * stands for any value of the corresponding variable. The formula of the expected frequencies is:

$$\widehat{f}(xy) = f(x*)f(*y)/N \tag{1}$$

However, the results of different objective interestingness measures are always different [11]. It is difficult to select an appropriate measure for all fields. Piatetsky-Shapiro [18] proposed three key properties that a good measure should satisfy. These properties have been extended by many authors [11, 19, 20]. Nonetheless, it was difficult to decide which properties must be satisfied in this study, and none of the interestingness measures satisfied all of the properties. Therefore, we tried a new method of selecting a proper interestingness measure, as shown in the following.

3   Hypothesis for Choosing an Interestingness Measure

The difference between two syndrome elements can be measured by the distance of the interestingness values from the symptoms. A hypothesis for choosing a good measure was proposed on the basis of the distance between two syndrome elements.

Hypothesis: A good objective interestingness measure should satisfy these three properties:

Property 1: If two syndrome elements are completely different then the distance between them should be large;

Property 2: If two syndrome elements are same then the distance between them should be small;

Property 3: If two syndrome elements are the same but the datasets are different then the distance between them obtained from different datasets should be similar.

The method for evaluating interestingness measures is based on these three properties. Section 4 shows the selection process.

4   Selection of an Interestingness Measure Based on the Hypothesis

4.1   Dataset

A series of inclusion and exclusion criteria were established by referring to the "National Union Catalog of the Works on Chinese Medicine" [21], and 229 works in the library of Shandong University of Chinese Medicine were screened. These books were first scanned into images and then transformed into a text format by using optical character recognition. A database including 51,186 records was established, which referred to 1484 famous physicians of the Song, Yuan, Ming and Qing dynasty and the twentieth century. The dataset used in this paper was extracted from this database.

We proposed a Symptomatic Unit hypothesis in order to standardize the symptoms into independent symptoms with the least intension [22]. Finally, we obtained 427 symptomatic units from the clinical record database. We screened the syndrome elements from the syndrome description and obtained 55 syndrome elements. The records that corresponded to any of the 427 symptomatic units and any of the 55 syndrome elements were selected as the dataset in this study. The selected dataset had 44,600 records.

4.2   Selection of Samples of Different Syndrome Elements

The 55 syndrome elements screened from the database were not standardized. If two syndrome elements are similar or dependent, some of the same symptoms may be closely related to both of them. For example, both the deficiency of Jin and the deficiency of Ye are closely related to some of the same symptoms, such as dry mouth and dry tongue. Therefore, the distance between two similar or dependent syndrome elements should be smaller than the distance between two completely different syndrome elements. According to the hypothesis in section 3, the distance between two different syndrome elements can be regarded as an indicator and it is expected to be large. Therefore, two completely different syndrome elements should be selected. Han and Re were selected from the 55 syndrome elements from a professional viewpoint.

4.3   Calculate the Interestingness of the Syndrome Elements and Symptoms

The frequency of each syndrome element and each symptom formed one 2×2 contingency table as follows:

Table 2. 2×2 contingency table of syndrome elements and symptoms

		Symptom	
		+	-
Syndrome	+	a	b
element	-	c	d

We compiled the program and calculated each frequency in the 2×2 contingency table. Then we calculated the interestingness of each syndrome element (Han and Re) and each symptom according to the formula of the 60 measures in section 2. Finally, we had 60 by 2 by 427 data points.

If a, b, c or d is zero, the calculation of some measures, such as *Mutual Information* and *Positive Likelihood Ratio*, cannot be continued since zero is a denominator or an antilogarithm. In order to complete the calculation, we added 1 to a, b, c and d and, as 1 is a very small number compared to the total number of 44,600, this change should not have obviously influenced the results.

4.4   Calculating the Distance between Han and Re

As the syndrome elements, symptoms and interestingness measures were expressed as $\{sd_1, sd_2\}$, $\{sp_1, sp_2, sp_3, \cdots, sp_{427}\}$ and $\{me_1, me_2, me_3, \cdots, me_{60}\}$, respectively, the results of the above step were expressed as shown in the following table:

Table 3. Interestingness values of syndrome elements and symptoms using the interestingness measure me_i

me_i	sp_1	sp_2	sp_3	...	sp_{427}
sd_1	x_{11}	x_{12}	x_{13}	...	x_{1427}
sd_2	x_{21}	x_{22}	x_{23}	...	x_{2427}

In this table, x_{11} corresponds to the interestingness value of syndrome element sd_1 and symptom sp_1 using the interestingness measure me_i. The value of interestingness can be regarded as being an interval-scaled variable. The appropriate distance measures for an interval-scaled variable include Euclidean, Manhattan and Minkowski distances. Among these, the Euclidean distance is the most popular distance measure; therefore, it was used in this study.

Before distance calculating the data should be standardized. The standardization used in this paper can be viewed as a two-step process:

1. The mean absolute deviation was calculated:

$$s_f = \frac{1}{n}(|x_{f1} - m_f| + |x_{f2} - m_f| + \cdots + |x_{fn} - m_f|) \qquad (2)$$

Where $x_{f1}, \ldots, x_{fn}$ are n measurements of the interestingness value, and m_f is the mean interestingness value , that is:

$$m_f = \frac{1}{n}(x_{f1} + x_{f2} + \cdots + x_{fn}) \qquad (3)$$

2. The standardized measurement, or z-score, was calculated:

$$z_{fi} = \frac{x_{fi} - m_f}{s_f}. \qquad (4)$$

The mean absolute deviation, s_f, is more robust to outliers than the standard deviation, σ_f; therefore, the mean absolute deviation was used in this study [23].

Finally, we calculated the Euclidean distance of Han and Re according to the following formula:

$$d(i, j) = \sqrt{(x_{i1} - x_{j1})^2 + (x_{i2} - x_{j2})^2 + \cdots + (x_{in} - x_{jn})^2}. \qquad (5)$$

These distances represent the distances of the completely different syndrome elements corresponding to property 1 in the hypothesis. They denoted the differentiating capacity of the interestingness measures. Thus, the larger the difference, the better the measure.

4.5   Calculating the Distance of the Same Syndrome Elements

There were no completely the same syndrome elements within the 55 syndrome elements. However, by dividing the clinical cases corresponding to one syndrome element into two parts randomly, the two parts may then be regarded as corresponding to the same syndrome elements. According to the method in section 4.3 and section 4.4, we calculated the distance between the same syndrome elements. The division method is shown in Figure 1.

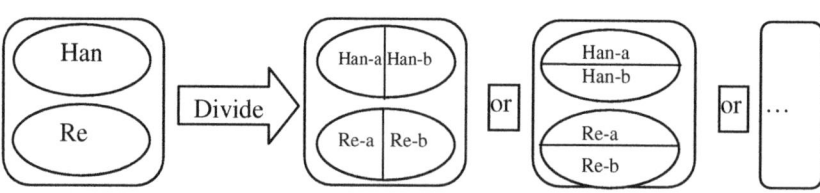

Fig. 1. Different division methods of syndrome elements Han and Re

As shown in Figure 1, clinical cases corresponding to syndrome element Han were randomly divided into two parts that corresponded to Han-a and Han-b, respectively, and the same was performed for syndrome element Re. Then, we calculated the distances between Han-a and Han-b, and Re-a and Re-b, and calculated the average of the two distances. The division and calculation were repeated 100 times. Ideally, the distances between the same syndrome elements should be small and similar. Then, we calculated the means and coefficients of variation of the 100 distances. The mean distance and coefficient of variation corresponded to properties 2 and 3 respectively in the hypothesis. The smaller the distance and coefficient of variation, the more ideal the measure.

4.6   Comparison of Different Interestingness Measures

There were now three indicators corresponding to each interestingness measure. They were the distance between different syndrome elements (DD), the distance between the same syndrome elements (DS) and the coefficient of variation of the distances between the same syndrome elements (CV). They corresponded to the three properties in the hypothesis, respectively. An interestingness measure is a good measure when DD is large and DS and CV are small. Since DD and DS were both distances between two syndrome elements, we could calculate the ratio of them (RDS) as one indicator. Now there were two indicators: RDS and CV. These two indicators were of different dimensions so they cannot participate in the calculation directly. Therefore, the measures were sorted by RDS into descending order to obtain the rank of each measure. Then they were sorted by CV in ascending order to obtain another rank of each measure. Finally, we calculated the average of the two ranks and regarded the average (AR) as the final indicator. We easily verified two characteristics of AR as follows: AR monotonically increased with DD while DS and CV remained the same; AR monotonically decreased with DS (CV) while DD and CV (DS) remained unchanged. So we regarded AR as being a reasonable indicator.

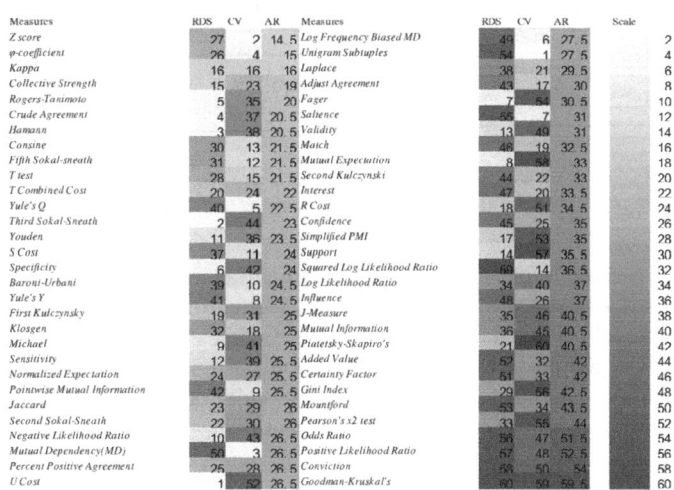

Fig. 2. RDS, CV and AR for all of the 60 interestingness measures

Figure 2 shows the calculated results of RDS, CV and AR for all of the 60 interestingness measures. It can be seen from Figure 2 that the ranks of each measure corresponding to RDS and CV were mostly inconsistent. A small RDS was always accompanied by large CV and vice versa. Among the 60 interestingness measures, the ARs of the *Z score*, *ϕ-coefficient* and *Kappa* were smaller. Therefore, in this section, the *Z score*, *ϕ-coefficient* and *Kappa* measures were selected as the best interestingness measures.

5   Confirmation of the Best Interestingness Measure

5.1   Confirmation by Subjective Interestingness Analysis

Whether the results are useful or not is finally decided by the user. Confirmation of interestingness measure selection by subjective interestingness analysis is an important step. The *Z score*, *ϕ-coefficient* and *Kappa* measures were selected in section 4. Next, the results of these three measures were compared to determine which measure was the best.

Ranks of symptoms are paid more attention than the data points of interestingness and more attention is paid to the symptoms of previous ranks. Moreover, it is easier to professionally compare the rankings if the difference between the rankings is large. So we selected some symptoms from the results. The selection process is shown in Figure 3.

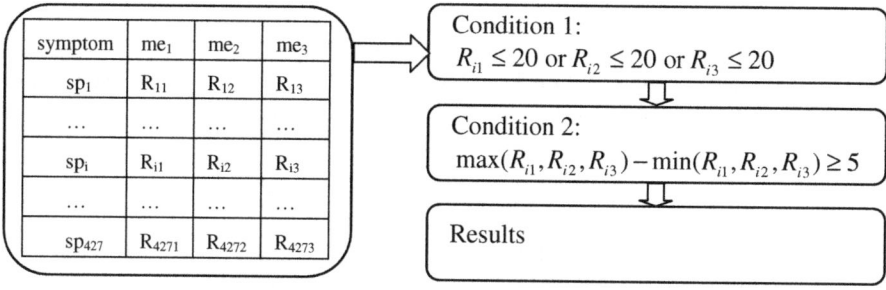

Fig. 3. The process of selection. R_{ij}: Ranking of symptom sp_i when the symptoms were sorted by the interestingness value of the interestingness measure me_j in descending order.

The selected symptoms and their corresponding rankings of the three measures are shown in Figures 4(syndrome element Re) and 5 (syndrome element Han).

In Figure 4, yellow sclera, pain on urination, red eyes, burning on urination, cloudy urine, brightly coloured skin and itchy skin were ranked behind by *Kappa* and fever, dry mouth, small amount of urine, vexation, difficulty in defecation, difficulty in urination and strong pulse were ranked behind by *ϕ-coefficient* and *Z score*. The former

Symptom	φ-coefficient	Kappa	Z score
Yellow sclera	4	12	4
Pain on urination	9	22	8
Red eyes	12	21	11
Burning on urination	16	31	15
Cloudy urine	17	24	17
Brightly colored skin	18	56	18
Itchy skin	20	26	19
Fever	11	7	13
Dry mouth	15	9	16
Small amount of urine	24	14	24
Vexation	25	13	26
Difficulty in defecation	26	18	25
Difficulty in urination	31	19	33
Strong pulse	37	11	40

Fig. 4. Symptom's ranks of three measures corresponding to Re

Symptom	φ-coefficient	Kappa	Z score
Cold feeling in abdomen	4	12	3
Swollen joints	15	25	15
Cold feeling in waist	16	34	16
Nasal obstruction	20	24	19
Headache	24	18	24

Fig. 5. Symptom's ranks of three measures corresponding to Han

group of symptoms had a closer correlation with Re than the latter group of symptoms from a professional viewpoint. In Figure 5, cold feeling in abdomen, swollen joints, cold feeling in waist and nasal obstruction were ranked behind by *Kappa* and headache was ranked behind by *φ-coefficient* and *Z score*. The former group of symptoms had a closer correlation with Han than the latter group of symptoms from a professional viewpoint. Therefore, *φ-coefficient* and *Z score* were considered as more idea measures than *Kappa*.

5.2   Confirmation by Computational Complexity

Two interestingness measures still remained. The formulae for these were rewritten as follows:

$$\phi - coefficient = \frac{ad - bc}{\sqrt{(a+b)(a+c)(c+d)(b+d)}} \tag{6}$$

$$z\ score = \frac{\sqrt{n}(ad - bc)}{\sqrt{(a+b)(a+c)(n^2 - (a+b)(a+c))}} \tag{7}$$

It can easily be seen from the two formulas that calculation of the *φ-coefficient* is easier. Therefore, *φ-coefficient* was regarded as the best measure.

6   Conclusion

This paper presents a method for selecting the most ideal objective interestingness measure for evaluating the correlation between syndrome elements and symptoms. Firstly, a hypothesis for a good interestingness measure was proposed and *Z score*, *φ-coefficient* and *Kappa* measures were selected from 60 interestingness measures based on this hypothesis. Secondly, *φ-coefficient* and *Z score* were selected from these three measures based on subjective interestingness. Finally, *φ-coefficient* was selected as the best measure because of its low computational complexity. We concluded that *φ-coefficient* is the most appropriate measure for evaluating the correlation between syndrome elements and symptoms. The method used in the first step of the selection process was based on existing data and was little interfered with by subjective factors. Thus, the result satisfied the characteristics of the data and was objective. It is expected that this method could be extended to other similar territories.

In the future, we plan to study the correlation between syndrome elements and symptoms using the *φ-coefficient* measure. Then we will be able to standardize the syndrome elements according to their correlation with symptoms and establish the diagnosis criteria of syndrome elements. The method of selecting the best measure will be used to resolve other problems in traditional Chinese medicine, such as the correlation between symptoms and syndromes, symptoms and herbs, syndromes and herbs and other such correlations. Based on the selected measure, other algorithms of data mining, such as classification, clustering and so on, could be developed to obtain more regularity in traditional Chinese medicine.

Acknowledgements. This study was supported by the National Science and Technology Major Projects of China (No: 2009ZX10005-019) and the National Natural Science Fund Project of China (No. 81001500). We are thankful to Xue-zhong Zhou, School of Computer and Information Technology, Beijing Jiaotong University, Beijing, China, for his instructions in this study.

References

1. Zhu, W.: Standardization Research of Differentiation System of Symptoms and Signs and Syndrome in TCM. Tianjin Journal of TCM 19, 1–3 (2002)
2. Wang, Y., Zhang, Q., Zhang, Z.: Extraction of Syndrome Elements and Destination. Journal of Shandong University of Chinese Medicine 30, 6–7 (2006)
3. Sun, Z., Xi, G., Yi, J., Zhao, D.: Select informative symptoms combination for diagnosing syndrome. Journal of Biological Systems 15, 27–38 (2007)
4. Wang, J., Chu, F., Li, J., Yao, K., Zhong, J., Zhou, K., He, Q., Sun, X.: Study on syndrome element characteristics and its correlation with coronary angiography in 324 patients with coronary heart disease. Chinese Journal of Integrative Medicine 14, 274–280 (2008)
5. Tan, S., Tillisch, K., Bolus, S., Olivas, T., Spiegel, B., Naliboff, B., Chang, L., Mayer, E.: Traditional Chinese medicine based subgrouping of irritable bowel syndrome patients. Am. J. Chin. Med. 33, 365–379 (2005)
6. Agrawal, R., Imielinski, T., Swami, A.: Mining association rules between sets of items in large databases. ACM SIGMOD Record 22, 207–216 (1993)

7. Aggarwal, C.C., Yu, P.S.: A new framework for itemset generation. Association for Computing Machinery, Inc., New York, 10036-5701 (1998)
8. Brijs, T., Vanhoof, K., Wets, G.: Defining interestingness for association rules. International Journal of Information Theories and Applications 10, 370–376 (2003)
9. Brin, S., Motwani, R., Silverstein, C.: Beyond market baskets: Generalizing association rules to correlations. ACM SIGMOD Record 26, 265–276 (1997)
10. McGarry, K.: A survey of interestingness measures for knowledge discovery. The Knowledge Engineering Review 20, 39–61 (2005)
11. Tan, P., Kumar, V., Srivastava, J.: Selecting the right objective measure for association analysis. Information Systems 29, 293–313 (2004)
12. Luo, K., Wu, J.: Evaluating Criterion of Association Rules. Control and Decision 18, 277–280 (2003)
13. Yi, W., Wei, J., Wang, M.: Mining Efficient Association Rules. Computer Engineering & Science 27, 91–94 (2005)
14. Chen, J., Gao, Y.: Evaluating Criterion of Association Rules Using Efficiency. Computer Engineering and Applications 45, 141–142 (2009)
15. Huang, Y.: Clinical Epidemiology. People's Medical Publishing House, Beijing (2006)
16. Pecina, P.: A machine learning approach to multiword expression extraction. In: Towards a Shared Task for Multiword Expressions (MWE 2008), pp. 54–57 (2008)
17. Zhou, X., Liu, B., Wu, Z., Feng, Y.: Integrative mining of traditional Chinese medicine literature and MEDLINE for functional gene networks. Artificial Intelligence in Medicine 41, 87–104 (2007)
18. Piatetsky-Shapiro, G.: Discovery, analysis, and presentation of strong rules. Knowledge Discovery in Databases, 229–248 (1991)
19. Lenca, P., Meyer, P., Vaillant, B., Lallich, S.: A multicriteria decision aid for interestingness measure selection. Departement LUSSI, ENST Bretagne, Technical Report LUSSI-TR-2004-01-EN (2004)
20. Geng, L., Hamilton, H.: Interestingness measures for data mining: A survey. ACM Computing Surveys (CSUR) 38, Article 9 (2006)
21. Zhang, Q., Wang, Y., Zhang, Z., Zhang, Q., Song, G.: The Establishment and Statistics on the Clinical Records Database of the Past Dynasties. Journal of Shandong University of Chinese Medicine 29, 298–299 (2005)
22. Zhang, Q., Wang, Y., Zhang, L., Yu, D., Wang, Y.: Independent Symptoms with the Least Intension. Journal of Beijing University of Traditional Chinese Medicine, 5–10 (2010)
23. Han, J., Kamber, M.: Data Mining: Concepts and Techniques, 2nd edn. The Morgan Kaufmann Series in Data Management Systems. Morgan Kaufmann Publishers, San Francisco (2006)

The Impact of Feature Representation to the Biclustering of Symptoms-Herbs in TCM

Simon Poon[1], Zhe Luo[1], and Runshun Zhang[2]

[1] School of Information Technologies, University of Sydney, Sydney, Australia
[2] Guananmen Hospital, China Academy of Chinese Medical Sciences, Beijing, China

Abstract. Traditional Chinese Medicine (TCM) is a holistic approach to medical treatment. Analysis and decision cannot be made in isolation, hence, the extraction of symptoms-herbs relationship is a crucial step to the research of the underlying TCM principle. Since this kind of relationship bears a lot of similarity with the gene-expression study in the microarray analysis, where the use of biclustering algorithms is common, it is logical to apply biclustering algorithms to the study of symptom-herb relationship. However, the choice of feature representation is a dominant factor in the success of any machine learning problem. This paper aims to understand the impact of different representation schemes in the biclustering of symptoms-herbs relationship. A bicluster is not helpful if the number of features is too large or too small. In order to get a desirable size for the biclusters, modified relative success ratio is considered to be the most appropriate one among the other four schemes. Some of the biclusters (using modified relative success ratio) do follow the therapeutic principle of TCM, while some biclusters with interesting feature combination that are worthwhile for clinical evaluation.

Keywords: Representation, Biclustering, Traditional Chinese Medicine, Symptoms-Herbs Relationship.

1 Introduction

Traditional Chinese Medicine (TCM) has started to become the attention in the western world. It is widely adopted as the supplementary therapeutics for disease treatment [1-4]. The use of herbs plays an important role in TCM. There are more than 8000 typical herbs being used. Each herbal treatment is comprised of multiple herbs. In contrast to typical Western Medicine, TCM is based on a huge accumulation of knowledge developed by observation, investigation and clinical practice over thousands of years. It follows a holistic and integrative principle in viewing the human body and its functioning, i.e. the body parts or symptoms can only be understood as a whole [5][6]. Since TCM is a highly empirical activity, it has created a tremendous data warehouse which is unparalleled and unsurpassed in the world's medicine history. To carry out a scientific analysis over these data is crucial for the future development of TCM. Feng et al. [2] provided a systemic literature review of recent researches in TCM data mining area, and propose its future direction. Zhou et al. [3] developed a clinical

L. Cao et al. (Eds.): PAKDD 2011 Workshops, LNAI 7104, pp. 384–394, 2012.
© Springer-Verlag Berlin Heidelberg 2012

reference information model and physical data model to deal with the various information entities and their relationship in TCM clinical data.

Discovering complex relationship between symptoms and herbs fascinate both TCM practitioners and data analysts. Clustering (or grouping) the subsets of symptoms and herbs that exhibit clinical significance together are an essential task in the data analysis. However, there are two characteristics which make the conventional clustering algorithms inappropriate: First, herbs are effective only over some but not all symptoms. Second, some herbs may be effective for more than one symptom. Unfortunately, there is a lack of literature in discovering of herbs-symptoms correlations by using data mining technologies.

On the other hand, diverse biclustering algorithms are reported extensively as a remarkable breakthrough in analysing gene expression data [4-10]. First, each gene is usually assigned to a single cluster by commonly used clustering methods, whereas a gene should be in other clusters because it may participate in multiple functions. Second, the expression of genes is usually considered over all experimental conditions, in fact, genes are typically affected only by specific experimental context that is the subset of these conditions. These limitations are so similar to TCM that the biclustering algorithm is proposed to address these issues in this paper. Also, Prelic et al. [12] conducted a systematic comparison and evaluation of five well-known biclustering algorithms for gene expression data. These algorithms have been used to evaluate the same problem respectively by using synthetic and real data. There are two conclusions of this paper. First, biclustering in general has advantages over a traditional hierarchical clustering approach for gene expression data. Second, for a specific problem, there are considerable performance differences among the different methods, that is, biclustering method strongly depends on problem formulation and algorithm, which may work well in certain scenarios and fail in others.

The aim of this work is to find the subset of symptoms-herbs that seem related and to explore the appropriate representation for the biclustering task. Section 2 briefly introduces our biclustering algorithm. The characteristics and the pre-processing of the dataset for our experiments are highlighted in Section 3. Different representation schemes used in different experiments are described in Section 4. Section 5 looks at and gives an interpretation to the results. The paper is finally closed with a conclusion and future work in Section 6.

2   Symptom-Herb Biclustering Algorithm

2.1   General Representation Pattern for Symptom-Herb Biclustering Algorithm

In addition to the algorithmic difference and the nature of the result, one significant difference between traditional clustering algorithm and biclustering algorithm is the representation scheme over their respective datasets. Data are usually organized in a tabular form in the traditional clustering algorithm, where the columns represent the features and each row represents an instance/object. Instead, the rows and columns for a biclustering algorithm represent the two distinct feature types, features and their manifestations (or vice versa). The value at the intersection of a row and a column represents the relationship between a gene and its appropriate expression in the microarray analysis.

In our symptoms-herbs analysis, the rows represent the symptoms and the columns are the herbs, the values of junctions of rows and columns represent the degree of relationship between the corresponding symptoms and herbs.

2.2   Symptom-Herb Biclustering Algorithm

As argued by Prelic et al. [12], a biclustering algorithm strongly depends on problem formulation. As a result, there is no universal framework for biclustering algorithm design as long as simultaneously clustering of both row and column sets in a data matrix. According to the general representation pattern of symptom-herb dataset, a biclustering algorithm in addressing the discovery of the correlations between subsets of symptoms and herbs is proposed and its basic principle is shown as follow.

Symptom-Herb Biclustering Algorithm

Input: Z, the data matrix of a certain representation pattern of
    symptom-herb dataset; S, a subset of symptoms that is initialized
        to input symptom(s); δ, the minimum acceptable mean score.
Iteration:
   1. Compute the mean score M of input symptom(s) for each
     herb and choose the herb(s) with a mean score no less than δ as
     input herb(s).
   2. Compute the mean score N of input herb(s) for each symptom and
     update the symptom(s) with a mean score no less than δ as
     input symptom(s).
   3. Repeat 1 to 2 until both input symptom(s) and herb(s) are not
     changed, return such subset of symptom(s) and herb(s) as a
     bicluster.
Output: A, a bicluster that is a submatrix of Z that includes a subset of
    symptoms and a subset of herbs with a mean score no less than δ.

Fig. 1. Symptom-Herb Biclustering Algorithm

The proposed biclustering algorithm is similar to the Iterative Signature Algorithm [9]. It receives a data matrix that is a certain data representation pattern of targeted dataset and a subset of symptoms as input symptom(s). The process of this biclustering algorithm can be divided into two main steps. In the first step, by using the input symptom(s) to select which herb(s) has (have) strong connection to such symptom(s), which indicates the mean score should be no less than a predefined threshold δ, and then update the input herb(s). In the second step, the herb(s) that was (were) obtained in first step is used as input herb(s) to select which symptom(s) has (have) strong connection to such herb(s) and then update the input symptom(s). These two steps will be executed iteratively until the convergence appears that is the input symptoms and herbs are not changed anymore. Finally, we return a subset of symptoms and herbs as a bicluster which suggest there are some strong connection between the subset of symptoms and herbs. The procedure of proposed biclustering algorithm is briefly described in Figure 1.

In this paper, all the symptoms in the input data matrix will be used as the input symptom, for each time only one symptom will be used and then execute the symptom-herb biclustering algorithm. The result must either be one bicluster that may or may not contains the input symptom(s) or be null, because any input symptom may be discarded during the iteration process of computing which depends on the value of predefined threshold δ. If $\delta = 0$, no input symptom will be discarded but the result bicluster will contain all symptoms and herbs which is equal to the input data matrix. In other words, the size of biclusters strongly depend upon the predefined threshold δ, specifically, the number of rows and columns of a bicluster, that is, the number of input symptoms and herbs will increase with the reduction of δ.

3   Dataset and Pre-processing

The original clinical data is about insomnia provided by *Guanganmen* hospital. The insomnia dataset includes the clinical information of outpatients in terms of the symptoms and the prescriptions. They were stored in separate Microsoft Office Excel worksheets and linked by a unique *Hospital_ID*.

Each row in the symptoms spreadsheet represents a medical record of an outpatient instance. Its columns represent some distinct symptoms which are related to the insomnia disease. The value of a symptom (*Symptom_Code*) is set to "1" to indicate that the corresponding symptom appears in the related instance, otherwise it is set to "0". There are 457 instances and 155 symptoms in the insomnia dataset. All the symptoms relating to the features of tongue and pulse are not included because the objectivity of these features is in doubt. As a result, only 103 symptoms remained.

The prescriptions worksheet has similar pattern as with the symptoms worksheet. The *Symptom_Code* in the previous one is replaced with *Herb_Code* that is used to depict some distinct TCM herbs. The value of a cell is set in a similar fashion as the symptom. This file has a column that represents the Outcome of each prescription. The outcome is annotated by TCM clinical experts. They evaluated the treatment efficacy based on the information from the subsequent consultation [13]. The outcomes are classified into two categories, good and bad. If the prescription of the corresponding instance is effective for its clinical treatment, it can be seen as good outcome ("1") in *Outcome*. Otherwise, the prescription will be classified as bad outcome and reported as "0". There are 460 instances and 111 herbs in the insomnia dataset, 68 instances have been labeled as bad outcome, and others are good.

It was noted that there was a discrepancy between the numbers of instances in the two spreadsheets. As a result, 457 instances remained after correction. Each table is then subdivided into two categories according to the *Outcome* of each instance. To be specific, the prescriptions with the good outcome were organized in the table *prescriptions_outcome_*1. In the same way, instances with bad outcome were put in the table *prescriptions_outcome_0*. The *symptoms* table has a similar process.

4   Experiments

The choice of representation is always a challenging task in machine learning research. There are no theoretical guidelines to select a representation. The user plays a significant role in choosing a suitable data representation for a specific situation by gathering the facts and conjectures about the data [11]. However, an appropriate data representation, which is yielded by careful investigation of the available features and any available transformations, usually has a simple and easily understandable result [11]. In this work, four data representation patterns were proposed for the input data matrixes in this work.

4.1   Effective Count

The effective count indicates that the effective successful count of a particular herb in the treatment a particular symptom, i.e. *good_outcome – bad_outcome*. The table *prescriptions_outcome_1* and the table *symptoms_outcome_1* are further integrated into a new table, say, *frequencyCount_outcome_1*. The corresponding entry in this new table was incremented according to each symptom-herb pair for each instance. A similar process was applied to the table's *prescriptions_outcome_0* and *symptoms_outcome_0* to create a new table of *frequencyCount_outcome_0*.

Based on the effectiveness of the prescriptions that are reported as the outcome of instances, the weight "1" and "-1" is assigned to table *frequencyCount_outcome_1* and table *frequencyCount_outcome_0* respectively, then simply add two tables together to obtain the table *Frequency Count*.

4.2   Binary Value

In contrast to the abovementioned effective count, the binary value concentrates on the "presence" rather than the "frequency" (value = 1, otherwise value = 0). To be more exact, if the cell value of *frequency Count* is equal or larger than 1, the entry in the table *Binary Value* is set to 1, otherwise set it as 0.

4.3   Relative Success Ratio

Effective count is not a good indicator to reveal the validity and reliability of an herb in treating a particular symptom. Relative success ratio is proposed as one of the data representation patterns; it is defined as the ratio of occurrence good outcome to the total occurrence. The occurrence of good outcome comes from the table *frequencyCount_outcome_1*, while the *total occurrence* of this dataset is the sum of table *frequencyCount_outcome_1* and *frequencyCount_outcome_0*.

4.4   Modified Relative Success Ratio

Modified relative success ratio is proposed as the combination of relative success ratio and frequency count. In this pattern, the frequency count is used as the pruning tool for modifying the relative success ratio table. For insomnia dataset, if the cell values within table *Frequency Count* are less than a predefined threshold, the cell values with same position in table *Relative Success Ratio* will be reset as "0". This predefined threshold is based on an arbitrary assumption that if the frequency of an herb-symptom pair is less than the threshold, it is considered as uninteresting. Since there are 457 instances in this dataset, we assume that the value of frequency count of a certain herb-symptom pair should be at least 2% of the amount of the cases (457 * 2% ≈ 9), otherwise it is considered as an insignificant herb-symptom pair in TCM.

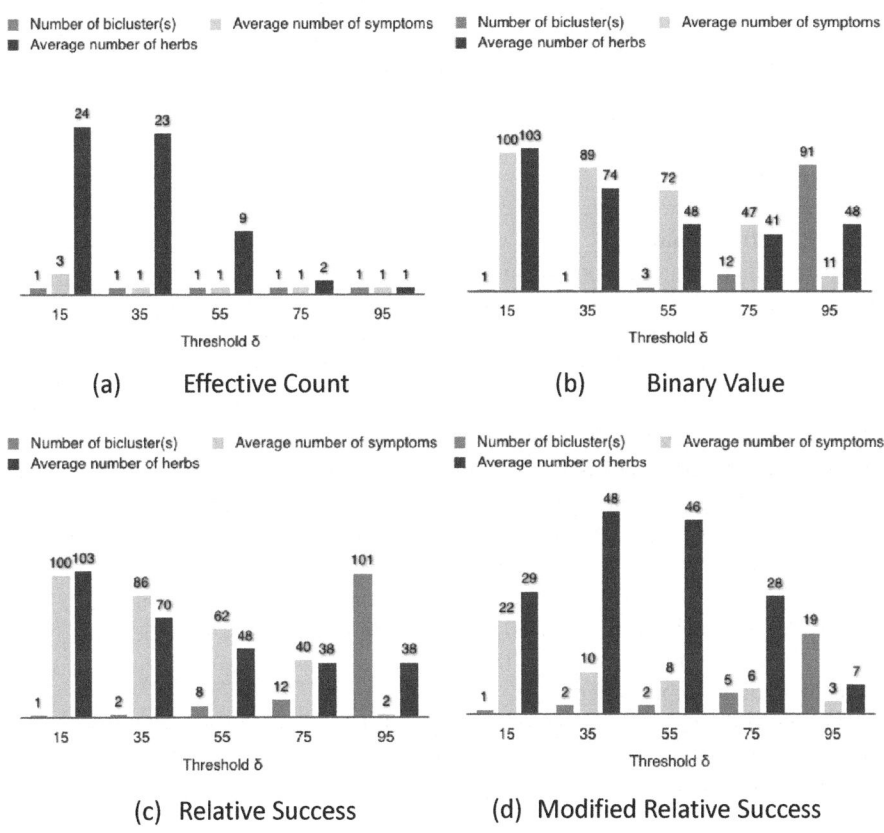

Fig. 2. The impact of the threshold

5   Results and Discussion

Each bicluster contains a subset of symptoms and a subset of herbs that exhibit certain significance, which refers to the effectiveness when a subset of herbs is prescribed for treating a subset of symptoms.

A bicluster is only useful when it has a reasonable size. If the size is too large, i.e. a bicluster contains too many symptoms and herbs, it is obviously difficult for a practitioner to comprehend the complex relations. However, if the size is too small, it may be too simplistic and uninteresting. Though there is no strict and specific definition for the "reasonable size", the domain expert suggested 5 to 7 as the ballpark figures for the number of symptoms and herbs.

As stated in Section 2.2, the threshold δ determines the size of a bicluster. It is quite important to show how the threshold will affect the size of biclusters in detail. As a result, for each representation scheme, the impact of threshold on the number and size of bicluster(s) is provided in the bar chart and discussed, which is followed by the discussion on the validity of its assumption. Meanwhile, some biclusters are shown and discussed in TCM field.

5.1   Results

5.1.1   Effective Count

In terms of effective count, one bicluster was found regardless of the threshold (Fig 2a). The bicluster barely changed. It only contained the subset of symptoms and herbs that frequently occurred; the less frequent symptoms and herbs were simply excluded because of δ. When the threshold was specified as 95, the bicluster only contained one symptom and herb, "Sym133" (失眠) and "Herb103" (炒酸枣仁 stir-frying spine date seed), which is too simple to lead to any interesting investigation.

This representation assumes that the higher the value of a symptom-herb pair, the more effective this pair is in the treatment. However, this assumption may not be true when it is not a balanced dataset, i.e. having roughly the same number of good and bad outcome in the dataset. Also, let say we have two symptom-herb pairs A and B, A appears 100 times with 80 bad outcomes, B appears only 20 times but with 2 bad outcome, A will receive a score of 20 while B only has 18, which implies A is more effective than B. It may be doubted that whether the simple count truly reflects the effectiveness of a symptom-herb pair.

Because of the dubious assumption and the barely changed bicluster pattern, using effective count does not seem to be a proper representation scheme for the proposed biclustering algorithm. Thus, the biclusters will not be shown and discussed.

5.1.2   Binary Value

There is no dominated symptom and herb in this representation scheme. Binary approach is based on an apparently inaccurate assumption that all the symptom-herb pairs are of equal importance, as soon as the number of good outcomes exceed the bad outcomes by one. In normal sense, a symptom-herb pair with 100 good outcomes and no bad outcome should be more interesting than a pair with 51 good outcomes and 49 bad outcomes in a clinical study. Refer to Fig 2(b), the average numbers of symptoms and herbs in a bicluster were as high as 100 and 108 in a low threshold value, this is far beyond any person to comprehend and to conduct a proper study. Even though δ was changed to a high 95, the average size of herbs is still 48 in a bicluster. As a result, this representation scheme is considered inappropriate, and no further evaluation was carried out.

Table 1. Interpretation of the biclusters using TCM principle

	Bicluster1	Bicluster2	Bicluster3
Symptoms	Sym34 (疲乏) Sym87 (失眠多梦)	Sym15 (入睡困难) Sym78 (头痛) Sym87 (失眠多梦)	Sym5 (心悸) Sym15 (入睡困难) Sym20 (汗出) Sym41 (醒后不易入睡)
Syndrome	deficiency of both the heart and spleen 心脾两虚证	heart disorder due to liver-fire 肝火扰心证	deficiency of vital energy of cholecyst and hear 心胆气虚证
Herbs	Herb76 (当归 Chinese angelica) Herb90 (炒白术 stir-frying largehead atractylodes rhizome) Herb101 (制远志 prepared thinleaf milkwort root) Herb103 (炒酸枣仁 stir-frying spine date seed)	Herb76 (当归 Chinese angelica) Herb104 (柴胡 Chinese thorowax root)	Herb15 (茯苓 Indian bread) Herb22 (石菖蒲 grassleaf sweetflag rhizome) Herb88 (生甘草 fresh liquorice root) Herb101 (制远志 prepared thinleaf milkwort root) Herb102 (知母 common anemarrhena rhizome) Herb103 (炒酸枣仁 stir-frying spine date seed)

5.1.3   Relative Success Ratio

In Fig 2(c), the result patterns using relative success ratio is quite similar to the binary value (Fig 2b). "The symptom-herb pair with the higher success ratio is considered to have higher clinical significance" is the assumption behind this representation; it is fairly convincing, especially when TCM is seen as a highly empirical activity. However, many symptom-herb pairs show the perfect score of 1, especially for those infrequently used herbs, i.e. they do not have bad outcome. As a result, some insignificant symptom-herb pairs were presented as highly effective under this scheme. For instance, there are two symptom-herb pairs A and B, A has 99 good outcomes and 1 bad outcome, B has only 1 good outcome but without any bad outcome, and the score of A is 0.99 and that of B is 1 in this representation pattern.

The tricky thing is that *B* is more interesting than *A* theoretically, but it may not be true. There is a serious possibility that the herb of *B* is accidentally added as an adiaphorous component in the prescription for treating the symptom of *A*. This scheme, though having a fairly persuasive assumption, had a disappointing performance.

5.1.4 Modified Relative Success Ratio

Compared with other representation schemes, the average number of biclusters is relatively small in modified relative success ratio (Fig 2(d)). Some biclusters with reasonable size were found in the results by proper thresholds, for example, when the threshold is around 95. The average number of herbs in this representation pattern was far less than that in any other schemes. The subsets of symptoms and herbs that have been narrowed down are desirable to facilitate the further study in TCM field.

The assumption of this representation pattern integrates the assumption of frequency count and that of relative success ratio. This heterogeneous assumption states a rule that is even more convincing, *a symptom-herb pair with higher success ratio and used more frequently* is considered to be more interesting. This scheme is able eliminate those insignificant symptom-herb pairs that show virtual effectiveness from relative success ratio. Owing to a more rational assumption and a desirable cluster size, this scheme is considered appropriate for the TCM biclustering.

5.2   Discussion of Biclusters in TCM Field

As mentioned above, the symptoms can only be understood as a whole in TCM. A certain subset of symptoms is related to a certain syndrome ("ZHENG"), and the typical treatment of a syndrome usually observes a therapeutic principle, which refers to the compound use of a certain subset of herbs.

Table 1 shows three of the biclusters that have been analysed and annotated by the domain expert. Besides those biclusters that follow/confirm the therapeutical principle of TCM, some biclusters were interesting and worthy for further study. One such bicluster (**bicluster4**) contained Sym2 (耳鸣), Sym15 (入睡困难) and Sym20 (汗出). They are mainly related to the syndrome "imbalance between heart-yang and kidney-yin". Nevertheless, the three herbs, Herb15 (茯苓 Indian bread), Herb102 (知母 common anemarrhena rhizome) and Herb103 (炒酸枣仁 stir-frying spine date seed), are not covered in the standard prescription. This may be attractive to the TCM practitioners for curing the syndrome "imbalance between heart-yang and kidney-yin" as it provides a novel herbal composition.

The first three biclusters (bicluster1-bicluster3) are consistent with the TCM principle. Though bicluster4 is inconsistent with the expert's clinical experience, but the expert reckoned that Herb103 (炒酸枣仁) has a wide range of applications. This herb is not only appropriate for the insomnia due to blood deficiency with heat, it is also relevant to the syndrome of deficiency of both heart and spleen, syndrome of disharmony between heart and kidney but not much use for the syndrome of liver-fire and heart-heat.

6   Conclusion and Future Work

This research was an attempt to understand the impact of different representations to discover the complex relationship between symptoms and herbs by using biclustering algorithm. In comparing the results under different representation schemes, the modified relative success ratio was found to be the most appropriate one for the further study in TCM field. Some of the biclusters, which derived from the modified relative success ratio, do follow the therapeutical principle of TCM. There were some other biclusters that displayed interesting pattern appear, which required evaluation in future clinical trials. Moreover, only one dataset (insomnia) was carried out in this paper, the implementation of this method on other datasets should be explored to understand its generality.

In this present work, symptoms relating to the features of tongue and pulse have been removed from the analysis, putting them together and/or considering separately will be another good alternative. Meanwhile, the dataset of this research does not include the herb dosage that is extensively recognized as an indispensable part of a prescription, which should be taken into consideration in the future.

References

1. Hsiao, C.F., Tsou, H.H., Wu, Y.J., Lin, C.H., Chang, Y.J.: Translation in different diagnostic procedures—traditional Chinese medicine and Western medicine. Journal of the Formosan Medical Association 107(12 suppl.), 74–85 (2008)
2. Feng, Y., Wu, Z., Zhou, X., Zhou, Z., Fan, W.: Knowledge discovery in traditional Chinese medicine: state of the art and perspectives. Artificial Intelligence in Medicine 38(3), 219–236 (2006)
3. Zhou, X., Chen, S., Liu, B., Zhang, R., Wang, Y., Li, P., Guo, Y., Zhang, H., Gao, Z., Yan, X.: Development of traditional Chinese medicine clinical data warehouse for medical knowledge discovery and decision support. Artificial Intelligence in Medicine 48(2-3), 139–152 (2010)
4. Ung, C.Y., Li, H., Cao, Z.W., Li, Y.X., Chen, Y.Z.: Are herb-pairs of traditional Chinese medicine distinguishable from others? Pattern analysis and artificial intelligence classification study of traditionally defined herbal properties. Journal of Ethnopharmacology 111(2), 371–377 (2007)
5. Yan, X., Milne, G.W.A., Zhou, J., Xie, G.: Traditional Chinese medicines: molecular structures, natural sources, and applications. Ashgate, Aldershot (1999)
6. Shao, L.: Network Systems Underlying Traditional Chinese Medicine Syndrome and Herbs Formula. Current Bioinformatics 4, 188–196 (2009)
7. Cheng, Y., Church, G.: Biclustering of Expression Data. In: Proceedings of the Eighth International Conference on Intelligent Systems for Molecular Biology, pp. 93–103. AAAI Press (2000)
8. Tanay, A., Sharan, R., Shamir, R.: Discovering Statistically Significant Biclusters in Gene Expression Data. Bioinformatics 18, S136–S144 (2002)
9. Ihmels, J., Friedlander, G., Bergmann, S., Sarig, O., Ziv, Y., Barkai, N.: Revealing Modular Organization in the Yeast Transcriptional Network. Nature Genetics 31, 370–377 (2002)

10. Ihmels, J., Bergmann, S., Barkai, N.: Defining Transcription Modules Using Large-Scale Gene Expression Data. Bioinformatics 20, 1993–2003 (2004)
11. Jain, A.K., Murty, M.N., Flynn, P.J.: Data Clustering: A Review. ACM Computing Surveys 31(3), 264–323 (1999)
12. Prelic, A., Bleuler, S., Zimmermann, P., Wille, A., Buhlmann, P., Gruissem, W., Hennig, L., Thiele, L., Zitzler, E.: A Systematic Comparison and Evaluation of Biclustering Methods for Gene Expression Data. Bioinformatics 22(9), 1122–1129 (2006)
13. Zhou, X., Poon, J., Kwan, P., Zhang, R., Wang, Y., Poon, S., Liu, B., Sze, D.: Novel Two-stage Analytic Approach in Extraction of Strong Herb-herb Interactions in TCM Clinical Treatment of Insomnia. In: International Conference on Medical Biometrics (ICMB 2010), pp. 28–30 (2010)

Second Workshop
on Data Mining
for Healthcare Management
(DMHM 2011)

Usage of Mobile Phones for Personalized Healthcare Solutions

M. Saravanan[1], S, Shanthi[2], and S. Shalini[2]

[1] Ericsson R & D, Ericsson India Pvt. Ltd, Chennai, India
m.saravanan@ericsson.com
[2] Department of ECE, Meenakshi Sundararajan Engg. College, Chennai, India
{shanthi.s89,shalini.msec}@gmail.com

Abstract. One of the greatest hurdles in providing the appropriate healthcare is the availability of proper information at the point of individual's care. Mobile phone-based health solutions can bridge this gap and can support with right information at the right time. In order to overcome some of the prevalent issues and hence provide the necessary healthcare, we here introduce one such mobile based system known as the Personalized Mobile Health Service System for Individual's Healthcare which caters to the specific needs of the user without the constraint on mobility. This system helps in guiding the user with regard to the food they consume, the precautionary measures to be taken in case of any ailments, and when they travel to a new location etc. This system also proves to be supportive in situations when the user is in a traumatic condition suffering alone. The main advantage of the system is that it will keep updating the details to the user on a regular basis.

Keywords: Personalized Mobile Health Service System, Ontological Framework, Cloud Computing, Wearable Sensors, Healthcare solutions.

1 Introduction

Mobile technologies are increasingly growing in developing countries. By the end of 2010, more than three quarters of any nation population will be covered by a mobile network. The availability of low-cost mobile phones and the existing broad coverage of GSM network availabilities are huge opportunities to provide services that would trigger development and improve people's lives. And the fact that mobile phones only require basic literacy and that they allow the transfer of data in a simple and faster manner, makes it suitable for various applications among which healthcare is benefited to a great extent.

The reason for choosing mobile phones as a tool for implementing the Personalized Mobile Health Service System is mainly because of the fact that the usage of mobile phones has increased tremendously over the past few decades, mainly in the developing countries. Today, people tend to be occupied or rather pre-occupied due to their busy daily schedule and mobile phones play a major role in assisting the users

L. Cao et al. (Eds.): PAKDD 2011 Workshops, LNAI 7104, pp. 397–407, 2012.
© Springer-Verlag Berlin Heidelberg 2012

with this regard. Allocating time for all our activities becomes an issue of great concern which has considerable impact when it comes to monitoring our health. Thus, we came up with this idea, which facilitates the mobile users with a system that would not only help them to monitor their health, but also recommend the right diet, alert them with respect to their activities and also help in the diagnosis measures when they are not well.

In the mobile health scenarios, the physician requires the patient's details, in order to understand and hence provide the required medication. There are various factors like mobility, availability of data etc. that affect this basic norm that is necessary for providing proper healthcare. Thus, our model proposes an overall healthcare service system, which helps to provide solutions; overcoming the various problems that prove to be a hindrance in the timely treatment of the user (patient), like for instance, when the user moves from one location to another, it is not usually possible for the user to be precisely aware about the prevailing diseases in the new location. Our system is designed to give an intimation regarding the health hazards prevalent in the new location along with various precautionary and remedial measures which will prove to be helpful to the user.

Our model basically deals with three major scenarios. The first situation is wherein the user uses the mobile application to know about the food that they consume, in the form of food's calories value etc. It also intimates the acceptable alternative intake and the prevailing diseases, during change of location. The second aspect helps the user in monitoring and following their regular medications and other health related activities and also it deals with the dynamic movement of the doctor in treating the user while both of them are in remote locations. The final situation involves the use of an IR camera and biosensor, for the purpose of detecting the ailment by means of an IR image and bio-signal, mainly when the user is alone.

The use of domain-specific ontology and adopting a cloud environment are two important technologies to be considered for implementing our system for faster access of required data. Ontology is a formal, explicit specification of a shared conceptualization of terms and their relationships for a domain [1]. It has the advantage of sharing of knowledge, logic inference and the reuse of knowledge. While, cloud computing is a style of computing in which on-demand resources are provided as a service over the Internet. Users do not need to have knowledge of, expertise in, or control over the technology infrastructure in the "cloud" that supports them [2]. Also cloud computing could be seen as a boon to healthcare IT services as a number of health centers could share infrastructure with vast number of system linked together for reducing cost and increasing efficiency.

The health cloud [3] is a structure that helps in the storage of large amounts of data, for the purpose of processing and future retrieval of the same. Our model is designed to provide an interface such that the user can store the data in this cloud and can also view and modify it anytime in the future. The application expects the user to login (for free) to a predesigned website, for the purpose of entering the required data in order to generate the database. This personalized database can be modified and updated by the user, can be a very useful resource for different scenarios.

2   Related Work

Mobile phones have become an integral part of communication, for the past few years. They were previously used for the purpose of communication, using voice as well as text. This made their necessity grow drastically, and thereby led to the development of mobiles, that basically supported fast communication as well as storage and transfer of data. Thus, this brought in the mobile phones, to play an active role in the field of healthcare.

Research and Development for Health Management is a continuous process. Prior to this system which we have proposed, there exist various other systems that have been designed for the purpose of facilitating different health services using advanced technologies. To begin with, there are various ontological frameworks available for the purpose of classification and arrangement of data [4]. There have been models proposed related to Mobile Telecommunications System combined with the ever-advancing miniaturization of sensor devices and computers [5]. Patients first hour following the trauma is of crucial importance in trauma care. The sooner treatment begins, the better the ultimate outcome for the patient [6]. The existing system [7] allows the incorporation of diverse medical sensors via wireless connections, and the live transmission of the measured vital signals over public wireless networks to healthcare providers. The above facts clearly highlights that there is an immediate need of a sophisticated system which can handle all the above scenarios combined together in a single solution environment.

Our system comprises of various scenarios which are based on certain user oriented health related parameters, in order to help the user to get a fast and precise solution. As for the generation of prescription to patient, we enable the application to facilitate the doctor with the necessary updated data of the user for the purpose of treatment. This information is then referred to perform multiple operations. The important features we consider for this overall system are food, location and give a meticulous solution to trauma care. The main advantage of our system is that we use the cloud environment to store the data, by which the mobile's memory space is not overloaded. Above all, the applications are designed such that, it performs multiple operations with the help of a single ontological framework [8], thereby making the system much more simple, elegant and fast. The ontology promotes the rapid development of mobile applications, more efficient use of resources, as well as reuse and sharing of information between communicating entities [9, 10]. Moreover, it includes machine-interpretable definitions of basic concepts in the domain and relations among them.

Using cloud computing in the health information technology sector makes sense from a cost basis. The SMArt program has a goal of making health-care IT software applications as easy to make and as interoperable as iPhone apps [11]. SMArt, which stands for Substitutable Medical Applications, reusable technologies, is a platform that will contain open APIs that would allow any application to connect with information in any health-care IT system, regardless of what legacy enterprise system it was based on. A strong case can be made for cloud computing today is on the research side of healthcare.

We have also made use of a list that consists of the name and photos of various pills, for the purpose of reminding patients regarding their medication [12]. The usage of IR imaging has been a useful and non-invasive approach to the diagnosis and treatment (as therapeutic aids) of many disorders, in particular in the areas of rheumatology, dermatology, orthopaedics and circulatory abnormalities [13], which basically give a relationship between body temperature and various diseases.

In order to implement our proposed mobile phone-based system, we have used certain basic concepts that have been considered from prevailing works. Most of the scenarios make use of ontology environment to classify data as per the requirement of the application.   While cloud computing is preferred in order to organize large amounts of data for the purpose of future retrieval and usage.

3   System Functional Architecture

The system we have developed is designed in such a way that it can be used for different healthcare purposes, like the calculation of calories value of food, intimation regarding prevailing disease while changing to a new location etc. The functional architecture of the proposed system consists of seven basic components which were shown in Fig 1, as discussed below:

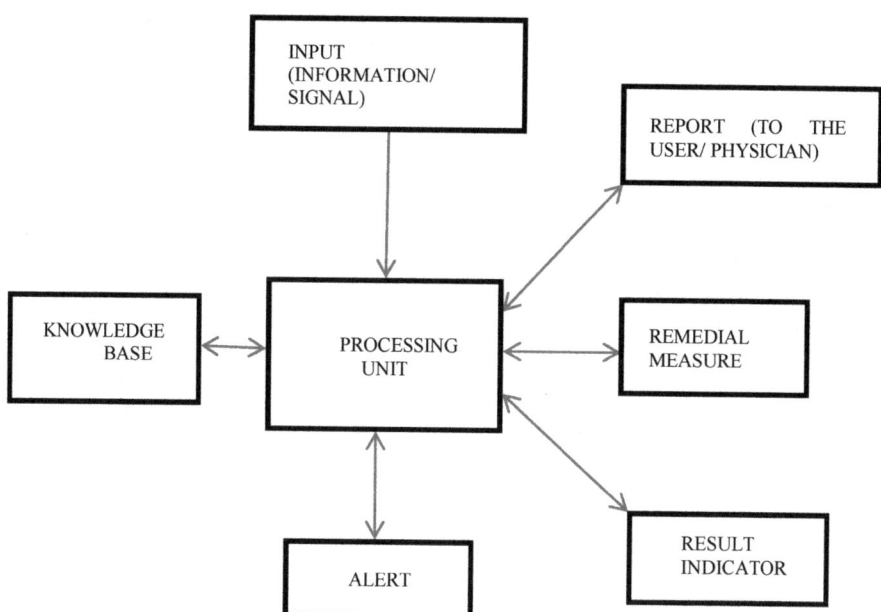

Fig. 1. Functional units for Personalized Healthcare Service System

1. Knowledge Base - Stores the information about the user, which includes their complete medical history stating information like major ailments, specific allergies, vaccinations etc., their current health status as well as their general personal information.

2. Input (Information/Signal) - The input plays a major role in deciphering the appropriate result, thus the input can be in the form of information or signal. Information is given as input in the case of calculating the calories value of a particular dish (the ingredients of the dish are the inputs) and a signal is given as input while using biosensors to detect bio-signals, in order to know the variations in the signal, with regard to a reference signal.

3. Processing Unit - The data in the database, along with the available input is processed and this processed information is either formatted to give alerts, reports, remedial measures or result.

4. Alerts - The alerts are basically used in situation wherein the person has skipped their morning exercise, yet follows the normal routine of intake. The alerts will thus help prevent the user from doing activities that will cause a health problem.

5. Reports - The reports are usually generated based on the processed data. These reports can be made to the user as well as to the physician. Generally the reports generated by the application are displayed in the mobile to notify the user. While in the case wherein the bio- signals are analyzed by the mobile, the reports that are generated are sent to the doctor for the purpose of diagnosis.

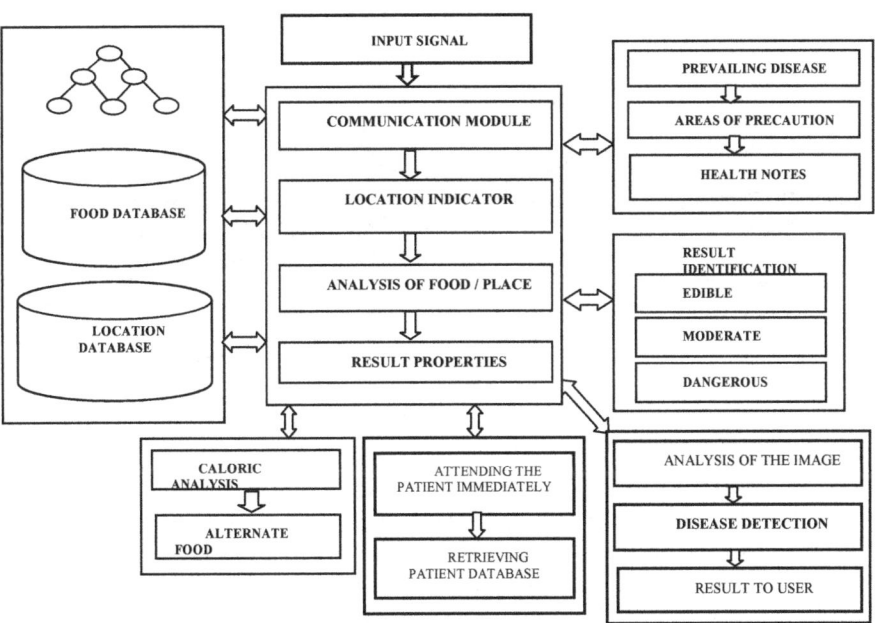

Fig. 2. The basic system details for healthcare scenarios 1 and 2

6. Remedial Measures - The application is designed to give remedial measures in situations like, when the user is in a different location and there prevails a particular epidemic, then the user is given certain measures in order to safeguard oneself, from getting affected.

7. Result Indicator - There are various cases where the results can be indicated, say when the calories value for a particular dish is calculated, the application is designed to indicate the results in three ways in order to indicate the user as to whether it is advisable to consume that particular dish.

4   Mobile Phone-Based Healthcare Scenarios

The Personalized Mobile Health Service System is designed to be applied for two different situations which have been given in Fig 2 were discussed as follows.

Scenario 1: Healthcare Solutions Based on the Analysis of Food and Location of the User

In this scenario, we use food and location ontology. There are three main aspects for this case. First, we consider the case wherein the user gives the input of the food they would like to consume, based on the details in the database and the input given, the data (say calories value) is extracted from the food database in order to be processed using the ontological framework which we have created (Fig 3). The results of the processing are indicated using the result indication block (Fig 1) of the system.

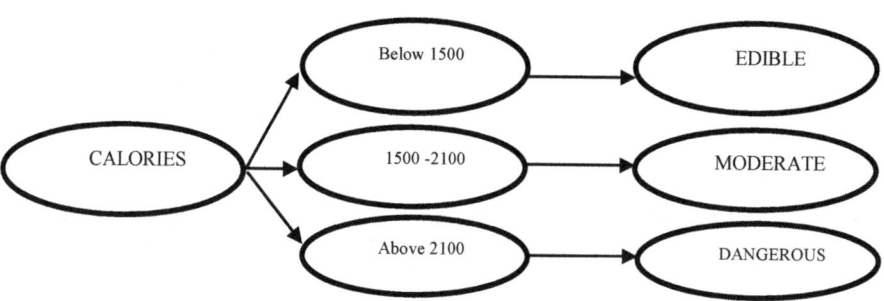

Fig. 3. Food Ontology

Next we have the scenario wherein the user moves from one location to another. In this case, the user is automatically intimated regarding the prevailing health related diseases in that location and also provided with the necessary remedial measures. In the final case, we use both the databases. The application takes the input from the food database (say calories value) and computes the total (calories) value for the food consumed by the user in the previous location and hence gives an alternative and equivalent food that is preferable for the user to consume in the new location. This is given in the form of report for the user's reference.

Scenario 2: Healthcare at Traumatic Situations

Healthcare plays a major role mainly when the user is alone. There exists lots of way to provide health services in usual conditions, but when the user is alone, it becomes difficult for doing the same. And mainly during critical conditions, the user is in a situation to help oneself.

In our model, we propose to use bio-sensors and IR camera as an intermediary device, in order to give input to the system, in the form of bio-signals or IR images. The system uses the input to compare it with a pre-loaded reference image. After comparison, the data is sent to the nearest hospital and family physician. We make use of the ontological framework to find the nearest hospital [8]. Then the necessary steps are taken by the hospital. A template is created during the installation of the application, which is sent along with the results obtained after the processing. Thus, this enables in quicker response.

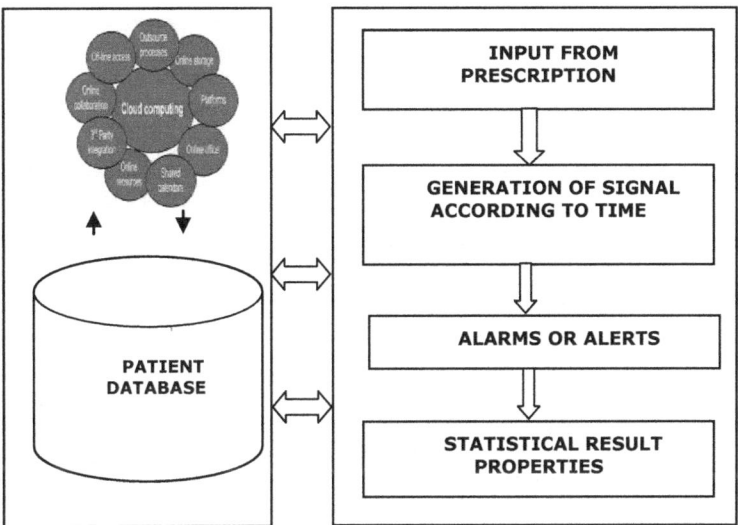

Fig. 4. Basic scenario involving extracting of data from patient database

Scenario 3: Monitoring User's Health Conditions

This scenario is designed based on two conditions, thus it can be split into two cases. The first case is when the user inputs all the details, like personal details, prescription details, recent medications, health related activities etc. and stored in the mobile itself. If the details to be stored are vast, then a cloud environment is set up in order to store the information. The application is designed to alert the user in case of missing an appointment, morning exercise or tablet etc. It will also generate a monthly report for the user, based on the various activities performed over the month which is shown in Fig 4.

We consider the other case to be when, the user as well as the physician are in remote places. In such a case, the database of the user is made available in the hospital, by means of creating a cloud environment. When the user contacts the physician, in case of any ailment, then physician collects the relevant details of a user from his clinic desktop, and suggest the appropriate medication to the user. Fig 4 can be modified and used for implementation in the second case of scenario 3, based on the requirements discussed above.

5   System Implementation

The implementation part of our system basically involves the setting up of a cloud environment that helps store the information of the user, required for the purpose of diagnosis and also for further retrieval. The mobile application actually links to the network where the cloud is set up for the purpose of entering the data. The user needs to login in order to input, modify, view and retrieve the data in future. The login can be done free of cost. Fig 5 illustrates the usage of cloud environment in our system.

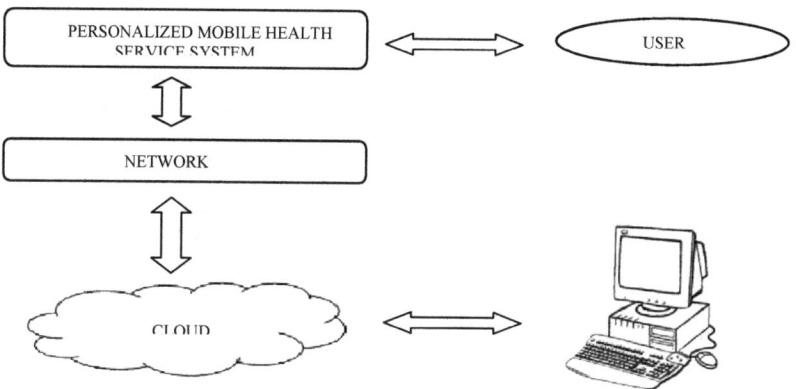

Fig. 5. Usage of cloud environment

The cloud environment is designed in such a way that the user is allowed to enter the preliminary data for further use. This data is stored in the cloud, which is based on the concept – Software as a Service (SaaS). We prefer using PUBLIC CLOUD as it allows flexible resource allocation and also allows further modifications that make the cloud to support mobile applications.

The other major concept used is ontology. An ontological framework is created using Protégé, which helps create the required domains and system-classes. Fig 6 shows a sample of the domains we have used, based on Scenario 3. The fields given as Vegetarian and Non-Vegetarian are system-classes while the various dishes listed below them are the sub-classes.

The ontology basically consists of various classes which are based on the type of food, like North-Indian, South-Indian, Vegetarian, Non-Vegetarian etc. Under each class we have the various dishes given as the sub-classes which are characterized based on its caloric value. Rule based approach is used for further computational purpose. This ontological framework helps in an elegant and simple way of classifying the data and hence processing it based on our conditions for any particular scenario.

Fig. 6. Sample Class Hierarchy

6   Applications

The system has various facilities to help the mobile users in the present scenario. As for the 1st user case, it helps in analyzing if the food consumed by the user is right for their health condition. It suggests the user an alternative food items while the user travels to a new place. It also helps in alerting the user about the diseases prevalent in a location, when the user travels to new location. Along with the alert, it also gives the remedial measures in order to help the user to prevent them from getting affected by the disease.

In our 2nd scenario the system help the user to get immediate attention during traumatic conditions like heart attack. In this case, the signal from the bio-sensor is given to the GSM module which in turn gives an alert to the nearby hospital so that they can attend to the user immediately. The hospitals are classified based on their distance using an ontological framework discussed in our earlier study [8].

As for the 3rd scenario, the system helps in monitoring the daily activities of the user and giving remainders with regard to the user's medication and exercises. It prepares detailed report at the end of every month which helps the user to decide on their future activities. Hence, we added the applications with the hand-held device which can act intelligently to serve the users in a more productive manner.

7   Conclusion

In this paper we have discussed the functional details of the proposed system, based on different scenarios that are most prevalent when we take into account an individual's healthcare. This system is basically an integration of various situations and technologies wherein the user is unable to get immediate treatment due to various reasons, most important of which is the location of the user as well as the location of the physician. This application not only intimates the user, but it also helps in the medication and continuously keeps updating the user with regard to their health conditions. The other benefit of the proposed system is that with the use cloud environment, the network, server and security headaches that exist for locally-installed systems are totally eliminated. We are sure that this system would be a great boon to the mobile user and in turn to the society, when implemented in a real-time.

References

1. Gruber, T.R.: Toward principles for the design of ontologies used for knowledge sharing. International Journal of Human-Computer Studies 43(4-5), 907–928 (1995)
2. Rowley, R.: Is Cloud Computing Right for Health IT?, Healthcare IT News E-Newsletters, Himss (2009)
3. Health Cloud details, https://healthcloud.net/
4. Schlenoff, C., Libes, R.D., Denno, P., Szykman, S.: An Analysis of Existing Ontological Systems for Applications in Manufacturing and Healthcare. Artificial Intelligence for Engineering Design, Analysis, and Manufacturing 14(4), 257–270 (2000)
5. Jones, V., van Halteren, A., Konstantas, R.B.D., Widy, I., Herzog, R.: Mobile Patient Monitoring: The MobiHealth System. The Journal on Information Technology in Healthcare 2(5), 365–373 (2004)
6. Jones, V.M., Bults, R.G.A., Konstantas, D., Vierhout, P.A.M.: Healthcare PANs: personal area networks for trauma care and home care. In: Proceedings Fourth International Symposium on Wireless Personal Multimedia Communications (WPMC), Aalborg, Denmark (2001)
7. Van Halteren, A.T., Bults, R.G.A., Widya, I.A., Jones, V.M., Konstantas, D.: Mobihealth-Wireless body area networks for healthcare. In: Wearable eHealth Systems for Personalised Health Management, Il Ciocco Castelvecchio Pascoli Lucca, Tuscany, pp. 11–14 (2003)
8. Saravanan, M., Prasad, G., Yeshwanth, K.S., Venkatesh, B.: Development of Ontological Framework for Mobile Health Services. In: PAKDD 2010: Workshop on Data Mining for Healthcare Management, DMHM 2010 (2010)

9. Weibenberg, N., Voisard, A., Gartmann, R.: Using ontologies in personalized mobile applications. In: Proceedings of the 12th Annual ACM International Workshop on Geographic Information Systems, November 12-13, Washington DC, USA (2004)
10. Korpipää, P., Mäntyjärvi, J.: An Ontology for Mobile Device Sensor-Based Context Awareness. In: Blackburn, P., Ghidini, C., Turner, R.M., Giunchiglia, F. (eds.) CONTEXT 2003. LNCS, vol. 2680, pp. 451–458. Springer, Heidelberg (2003), doi:10.1007/3-540-44958-2_37
11. Connolly, J.M.: Health IT in the Cloud: A Long Road, Mass High Tech. The Voice of New England Innovation (June 24, 2010)
12. My Pillbox - Medication List, http://www.mypillbox.org/med_list.php
13. Jiang, L.J., Ng, E.Y., Yeo, A.C., Wu, S., Pan, F., Yau, W.Y., Chen, J.H., Yang, Y.: A Perspective on Medical Infrared Imaging. J. Med. Eng. Technol. 29(6), 257–267 (2005)

Robust Learning of Mixture Models and Its Application on Trial Pruning for EEG Signal Analysis

Boyu Wang, Feng Wan, Peng Un Mak, Pui In Mak, and Mang I Vai

Department of Electrical and Electronics Engineering,
Faculty of Science and Technology,
University of Macau

Abstract. This paper presents a novel method based on deterministic annealing to circumvent the problem of the sensitivity to atypical observations associated with the maximum likelihood (ML) estimator via conventional EM algorithm for mixture models. In order to learn the mixture models in a robust way, the parameters of mixture model are estimated by trimmed likelihood estimator (TLE), and the learning process is controlled by temperature based on the principle of maximum entropy. Moreover, we apply the proposed method to the single-trial electroencephalography (EEG) classification task. The motivation of this work is to eliminate the negative effects of artifacts in EEG data, which usually exist in real-life environments, and the experimental results demonstrate that the proposed method can successfully detect the outliers and therefore achieve more reliable result.

Keywords: Deterministic annealing, mixture models, robust learning, trial pruning, EEG signals.

1 Introduction

A brain computer interface (BCI) is a system forming a direct connection between brain and machine, which enables individuals with severe motor disabilities to have effective control over external devices without using the traditional pathways as peripheral muscle or nerves [1-3]. The brain activities are often recorded noninvasively by electroencephalogram (EEG), which has excellent temporal resolution and usability, and the EEG signal is therefore a popular choice for BCI research. In order to control an EEG-based BCI, the user must produce different brain activity patterns, which are recorded by electrodes on the scalp, and then features are extracted from the EEG signals and translated into the control commands. In most existing BCIs, this translation relies on a classification algorithm [4], [5].

Finite mixture models, in particular Gaussian mixture models (GMMs) [6] have been applied to EEG signal analysis in BCI system due to their computational tractability, ease to implement, and capability of representing arbitrarily complex probability density function with high accuracy. In [7] and [8], the mixture of Gaussian was introduced as the online classifier and the parameters were updated in a simulated online scenario. In [9] a GMM-based classifier was used to separate the signal into different classes of mental task, where adaptation is concerned by using a

L. Cao et al. (Eds.): PAKDD 2011 Workshops, LNAI 7104, pp. 408–419, 2012.

supervised method. Similarly, [10] and [11] proposed an online GMM classifier via the decorrelated least mean square (DLMS) algorithm. On the other hand, GMMs can be also applied to model the features extracted from EEG data in which the rest or active state of brain signals are modeled so that the changes in EEG signal can be detected rather than classified [12], [13].

The conventional approach to learning the parameters of mixture models is maximum likelihood (ML) estimator via EM algorithm [14]. However, a well-known problem of the ML estimator via conventional EM algorithm for GMMs is its sensitivity to atypical observations. On the other hand, noise is ubiquitous in EEG signals due to the factors such as measurement inaccuracies, physiological variations in background EEG, muscle and eyes blink artifacts. Therefore, contaminated samples in EEG data should be pruned to achieve a reliable classification result. Unfortunately, none of the GMM-based EEG analysis algorithm, to our best knowledge, considered the negative effects of the outliers in EEG data.

In machine learning community, one approach to detect the outliers is to fit the model based on maximum trimmed likelihood (MTL) to select a subset of the data, on which the mixture models are trained to the majority of the data, whereas the remaining data which do not follow the models are viewed as anomalous data. The ML estimator can be viewed as a special case of MTL estimator. The resulting estimation of the parameters obtained by trimmed likelihood estimator (TLE) is usually more robust, and therefore can be used for outlier detection [17] [18]. One drawback of this approach, however, is that it is a local algorithm, and therefore usually gets trapped in local optima with a poor estimation.

The motivation of this paper is to go a step further along in this research direction, that is, to develop a robust learning algorithm for mixture models. In particular, we propose a deterministic annealing (DA) learning approach for robust fitting of GMMs. The GMMs are learned based on MTL via EM algorithm and the learning process is controlled by annealing temperatures, leading gradual optimization of the objective function, so that the local optima problem of MTL can be avoided. As a result, the outliers can be automatically detected so that the estimation of parameters of GMMs is more robust and reliable.

The reminder of this paper is organized as follows. The robust learning algorithm is developed in Section 2, and experiments on both synthetic and benchmark real data sets are reported in Section 3. In Section 4, we apply our method on EEG signal analysis. The conclusions are provided in Section 5.

2   Deterministic Annealing for Robust Learning

2.1   Mixture Models, Trimmed Likelihood Estimator and FAST-TLE

Given a data set $\mathbf{X} = \{\mathbf{x}_1, ..., \mathbf{x}_N\}$, consisting of N independent identical distributed (i.i.d.) observations of a random d-dimensional variable $\mathbf{x}$. If it follows a K-component finite mixture distribution, its probability density function (pdf) can be given by:

$$p(\mathbf{x}) = \sum_{k=1}^{K} \pi_k p(\mathbf{x} \mid \boldsymbol{\theta}_k), \text{ with } 0 \le \pi_k \le 1 \text{ and } \sum_{k=1}^{K} \pi_k = 1 \tag{1}$$

where π_k is the mixing coefficient, and $\boldsymbol{\theta}_k$ is the parameter set for the kth component.

Define $\boldsymbol{\Theta} \equiv \{\pi_1, ..., \pi_K, \boldsymbol{\theta}_1, ..., \boldsymbol{\theta}_K\}$ as the complete set of the parameters specifying the mixture model. The ML estimate of the optimal set of the parameters is defined as a maximum of the log-likelihood function:

$$\log p(\mathbf{X} \mid \boldsymbol{\Theta}) = \sum_{n=1}^{N} \log p(\mathbf{x}_n \mid \boldsymbol{\Theta}) = \sum_{n=1}^{N} \log \sum_{k=1}^{K} \pi_k p(\mathbf{x}_n \mid \boldsymbol{\theta}_k) \tag{2}$$

It is well known that the ML estimator cannot be obtained in a closed form. Hence, we need to resort the optimization techniques, and one common choice is the EM algorithm, which is an iterative procedure to find the ML estimator of the parameter set of a probability. For more detailed description of the EM algorithm see [6], [14].

To estimate GMMs in a robust way, one approach is to calculate the MTL solution, which is given by

$$\log p_{TL}(\mathbf{X} \mid \boldsymbol{\Theta}) = \sum_{n=1}^{N} \omega_n \log p(\mathbf{x}_n \mid \boldsymbol{\Theta}) = \sum_{n=1}^{N} \omega_n \log \sum_{k=1}^{K} \pi_k p(\mathbf{x}_n \mid \boldsymbol{\theta}_k) \tag{3}$$

where $\omega_n \in \{0,1\} \ \forall \ n = 1, ..., N$, and $\sum_{n=1}^{N} \omega_n = M$ is the indicator describing which of the N observations is viewed as a typical sample, so that $\mathbf{x}_n$ is detected as an outlier when $\omega_n = 0$, and $\omega_n = 1$ for a typical sample, which contributes to the log-likelihood function. M is the trimming parameter indicating the removal of $(N\text{-}M)$ samples which cannot be fitted well by any component of the mixtures, and MTL degenerates to ML when $N=M$.

The FAST-TLE algorithm was proposed in [19] to get an approximation of TLE solution for generalized linear models (GLMs), and was extended for robust fitting of mixture model in [18]. The FAST-TLE algorithm consists of two steps called the trial step and a refinement step. However, this algorithm is a local method, and therefore may get trapped in local optima resulting in a poor estimation.

2.2   Deterministic Annealing Outlier Detection

To avoid the local optima problem with FAST-TLE, we resort the deterministic annealing (DA) approach in which the optimization problem is reformulated as that of seeking the probability distribution that minimizes the application-specific cost subject to a constraint of randomness of the solution. During the annealing process, the algorithm tracks the minimum while the temperature is gradually lowered so that many shallow local optima can be avoid, and finally achieves the hard (nonrandom) solution as the temperature approaches to zero [20], [21].

In the light of TLE and DA algorithm, we consider the following objective function:

$$F(\Theta, \omega) = -\sum_{n=1}^{N} \omega_n \log p(\mathbf{x}_n \mid \Theta) - TH_\omega \tag{4}$$

under the constraints $\sum_{n=1}^{N} \omega_n = M$, $\sum_{k=1}^{K} \pi_k = 1$, and

$$H_\omega = -\sum_{n=1}^{N} \omega_n \log \omega_n \tag{5}$$

where T is the Lagrange multiplier, which is analogous to the temperature in statistics physics, H_ω is the Shannon entropy, which represents a specified level of randomness. At high value of T, the objective function is very smooth and we mainly maximize the entropy, with $\sum_{n=1}^{N} \omega_n = M$, yielding $\omega_n = M/N$, i.e., each samples is equally treated, and the MTL is therefore equivalent to ML. Hence, we have

$$\min_{\Theta, \omega} \lim_{T \to \infty} F(\Theta, \omega) = \max_{\Theta} \log p(\mathbf{X} \mid \Theta) = \max_{\Theta} \sum_{n=1}^{N} \log p(\mathbf{x}_n \mid \Theta) \tag{6}$$

As T is gradually lowered, the influence of log-likelihood function is increasing, which makes the solution of ω_n harder and harder. Finally, as T approaches to zero, the optimization is carried out directly on the trimmed log-likelihood function, forcing ω_n to either zero or one, which yields the MTL

$$\min_{\Theta, \omega} \lim_{T \to 0} F(\Theta, \omega) = \max_{\Theta} \log p_{TL}(\mathbf{X} \mid \Theta)\big|_{\omega_n \in \{0,1\}} = \max_{\Theta} \sum_{n=1}^{N} \omega_n \log p(\mathbf{x}_n \mid \Theta)\big|_{\omega_n \in \{0,1\}} \tag{7}$$

The motivation of the DA based learning procedure is that there is no guarantee that the selected subset in the early stage of learning is near the true one. Therefore, all of the samples should be equally treated at early stage, and the constraint is gradually relaxed during the learning process to increase the effect of the selection of subset, so that the global (at least a better local) optimal solution could be achieved.

For GMMs, to maximize (4), given the fixed $\{\omega_n\}$, we have

$$\pi_k = \frac{1}{M} \sum_{n=1}^{N} \omega_n p(k \mid \mathbf{x}_n), \quad \mu_k = \frac{\sum_{n=1}^{N} \omega_n p(k \mid \mathbf{x}_n) \mathbf{x}_n}{\sum_{n=1}^{N} \omega_n p(k \mid \mathbf{x}_n)} \tag{8}$$

$$\Sigma_k = \frac{\sum_{n=1}^{N} \omega_n p(k \mid \mathbf{x}_n)(\mathbf{x}_n - \mu_k)(\mathbf{x}_n - \mu_k)^T}{\sum_{n=1}^{N} \omega_n p(k \mid \mathbf{x}_n)}$$

where

$$p(k \mid \mathbf{x}_n) = \frac{\pi_k N(\mathbf{x}_n \mid \boldsymbol{\mu}_k, \boldsymbol{\Sigma}_k)}{\sum_{j=1}^{K} \pi_j N(\mathbf{x}_n \mid \boldsymbol{\mu}_j, \boldsymbol{\Sigma}_j)} \tag{9}$$

It can be observed from (8) that the larger the value of ω_n is, the more the corresponding sample contributes to the estimation of the parameters. As T approaches to zero, ω_n skews either to one or zero, indicating whether the sample is viewed as typical or eliminated as an outlier.

For the parameters $\{\omega_n\}$, we minimize the following objective function

$$\omega_n = \arg\min_{\omega_n} \left\{ \begin{array}{l} F_{RDAEM}(\boldsymbol{\Theta}, \boldsymbol{\omega}, \boldsymbol{\upsilon}) \\ \text{s.t. } \sum_{n=1}^{N} \omega_n = M, \ \omega_n \in [0,1] \end{array} \right\} \tag{10}$$

This bound-constrained convex optimization can be solved by cvx, a Matlab package for solving convex program [22].

A description of the proposed algorithm for GMMs is summarized in Fig. 1.

Algorithm: *Deterministic Annealing Based Approach for Outlier Detection*

Input: Data Matrix $\mathbf{X} = \{\mathbf{x}_1, ..., \mathbf{x}_N\}$, scaling factor α, T_{max}, and T_{min}

Output: Optimal mixture model $\boldsymbol{\Theta}$, and outlier indicators $\{\omega_n\}$

Procedure:

    Initialize the parameter set of the mixture models $\boldsymbol{\Theta}$, the indicator $\omega_n = M / N$, and minimum temperature T_{min}. Set $T = T_{max}$, $t = 0$.

    Repeat

        Repeat

            $t = t+1$

            E-Step:

                Calculate $p(k \mid \mathbf{x}_n)$ according (9)

            M-Step:

                Update $\{\pi_k\}, \{\boldsymbol{\mu}_k\}$, and $\{\boldsymbol{\Sigma}_k\}$ according to (8)

                Update $\{\omega_n\}$

        Until a stop criterion is met.

        $T = \alpha T$ ($0 < \alpha < 1$).

    Until $T < T_{min}$,

    Return the model parameter set $\boldsymbol{\Theta}$ and $\{\omega_n\}$

Fig. 1. Deterministic Annealing Based Approach for Outlier Detection

In the proposed algorithm, T_{max} is set to 100, and T_{min} is set in the range of [0.005 0.01]. The choice of scaling factor α involves a tradeoff between execution time and the risk of poorer performance. In practical application, $\alpha \in [0.8\ 0.9]$ can achieve satisfactory results.

The stop criterion can be a maximum number of EM cycles or a convergent indicator. In general, if the algorithm is executed until convergence or the maximum number is set to a large value, the computation time is longer, and the algorithm may get trapped in local minima in the early stage of learning. Therefore, the criterion is set to be execution of 10 to 20 EM cycles at each temperature in our experiment.

In summary, the learning process consists of repeated E-step and modified M-step while gradually lowering the temperature, and its monotonicity in objective function is obvious. When the temperature approaches to zero, the method degenerates to FAST-TLE algorithm, of which the monotonicity has been proved in [19]. The convergence property of deterministic annealing has also been discussed in [21], [23].

3   Experiments

3.1   Synthetic Data Sets

The first example is a synthetic dataset which consists of 100 samples from three Gaussian components with equal mixing coefficients, and the parameter set of each component is given by

$$\mu_1 = [0,\ 3]^T,\ \mu_2 = [3,\ 0]^T,\ \mu_3 = [-3,\ 0]^T$$

$$\Sigma_1 = \begin{bmatrix} 2 & 0.5 \\ 0.5 & 0.5 \end{bmatrix},\ \Sigma_2 = \begin{bmatrix} 1 & 0 \\ 0 & 1 \end{bmatrix},\ \Sigma_3 = \begin{bmatrix} 2 & -0.5 \\ -0.5 & 0.5 \end{bmatrix}$$

In addition, 50 noise points generated from a uniform distribution within [-10, 10] on each dimension are added to the typical samples, which is similar to the data set discussed in [1], [15], [18]. Thus, the total number of samples $N = 150$, and the number of typical samples $M = 100$. The obtained samples, as well as the Gaussian components are shown in Fig. 2(a). The typical samples are marked by magenta dots, whereas the outliers are marked by cyan crosses. On the other hand, the colors of the observations also indicate the values of weights $\omega_n \in [0,1]$ which are represented by cyan when $\omega_n = 0$ and by magenta when $\omega_n = 1$. The colors of samples vary smoothly from cyan to magenta as the values of $\{\omega_n\}$ approach from zero to one.

Fig. 2(b)-(f) demonstrate the learning process of proposed deterministic annealing based outlier detection method (we refer it as "DAOD" here). At the beginning, three components are randomly initialized among the samples, and the values $\{\omega_n\}$ at high temperature are almost same. As T is lowered, the components converge to the true model, and the atypical observations are gradually detected and eliminated (depicted by the smooth varying of the colors from purple to cyan). Fig. 2(e) also shows the result of conventional EM, which is marked by dashed line. It can be observed that it fails to fit the samples due to the existing of outliers. The proposed method is also tested with different levels of trimming (1- M/N = 0.25, 0.35, 0.45), which is presented in Fig. 2(f). We can observe that the proposed method can still identify the clusters with higher or lower trimming level, which indicate that our algorithm is also robust to the trimming percentage. In other words, even when the prior knowledge of noise level is not consistent with the true one, our method can still give reasonable results which fit the samples appropriately.

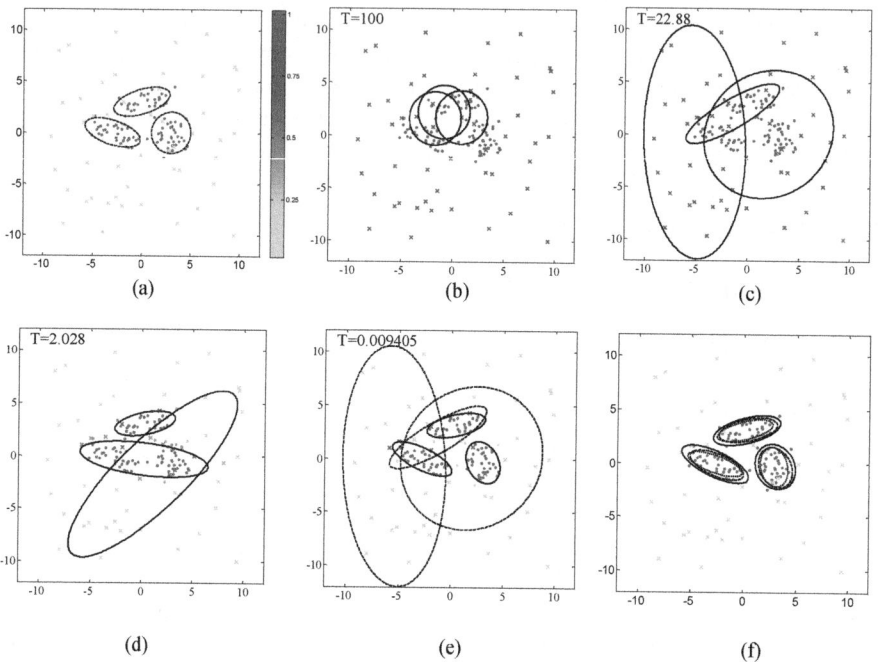

Fig. 2. Fitting Gaussian mixtures with noise: (a) the true model shown by dashed line; (b)-(d) the learning process of DAOD; (e) the results of DAOD (solid line) and the conventional EM algorithm; (f) final estimates with different trimming levels

The second example is a more complicated one since it has more components and a higher degree of overlap than the first one. This data set consists of 1000 samples from a mixture of eight two-dimensional Gaussian components with equal mixing coefficients (see also [29]), to which 250 outliers are added from a uniform distribution within [-3, 3] on each dimension, and the parameter set of each component is given by

$$\mu_1 = [1.5,\ 0]^T \ \mu_2 = [1,\ 1]^T \ \mu_3 = [0,\ 1.5]^T \ \mu_4 = [-1,\ 1]^T$$

$$\mu_5 = [-1.5,\ 0]^T \ \mu_6 = [-1,\ -1]^T \ \mu_7 = [0,\ -1.5]^T \ \mu_8 = [1,\ -1]^T$$

and

$$\Sigma_1 = \Sigma_5 = \begin{bmatrix} 0.01 & 0 \\ 0 & 0.1 \end{bmatrix} \quad \Sigma_3 = \Sigma_7 = \begin{bmatrix} 0.1 & 0 \\ 0 & 0.01 \end{bmatrix} \quad \Sigma_2 = \Sigma_4 = \Sigma_6 = \Sigma_8 = \begin{bmatrix} 0.1 & 0 \\ 0 & 0.1 \end{bmatrix}$$

Fig. 3 illustrates the data set, the results obtained by conventional EM, as well as DAOD with different trimming percentages. Again, it can be observed that the conventional EM cannot fit the typical observations correctly since the outliers are fitted by some components whereas some samples generated by more than one component are fitted by a single Gaussian. On the contrary, the results of our robust algorithm with different trimming percentages in Fig 3(c) indicate that our robust algorithm can locate the components correctly. The change of the trimming level only affects the estimation of covariances.

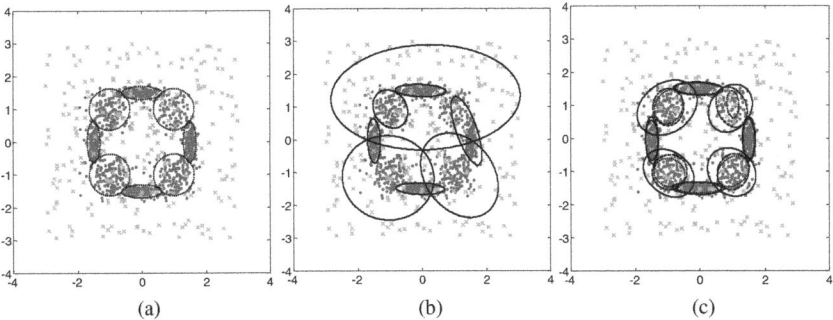

(a)   (b)   (c)

Fig. 3. Fitting an eight-components Gaussian mixture with noise (the typical samples are marked by dots, and the outliers are marked by crosses): (a) the true model presented by dashed line; (b) result of the conventional EM; (c) final estimates of DAOD with different trimming levels

To further evaluate and compare the performances of conventional EM algorithm, FAST-TLE and DAOD, we repeat the second example with different noise levels, and then compare their classification accuracies. To check the dependence of the algorithms on the initial conditions, we repeat the experiments 50 times for each noise level. In unsupervised learning scenario, labels of samples are not needed, and the classification accuracies are evaluated as below. After fitting mixtures, the samples are first partitioned into different clusters according their posterior probabilities to each component. Since the true label of each sample is known in prior, the label of each estimated cluster is assigned as the label that most samples in this cluster have. Then the sample of which the label does not agree with the cluster label is considered as misclassified, and therefore the classification accuracy can be calculated. Fig. 4 demonstrates the classification accuracies of different algorithms. Notice that the all of the algorithms have high accuracies when there is no outlier. The conventional EM algorithm is very sensitive to outliers. Both FAST-TLE and DAOD perform well when the noise level is not high. However, as the number of outliers increases, performance of FAST-TLE degrades significantly. One the other hand, DAOD can mitigate the local optima problem with FAST-TLE and therefore is more robust than the other methods.

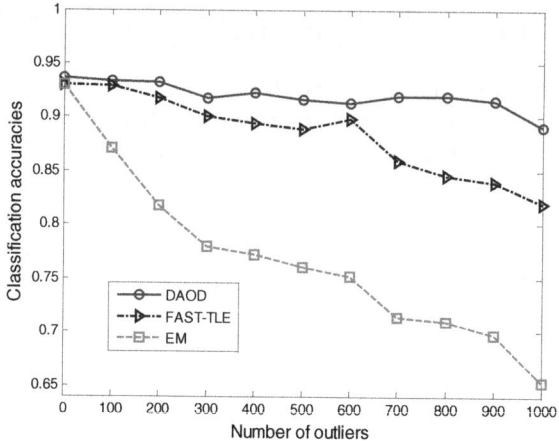

Fig. 4. Classification accuracies of various algorithms as a function number of outliers (the number of typical samples = 1000)

3.2   Real World Data Sets

We now consider the classification task for two real world data sets, i.e., Iris data set and waveform data set, which are all available from the UCI machine learning repository [24]. The Iris data set contains 150 four-dimensional samples from three classes, with each class consists of 50 samples. The waveform data set contains 5000 instances of three classes of waves, and each sample consists of 40 attributes. Therefore, we use three Gaussian components to fit each data set. The label for each component is set to dominant label of the samples, and each sample is classified according to its corresponding posterior probability.

The average classification accuracies and the standard deviations over 50 runs are demonstrated in Table I, from which it can be observed that the classification accuracy can be improved by DAOD with different trimming levels. In summary, the performance of conventional EM algorithm can be improved by gradually pruning off some samples which are located at the boundary of the components, so that the estimates will be more robust and reliable.

Table 1. Classification Accuracy (%) for Two Data Sets with Different Trimming Levels

Data Set	Proposed Robust Approach (with different trimming levels)			Conventional EM
	0.03	0.05	0.1	
Iris	94.03±4.92	95.52±2.74	94.10±4.38	93.21±4.66
Waveform	81.66±0.49	81.47±0.46	80.74±0.50	79.46±0.48

4   EEG Data Set

Finally, the proposed approach is evaluated on a more realistic application – the classification task of EEG signals. We applied our algorithm to the data set IIa from BCI competition IV [25], which consists of EEG data sets from 9 subjects. For each subject, two sessions were recorded, each of which consists of 288 trials with duration of 7s. In addition, the data set for each subject also contains some rejected trials, which are contaminated by noise or artifacts. For detailed description of this data set, see [25]. Before feature extraction, the EEG signals are filtered by 8-30Hz band pass filter. Then we applied common spatial pattern (CSP), a discriminative approach decomposing the signals into spatial patterns, to extract the features from multichannel EEG signals, and three Gaussian components are used for each class. Since the CSP is a data-driven feature extraction approach, after the elimination of noise samples, we re-train the CSP and GMM classifiers.

Fig.5 illustrates the average classification accuracies of the EEG signals of nine subjects obtained by conventional EM algorithm and the proposed robust approach with different trimming levels. The classification accuracy of the EM algorithm is not improved significantly for the original EEG data (solid line). However, when the signals are contaminated by the noise samples (rejected trials), the performance of conventional EM algorithm deteriorates obviously, whereas our proposed approach can detect and eliminate the outliers, so that more robust and reliable results (dashed line) can be obtained. We further investigate the performances of DAOD for each subject. On a whole, seven out of nine subjects benefit from DAOD. In addition, for the subjects with high classification accuracies (>80%, with fewer noise samples), the improvements are not remarkable (<2%); for the subjects with lower accuracies (<80%, with more noise samples), however, the classification accuracies of four out five subjects are improved significantly (>5%). Therefore, the proposed algorithm can successfully reduce the negative effects of EEG signals contaminated by artifacts and noise.

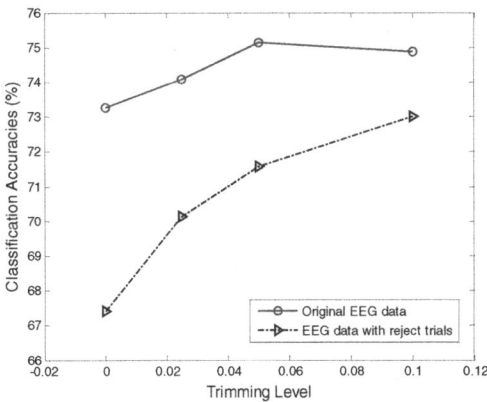

Fig. 5. The Comparison of the classification accuracies for the EEG data sets with and without rejected trials

5   Conclusion

In this paper, we proposed a DA based EM algorithm to detect outliers for mixture models. The experiments demonstrate that the performances of conventional learning approaches are significantly deteriorated due to the outliers while our method can successfully alleviate the negative effects of outliers. In addition, the proposed method can automatically prune off the EEG signals contaminated by artifacts and noise without any additional channel rejection operation (i.e., independent component analysis) or visual inspection of an expert. Since the noise is ubiquitous in EEG signals, it is necessary to prune off a small account of samples to achieve reliable result even though the noise level is unknown.

The future work will focus on the reduction on the dependence on the prior knowledge of the trimming level. It should be noted that although our method is applied to Gaussian mixtures, it can be extended to non-Gaussian cases, which will be also considered in the future work.

Acknowledgment. The authors gratefully acknowledge the support from the Macau Science and Technology Department Fund (Grant FDCT/036/2009/A) and the University of Macau Research Fund (Grants RG059/08-09S/FW/FST, RG080/09-10S/WF/FST and MYRG139 (Y1-L2)-FST11-WF).

References

1. Wolpaw, J.R., et al.: Brain–computer interface technology: a review of the first international meeting. IEEE Transactions on Rehabilitation Engineering 8(2), 164–173 (2000)
2. Wolpaw, J.R., Birbaumer, N., McFarland, D.J., Pfurtscheller, G., Vaughan, T.M.: Brain-computer interface for communication and control. Clinical Neurophysiology 133(6), 767–791 (2002)
3. Bashashati, A., Fatourechi, M., Ward, R.K., Birch, G.E.: A survey of signal processing algorithms in brain-computer interfaces based on electrical brain signals. Journal of Neural Engineering 4(2), R32–R57 (2007)
4. McFarland, D.J., Anderson, C.W., Müller, K.-R., Schlogl, A., Krusienski, D.J.: BCI meeting 2005-workshop on BCI signal processing: feature extraction and translation. IEEE Transactions on Neural Systems and Rehabilitation Engineering 14(2), 135–138 (2006)
5. Lotte, F., Congedo, M., Lecuyer, A., Lamarche, F., Arnaldi, B.: A review of classification algorithms for EEG-based brain-computer interfaces. Journal of Neural Engineering 4(2), R1–R13 (2007)
6. McLachlan, G., Peel, D.: Finite Mixture Models. Wiley, New York (2000)
7. Millán, J.R., Renkens, F., Mouriño, J., Gerstner, W.: Brain-actuated interaction. Artificial Intelligence 159, 241–259 (2004)
8. Millán, J.R.: On the need for on-line learning in brain–computer interfaces. In: Proceedings of International Joint Conference on Neural Networks, Budapest, Hungary, pp. 2877–2882 (2004)
9. Buttfield, A., Millán, J.R.: Online classifier adaptation in brain-computer interfaces. Techical Report, IDIAP–RR 06-16 (2006)

10. Sun, S., Zhang, C., Lu, N.: On the On-line Learning Algorithms for EEG Signal Classification in Brain Computer Interfaces. In: Wang, L., Jin, Y. (eds.) FSKD 2005. LNCS (LNAI), vol. 3614, pp. 638–647. Springer, Heidelberg (2005)

11. Sun, S., Zhang, C.: Learning On-line Classification via Decorrelated LMS Algorithm: Application to Brain–Computer Interfaces. In: Hoffmann, A., Motoda, H., Scheffer, T. (eds.) DS 2005. LNCS (LNAI), vol. 3735, pp. 215–226. Springer, Heidelberg (2005)

12. Schalk, G., Brunner, P., Gerhardt, L.A., Bischof, H., Wolpaw, J.R.: Brain-computer interfaces (BCIs): Detection instead of classification. Neuroscience Methods 167(1), 51–62 (2008)

13. Fazli, S., Danóczy, M., Popescu, F., Blankertz, B., Müller, K.-R.: Using Rest Class and Control Paradigms for Brain Computer Interfacing. In: Cabestany, J., Sandoval, F., Prieto, A., Corchado, J.M. (eds.) IWANN 2009. LNCS, vol. 5517, pp. 651–665. Springer, Heidelberg (2009)

14. McLachlan, G., Krishnan, T.: The EM Algorithm and Extensions. Wiley, New York (1997)

15. Nguyen, D.T., Chen, L., Chan, C.K.: An outlier-aware data clustering algorithm in mixture model. In: Proceedings of 7th IEEE International Conference on Information, Communication and Signal Processing, Macau, China, pp. 1–5 (2009)

16. Blankertz, B., Tomioka, R., Lemm, S., Kawanabe, M., Müller, K.-R.: Optimizing spatial filters for robust EEG single-trial analysis. IEEE Signal Processing Magazine 25(1), 41–56 (2008)

17. Hadi, A.S., Luceño, A.: Maximum trimmed likelihood estimators: a unified approach, examples, and algorithms. Computational Statistics & Data Analysis 25(3), 251–272 (1997)

18. Neykov, N., Filzmoser, P., Dimova, R., Neytchev, P.: Robust fitting of mixtures using trimmed likelihood estimator. Computational Statistics & Data Analysis 52(1), 299–308 (2007)

19. Neykov, N., Müller, C.: Breakdown point and computation of trimmed likelihood estimators in generalized linear models. In: Developments in Robust Statistics, pp. 277–286. Physica-Verlag, Heidelberg (2003)

20. Rose, K., Gurewitz, E., Fox, G.C.: Statistical mechanics and phase transitions in clustering. Physical Review Letters 65(8), 945–948 (1990)

21. Rose, K.: Deterministic annealing for clustering, compression, classification, regression, and related optimization problems. Proceedings of the IEEE 86(11), 2210–2239 (1998)

22. Grant, M., Boyd, S.: CVX: Matlab software for disciplined convex programming, version 1.21, http://cvxr.com/cvx

23. Rose, K., Gurewitz, E., Fox, G.C.: Constrained clustering as an optimization method. IEEE Transactions on Pattern Analysis and Machine Intelligence 15(8), 785–794 (1993)

24. Machine Learning Repository website, http://archive.ics.uci.edu/ml/index.html

25. BCI competition IV website, http://bbci.de/competition/iv/

26. Ueda, N., Nakano, R.: Deterministic annealing EM algorithm. Neural Networks 11(2), 271–282 (1998)

27. Zhao, Q., Miller, D.J.: A deterministic, annealing-based approach for learning and model selection in finite mixture models. In: Proceedings of 29th IEEE International Conference on Acoustics, Speech, and Signal Processing, Montreal, Canada, pp. V-457–V-460 (2004)

28. Figueiredo, M.A.T., Jain, A.K.: Unsupervised learning of finite mixture models. IEEE Transactions on Pattern Analysis and Machine Intelligence 24(3), 381–396 (2002)

29. Zhang, B., Zhang, C., Yi, X.: Competitive EM algorithm for finite mixture models. Pattern Recognition 37(1), 131–144 (2004)

An Integrated Approach to Multi-criteria-Based Health Care Facility Location Planning

Wei Gu, Baijie Wang, and Xin Wang

Department of Geomatics Engineering, University of Calgary,
2500 University Drive, Calgary, AB, Canada
{wgu,baiwang,xcwang}@ucalgary.ca

Abstract. Optimal location of health care facilities is critical to the success of health care services. Given its importance, this is an active research topic in health informatics, operational research and GIS. This paper presents an integrated approach to health care facility planning whereby the methods from three research topics are combined. The integrated approach is applied in order to solve preventive health care facility location planning problems. In this approach, a new health accessibility estimation method is developed in order to capture the current characteristics of preventive health care services. Based on this, the preventive health care facility location planning problem is formalized as a multi-criteria facility location model. A new algorithm is proposed in order to solve the model. Experiments on synthetic datasets and on the Alberta breast cancer screening program data are conducted and the results support our analysis.

Keywords: GIS, Health Accessibility Estimation, Multi-criteria Health Care Facility Location Planning, Heuristic Algorithm.

1 Introduction

The location of health care facilities is critical to the success of health care services. Consequently this has been an active research topic in health informatics, operational research and Geographical Information Systems (GIS) [1, 2, 3].

In this paper, we categorize the current research of health care facility locations into three main groups: (1) accessibility estimation, (2) facility location planning, and (3) locational evaluation. *Accessibility* in health care is "a concept representing the degree of 'fit' between the clients and the service supply" [4]. *Accessibility estimation* research aims to find the best method for estimating accessibility based on current characteristics of health care services. The purpose of *facility location planning* is to find those locations for health facilities that optimize social welfare. The planning procedure includes identifying location criteria, formalizing the facility location model [2] and implementing the spatial analysis algorithm for solving the model [5]. The *locational evaluation* is to measure the efficiency of the health care service's configurations in large time series [6]. By comparing the configurations of the health care service in different years, the locational evaluation aims to connect the locational pattern with the health outcome.

L. Cao et al. (Eds.): PAKDD 2011 Workshops, LNAI 7104, pp. 420–430, 2012.

A large number of facility location models and spatial analysis algorithms have been proposed for improving the accuracy of the planning procedure [2]. However, none of the existing research discusses the potential benefits of incorporating the accessibility estimation research and the locational evaluation research into the planning procedure. In this paper, we believe that the integration of these three research areas can improve the accuracy of facility location planning for three reasons. First, most current facility location models formalize the facility location problem in terms of distance, which can be seen as a basic accessibility measurement. A number of advanced accessibility estimation methods [1, 8] have been developed in accessibility estimation research in order to capture the characteristics of the specific health service. These methods more precisely reflect the real world demand for health care than does the basic accessibility measurement. Second, locational evaluation uses several measures, such as facility characterization and proximity analysis for comparing the distribution of facilities. By using the locational evaluation method to compare the hypothetical optimal planning result with the existing spatial arrangement of that health service, the advantages and disadvantages of the hypothetical optimal result could be ascertained clearly and the facility location model could be modified accordingly.

Integrating the methods from the three research approaches to facility location planning is challenging for the following reasons. First, the assumptions behind each of the methods need to be clear and conflict-free. A method that is good when used on its own may not necessarily be good when used in conjunction with others, since its assumptions may be in conflict with those of the other methods. For example, the regional availability method [13] is good for estimating the accessibility of preventive health care services. However, it cannot be combined with the preventive health care facility model [11] because the regional availability method assumes that people may go to any facility within an acceptable travelling distance, whereas the model assumes that people choose the closest one. Second, some accessibility estimation methods cannot be applied directly to the facility location model, which aims to optimize the global value for the whole area. For example, a ratio between patients and health facilities within a region is used to define accessibility. As such, it is useful to identify those areas where service shortage, in other words the ratio, is high. However, this method cannot be used for facility location planning because a change in facility locations would not influence the total ratio between patients and health facilities. Third, spatial analysis algorithms should be modified in order to accelerate the process of solving the compound facility location model based on an advanced accessibility estimation method. By analyzing the structure of data produced by the advanced accessibility estimation method, the modified algorithm can reduce the high computing complexity that is brought about by replacing the basic accessibility estimation method with the advanced one.

The main contributions of this paper are summarized: First, the paper gives an integrated approach to summarizing the relations among the three research topic methods. The most used methods in each research topic and the assumptions behind these are also discussed. Second, as an example, this paper applies the integrated approach to solving the challenge of preventive health care facility location planning. In this approach, we formalize the preventive health care facility location planning problem as a multi-criteria facility location model. By adding the distance factor to the regional availability accessibility estimation method [13], a new health accessibility estimation method is developed and added into the model. Third, a new algorithm, named AFI is proposed in order to solve the model in the approach efficiently. Finally, the approach is evaluated using a real application: optimizing the configuration of breast cancer screening clinics in Alberta, Canada. The locational evaluation performed among the existing configuration and optimal solutions proves the efficiency of the approach.

The remainder of the paper is organized as follows. Section 2 proposes an integrated approach to health care facility location planning. Section 3 describes the implementation of the approach for preventive health care facility location planning. Section 4 evaluates the algorithm using synthetic datasets and describes the procedure of using the approach for the Alberta breast cancer screening program. Finally, Section 5 discusses major conclusions of the paper.

2   An Approach to Planning Health Care Facility Locations

An integrated approach is proposed for health care facility location planning (as shown in Fig. 1), which includes three main components: *health accessibility estimation*, *facility location planning* and *locational evaluation*. In the rest of the section, we elaborate on the associations between the problem and components and on the dependence among different components, emphasizing the assumption behind each component, and introducing the best methods to be used.

(1) Health Accessibility Estimation Method. The health accessibility estimation describes the level of ease with which patients can get health care service. As shown in Fig. 1, it is implemented according to the characteristics of the health facility location problem and needs to be considered while implementing the facility location planning method. Two types of GIS-based accessibility measures are introduced below.

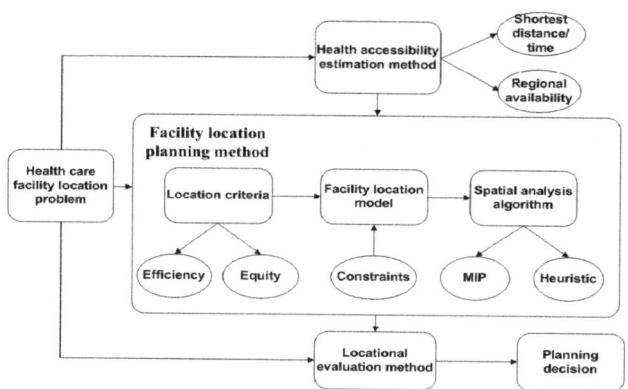

Fig. 1. The structure of the integrated approach to health care facility location planning

The most common way to measure health accessibility is the *shortest distance/time* method [9]. In this method, accessibility is defined as the distance from the patient to the closest health facility. The distance can be Euclidean distance, travel distance or travel time. The method is popular because it is easy to implement and satisfies the well-known fact that distance affects access to health care services [10]. The method is appropriate for describing the service, which is usually offered to the patient by the closest facility, such as emergency medical services [2].

However, the shortest distance/time method cannot be used to measure all kinds of the health services. For example, it is not a good measurement for the accessibility of preventive health care services, such as screening services and flu shot services. Since preventive health care services are given to people with no clear symptoms of illness, people who seek preventive health care services are more flexible and don't necessarily always have to go to the closest facility [11, 12]. The *Regional availability* method [13] is another popular method for measuring the accessibility of health care services. It is a good supplement to the shortest distance/time method. The method generally assumes that given a specific range for a service being offered at a facility, every resident within the range is a potential patient of the service. The regional availability method defines the accessibility of health care services as the ratio of the number of people living in the region to the number of health care facilities. People living in a lower ratio region have more convenient to access to the service.

(2) Facility Location Planning Method. As shown in Fig. 1, the facility location planning method includes three parts: *location criteria*, *facility location model* and *spatial analysis algorithm*.

Location criteria are the objectives of facility location planning as extracted from the health care facility location problem. Most health care services should be efficient and equitable [14] and the planning of health care facility locations is ideally based on multiple-criteria. The *efficiency* criterion aims to optimize the accessibility values of the target region by optimally arranging the health care facilities. The *equity* criterion dictates that accessibility values should not be too unevenly distributed.

The facility location model aims to formalize the health care facility location problem into objective equations, which depend on the location criteria, the characteristics of the problem, and the health accessibility estimation method. First, the facility location model must be constructed in order to balance multiple criteria. Mehretu et al. [15] propose a multi-criteria model in terms of efficiency and equity to locate rural health clinics in the Eastern Region of the Upper Volta. They implement the equity criterion as a constraint on the efficiency criterion and formalize the problem to maximize efficiency. Recently, Mitropoulos et al. [14] proposed a biobjective mathematical programming model for simultaneously achieving the criteria of efficiency and equity. In the model, the two criteria are transferred to one criterion by defining a coefficient α. The value of the coefficient is determined by the importance of each criterion according to the need of real applications. The biobjective model can be simplified as Eq. (1).

$$\text{Optimize (efficiency} + \alpha \times \text{equity)} \tag{1}$$

Second, the facility location model should also be built based on the characteristics of the health care facility location problem, which can be extracted as specific constraints and added into the objective equations. For example, the authors in [12] build the capacitated model to optimize the locations of preventive health care facilities. They assume each facility can only serve a maximum number of patients. When doing capacitated constraint assignment, the model assigns the closer patient to the facility until the facility's capability is reached. (The implied assumption behind the assignment is that people would go to the closest facility.)

Third, the facility location model is also influenced by the health accessibility estimation method. In order to let the health accessibility estimation method fit for the model, two conditions should be satisfied: (1) The assumptions behind the health accessibility estimation method cannot conflict with that of the model. For example, although the regional availability method [13] and the capacitated model [12] perform well when used separately in a preventive health care facility location problem, their conflicting assumptions make it impossible for them to be used in conjunction with one another. (2) After adopting the new accessibility estimation method, the facility location model can still reflect the change of facility locations. We will discuss a new estimation method in the next section.

The Spatial analysis algorithms aim to provide optimal facility locations by solving the equations formalized by the facility location model. Because the facility location problem is NP-complete, attempting a solution consumes a large amount of computational resources. There are two types of algorithms for solving this: The Mixed-Integer Programming algorithm (MIP) [11] and the heuristic algorithm [5]. As shown in Fig. 1, the selection of the spatial analysis algorithm depends on the number of constraints in the model. The MIP algorithm is recommended for solving problems with a small number of constraints in the model and with small sized datasets, while heuristic algorithms are used under the opposite conditions. In addition, by analyzing the structure of data produced by the health accessibility estimation method, the new spatial analysis algorithm could be developed in order to reduce execution time and improve the accuracy.

(3) Locational Evaluation. By comparing the existing configuration of health care facilities with the optimal solution given by the facility location planning method, the locational evaluation method can be used to forecast the service pattern of the facility location in the optimal solution based on the existing service pattern. Next, the forecasting results can be used to modify the optimal solution. Take one of the most used locational evaluation methods, proximity analysis [3], as an example. By using proximity analysis, the features of existing facility locations can characterized, such as proximity of a facility to a highway. Such location features can be used to guide planning.

3   Preventive Health Care Facility Location Planning Approach

The characteristics of preventive health care services are inherently different from other health care services (such as health care for acute diseases), which requires a different location decision approach. The first characteristic is that patients might not seek services from the closest preventive health care facility since preventive services are given to people with no clear symptom of illness [12]. The second characteristic is that each preventive health care facility needs to have a minimum number of patients in order to retain accreditation. In this section, we describe the implementation of the integrated approach for preventive health care facility location planning. A new accessibility estimation method for preventive health care is implemented and the preventive health care location criteria are summarized in 3.1. In 3.2, a multi-criteria preventive health care facility location model is formalized. Finally, the algorithms for solving the model are proposed in 3.3. In this paper, the locational methods only compare the existing facility locations with the optimal solutions, while the comparison results are not used to guide the planning.

3.1   Accessibility Estimation and Location Criteria

In order to capture the characteristics of preventive health care services, we define the accessibility of preventive health care service by combining the regional availability and the distance factor. This can be calculated using the following two steps:

Step 1. For each candidate site of preventive facilities j, search all population centers i that are within a travelling distance threshold d_0 from location j (that is, catchment area of j), and compute the facility-to-patient ratio R_j, within the catchment area:

$$R_j = 1 \bigg/ \sum_{i \in I \cap d_{ij} \leq d_0} P_i \tag{2}$$

Where P_i is the number of patients in population center i whose centroid falls within the catchment area $d_{ij} \leq d_0$. d_{ij} is the travelling distance between population center i and candidate site j.

Step 2. Calculate the accessibility value A_i for each population center i by searching all facilities whose locations that are within the travelling distance threshold d_0 from i and summing up the inverse distance weighted facility-to-patient ratio R_i.

$$A_i = \sum_{j \in J \cap d_{ij} \leq d_0} \frac{R_j}{d_{ij}} \times y_j = \sum_{j \in J \cap d_{ij} \leq d_0} \frac{y_j}{d_{ij} \times \sum_{i \in I \cap d_{ij} \leq d_0} P_i} \tag{3}$$

Where $y_j = 1$ if a facility opens at the candidate site j; Otherwise, $y_j = 0$.

The reasons that we combine the regional availability and the distance factor is that: First, the regional availability considers all the facilities within an acceptable travelling distance of a patient when calculating the accessibility of preventive health care service to that patient. Thus, the assumption behind the regional availability is that people may go to any facility within an acceptable travelling distance, which satisfies the first characteristic of preventive health care services. A_i can be seen as the value of the regional availability if d_{ij} is removed from Eq. (3). Second, the regional availability can not be used in the location decision directly. Changing facility locations would only result in the change of the ratio between facilities and patients for each facility, while the total ratio between facilities and patients would not change. That is because the regional availability considers facilities to have the same attraction to the patients within their catchment areas regardless of the actual travelling distance. Thus, by adding the distance factor into the regional availability, the sum of A_i varies under the change of the facility locations. Then the facility location model can be built based on the sum of A_i.

Since we are dealing with public services, the preventive health care facility location criteria should take into account efficiency and equity. According to the definition above, patients in a population center i can access the service as long as the value A_i is not zero and a larger value of A_i indicates a better accessibility at a population center i. Thus, the efficiency criterion is achieved by maximizing population weighted accessibility values. The efficiency of the service is shown as Eq. (4) and the equity of the service is shown as Eq. (5).

$$efficiency = \sum_{i \in I} A_i \times P_i \left/ \sum_{i \in I} P_i \right. \tag{4}$$

$$equity = \sum_{i \in I \cap A_i \neq 0} P_i \left/ \sum_{i \in I} P_i \right. \tag{5}$$

3.2   A Multi-Criteria Preventive Health Care Facility Location Model

Given a set of population centers ($i = 1,...,|I|$) and a set of candidate sites for facilities ($j = 1,...,|J|$), the purpose of the multi-criteria preventive health care facility location model is to identify optimal locations for the number of preventive health care facilities n that maximize the value of the objective equation below:

Where n is the predefined number of preventive health care facilities; W_{min} is the minimum required workload of a facility. Constraint (a) requires the number of facilities be equal to the predefined number of preventive health care facilities. Constraint (b) ensures that the population covered by each facility is beyond the minimum workload. The model we have defined here is based on the accessibility of the preventive health care service...which is the preventive health care facility location criterion. It uses the same strategy as [14], where α is defined as a co-efficient for incorporating the criteria of efficiency and equity.

$$Max\ (\sum_{i \in I} A_i + \alpha \frac{\sum_{i \in I \cap A_i \neq 0} P_i}{\sum_i P_i}) = Max\ (\sum_{i \in I} \sum_{j \in J \cap d_{ij} \leq d_0} \frac{R_j}{d_{ij}} \times y_j + \alpha \frac{\sum_{i \in I \cap A_i \neq 0} P_i}{\sum_i P_i}) \tag{6}$$

Subject to: $\sum_{j \in J} y_j = n$ **(a)**   and $\frac{1}{R_j} \geq W_{min} y_j$ **(b)**

3.3   Algorithm

Interchange algorithm [7] is a basic heuristic algorithm for solving the facility location model, which can produce acceptable solutions on large datasets and can be implemented easily. This subsection proposes an algorithm, based on the interchange algorithm, for solving the multi-criteria preventive health care facility location model efficiently and accurately.

The flow of the Interchange algorithm is shown in Fig. 2 (a). It first randomly selects the initial locations for predefined numbers of facilities from all candidate sites. All selected candidate sites are kept in a pool and unselected ones are kept in another pool. Then the algorithm looks for a pair of facilities (one from the selected pool, another from the unselected pool) that would lead to a larger value of the objective equation if swapped. The algorithm terminates if no swap can produce a larger value of the objective equation.

There is a large number of research devoted to accelerate Interchange algorithm. However, none of them can be used in this approach directly because they reduce the execution time by exploiting the structure of distance data. We implement a new heuristic algorithm called Accessibility Fast Interchange (AFI) algorithm which is based on the structure of data produced by the health accessibility estimation method. The AFI algorithm accelerates Interchange algorithm by calculating distribution feature and reuse the distribution feature whenever calculating the objective equation. The Distribution feature is a data structure that aggregates population centers into different groups based on candidate sites' catchment areas, where each group is covered by the same group of candidate sites. Fig. 3 gives an example to illustrate the distribution feature. In this case, the distribution feature has six groups of population centers, which are {1}, {2, 3}, {4}, {5}, {6} and {7} (i.e., 1 is only covered by the catchment area of a; 2 and 3 are covered by the catchment areas of a and b). Distribution feature speeds up calculating the value of the objective equation by avoiding scanning all of the population centers. The population centers in the same group will have the same change of accessibility values whenever they do a swap. Thus, when calculating the value of the objective equation, the AFI algorithm can update the accessibility value of population centers, which are aggregated together at the same time. As shown in Fig. 2 (b), the AFI algorithm starts by building the distribution feature. Then the AFI algorithm uses the same strategy as the Interchange algorithm to find potential swaps but uses the distribution feature before calculating the objective equation. The AFI algorithm is faster than the Interchange algorithm and produces the same output as that of the Interchange algorithm.

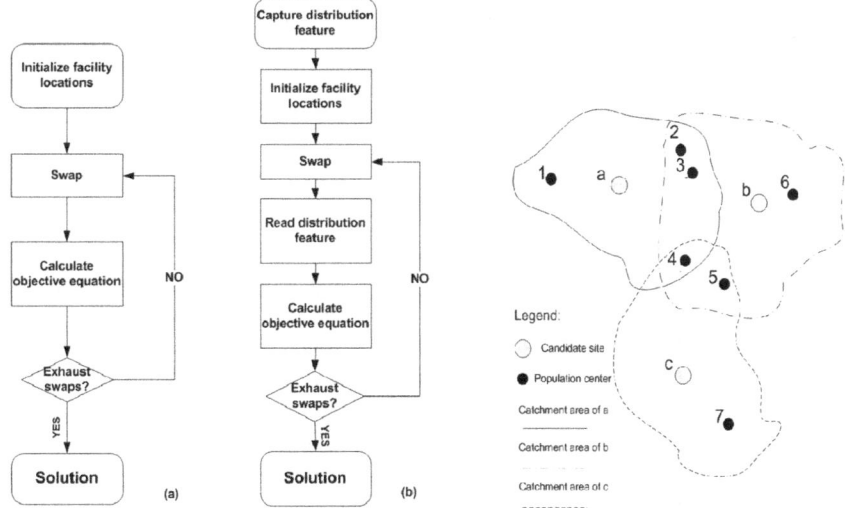

Fig. 2. Flow charts to: (a) the Interchange and (b) AFI algorithms

Fig. 3. An example of a distributiofeature

4   Experiments

4.1   Computational Experiments on Synthetic Datasets

In this subsection, we compare the performance of the Interchange and the AFI algorithms in terms of efficiency, equity and execution time. Efficiency is measured by Eq. (4) and equity is measured by Eq. (5) in section 3.1. Synthetic datasets for population centres and candidate sites of facilities were created in a 300×300 area. The number of candidate sites is set to 20, 40, 60, 80 or 100 while the number of population centres is fixed at 10,000. We use Euclidean distance between two locations to set travelling distance threshold d_0. The number of patients in each population centre ranges from 10 to 100. The algorithms were implemented in Java and experiments were performed on a Core 2 Duo 2.40GHz PC with 3GB memory, running on Windows XP platform.

The results of using two algorithms are summarized in Tables 1 (*Ef* denotes *Efficiency*). When the number of candidate sites increases from 20 to 100, the execution time of the Interchange algorithm increases from 4.39 seconds to 35.08 seconds. The AFI algorithm has the same efficiency and equity as the Interchange algorithm but reduces the execution time, which ranges from 0.14 seconds to 0.21 seconds.

Table 1. The performance of the two algorithms ($d_0 = 30$, $n = 10$, $\alpha = 30$, $W_{min} = 1000$)

No. of candidate sites	Interchange			AFI		
	Ef (10^{-2})	Equity (%)	Time (s)	Ef (10^{-2})	Equity (%)	Time (s)
20	1.00	96.8	4.39	1.00	96.8	0.14
40	1.73	98.9	16.83	1.73	98.9	0.18
60	1.74	99.7	27.89	1.74	99.7	0.19
80	1.96	100	33.06	1.96	100	0.19
100	2.86	100	35.08	2.86	100	0.21

4.2   A Real Application

In this subsection, we apply the preventive health care facility location planning approach to the breast cancer screening program in Alberta, Canada.

4.2.1   Problem Statement and Data Issues

The Alberta Breast Cancer Screening Program (ABCSP) started regular screening mammograms for women in this age range. The research considers the patients for service, as measured by population in target groups in various locations. The population data is essentially the census data in the Dissemination area (DA) level, which were derived from 2006 Statistics Canada. There are 327830 women within the target age in Alberta. In order to calculate the distance between the DAs and facilities, we use the Postal Code Conversion File to estimate the location of the DAs. A total of 5180 DAs with location and population values were used in the research. In addition, data from the 53 existing Alberta screening clinics were extracted from the ABCSP. The other 92 candidate screening clinics data in Alberta ware extracted from Alberta Health Services. The locations of clinics are geocoded to point locations using

ArcGIS address matching techniques. Travel distance and time are calculated from real world road networks by calling these functions in the Google Maps API (http://code.google.com/apis/maps/).

4.2.2   Optimal Solution and Locational Evaluation

The AFI algorithm is used to optimize the locations of screening clinics. Since the number of current screening clinics is 53 in Alberta, the predefined number of preventive health care facilities n is set to 53. The threshold travelling distance d_0 of each facility's service radius is defined as thirty minutes driving time distance (American standard [8]). Minimum required workload at each clinic W_{min} is set to 4000 according to the policy decision made by the Ministry of Health [11]. Table 2 compares the solutions given by the AFI algorithm with the current situation on efficiency and equity. The result shows that the AFI algorithm increases efficiency from 0.35 to 0.40 and improves equity from 78.42% to 81.86%. Comparing the number of people in different accessibility value segments, the AFI algorithm achieve better results than that of the current situation in that it increases the number of people on every non zero segment and reduces the number of people not covered by any facility (the zero segment).

Table 2. Comparison between the current situation and the optimal solution

	Efficiency	Equity (%)	Accessibility value segment			
			0	(0,0.5)	[0.5,1)	[1,max]
Current	0.35	78.42	70745	233700	7855	15530
AFI	0.40	81.86	59460	233720	14880	19770

Compared with traditional methods for preventive health care facility location planning [11, 12], the integrated approach has the following advantages: first the new accessibility estimation method precisely reflects the real world demand for preventive health care because it considers two unique characteristics of preventive health care services; Second, based on the multi-criteria facility location model, the planning procedure considers the efficiency and equity of the service simultaneously. Finally, we do the locational evaluations of the optimal solutions given by the AFI algorithm using different parameters, and by the current solution. The conclusions are: (1) for the multi-criteria model it is impossible to improve any criterion without deteriorating the other. (2) Both efficiency and equity are improved with increases in the number of open facilities, and vice versa. We also find that the AFI algorithm can produce a better solution with even 50 facilities, rather than with the current solution of 53 open facilities.

5   Conclusions

This paper presents an integrated approach to health care facility location planning. To achieve the best planning result, the integrated approach incorporates the methods from three health care facility location research areas and indentifies the relations among the methods to be combined. The advantages of the integrated approach are proven by implementing it for use in preventive health care facility location planning.

In the approach, we modify the regional availability method by adding the distance factor, which captures the characteristics of preventive health care services and is also suitable for the rest of the planning procedure. Based on this, we formalize the preventive health care facility location planning problem as a multi-criteria facility location model. The AFI algorithm is proposed to solve the model. Compared with the Interchange algorithm, the AFI algorithm can produces the same results but dramatically reduces the execution time. Experiments on synthetic datasets and on a real application for the Alberta breast cancer screening program support our analyses of the algorithms and show that our work can improve the performance of the existing preventive health care program. Future efforts will be made on comparing the newly defined accessibility measurement with traditional measurements (e.g., the shortest distance method) by an individual level health survey in the Alberta, Canada.

References

1. Wang, F., Luo, W.: Assessing spatial and nonspatial factors for healthcare access: Towards an integrated approach to defining health professional shortage areas. Health and Place 11, 131–146 (2005)
2. Daskin, M.S., Dean, L.K.: Location of Health Care Facilities. In: Sainfort, F., Brandeau, M., Pierskalla, W. (eds.) Handbook of OR/MS in Health Care: A Handbook of Methods and Applications, pp. 43–76. Kluwer (2004)
3. Birkin, M., Clarke, G., Clarke, M., Wilson, A.: Intelligent GIS: Location decisions and strategic planning. Wiley, New York (1996)
4. Penchansky, R., Thomas, J.W.: The concept of access. Medical Care 19, 127–140 (1981)
5. Pacheco, J., Casado, S., Alegre, J.F.: Heuristic Solutions for Locating Health Resources. IEEE Intelligent Systems 23, 57–63 (2008)
6. Kumar, N.: Changing geographic access to and locational efficiency of health services in two Indian districts between 1981 and 1996. Social Science & Medicine 10, 2045–2067 (2004)
7. Teitz, M.B., Bart, P.: Heuristic methods for estimating the generalized vertex median of a weighted graph. Operations Research 16, 955–961 (1968)
8. Wang, F., McLafferty, S., Escamilla, V., Luo, L.: Late-stage breast cancer diagnosis and health care access in Illinois. Prof. Geographer 60, 54–69 (2008)
9. Brabyn, L., Gower, P.: Mapping accessibility to general practitioners. In: Khan, O., Skinner, R. (eds.) Geographic Information Systems and Health Applications, pp. 289–307. Idea Group Publishing, Hershey (2003)
10. Weiss, J.E., Greenlick, M.R., Jones, J.F.: Determinants of Medical Care Utilization: The Impact of Spatial Factors. Inquiry 8, 50–57 (1971)
11. Verter, V., Lapierre, S.D.: Location of preventive health care facilities. Annals of Operations Research 110, 123–132 (2002)
12. Zhang, Y., Berman, O., Verter, V.: Incorporating congestion in preventive healthcare facility network design. European Journal of Operational Research 198, 922–935 (2009)
13. Joseph, A.E., Phillips, D.R.: Accessibility and Utilization—Geographical Perspectives on Health Care Delivery. Happer & Row Publishers, New York (1984)
14. Mitropoulos, P., Mitropoulos, I., Giannikos, I., Sissouras, A.: A biobjective model for the locational planning of hospitals and health centers. Health Care and Management Science 9, 171–179 (2006)
15. Mehretu, A., Wittick, R.I., Pigozzi, B.W.: Spatial design for basic needs in eastern Upper Volta. The Journal of Development Area 7, 383–394 (1983)

Medicinal Property Knowledge Extraction from Herbal Documents for Supporting Question Answering System

Chaveevan Pechsiri[1], Sumran Painuall[1], and Uraiwan Janviriyasopak[2]

[1] Dept. of Information Technology, DhurakijPundit University, Bangkok, Thailand
itdpu@hotmail.com, sumran@it.dpu.ac.th
[2] Eastern Industry Co.ltd., Bangkok, Thailand
uraiwanjan@hotmail.com

Abstract. The aim of this paper is to automatically extract the medicinal properties of an object, especially an herb, from technical documents as knowledge sources for health-care problem solving through the question-answering system, especially What-Question, for disease treatment. The extracted medicinal property knowledge is based on multiple simple sentence or EDUs (Elementary Discourse Units). There are three problems of extracting the medicinal property knowledge: the herbal object identification problem, the medicinal property identification problem for each object and the medicinal property boundary determination problem. We propose using NLP (Natural Language Processing) with statistical based approach to identify the medicinal property and also with machine learning technique as Naïve Bayes with verb features for solving the boundary problem. The result shows successfully the medicinal property extraction of the precision and recall of 86% and 77%, respectively, along with 87% correctness of the boundary determination.

Keywords: Medicinal Property Knowledge, Elementary Discourse Unit, Medicinal Property Boundary.

1 Introduction

The objective of this research is to develop a system of automatic knowledge extraction of the medicinal properties of herbs from technical documents for constructing knowledge base of the herbal effect property network which is beneficial for solving health problems through an automatic what-question answering system at the service center. According to Wordnet (http://wordnet.prince ton.edu/), 'property' is 'a basic or essential attribute shared by all members of a class'. Then, an object class can contain several attributes or properties, e.g. color, size, weight, contour, effect, etc. This research concerns only the medicinal property or effect property of herbs for contributing the healthcare knowledge. An example of herbal effect property is as follows:

"ขมิ้น/*Tumaric* ลด/*reduce* กรด/*acid* ใน/*in* กระเพาะ/*a stomach*" ("Tumaric reduces acid of a stomach.")

L. Cao et al. (Eds.): PAKDD 2011 Workshops, LNAI 7104, pp. 431–443, 2012.

where 'reduces acid' is the herbal effect property. According to [1], the analysis of drug effect properties requires the consideration of 3 main entities: the drug (A), the physiological effect (B), and the disease (C). The physiological effect relationships of A → B and B → C are extracted from medical abstracts to obtain A → C as in [1] where each A, B, and C is expressed in the term of a noun phrase (NP) entity. In our research, the entities used by [1] are adapted to automatically extract only the high occurrence relations on the herbal documents as

$$\text{herb}_i \rightarrow \text{S-medicinal-activity}$$
$$\text{herb}_i \rightarrow \text{S-disease}$$

where $\text{herb}_i \rightarrow$ Herb (Herb is the medicinal herb set and equivalents to A) and i is an iterate number with >=1, as shown in the following.

$$\text{Herb} = \{\text{herb}_1, \text{herb}_2,..\text{herb}_n\}$$

S-medicinal-activity is the subset of the herbal-activity set (H-activity) which is the physiological effect or B.

$$\text{H-activity} = \{\text{act}_1, \text{act}_2,..\text{act}_m\}$$
$$\text{S-medicinal-activity} \subseteq \text{H-activity}$$

D is the disease set (which equivalents to C). S-disease is the subset of D.

$$D = \{d_1, d_2,..d_p\}$$
$$\text{S-disease} \subseteq D$$

In addition, herb_i exists in the documents as a noun phrase of a name entity which is an object entity, e.g. 'ใหระพา/basil', 'ขิง/ginger', 'ตะไคร้/lemon grass', 'กระเทียม/garlic', etc. Both act_i and d_i are mostly expressed in the documents as either a verb phrase (VP) of EDU (Elementary Discourse Unit which is a simple sentence or a clause, [2] or VP of an EDU-like name entity, as shown in the following example.

a. VP of EDU
   EDU1: "กระเทียม/***Garlic*** ระงับ/***stops*** การแข็งตัวของเลือด/***blood clotting***" ("Garlic stops blood clotting.")
   where 'stops blood clotting' is the VP expression of act_i in EDU1.
   EDU2: "[กระเทียม/***The garlic***] รักษา/***cures*** โรคหัวใจ/***heart disease***" ("[The garlic] cures heart disease.")
   where 'cures heart disease' is the VP expression of d_i in EDU2 and the [..] symbol means ellipsis.

b. VP of EDU-like name entity
   EDU1: "ใบกระเพรา/***A basil leaf*** ใช้เป็น/***is used as*** ยา/***a medicine*** ขับ/***releases*** ลม/***gas***" ("A basil leaf is used as a medicine releasing gas.")
   where '***a medicine*** ขับ/***releases*** ลม/***gas***' is the EDU-like name entity and 'releases gas' is the VP expression of the event act_i which is releasing gas inside of the human body system'.

The expression of herb_i, act_i, and d_i can be represented by the following regular expression.

EDU → NP1 VP

VP → V_1 NP2 | V_2 Name-entity

NP_1 → φ | Herb-name | N_1 Modifier$_1$ | N_1

NP_2 → N_2 Modifier$_2$ | N_3

Name-entity → EDU

Modifier1 → Herb-name | Prep Herb-name | Adj

Modifier2 → V_3 | V_3 N_3 |

Prep → 'ของ/of' | ...

Herb-name → 'โหระพา/basil' | 'กระเทียม/garlic' | ....

Adj → 'อ่อน/young' | 'แก่/old' | 'สูง/be high' | ....

N_1 → 'ใบ/leaf'' | 'ราก/root' | 'ยา/medicine' | ...

N_2 → φ | 'อาการ/symptom'

N_3 → 'ไข้/fever' | 'ท้องเสีย/diarrhea' | 'ลม/gas' | 'โรคหัวใจ/heart disease' | 'หัว/head' | 'ท้อง/
stomach' | 'กล้ามเนื้อ/muscle' | 'โลหิต/blood' | 'แผล/scar' | 'ท้องอืดท้องเฟ้อ/ flatulence' | ...

V_1 → φ | 'แก้/stop' | 'บรรเทา/relieve' | 'ขับ/release' | 'รักษา/remedy' | 'บำรุง/nourish' | 'ลด/
reduce' | ...

V_2 → 'เป็น/be' | 'ใช้เป็น/be used as'

V_3 → 'ปวด/ache' | 'เจ็บ/pain' | 'อักเสบ/inflame' | 'คลื่นไส้/nausea' | 'วิงเวียน/be dizzy' | ...

where NP is a noun phrase, Prep is a preposition, Adj is an adjective, V_1 is an medicinal-property verb concept, V_2 is a general verb showing is-a relation, V_3 is a symptom verb concept, N_1 is an object noun concept, N_2 is a head-noun key word of the 'symptom' concept, N_3 is a noun concept of the symptom name, the symptom location, or the disease, and φ means ellipsis.

There are several techniques ([1], [3], [4], [5]) having been used to extract property knowledge from texts (see section 2). In our research, we emphasize on extracting the medicinal property knowledge of herbs from the Thai herb websites (http://www.thaihealth.or. th/node/5173 ; http://www.rspg.or.th/plants_data/herbs /herbs_200.htm). Most of the medicinal properties of each herb occur sequentially in the documents as shown in the following example.

EDU1 "ขมิ้น/**Turmeric** ใช้เป็น/**is used as** ยา/**medicine** ลดกรด /**reduces acid**" ("Turmeric is used as a antacid medicine.")

EDU2 "[**Turmeric**] ขับ/**releases** ลม/**gas**" (" [Turmeric] releases gas.")

EDU3 "[**Turmeric**] แก้/**stops** ปวดท้อง/ **abdominal pain**" ("[Turmeric] stops paining abdominal pain.")

EDU4 "[**And Turmeric**] คลาย/**relaxes** อาการปวดเกร็งช่องท้อง /**muscle cramps**" ("[And Turmeric] relaxes muscle cramps.")

Moreover, Thai has several specific characteristics, such as the existence of sentence-like name entity, zero anaphora or the implicit noun phrase, without word and sentence delimiters, and etc. All of these characteristics are involved in three main problems of extracting the medicinal property knowledge of an object, especially a herb (see section 3): the first problem is how to identify an object entity or herb entity, the second one is how to identify the object's interesting medicinal properties, and the third one is how to identify the medicinal property boundary, where problem of implicit delimiter of the boundary is involved. From all of these problems, we need to develop a framework which combines the linguistic phenomena and the machine learning technique as Naïve Bayes [6] to learn the object's medicinal property behavior and the medicinal property boundary.

Our research will be separated into 5 sections. In section 2, related work is summarized. Problems in medicinal property knowledge extraction from Thai documents will be described in section 3 and in section 4 our framework for extracting the medicinal property knowledge. In section 5, we evaluate and conclude our proposed model.

2   Related Work

Several strategies ([1], [3], [4], [5], [6]) have been proposed to extract knowledge from the textual data.

In 1998, [1] suggested that the understanding of drug effect properties require the consideration of 3 main components, the drug A, the physiological effect B, and the disease C, where physiological relationships A→B and B→C are common knowledge. [1] proposes to extract medical knowledge, A→C, by applying Intelligent Data Analysis in Medicine to extract the medical knowledge on textual databases, and comparing to the traditional statistical based computer linguistics approach using log likelihood and Chi-square to find the association between words including drugs and side effects which mostly are noun phrase expression. Meanwhile, Intelligent Data Analysis approach uses frequency based variability measures to classify side-effect-related words in medical abstracts. The knowledge extraction based on the Intelligent Data Analysis to the computer linguistic approach (achieved the averages of 0.57recall and 0.08 precision.) show that the burstiness rule derived from classification tree analysis is more powerful than the one using traditional statistical based approach. The reason of low precision and low recall is due to sparse data. However, our corpus contains a lot of implicit noun phrases and most of the medicinal property being the verb phrase expression; their method presented by [1] cannot be applied.

In 2005, [3] extracted object-property knowledge, especially physical properties i.e. color and size, by applying the linguistic pattern as follows:

[object/NP ['มี(have)'/vt [[prop]/NP value]/NP]/VP]/S

prop   = property of object in the Sentence (S)
value   = qnum | num | measure
qnum   = quantifier of numerical value
num   = number of numerical values
measure   = measurement of numerical value

Their object-property knowledge is expressed in terms of NP patterns. The achieved precision and recall are 0.88 and 0.47 respectively.

In 2008, [4] presented TCMGeneDIT, a database that provides associations on Traditional Chinese Medicine (TCM), genes, diseases, effects of TCM, ingredients, and the TCM effect and effecter relationships, which are mined and extracted from literature. According to [4], association discovery on noun phrases (TCM, disease, gene, ingredient, and effect) was conducted by using hypothesis testing and colloca-tion analysis on annotated documents where a rule-based information extraction was performed. The Swanson's ABC model was also applied where [4] suggested that gene (A), may be regulated by ingredients (B), which are isolated from TCMs (C). Transitive association was used to imply A→C when there are significant A→B and B→C. [4]'s results shown a precision of 0.91 with 1185 relations from the associa-tions between effects and effecters. The analysis of errors has indicated incomplete, incorrect, and too broad effecter NPs as the cause of errors. However, [4]'s technique cannot be applied to this research due to the use of NPs.

In 2008, [5] identify factual knowledge of object class by using the keyword queries with Is-A pattern to extract hierarchical class attributes (which are the class properties) from web text and query log. In order to identify the object class, five attributes are used in the form of NP. A precision of 0.8 is achieved for 100 classes from 9,537 extracted classes. In our research, the attribute/property extraction is adapted where the relations between the instances (i.e. basil, ginger, garlic, etc.) and properties (i.e. release gas, stop pain, etc.) are focused on the herb class. And, most of our properties are expressed in terms of verb phrases (VP).

However, unlike our research, the property knowledge extracted by previous researches is expressed in the form of NP without the boundary determination of properties. This research proposes using NLP with statistical based approach to identify the medicinal properties and using machine learning technique of Naïve Bayes to determine the medicinal property boundary.

3   Problems of Medicinal Property Knowledge Extraction

To extract the medicinal properties, there are three main problems that must be solved. The first problem is how to identify an interesting object entity which is an herb. The second problem is how to identify the medicinal property of the interesting object, and the third one is how to identify the medicinal property boundary with the implicit boundary delimiter.

3.1   Object Identification Problems

There are two major linguistic-phenomenon problems, the zero anaphora and the textual ellipsis.

Zero Anaphora. Zero anaphora is the relation in which a phonetically null element is seen as linked by anaphora to an antecedent (http://www.encyclopedia.com/doc/1O36-zeroanaphora .html). The use of an ana-phoric pronoun or zero anaphora to indicate the same participant forces the reader to "remember" the participant. In our corpus, there are high occurrences of zero anapho-ra including the herb entity which mostly occurs at the EDU subject. For example:

EDU1 *"กระเทียม/**Garlic** ใช้เป็น/**is used as** ยา/**medicine** ขับ/**releases** ลม/**gas**"* ("Garlic is used as a medicine releasing gas.")

EDU2 *"φ แก้ /**stops** ไอ/ **cough**"* (" φ stops coughing.") where φ (or[..] from section 1) represents zero anaphora.

In generally, φ refers to the previous EDU subject which is "Garlic" from EDU1.

Textual Ellipsis. Most of both semi-structure and non structure herb corpora contain ellipses of preposition phrases in the object-entity noun phrases, as shown in the following example.

"ทับทิม/ Pomegranate

....

EDU1: *"ราก/**Roots** ใช้เป็น/**are used as** ยา/**a medicine** ขับ/**releases** ปัสสาวะ/**urine**"* ("Roots [of pomegranates] are used as diuretic medicine.")

..............................*"*

where "[of pomegranates]" is omitted. This problem can be solved by referring the document topic name.

3.2 Medicinal Property Identification Problem

In general, the medicinal property can be identified by a property cue word, e.g. "สรรพคุณ/property" "คุณสมบัติ/property", existing in a subtopic and/or a document body. However, there are some document containing the absent of the property cue word. For example:

EDU1: *"น้ำต้ม/**Boil** ใบบัวบก/**Asiatic Pennywort leaves**"*
EDU2: *"แช่เย็น/**Cool** [**it**]"*
EDU3: *"ดื่ม/**Drink** [**it**]"*
EDU4: *"ลด/**Reduce** ไข้/ **fever**"*
EDU5: *"แก้/**Relief** เจ็บคอ/**sore throat**"*

This problem can be solved by learning the relatedness of verb order pair (v_{pmp}, v_{mp}) from two consecutive EDUs, where $v_{pmp} \in V_{pmp}$ (which is a pre-medicinal-property verb concept set) in one EDU followed by $v_{mp} \in V_{mp}$ (which is a medicinal-property verb concept set) in another EDU. Both concepts of V_{pmp} and V_{mp} from the annotated corpus are obtained from Wordnet. Moreover, V_{mp} contains all V_1 from the regular expression in section 1 along with concept from Wordnet.

V_{pmp} = { 'เป็นยา (be medicine)', 'ใช้เป็นยา (used as medicine)', 'ทา (apply)', 'ใช้(apply)', 'ดื่ม (consume)', 'รับประทาน (consume)', 'ตำ(grind)', 'บด(grind)', 'มีรส(have taste)' …}

V_{mp}={ 'แก้(stop/prevent)', 'แก้ (relieve/ treat)', 'บรรเทา (relieve/treat)', 'ขับ(release /discharge)', 'รักษา(remedy/treat)', 'บำรุง(nourish/supply)', 'ลด(reduce/decrease)', 'คลาย(relax/loosen up)', 'ฆ่า(kill)'…}

3.3 Medicinal Property Boundary Determination Problems

There are two problems of the verb ellipsis and the ending-boundary-cue ellipsis.

Verb Ellipsis. There are some EDUs in the corpus containing the v_{mp} ellipsis (where $v_{mp} \in V_{mp}$) along with the zero anaphora problems, for example:

EDU1: "*กระเพรา/The basil* แก้/*relieves* ปวดท้อง/ *abdominal pain*" ("The basil relieves abdominal pain.")

EDU2: "[*The basil*] [*relieves*] ท้องเสีย/ *diarrhea*" ("[The basil] [relieves] diarrhea.")

EDU3: "*และ/And* [*The basil*] [*relieves*] คลื่นไส้/ *nausea*" ("And [The basil] [relieve] nausea.")

EDU4: "[*The basil*] [*relieves*] อาเจียน/ *vomiting*" ("[The basil] [relieves] vomiting.")

Each EDU2 and EDU3 contains the implicit of v_{mp} and noun phrase (zero anaphora) of the herbal object. These problems of the v_{mp} ellipsis can be resolved by using the previous explicit v_{mp}.

Ending-Boundary-Cue Ellipsis. In general the medicinal properties of herbs occur sequentially in the document. The medicinal property boundary determination is necessary when the ending boundary cue, e.g. "และ/and", is absent. For example:

EDU1: "*ดอกกระเจี๊ยบ/A Roselle flower* เป็นยา/*is a medicine* ละลาย/*dissolves* เสมหะ/ *phlegm*" ("A Roselle flower is a medicine dissolves phlegm.")

EDU2: "[*The Roselle flower*] ละลาย/*dissolves* คอเลสเตอรอล/*cholesterol*" ("[The Roselle flower] dissolves cholesterol".")

EDU3: "[*The Roselle flower*] แก้/*treats* โรคนิ่วในไต/ *a kidney stone disease*" (" [The Roselle flower] treats a kidney stone disease.")

EDU4: "*ดอก*[*Rosselle*]*Flowers* บด/*are ground*….("[Rosselle]Flowers are ground….")

where EDU3 is the ending boundary of the medicinal properties of the Roselle flower. This problem can be resolved by using Naïve Bayes with a slide window size of two verbs from consecutive EDUs.

4   A Framework for Medicinal Property Knowledge Extraction

There are three steps in our framework. First is corpus preparation step followed by the medicinal property learning step and the medicinal property extraction step as shown in Fig. 1.

4.1 Corpus Preparation

This step is the preparation of corpus in the form of EDU from herbal medicine documents download from several Thai organization websites. The step involves with using Thai word segmentation tools [7], including Name entity [8].

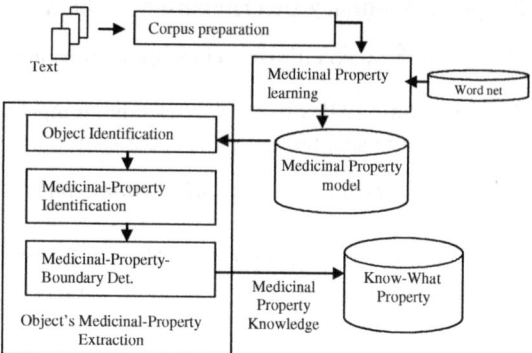

Fig. 1. System Overview

```
"พริกไทยดำ..............
พริกไทยดำมีรสเผ็ดอุ่น EDU1 เมื่อรับประทานเข้าไป EDU2 จะรู้สึกอุ่นวาบที่ท้อง EDU3 ช่วยขับลม EDU4 ขับปัสสาวะ
EDU5 แก้ท้องอืดท้องเฟ้อ EDU6 แก้ไข้มาลาเรีย EDU7 แก้อหิวาตกโรค , EDU8 ใช้ก้านพริกไทย 10 ก้าน"
"Black pepper..............
Black pepper has hot taste. EDU1 When [you] have it, EDU2 [you] would feel warm in the stomach.
EDU3 [Black pepper] will release gas, EDU4 discharge urine, EDU5 stop flatulence, EDU6 cure malaria,
EDU7 stop diarrhea. EDU8 Using 10 Black pepper stems..."
<Topic_name Entity-concept=black pepper/herb>พริกไทยดำ</Topic_name>..........

<EDU1> เมื่อ <NP1 concept=person>φ</NP1>
        <VP><Vmp concept=consume>รับประทาน </Vmp>เข้าไป</EDU>

<id =1 class=Medicinal Property>
<EDU2><NP1 concept= black pepper/herb>φ</NP1>
        <VP>จะ<Vmp concept=be warm>รู้สึกอุ่นวาบ </Vmp>ที่ท้อง</EDU>
<EDU3><NP1 concept= black pepper/herb>φ</NP1>
        <VP><Vmp concept=release>ช่วยขับ</Vmp>
        <NP2 concept=gas>ลม<NP2></VP></EDU>
<EDU4><NP1 concept= black pepper/herb>φ</NP1>
        <VP><Vmp concept=release>ช่วยขับ </Vmp>
        <NP2 concept=gas>ลม<NP2></VP></EDU>
<EDU5><NP1 concept= black pepper/herb>φ</NP>
        <VP>< Vmp concept=discharge/release >ขับ </ Vmp >
        <NP2 concept=urine>ปัสสาวะ<NP2></VP></EDU>
<EDU6><NP1 concept= black pepper/herb>φ</NP1>
        <VP>< Vmp concept=stop>แก้ </ Vmp >
        <NP2 concept=flatulence/symptom >ท้องอืดท้องเฟ้อ <NP2></VP></EDU>
<EDU7><NP1 concept= black pepper/herb>φ</NP1>
        <VP>< Vmp concept=cure>แก้ </ Vmp >
        <NP2 concept= malaria >ไข้มาลาเรีย<NP2></VP></EDU>
<EDU8><NP1 concept= black pepper/herb>φ</NP1>
        <VP>< Vmp concept=cure>แก้ </ Vmp >
        <NP2 concept= cholera >อหิวาตกโรค <NP2></VP></EDU>
</id>
Vmp is the medicinal property verb tag
```

Fig. 2. Medicinal Property Annotation

After the word segmentation is achieved, EDU segmentation is then to be dealt with [9]. These annotated EDUs will be kept as an EDU corpus. This corpus will contain 3000 EDUs and will be separated into 2 parts; one part is 2000 EDUs for learning the medicinal property knowledge and the other part of 1000 EDUs for themedicinal property knowledge extraction. In addition to this step of corpus preparation, we manually annotate the medicinal property EDUs, as shown in Fig.2., with a verb concept and a noun concept referred to Wordnet (http:// word-net.princeton. edu/obtain) after translating from Thai to English, by using Lexitron (the Thai-English dictionary) (http://lexitron.nectec.or.th/).

4.2 Medicinal Property Learning

There are two objectives of learning: to identify the medicinal property and to determine the medicinal property boundary.

Medicinal Property Identification Learning. According to [10], the relatedness, r, has been applied in this research by learning the verb order pair (v_{pmp}, v_{mp}) of two consecutive EDUs ($EDU_{pmp}+EDU_{mp}$ where EDU_{pmp} and EDU_{mp} contain v_{pmp} and v_{mp}, respectively) as shown in equation (1)

$$r(v_{pmp}, v_{pmp}) = \frac{fv_{pmp}v_{mp}}{fv_{pmp} + fv_{pmp} - fv_{pmp}v_{mp}} .$$

where $r(v_{pmp}, v_{pmp})$ is the relatedness of the verb order pair.

$v_{pmp} \in V_{pmp}$, $v_{mp} \in V_{mp}$

V_{pmp} is the pre$-$medicinal$-$ propertyverb conceptset.

V_{mp} is the medicinal$-$ propertyverb concept set.

fv_{pmp} is the numbers of v_{pmp} occurences.

fv_{mp} is the numbers of v_{mp} occurences.

$fv_{pmp}v_{mp}$ is the numbers of v_{pmp} and v_{mp} occurences.

(1)

We select each v_{pmp} with the higher relatedness of the verb order pairs in the medicinal concept than its own pair in the non-medicinal concept. All selected v_{pmp} as shown in the following $V_{pmp-selected}$ set will be used for the medicinal property knowledge identification in the medicinal property knowledge extraction step.

$V_{pmp-selected}$ ={'เป็นยา (be medicine)', 'ใช้เป็นยา (used as medicine)', 'ทา (apply)', 'ดื่ม/รับประทาน(consume)', 'ดม(smell)', 'มีรส(have taste)' }

Medicinal Property Boundary Learning. All annotated verbs and noun phrases with concepts from the corpus preparation step are extracted into a property-verb-concept vector (V_i) in matrix vector V of a herb entity.

$V_i = \{v_{i1}, v_{i2}....v_{ik}$ mp /non-mp} where mp is a medicinal-property-verb-concept vector and non-mp is non medicinal-property-verb-concept vector, existing in $EDU_1, EDU_2...EDU_k$, respectively.

$$V = \{V_i\} \quad \text{where } i=1..n$$

After we have obtained the extracted verb features, we then determine the probability of medicinal property relation and non medicinal property relation from a slide window size of two verbs from consecutive EDUs with one sliding EDU distance, shown in Table1, by using Weka (http://www.cs. wakato.ac.nz/ml/weka/).

Table 1. Show probability of v_{ij} concept and v_{ij+1} concept of a medicinal property verb and a non medicinal property verb

v_{ij}	Medicinal Property Verb	Non Medicinal Property Verb.
stop	0.4110	0.1731
release	0.1507	0.1346
relief	0.0069	0.0385
treat	0.1367	0.0096
discharge	0.0137	0.0385
be-drug	0.0205	0.0096
...	...	...
v_{ij+1}	Medicinal Property Verb	Non Medicinal Property Verb.
stop	0.375	0.1091
release	0.1447	0.0273
reduce	0.0197	0.0091
treat	0.0132	0.0182
discharge	0.0263	0.0091
be-drug	0.0132	0.0182
	....	.....

4.3 Medicinal Property Knowledge Extraction

The objective of this step is to recognize and extract the medicinal property knowledge from the testing EDU corpus after the herbal object entity has been identified by the document topic name. Then, the medicinal property knowledge extraction can be processed within 2 steps: the medicinal property identification step and the medicinal property boundary determination step.

Medicinal Property Identification. This step is using $V_{pmp\text{-selected}}$ to anchor the interesting location or EDU for notifying the following EDU as the medicinal property knowledge expressed by v_{mp}, for example:

EDU1: "กระเพรา/***Sweet basil*** เป็นยา/***is a medicine*** ขับ/***releases*** ลม/***gas***" ("Sweet basil is a medicine releaseing gas.")

EDU2: "แก้/***Stop*** ท้องอืด ท้องเฟ้อ/***flatulence***" ("[Sweet basil] stops flatulence.")

EDU3 ...

Medicinal Property Boundary Determination. After the medicinal property EDU identification, the medicinal property boundary is determined by using the Naive Bayes classifier [6] in the equation (2) to determine the consecutive medicinal property EDUs of the herbal documents with a slide window size of two verbs(in Table 1) from consecutive EDUs with one sliding EDU distance. As soon as the class 0 (non medicinal property relation) is determined, the medicinal property boundary is ended, as shown in Fig. 3 of the medicinal property knowledge extraction algorithm. In addition to the corpus study, the v_{mp} ellipsis EDU has to be solved by replacing it with the previous explicit v_{mp} before the medicinal property boundary determination.

$$MedicinalProperyBoundaryClass = \underset{class \in Class}{\arg\max} \ P(class \,|\, v_{ij}, v_{ij+1}).$$

$$= \underset{class \in Class}{\arg\max} \ P(v_{ij} \,|\, class) P(v_{ij+1} \,|\, class) P(class). \tag{2}$$

where $v_{ij} \in V_i$ and $v_{ij+1} \in V_i$ (V_i is a medicinal_ property_verb_concept vector)

$i = \{1,2,..n\} \qquad j = \{1,2,..k\}$

```
Assume that each EDU is represented by (NP VP).    L is a list of
EDU.  Vpmp is the pre-medicinal-property verb concept set Vmp is the
medicinal-property verb concept set
   Vij, Vij+1 are learned verb sets of  Vmp

1      i ← 1, j←1 R←∅       MEDPROPi← ∅
2      while i ≤ length[L] do
3      begin_while1
4        If vi∈Vpmp^vi+1∈Vmp  /*find the medicinal  property EDU
5           bd=yes ;  MEDPROPi ←  MEDPROPi ∪ {j}
7           while(vj ∈Vij )∧(vj+1∈Vij+1)∧ bd=yes do
8           begin_while2            /* Boundary determination
9             bd= arg max  P(vj |c)P(vj+1 |c)P(c)
                 c∈(yes,no)
10             if bd = yes    then
11                 MEDPROPi ← MEDPROPi ∪ {j+1}; j=j+1
13            end_while2
14            R = R ∪ {MEDPROPi }
15            j=1;  i=i+1
16       end_while1
17 : Return
```

Fig. 3. Medicinal Property Knowledge Extraction Algorithm

5 Evaluation and Conclusion

The Thai corpora used to evaluate the proposed medicinal property knowledge extraction algorithm consist of about 1,000 EDUs collected from several herbal web sites. The evaluation of the medicinal property knowledge extraction performance of this research methodology is expressed in terms of the precision and the recall as shown below, where R is the medicinal relation:

$$Precision = \frac{\#\,of\ samples\ correctly\ extracted\ as\ R}{\#\,of\ all\ samples\ output\ as\ being\ R} \tag{3}$$

$$Recall = \frac{\# \, of \, samples \, correctly \, extracted \, as \, R}{\# \, of \, all \, samples \, holding \, the \, target \, relation \, R} \qquad (4)$$

The results of precision and recall are evaluated by three expert judgments with max win voting. The precision of the extracted is 86% and 77% recall with 87% boundary correctness. Our methodology in this research aims to be cost effective by focusing on the verb concepts from two consecutive EDUs which leads to the 86% precision. For example:

EDU1:"หลังจาก/*After* ทาน/ **have** น้อยหน่า/*a sugar apple*" ("After[you] have a sugar apple.")
EDU2: "น้ำเมล็ดของมัน/*Its seed juice* สามารถ/*can* ฆ่า/ **kill** เห็บ/*ticks*" (Its seed juice can kill ticks.") (where 'ทาน/ have(consume)' $\in V_{pmp}$ and 'ฆ่า/ kill' $\in V_{mp}$)

However, EDU3 is not a medicinal property because ticks are not pathogens. These precisions can be further improved by taking in to consideration of the verb's contexts. Our boundary determination is able to obtain 87% correctness because of the lack of consideration of the context. For example:

EDU1:"ใบกระเจี๊ยบ/*Roselle leaf* มีรสเปรี้ยว/*has sour taste*" ("A Roselle leaf has sour taste.")
EDU2:"[*Roselle leaf*] ขับ/*dissolves* เสมหะ/*phlegm*" ("[The leaf] dissolves phlegm.")
EDU3:"ดอก/*A flower* แก้/*relieves* โรคนิ่ว/ *kidney stone disease*" ("A flower relieves the kidney stone disease.")

From this example, the boundary should end at EDU2 for Roselle leaf instead of EDU3. In conclusion, our methodology was able to handle questions that aim to describe medicinal properties of medicinal plants which are beneficial for supporting the question answering system that aim to provide information on medicinal plants, e.g. "ใบโหระพามีสรรพคุณอะไรบ้าง/What are the properties of Basil leaf?".

References

1. Weeber, M., Vos, R.: Extracting expert medical knowledge from texts. In: Working Notes of the Intelligent Data Analysis in Medicine and Pharmacology Workshop (1998)
2. Carlson, L., Marcu, D., Okurowski, M. E.: Building a Discourse-Tagged Corpus in the Framework of Rhetorical Structure Theory. In: Current Directions in Discourse and Dialogue, pp. 85–112 (2003)
3. Kongwan, K., Kawtrakul, A.: Know-what: A Development of Object-Property Extraction from Thai Texts and Query System. In: Proceedings of SNLP 2005, Bangkok, Thailand, pp. 157–162 (2005)
4. Fang, Y.-C., Huang, H.-C., Chen, H.-H., Juan, H.-F.: TCMGeneDIT: a database for associated traditional Chinese medicine, gene and disease information using text mining. BioMed. Central Complementary and Alternative Medicine 8, 58 (2008)
5. Paşca, M.: Turning Web Text and Search Queries into Factual Knowledge: Hierarchical Class Attribute Extraction. In: Proceedings of the Twenty-Third AAAI Conference on Artificial Intelligence (2008)

6. Mitchell, T.M.: Machine Learning. The McGraw-Hill Companies Inc. and MIT Press, Singapore (1997)
7. Sudprasert, S., Kawtrakul, A.: Thai Word Segmentation based on Global and Local Unsupervised Learning. In: Proceedings of NCSEC 2003 (2003)
8. Chanlekha, H., Kawtrakul, A.: Thai Named Entity Extraction by incorporating Maximum Entropy Model with Simple Heuristic Information. In: IJCNLP 2004 Proceedings (2004)
9. Chareonsuk, J., Sukvakree, T., Kawtrakul, A.: Elementary Discourse unit Segmentation for Thai using Discourse Cue and Syntactic Information. In: Proceedings of NCSEC 2005 (2005)
10. Guthrie, J.A., Guthrie, L., Wilks, Y., Aidinejad, H.: Subject-dependent co-occurrence and word sense disambiguation. In: Proceedings of the 29th Annual Meeting on Association for Computational Linguistics (1991)

First PAKDD Doctoral Symposium on Data Mining (DSDM 2011)

Age Estimation Using Bayesian Process

Yu Zhang

Hong Kong University of Science and Technology
zhangyu@cse.ust.hk

Abstract. Age problems have attracted many researchers' attentions in recent years since they have many potential applications in human-computer interaction and other areas. Among all the age problems, automatic age estimation is one interesting problem and many methods have been proposed to solve this problem. In this paper, we use two Bayesian process regression algorithms, Gaussian process and t process, for age estimation. Different from previous regression methods on age estimation, which need to specify the form of regression functions or determine many parameters in regression functions in inefficient ways such as cross validation, in our methods, the form of regression function is implicitly defined by kernel function and almost all the parameters of our methods can be learnt from data automatically using efficient gradient methods. Moreover, our methods are very simple and easy to implement. Since Gaussian process is easy to be affected by outlier data points, t process can be viewed as a robust version of Gaussian process to solve this problem. Experiments on one public aging database FG-NET show our method is effective and comparable with the state-of-the-art methods on age estimation.

1 Introduction

Human face contains much information about a person, i.e., identity, gender, expression, ethnicity, pose and age. Due to this, many problems on face have emerged, i.e., face detection, face recognition, face verification, gender classification, expression recognition, ethnic classification, pose estimation and age estimation, and these problems have attracted many researchers' attentions over the past decades. Among these problems, age estimation maybe receives the least attentions. One reason for this is that age images require much longer time to collect. Because of the emergence of one public age database FG-NET[1], age estimation has become an active problem over the last decade.

In the pioneering work on age estimation [1], the authors treat age estimation as a regression problem and propose four aging functions by using different strategies or utilizing additional information. Since aging process is very complicate, different from the simple aging functions, i.e., quadratic or cubic regression function, used in [1], [2,3,4,5] propose more sophisticated regression functions as aging functions for age estimation. Since the age quantity accompanying with each age image in the age database is usually an integer but not a continuous real number, in some research work such as [6], age estimation problem is treated as a classification problem and many classification techniques such as nearest-neighbor classifier and neural network are adopted. [7,8,9]

---

[1] http://www.fgnet.rsunit.com/

L. Cao et al. (Eds.): PAKDD 2011 Workshops, LNAI 7104, pp. 447–458, 2012.

propose aging pattern which is a sequence of personal face images sorted in time order and principal component analysis [10] or its kernel extension is used to learn linear or nonlinear aging pattern subspace by EM-style algorithm. Different from the above method, [11,12] consider age quantity of each image contains some uncertainty, that is, the age of each image is represented as an age range but not an integer, and use ranking or regression method to make estimation. Moreover, in age database, there usually exists some other information, such as subject information, that is, we know which images belong to one person. Some of the methods described above utilize this information and others do not, i.e., [1,6,7,8,9] belong to the first class and others are the second class.

Besides age estimation, there are many other problems on age, such as [13,14,15,16,17]. [13], which studies an age classification problem, divides images into three age-groups: babies, young adults and senior adults, and uses some facial features to define some quantity to distinguish different age-groups. [14] studies the effect of aging progress on similarity measure which is used for face recognition. Similar to [14], [15] studies the effect of aging progress on defining similarity measure for face verification, and proposes a Bayesian age difference classifier to perform face verification. [16] uses a craniofacial growth model which can characterize growth related shape variations in young human faces and then the craniofacial growth model can be used to synthesize face images across years and perform face recognition across age progression. [17] proposes using the gradient orientation pyramid to extract facial features and then models face verification as a two-class problem in which a support vector machine is used as a classifier.

In this paper, we focus on age estimation, in which age labels are considered to contain no uncertainty. We also model this problem as a regression problem, since it is not a conventional classification problem in nature. That is, from performance measure, the errors that you classify a zero-age baby image to 3 or 5 years are different, but for conventional classification methods, these two errors are considered to be the same. Since previous regression methods used for age estimation have some limitations, i.e., the form of aging function in [1] and many model parameters in [2,3,4,5] need to be chosen in inefficient ways such as using cross validation method which needs to re-train model multiple times and is computational demanding, here we use two powerful Bayesian processes regressors, Gaussian Process [18] and t Process, for age estimation. These two regressors are very simple to implement and almost 'automatic', that is, almost all parameters in the models can be learnt from data automatically by using efficient gradient methods. Moreover, Gaussian process is easy to be affected by outliers, t process used here is to solve this problem by adding robustness control.

2   Age Estimation Using Gaussian Process

Suppose we are given a training set D which consists of n labeled data points $\{(\mathbf{x}_i, y_i)\}_{i=1}^{n}$ with the ith point $\mathbf{x}_i \in \mathbb{R}^d$ and its output $y_i \in \mathbb{R}$.

Before introducing Gaussian process, we first revisit probability linear regression which has much relationship with Gaussian process.

The model for probability linear regression is defined as

$$y_i = \mathbf{w}^T \mathbf{x}_i + \varepsilon_i$$
$$\mathbf{w} \sim \mathcal{N}(\mathbf{0}_d, \mathbf{I}_d)$$
$$\varepsilon_i \sim \mathcal{N}(0, \sigma^2),$$

where $\mathcal{N}(\mathbf{m}, \boldsymbol{\Sigma})$ denotes a multivariate Gaussian distribution with mean $\mathbf{m}$ and covariance matrix $\boldsymbol{\Sigma}$, $\mathbf{0}_d$ denotes a $d \times 1$ zero vector and $\mathbf{I}_d$ is the $d \times d$ identity matrix. According to the property of Gaussian distribution, we can get

$$\mathbf{w}^T \mathbf{x}_i \sim \mathcal{N}(0, \mathbf{x}_i^T \mathbf{x}_i).$$

So y_i can be viewed as two parts: the first one is a variable with zero mean and its linear kernel value or dot product as the variance and the second one is a noise term with zero mean and variance σ^2. If written in matrix form, then we can get

$$\mathbf{y} = \mathbf{f} + \boldsymbol{\varepsilon}$$
$$\mathbf{f} \sim \mathcal{N}(\mathbf{0}_n, \mathbf{X}^T \mathbf{X})$$
$$\boldsymbol{\varepsilon} \sim \mathcal{N}(\mathbf{0}_n, \sigma^2 \mathbf{I}_n),$$

where $\mathbf{X} = (\mathbf{x}_1, \ldots, \mathbf{x}_n)$, $\mathbf{y} = (y_1, \ldots, y_n)^T$, $\mathbf{f} = (f_1, \ldots, f_n)^T$, $\boldsymbol{\varepsilon} = (\varepsilon_1, \ldots, \varepsilon_n)^T$. So this model can be viewed as a prior of $\mathbf{f}$, which is Gaussian distribution with zero mean and linear kernel matrix as its covariance matrix, and a likelihood term $p(\mathbf{y}|\mathbf{f})$ which is also a Gaussian distribution. Since this model can only handle data with linear structure, it is natural to generalize to nonlinear extension using kernel trick, which is just Gaussian process.

For Gaussian process, we define a latent variable f_i for each data point $\mathbf{x}_i$. The prior of $\mathbf{f} = (f_1, \ldots, f_n)^T$ is defined as

$$\mathbf{f} \,|\, \mathbf{X} \sim \mathcal{N}(\mathbf{0}_n, \mathbf{K}), \tag{1}$$

where $\mathbf{X} = (\mathbf{x}_1, \ldots, \mathbf{x}_n)$ denotes data matrix and $\mathbf{K}$ denotes the kernel matrix defined on $\mathbf{X}$ using a kernel function $k_{\boldsymbol{\theta}}(\cdot, \cdot)$ parameterized by $\boldsymbol{\theta}$. One example for kernel function $k_{\boldsymbol{\theta}}(\cdot, \cdot)$ is RBF kernel whose form is $k_{\boldsymbol{\theta}}(\mathbf{x}_1, \mathbf{x}_2) = \theta_1 \exp(-\frac{||\mathbf{x}_1 - \mathbf{x}_2||_2^2}{2\theta_2^2})$ where $|| \cdot ||_2$ denotes the 2-norm of a vector and θ_i is the ith element of $\boldsymbol{\theta}$.

The likelihood for each data point is defined based on the Gaussian noise model:

$$y_i \,|\, f_i \sim \mathcal{N}(f_i, \sigma^2),$$

where σ^2 defines the noise level. Since all y_i's are independent when given f_i. So we can write in a vectorial form:

$$\mathbf{y} \,|\, \mathbf{f} \sim \mathcal{N}(\mathbf{f}, \sigma^2 \mathbf{I}_n), \tag{2}$$

where $\mathbf{y} = (y_1, \ldots, y_n)^T$.

The graphical model for Gaussian process is depicted in Figure 1.

In summary, Eqs. (1) and (2) compose the whole model for Gaussian process which is simple and powerful demonstrated in many applications. In next two sections, we will show how to learn model parameters and make prediction.

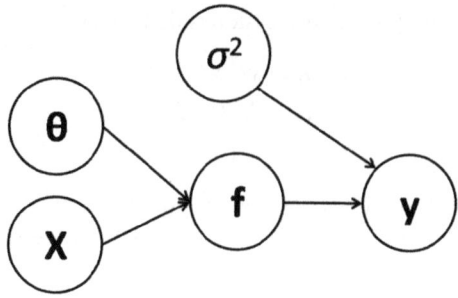

Fig. 1. Graphical model for Gaussian Process

2.1 Learning Model Parameters

The noise level σ^2 and kernel parameters θ are model parameters in Gaussian process.

In Bayesian statistics, marginal likelihood (also called model evidence) is usually used to learn model parameters [19]. In Gaussian process, since the marginal likelihood has close form, so it is a good choice for model selection.

Since

$$p(\mathbf{y}|\mathbf{X}) = \int p(\mathbf{y}|\mathbf{f})p(\mathbf{f}|\mathbf{X})df = \mathcal{N}(\mathbf{y}|\mathbf{0}_n, \mathbf{K} + \sigma^2 \mathbf{I}_n), \tag{3}$$

the negative log-likelihood of all data points $\mathbf{X}$ can be computed as

$$l = \frac{1}{2} \left[\mathbf{y}^T (\mathbf{K} + \sigma^2 \mathbf{I}_n)^{-1} \mathbf{y} + \ln |\mathbf{K} + \sigma^2 \mathbf{I}_n| \right] + \text{Const},$$

where $|\mathbf{A}|$ denotes the determinant of a square matrix $\mathbf{A}$. Here we use gradient method to minimize l to estimate the optimal values of θ and σ. Since each element of θ and σ is positive, we instead treat $\ln \theta$ and $\ln \sigma$ as variables where each element of $\ln \theta$ is the logarithm of the corresponding element in θ. The gradients of the negative log-likelihood with respect to $\ln \theta$ and $\ln \sigma$ can be computed as:

$$\frac{\partial l}{\partial \ln \sigma} = \frac{\partial l}{\partial \sigma^2} \frac{\partial \sigma^2}{\partial \ln \sigma} = \sigma^2 \text{tr} \left(\tilde{\mathbf{K}}^{-1} - \tilde{\mathbf{K}}^{-1} \mathbf{y} \mathbf{y}^T \tilde{\mathbf{K}}^{-1} \right), \tag{4}$$

$$\frac{\partial l}{\partial \ln \theta} = \frac{1}{2} \text{diag}(\theta) \text{Tr} \left([\tilde{\mathbf{K}}^{-1} - \tilde{\mathbf{K}}^{-1} \mathbf{y} \mathbf{y}^T \tilde{\mathbf{K}}^{-1}] \frac{\partial \mathbf{K}}{\partial \theta} \right), \tag{5}$$

where $\text{tr}(\cdot)$ deontes the trace of a square matrix, $\tilde{\mathbf{K}} = \mathbf{K} + \sigma^2 \mathbf{I}_n$, $\text{Tr}(\mathbf{A} \frac{\partial \mathbf{K}}{\partial \theta})$ denotes a vector whose jth element is $\text{tr}(\mathbf{A} \frac{\partial \mathbf{K}}{\partial \theta_j})$ where θ_j is the jth element of θ, and $\text{diag}(\theta)$ denotes the diagonal matrix whose (j,j)th element is the jth element of θ.

2.2 Make Prediction

Suppose we are given a test data point $\mathbf{x}_\star$, we need to determine the corresponding output $y_\star$. From the marginal likelihood calculated in Eq. (3), we can get

$$\begin{pmatrix} \mathbf{y} \\ y_\star \end{pmatrix} \sim \mathcal{N} \left(\mathbf{0}_{n+1}, \begin{pmatrix} \mathbf{K} + \sigma^2 \mathbf{I}_n & \mathbf{k}_\star \\ \mathbf{k}_\star^T & k_\theta(\mathbf{x}_\star, \mathbf{x}_\star) + \sigma^2 \end{pmatrix} \right),$$

where $\mathbf{k}_\star = (k_\theta(\mathbf{x}_\star, \mathbf{x}_1), \ldots, k_\theta(\mathbf{x}_\star, \mathbf{x}_n))^T$. Then using the linear Gaussian model in [20], we can get the predictive distribution $p(y_\star \mid x_\star, \mathbf{X}, \mathbf{y})$ which is a Gaussian distribution with mean $m_\star$ and variance $(\sigma_\star)^2$ as

$$m_\star = (\mathbf{k}_\star)^T (\mathbf{K} + \sigma^2 \mathbf{I}_n)^{-1} \mathbf{y} \tag{6}$$
$$(\sigma_\star)^2 = k_\theta(\mathbf{x}_\star, \mathbf{x}_\star) + \sigma^2 - \mathbf{k}_\star^T (\mathbf{K} + \sigma^2 \mathbf{I}_n)^{-1} \mathbf{k}_\star. \tag{7}$$

Finally we can use $m_\star$ as the prediction for $\mathbf{x}_\star$.

2.3 Discussion

The computational complexity of our model is $O(n^3)$. When the data size n is small, then our model is quite efficient; otherwise we can use informative vector machine [21], a sparse extension of Gaussian process which selects data points based on information theory and use these selected data points to learn parameters and make prediction, for age estimation. The complexity of informative vector machine is $O(nm^2)$ where m is the number of selected data points and so informative vector machine is very efficient and also very powerful.

3   Age Estimation Using t Process

A well-known limitation of a Gaussian distribution is that it is not robust, since if the training data are contaminated by outliers, then the accuracy of estimated mean and covariance can significantly be compromised [22]. And this also holds for Gaussian process.

A more robust alternative for Gaussian distribution is the multivariate t distribution whose p.d.f. is defined as

$$t_\nu(\mathbf{m}, \boldsymbol{\Sigma}) = \pi^{-\frac{p}{2}} |\boldsymbol{\Sigma}|^{-\frac{1}{2}} \nu^{\frac{\nu}{2}} \frac{\Gamma(\frac{\nu+p}{2})}{\Gamma(\frac{\nu}{2})} (\nu + (\mathbf{z} - \boldsymbol{\mu})^T \boldsymbol{\Sigma}^{-1} (\mathbf{z} - \boldsymbol{\mu}))^{-\frac{\nu+p}{2}},$$

where $\mathbf{z} \in \mathbb{R}^p$ is a random variable, $\Gamma(\cdot)$ is the gamma function and ν is the degree of freedom in t distribution. The t distribution is known to have heavy tails in its p.d.f. compared to the corresponding Gaussian distribution with identical mean and covariance. Some useful properties for the multivariate t distribution are summarized in the following propositions.

Proposition 1. $\lim_{\nu \to +\infty} t_\nu(\boldsymbol{\mu}, \boldsymbol{\Sigma}) = \mathcal{N}(\boldsymbol{\mu}, \boldsymbol{\Sigma})$.

Proposition 2. *If* $\mathbf{z} \in \mathbb{R}^p \sim t_\nu(\boldsymbol{\mu}, \boldsymbol{\Sigma})$, *then for any matrix* $\mathbf{W} \in \mathbb{R}^{q \times p}$, *we have* $\mathbf{Wz} \sim t_\nu(\mathbf{W}\boldsymbol{\mu}, \mathbf{W}\boldsymbol{\Sigma}\mathbf{W}^T)$.

Proposition 3. *Assume a random variable* $\mathbf{z} \in \mathbb{R}^p$ *satisfies* $\mathbf{z} \sim t_\nu(\boldsymbol{\mu}, \boldsymbol{\Sigma})$. *Let* $\mathbf{z} = \begin{pmatrix} \mathbf{z}_1 \\ \mathbf{z}_2 \end{pmatrix}$, $\boldsymbol{\mu} = \begin{pmatrix} \boldsymbol{\mu}_1 \\ \boldsymbol{\mu}_2 \end{pmatrix}$ *and* $\boldsymbol{\Sigma} = \begin{pmatrix} \boldsymbol{\Sigma}_{11} & \boldsymbol{\Sigma}_{12} \\ \boldsymbol{\Sigma}_{21} & \boldsymbol{\Sigma}_{22} \end{pmatrix}$ *be the* $[p_1, p - p_1]$ *partition of the corresponding vectors and matrix. Then* $\mathbf{z}_1$ *and* $\mathbf{z}_2 | \mathbf{z}_1$ *are independently distributed, with*

$$\mathbf{z}_1 \sim t_\nu(\boldsymbol{\mu}_1, \boldsymbol{\Sigma}_{11}), \mathbf{z}_2 | \mathbf{z}_1 \sim t_{\nu+p_1}(\boldsymbol{\mu}_{\mathbf{z}_2|\mathbf{z}_1}, \boldsymbol{\Sigma}_{\mathbf{z}_2|\mathbf{z}_1}),$$

where

$$\boldsymbol{\mu}_{\mathbf{z}_2|\mathbf{z}_1} = \boldsymbol{\Sigma}_{21}\boldsymbol{\Sigma}_{11}^{-1}(\mathbf{z}_1 - \boldsymbol{\mu}_1) + \boldsymbol{\mu}_2$$

$$\boldsymbol{\Sigma}_{\mathbf{z}_2|\mathbf{z}_1} = \frac{\nu + (\mathbf{z}_1 - \boldsymbol{\mu}_1)^T\boldsymbol{\Sigma}_{11}^{-1}(\mathbf{z}_1 - \boldsymbol{\mu}_1)}{\nu + p_1}(\boldsymbol{\Sigma}_{22} - \boldsymbol{\Sigma}_{21}\boldsymbol{\Sigma}_{11}^{-1}\boldsymbol{\Sigma}_{12}).$$

From **Proposition 1**, we can find ν controls the robustness of t distribution, that is, when ν is large enough, then t distribution will degenerate to Gaussian distribution whose robustness is the smallest, and when ν is small, t distribution is more robust.

Similar to Gaussian process, we start from introducing probabilistic linear regression. The new model for probability linear regression is defined as

$$y_i = \mathbf{w}^T\mathbf{x}_i + \varepsilon_i$$
$$\mathbf{w} \sim t_\nu(\mathbf{0}_d, \mathbf{I}_d)$$
$$\varepsilon_i \sim t_\nu(0, \sigma^2).$$

According to **Proposition 2**, we can get $\mathbf{w}^T\mathbf{x}_i \sim t_\nu(0, \mathbf{x}_i^T\mathbf{x}_i)$. If written in matrix form, then we can get

$$\mathbf{y} = \mathbf{f} + \boldsymbol{\varepsilon}$$
$$\mathbf{f} \sim t_\nu(\mathbf{0}_n, \mathbf{X}^T\mathbf{X})$$
$$\boldsymbol{\varepsilon} \sim t_\nu(\mathbf{0}_n, \sigma^2\mathbf{I}_n),$$

where $\mathbf{f} = (f_1, \ldots, f_n)^T$, $\boldsymbol{\varepsilon} = (\varepsilon_1, \ldots, \varepsilon_n)^T$. Since $\mathbf{w}$ and $\boldsymbol{\varepsilon}$ are independent, we can get

$$\begin{pmatrix} \mathbf{w} \\ \boldsymbol{\varepsilon} \end{pmatrix} \sim t_\nu\left(\mathbf{0}_{n+d}, \begin{pmatrix} \mathbf{I}_d & \mathbf{0}_{d\times n} \\ \mathbf{0}_{n\times d} & \sigma^2\mathbf{I}_n \end{pmatrix}\right),$$

where $\mathbf{0}_{d\times n}$ denotes $d \times n$ zero matrix. Since $\mathbf{y} = (\mathbf{X}^T, \mathbf{I}_n)\begin{pmatrix} \mathbf{w} \\ \boldsymbol{\varepsilon} \end{pmatrix}$, we can calculate the marginal likelihood as by using **Proposition 2**

$$\mathbf{y} \sim t_\nu(\mathbf{0}_n, \mathbf{X}^T\mathbf{X} + \sigma^2\mathbf{I}_n).$$

So this new probabilistic linear regression is similar to that in Section 3. The nonlinear extension of this model leads to t process.

For t process, we also define a latent variable f_i for each data point $\mathbf{x}_i$. The prior of $\mathbf{f} = (f_1, \ldots, f_n)^T$ is defined as

$$\mathbf{f}\,|\,\mathbf{X} \sim t_\nu(\mathbf{0}_n, \mathbf{K}), \tag{8}$$

where $\mathbf{X} = (\mathbf{x}_1, \ldots, \mathbf{x}_n)$ denotes data matrix and $\mathbf{K}$ denotes the kernel matrix defined on $\mathbf{X}$ using a kernel function.

The likelihood is defined based on the t noise model:

$$\mathbf{y}\,|\,\mathbf{f} \sim t_\nu(\mathbf{f}, \sigma^2\mathbf{I}_n), \tag{9}$$

where $\mathbf{y} = (y_1, \ldots, y_n)^T$.

The graphical model for t process is depicted in Figure 2.

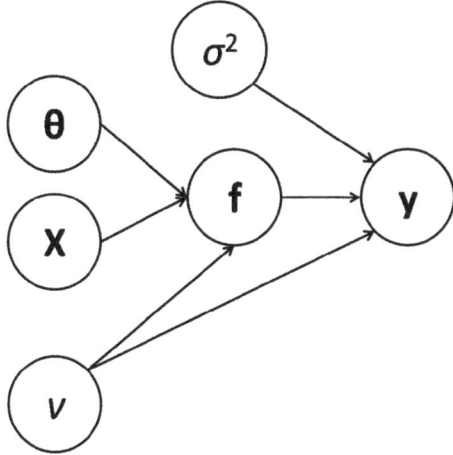

Fig. 2. Graphical model for t Process

3.1   Learning Model Parameters

The model parameters in t process are noise level σ^2, kernel parameters θ and ν. Due to possible pitfalls of empirically estimating ν reported in [23], we can first use cross validation method to determine ν and then learn other model parameters with ν fixed.

Similar to the case in probabilistic linear regression, the marginal likelihood can be computed as

$$p(\mathbf{y}|\mathbf{X}) = \int p(\mathbf{y}|\mathbf{f})p(\mathbf{f}|\mathbf{X})d\mathbf{f} = t_\nu(\mathbf{y}|\mathbf{0}_n, \mathbf{K} + \sigma^2\mathbf{I}_n). \tag{10}$$

Then the negative log-likelihood of all data points $\mathbf{X}$ can be computed as

$$l = \frac{1}{2}\left[(\nu + n)\ln\left(\nu + \mathbf{y}^T(\mathbf{K} + \sigma^2\mathbf{I}_n)^{-1}\mathbf{y}\right) + \ln|\mathbf{K} + \sigma^2\mathbf{I}_n|\right] + \text{Const.}$$

Here we still use gradient method to minimize l to estimate the optimal values of θ and σ. $\ln\theta$ and $\ln\sigma$ is treated as variable because σ and each element of θ are positive. The gradients of the negative log-likelihood with respect to $\ln\theta$ and $\ln\sigma$ can be computed as:

$$\frac{\partial l}{\partial \ln\sigma} = \frac{\partial l}{\partial \sigma^2}\frac{\partial \sigma^2}{\partial \ln\sigma} = \sigma^2\mathrm{tr}\left(\tilde{\mathbf{K}}^{-1} - \frac{\nu + n}{\nu + \mathbf{y}^T\tilde{\mathbf{K}}^{-1}\mathbf{y}}\tilde{\mathbf{K}}^{-1}\mathbf{y}\mathbf{y}^T\tilde{\mathbf{K}}^{-1}\right) \tag{11}$$

$$\frac{\partial l}{\partial \ln\theta} = \frac{1}{2}\mathrm{diag}(\theta)\mathrm{Tr}\left([\tilde{\mathbf{K}}^{-1} - \frac{\nu + n}{\nu + \mathbf{y}^T\tilde{\mathbf{K}}^{-1}\mathbf{y}}\tilde{\mathbf{K}}^{-1}\mathbf{y}\mathbf{y}^T\tilde{\mathbf{K}}^{-1}]\frac{\partial \mathbf{K}}{\partial\theta}\right), \tag{12}$$

where $\tilde{\mathbf{K}} = \mathbf{K} + \sigma^2\mathbf{I}_n$.

3.2   Make Prediction

Suppose we are given a test data point $\mathbf{x}_\star$, we need to determine the corresponding output $y_\star$. From the marginal likelihood calculated in Eq. (10), we can get

$$\begin{pmatrix} \mathbf{y} \\ y_\star \end{pmatrix} \sim t_\nu \left(\mathbf{0}_{n+1}, \begin{pmatrix} \mathbf{K} + \sigma^2 \mathbf{I}_n & \mathbf{k}_\star \\ \mathbf{k}_\star^T & k_\theta(\mathbf{x}_\star, \mathbf{x}_\star) + \sigma^2 \end{pmatrix} \right),$$

where $\mathbf{k}_\star = (k_\theta(\mathbf{x}_\star, \mathbf{x}_1), \dots, k_\theta(\mathbf{x}_\star, \mathbf{x}_n))^T$. Then using **Proposition 3**, we can get the predictive distribution $p(y_\star \mid x_\star, \mathbf{X}, \mathbf{y})$ is $t_{n+\nu}(m_\star, (\sigma_\star)^2)$ with

$$m_\star = (\mathbf{k}_\star)^T \tilde{\mathbf{K}}^{-1} \mathbf{y} \tag{13}$$

$$(\sigma_\star)^2 = \frac{\nu + \mathbf{y}^T \tilde{\mathbf{K}}^{-1} \mathbf{y}}{\nu + n} \left(k_\theta(\mathbf{x}_\star, \mathbf{x}_\star) + \sigma^2 - \mathbf{k}_\star^T \tilde{\mathbf{K}}^{-1} \mathbf{k}_\star \right), \tag{14}$$

where $\tilde{\mathbf{K}} = \mathbf{K} + \sigma^2 \mathbf{I}_n$. Finally we can use $m_\star$ as the prediction for $\mathbf{x}_\star$.

3.3   Discussion

When $\nu \to +\infty$, the prior Eq. (8), likelihood Eq. (9) and marginal likelihood Eq. (10) in t process become the prior Eq. (1), likelihood Eq. (2) and marginal likelihood Eq. (3) in Gaussian process. And this still holds for gradient of negative log-likelihood, predictive mean and variance, that is, Eqs. (11), (12), (13) and (14) become Eqs. (4), (5), (6) and (7) respectively when $\nu \to +\infty$. So in this sense, Gaussian process can be viewed as a special case of t process.

The prior Eq. (8) and the likelihood Eq. (9) in t process can have different freedom, but in this situation the marginal likelihood does not have close form and we need to resort to approximation methods such as variational Bayes method [20] which will increase computational cost. Here using identical freedom is for computational consideration.

4   Experiment

In this section, we will report some experimental results on age estimation to evaluate the performance of our methods.

4.1   Experimental Setting

FG-NET, one public aging database used in many research works, is used in our experiments. There are totally 1002 images from 82 persons, each of which has has 6-18 face images labeled with ground truth ages, in FG-NET database. The ages are distributed in a wide range from 0 to 69 but the age distribution in the number of images is highly uneven, i.e., the number of images above aged 50 is just 21. The sample images of one person in FG-NET database is present in Figure 3. From Figure 3, you can see that the images are collected under totally uncontrolled conditions, in which the images are

Fig. 3. Sample images of one person in the FG-NET database

corrupted by other facial variations such as pose, illumination, expression and so on. This increases the difficulty of age estimation on this database.

For fair comparison with methods reported in [1,6,7,8,9], the facial feature extractor used in our experiments is active appearance model, which is a combination of shape model and intensity model. Each face image is marked 68 landmark points manually and the shape model is trained on the coordinates of the 68 landmark points by principal component analysis. For the intensity model, each face image is aligned to the face shape template of the training set and principal component analysis is used to extract intensity features. In our experiments, we use 238 model parameters to describe 98% variance in the data.

For performance measure, similar to [7,8,9,2], we use two criterions to measure the performance of our methods. The first one is the Mean Absolute Error (MAE). Suppose there are m test images, the real age for kth image I_k is a_k and the predictive one is $\tilde{a}_k$, then the MAE can be calculated as

$$MAE = \frac{1}{m} \sum_{k=1}^{m} |a_k - \tilde{a}_k|,$$

where $|\cdot|$ applying to a scalar means the absolute value of the scalar. Another measure is cumulative score. Still suppose there are m test images, $m_{e \leq l}$ is the number of test images on which the age estimation makes an absolute error no higher than l (years), then the cumulative score at error level l can be calculated as

$$CumScore(l) = \frac{m_{e \leq l}}{m} \times 100\%.$$

The algorithms are tested through the Leave-One-Person-Out (LOPO) mode, i.e. in each fold, the images of one person are used as the test set and those of the others are used as the training set, which simulates the situation in real applications. The compared methods are WAS [1], AAS [6], AGES [7] and KAGES [9]. We use GP to denote Gaussian process and TP for t Process.

4.2   Experimental Results

The mean absolute errors of all algorithms for age estimation are recorded in Table 1. Note that the results of WAS, AAS, AGES and KAGES are just the results reported in

[7,9]. From Table 1, we can find the performance of our methods GP and TP are better than that of other compared methods and the performance of TP is better than that of GP, which shows adding robustness control is helpful. Even though the results reported in [2,5] on FG-NET database are better than those of our methods, however, the features used in [2,5] are different from ours and so direct comparison is not very meaningful. Moreover the difference of our best result 5.23 and the best result 4.95 in [2,5] is not very large.

Table 1. Mean Absolute Error (in Years) of the Algorithms in Age Estimation on the FG-NET Aging Database. GP here represents the Gaussian process and TP denotes the t process.

Method	WAS	AAS	AGES	KAGES	GP	TP
MAE	8.06	14.83	6.77	6.18	5.39	**5.23**

The cumulative scores of all algorithms at error levels from 0 to 10 are present in Figure 4. From Figure 4, we can see that the cumulative scores of our methods GP and TP are better than those of WAS and AAS at all levels. Compared with AGES and KAGES, our methods are comparable with them at error levels below 5 and better than them at error levels larger than 5(year). This shows the effectiveness of our methods.

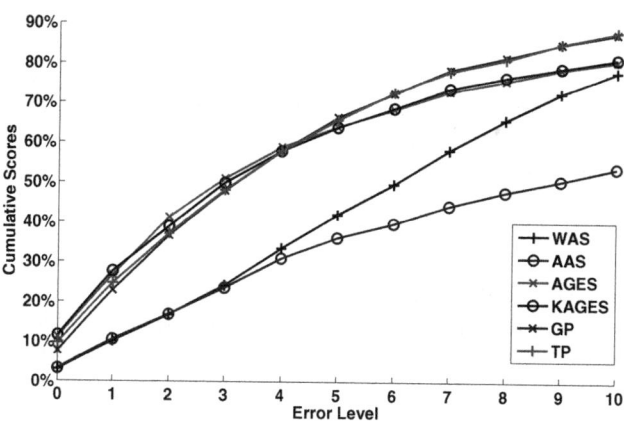

Fig. 4. Cumulative scores of the algorithms in age estimation at error levels from 0 to 10 (years)

The only free parameter in our methods is the freedom ν in t process. We investigate the sensitivity of performance of t process with respect to ν. The mean absolut errors and cumulative scores of t process with different ν are recorded in Table 2 and Figure 5. From Table 2 and Figure 5, we can see that the performance of t process is not very sensitive to the choice of ν.

Table 2. Mean Absolute Error (in Years) of t process on the FG-NET Aging Database with different ν

Method	TP($\nu = 1$)	TP($\nu = 5$)	TP($\nu = 50$)	TP($\nu = 120$)
MAE	5.24	5.23	5.24	5.24

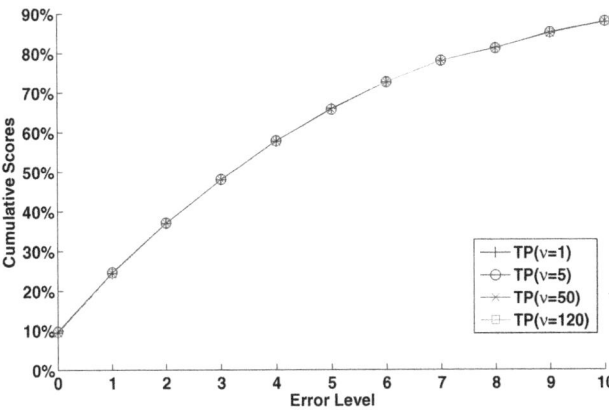

Fig. 5. Cumulative scores of t process in age estimation at error levels from 0 to 10 (years) with different ν

5 Conclusion

In this paper, we use two Bayesian regression models, Gaussian process and its robust version t process to study the age estimation problem. These two models are very simple and powerful especially for model parameter learning. In the future, we will investigate our models on age estimation using other feature extractors, such as the feature extractor present in [2]. Another interesting problem is to apply our models to other regression problems in computer vision, such as pose estimation.

References

1. Lanitis, A., Taylor, C.J., Cootes, T.F.: Toward automatic simulation of aging effects on face images. IEEE Transactions on Pattern Analysis and Machine Intelligence 24(4), 442–455 (2002)
2. Yan, S., Zhou, X., Liu, M., Hasegawa-Johnson, M., Huang, T.S.: Regression from patch-kernel. In: Proceedings of the IEEE Computer Society Conference on Computer Vision and Pattern Recognition, Anchorage, Alaska (2008)
3. Yan, S., Liu, M., Huang, T.S.: Extracting age information from local spatially flexible patches. In: Proceedings of the IEEE International Conference on Acoustics, Speech and Signal Processing, Las Vegas, Nevada, pp. 737–740 (2008)
4. Fu, Y., Huang, T.S.: Human age estimation with regression on discriminative aging manifold. IEEE Transactions on Multimedia 10(4), 578–584 (2008)

5. Guo, G., Fu, Y., Dyer, C.R., Huang, T.S.: Image-based human age estimation by manifold learning and locally adjusted robust regression. IEEE Transactions on Image Processing 17(7), 1178–1188 (2008)
6. Lanitis, A., Draganova, C., Christodoulou, C.: Comparing different classifiers for automatic age estimation. IEEE Transactions on Systems, Man, and Cybernetics–Part B: Cybernetics 34(1), 621–628 (2004)
7. Geng, X., Zhou, Z.H., Zhang, Y., Li, G., Dai, H.: Learning from facial aging patterns for automatic age estimation. In: Proceedings of the 14th ACM International Conference on Multimedia, Santa Barbara, CA, USA, pp. 307–316 (2006)
8. Geng, X., Zhou, Z.H., Smith-Miles, K.: Automatic age estimation based on facial aging patterns. IEEE Transactions on Pattern Analysis and Machine Intelligence 29(12), 2234–2240 (2007)
9. Geng, X., Smith-Miles, K., Zhou, Z.H.: Facial age estimation by nonlinear aging pattern subspace. In: Proceedings of the 16th ACM International Conference on Multimedia, Vancouver, British Columbia, Canada, pp. 721–724 (2008)
10. Jolliffe, I.T.: Principal Component Analysis, 2nd edn. Springer, New York (2002)
11. Yan, S., Wang, H., Huang, T.S., Yang, Q., Tang, X.: Ranking with uncertain labels. In: Proceedings of the 2007 IEEE International Conference on Multimedia and Expo., Beijing, China, pp. 96–99 (2007)
12. Yan, S., Wang, H., Tang, X., Huang, T.S.: Learning auto-structured regressor from uncertain nonnegative labels. In: Proceedings of the IEEE 11th International Conference on Computer Vision, Rio de Janeiro, Brazil (2007)
13. Kwon, Y.H., Lobo, N.V.: Age classification from facial images. Computer Vision and Image Understanding 74(1), 1–21 (1999)
14. Ramanathan, N., Chellappa, R., Chowdhury, A.K.R.: Facial similarity across age, disguise, illumination and pose. In: Proceedings of the 2004 International Conference on Image Processing, Singapore, pp. 1999–2002 (2004)
15. Ramanathan, N., Chellappa, R.: Face verification across age progression. IEEE Transactions on Image Processing 15(11), 3349–3361 (2006)
16. Ramanathan, N., Chellappa, R.: Modeling age progression in young faces. In: Proceedings of the IEEE Computer Society Conference on Computer Vision and Pattern Recognition, New York, NY, USA, pp. 387–394 (2006)
17. Ling, H., Soatto, S., Ramanathan, N., Jacobs, D.W.: A study of face recognition as people age. In: Proceedings of the IEEE 11th International Conference on Computer Vision, Rio de Janeiro, Brazil (2007)
18. Rasmussen, C.E., Williams, C.K.I.: Gaussian Processes for Machine Learning. The MIT Press, Cambridge (2006)
19. MacKay, D.J.C.: Bayesian interpolation. Neural Computation 4(3), 415–447 (1992)
20. Bishop, C.M.: Pattern Recognition and Machine Learning. Springer, New York (2006)
21. Lawrence, N.D., Seeger, M., Herbrich, R.: Fast sparse Gaussian process methods: The informative vector machine. In: Becker, S., Thrun, S., Obermayer, K. (eds.) Advances in Neural Information Processing Systems, Vancouver, British Columbia, Canada, vol. 15, pp. 609–616 (2003)
22. Gelman, A., Carlin, J.B., Stern, H., Rubin, D.B.: Bayesian data analysis. Chapman and Hall-CRC (1996)
23. Fernandez, C., Steel, M.F.J.: Multivariate student-t regression models: Pitfalls and inference. Biometrika 86(1), 153–167 (1999)

Significant Node Identification
in Social Networks

Chi-Yao Tseng and Ming-Syan Chen

Research Center for Information Technology Innovation,
Academia Sinica, Taipei, Taiwan, ROC
{cytseng,mschen}@citi.sinica.edu.tw

Abstract. Given a social network, identifying significant nodes from
the network is highly desirable in many applications. In different networks formed by diverse kinds of social connections, the definitions of
what are significant nodes differ with circumstances. In the literature,
most previous works generally focus on expertise finding in specific social networks. In this paper, we aim to propose a general node ranking
model that can be adopted to satisfy a variety of service demands. We
devise an unsupervised learning method that produces the ranking list
of top-k significant nodes. The characteristic of this method is that it
can generate different ranking lists when diverse sets of features are considered. To demonstrate the real application of the proposed method, we
design the system DblpNET that is an author ranking system based on
the co-author network of DBLP computer science bibliography. We discuss further extensions and evaluate DblpNET empirically on the public
DBLP dataset. The evaluation results show that the proposed method
can effectively apply to real-world applications.

1 Introduction

Due to the popularity of social-related applications, social network analysis has
become an emerging research topic. Social network is a graph formed by col-
lecting human social communication, and there is much interesting and useful
knowledge hidden in a network. Among them, identifying significant nodes from
a network is highly desirable. Note that in different networks formed by diverse
kinds of social connections, the definitions of what are significant nodes also differ with circumstances. Therefore, previous works generally focus on one specific
application, such as the expertise finding in web forums [1], the expertise ranking
in collaborative tagging systems [2], and so forth [3], [4], [5].

In this paper, we aim to propose a general node ranking model that can be
adopted to satisfy a variety of service demands. Observing that significant nodes
usually have particular social behavior which is distinct from that of normal
ones, our model is based on the features extracted from each node in a network.
Moreover, since each feature has its own underlying meaning, it is natural that
the top significant nodes will vary when different sets of features are considered.
To capture this idea, we investigate a general node ranking problem in social
networks. The problem description is as follows.

L. Cao et al. (Eds.): PAKDD 2011 Workshops, LNAI 7104, pp. 459–470, 2012.
© Springer-Verlag Berlin Heidelberg 2012

Problem Description. The goal of the node ranking problem is to produce the ranking list of top-k significant nodes. There is no training data needed, meaning that the unsupervised learning model is employed to resolve this problem. A set of selected features and the number of top nodes should be given. The concept of the ranking process is to find top-k nodes that predominate over others when all the selected features are simultaneously considered. It is worth mentioning that the resulting list changes as the different sets of features are considered. This effect corresponds to the phenomenon of real applications. The main purpose is that the proposed method can have a more flexible mechanism and can be generally applied to miscellaneous applications.

To demonstrate the real application of this problem, we design the system DblpNET that is an author ranking system based on the co-author network of DBLP computer science bibliography [6]. The analysis of the co-author network constructed from a bibliographic database has been of significant interest in many studies [7], [8], [9], [10], [11]. Co-author network is an undirected graph where each node represents an author, and each link between two nodes indicates that these two authors have collaborated on a publication. Previous literature has proposed various methodologies to evaluate the significance of authors. In general, these works use certain metrics or design ranking algorithms with specific features to generate a list of top-k authors. However, it is more reasonable that the ranking shall depend on what features are taken into account. For instance, the author who has the highest number of papers may not have the highest number of co-authors. Moreover, when more than one feature is considered, the result should also be changed. Accordingly, it is necessary in view of this fact to devise a more flexible mechanism.

In this paper, we present an author ranking system DblpNET [12] that can generate different ranking lists of top-k authors with diverse sets of features. This system uses the dataset of DBLP computer science bibliography to construct a co-author network. The distinguishing characteristic is that DblpNET is capable of generating the ranking list of top-k authors according to what features are taken into account. Namely, the user can select what features they really care about. This characteristic makes DblpNET more flexible and closer to real situations. For each feature, we sort the list of authors by the value of this feature. Note that for some features, such as the number of papers, the larger value indicates a more significant author. However, other types of features, such as the closeness centrality[1], have opposite nature. With a set of selected features, the primary principle of the ranking strategy is that when the overall rank of author A is higher than that of author B, it means that all feature ranks of author A are higher than the worst feature rank of author B. This strategy guarantees that top authors on average perform well with all considered features.

Note that there have been numerous indices (such as h-index, citation count, etc.) defined in the literature to evaluate the importance of authors in academic community. We can also incorporate any of these indices in our system to provide users more features to select. The rest of this paper is outlined as follows.

---

[1] The definition of the closeness centrality will be given in Section 3.1.

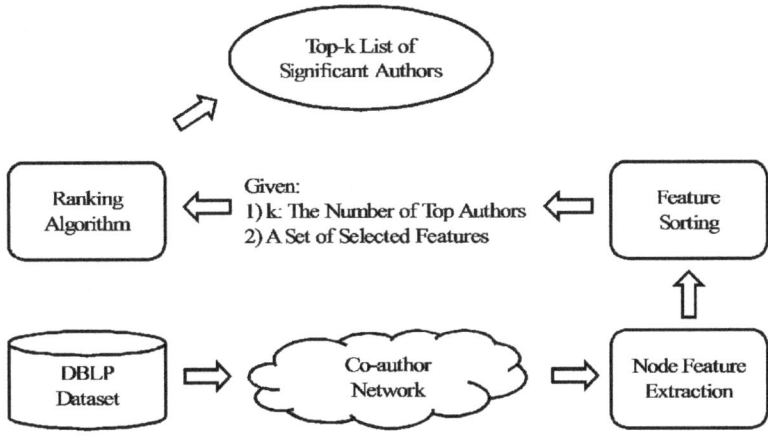

Fig. 1. System model of DblpNET

In Section 2, preliminaries including the system model of DblpNET and the related works are given. In Section 3, we elaborate the details of the DblpNET system. The experimental results are shown in Section 4. We discuss further extensions in Section 5, and finally, Section 6 summarizes this paper.

2   Preliminaries

In this section, the system model of the designed system DblpNET is presented in Section 2.1. We then review the related works on co-author network analysis in Section 2.2.

2.1   System Model of DblpNET

Figure 1 shows the system model of DblpNET. This system employs the DBLP dataset [6], which is the most prestigious and complete digital bibliography that contains bibliographic information on major computer science journals and proceedings, to construct a co-author network. Subsequently, several features are extracted from each author in the network. For each feature, we produce a ranking list of authors by sorting the feature values in descending or ascending order, which is determined according to whether or not the larger value indicates a more significant author. Until this phase is reached, all procedures are done off line.

DblpNET is ready for use when all sorted author lists of features are prepared. To generate the top list of significant authors, the number k of top authors and a set of selected features should be given. The designed ranking algorithm then proceeds with the sorted author lists of selected features. It is important to note that while diverse sets of features are considered, DblpNET can efficiently produce different top-k lists of significant authors. This distinguishing characteristic enables DblpNET to better satisfy practical requirements. The web version of DblpNET can be accessed through [12].

2.2   Related Works

Regarding the analysis of co-author networks, there are many studies investigating the hidden knowledge in this type of social network. In [10], the authors aim at evaluating the relevance between authors and discovering research communities by an iterative random walk algorithm. In [9], Newman analyzes the scientific collaboration patterns in bibliographic databases of biology, physics, and mathematics fields. With regard to the significance evaluation of authors, in [8], the closeness centrality score is used as the ranking criterion. Moreover, Liu et al. [7] define an alternative centrality metric based on a weighted directional network model. To rank not only authors but also documents, the authors in [11] propose a framework in a heterogeneous network connecting both researchers and publications.

In addition, Arnetminer[2] is a project developing the techniques of search and mining on academic social networks. Several indices[3] have been calculated or defined to produce the ranking lists of top authors, such as h-index, citation count, uptrend, etc. It can be observed that the result of each index differs from other ones. In this paper, we further consider combining any number of features to produce the ranking results. This flexible mechanism can better satisfy practical applications.

3   Top-k Author Ranking in Co-author Network

In this section, we present the designed author ranking system DblpNET. There is no training data needed, and thereby the proposed ranking algorithm is an unsupervised learning approach. Note that although we consider the co-author network in this section, the proposed algorithm can also be applied to similar applications in other social networks. We introduce the extracted features used to measure the significance of authors in Section 3.1. We elaborate the details of the ranking algorithm DblpRank in Section 3.2. Finally, the complexity analysis of the system DblpNET is given in Section 3.3.

3.1   Extracted Features

To identify significant authors from the co-author network of the DBLP database, four features of each author are extracted in DblpNET. Extracted features are as follows: 1) number of papers, 2) number of co-authors, 3) closeness centrality, and 4) component size. Though the significance of an author may not be assessed only by these features, they are still representatives related to the importance of authors. The details of each feature are depicted as follows.

---

[2] http://www.arnetminer.org/

[3] http://www.arnetminer.org/expertrank/

1) Number of papers. In general, the author who has published more papers is regarded as more important. This feature cannot be derived from the co-author network. We accumulate the number of papers of each author during the network construction phase.

2) Number of co-authors. Scientific collaboration is a common phenomenon in academic community. Thus, an author who has collaborated with more people is generally regarded as more important. This feature is equivalent to the degree of a node in the network.

3) Closeness centrality. As shown in [13], the scientific co-author network is a small world, meaning that most nodes can be reached from each other by a small number of steps. This phenomenon also exists in the DBLP dataset. In our experimental network constructed by the snapshot (as of December 31, 2010) of DBLP dataset, about 80 percent of authors compose the largest connected component. Moreover, important authors are usually situated near the center of the whole network. To capture this concept, closeness centrality is defined as the average shortest path length between this node and all other reachable nodes in the graph.

$$Centrality(x_i) = \frac{\sum\limits_{x_j \in V \setminus x_i} dist(x_i, x_j)}{|V| - 1},\tag{1}$$

where V is the set of nodes that can be reached from node x_i, and the function $dist$ returns the length of shortest path between node x_i and node x_j. An author who has smaller value of closeness centrality is normally viewed as more important. For ease of presentation, we use the term centrality to represent closeness centrality in this paper.

4) Component size. Component size of node x_i is defined as the number of nodes that can be reached from node x_i. This is an auxiliary feature for centrality. Concerning a group of authors that form a tiny and isolated network, their centrality values will be very small and close to 1. Consequently, since significant authors are usually not isolated and are in the largest connected component of the network, it is meaningful and reasonable to include the feature of component size when centrality is taken into account.

After the completion of the feature extraction phase, for each feature, we need to produce a ranking list of authors by sorting the feature values in descending or ascending order, which is determined according to whether or not the larger value indicates a more significant author. These sorted author lists are provided for the input of the ranking algorithm DblpRank.

3.2 Ranking Algorithm DblpRank

Once the sorted author lists are prepared, all off-line procedures of DblpNET are finished. As mentioned previously, DblpNET can adaptively generate different top-k ranking lists while diverse sets of features are considered. An intuitive approach is to give each selected feature a specific weight and define a weighting function to determine the ranking of authors. However, how to adequately set the weight of each feature is a difficult issue, and applying fixed weights to features

---

Algorithm DblpRank
Input: k: the number of top authors,
  S: the set of sorted author lists of selected features
Output: *topList*: top-k ranking list of authors
1 Create an empty list *topList*;
2 Create an empty candidate author set *candAuSet*;
3 $i = 0$; // i is the current processed rank index
4 **while** (*topList.size* < k)
5    $i = i + 1$;
6   Clear *candAuSet*;
7   **for** (each *list* in S)
8   Add authors who rank number i into *candAuSet*;
9   **for** (each *author* in *candAuSet*)
10   **if** (the rank of *author* in each *list* of S is equal to or smaller than i)
11   Append *author* to the end of *topList*;
12 **return** *topList*;
End

Fig. 2. Algorithmic form of DblpRank

will also limit the flexibility of the algorithm. Therefore, to provide a more objective assessment, the primary concept of the designed ranking algorithm is to find top-k authors that predominate over others when all the selected features are simultaneously taken into account. The core algorithm DblpRank is outlined in Figure 2, and an execution scenario of DblpRank is shown in Figure 3.

Initially, the number k of top authors and the set of selected features should be specified. In the example scenario of Figure 3, k is set as 5 and there are three features considered. In line 1 of Figure 2, an empty list *topList* is created to progressively store the top-k authors. In addition, an empty set *candAuSet* is also created in line 2 to store the candidate authors that are likely to be added into *topList* in each iteration. Subsequently, DblpRank starts to execute the main loop from line 4 to line 11. Given the set of sorted author lists of selected features, DblpRank simultaneously searches every list from top to down to find authors who rank high in all lists. When searching from top to down, the earlier an author appears in all lists, the higher this author ranks. For each iteration, we process one rank index. As shown in Iteration 1 of Figure 3, the authors who rank number 1 in each list are processed. From line 7 to line 8, DblpRank adds these authors into *candAuSet* (i.e., <B, D, E>). Then, from line 9 to line 11, for each author in *candAuSet*, if the rank of this author in each list is equal to or smaller than 1, this author will be appended to the end of *topList*. Since author E ranks number 1 in all three lists, after Iteration 1, author E is in *topList*. Similarly, in Iteration 2, the authors who rank number 2 are included in *candAuSet* (i.e., <C, A, B, D>). However, none of these authors rank equal to or smaller than 2 in all three lists, meaning that no one simultaneously appears in the rectangular area of each list. Thus, no author is appended to *topList* in this iteration. Note that as in Iteration 4 of Figure 3, both author A and author

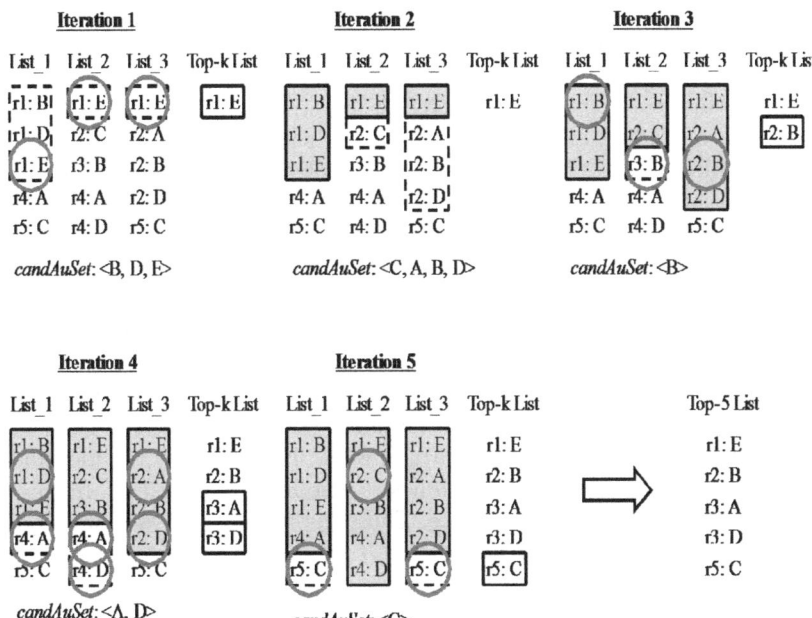

Fig. 3. Execution scenario of algorithm DblpRank

D are appended to *topList* in this iteration. For this situation, the final ranks of these two authors are set as the same number (i.e., 3). The main loop of DblpRank is repeated until the size of *topList* is equal to or lager than k.

As a whole, DblpRank provides the flexibility for users to choose the set of features that they really care about. Moreover, instead of determining the author ranking by defining a weighting function of feature values, DblpRank considers the ranks of authors in each selected feature. Furthermore, DblpRank does not take sides with any selected feature but extracts top-k authors that predominate over others when all selected features are simultaneously taken into account. These characteristics make DblpRank better satisfy practical requirements and suitable for the node ranking problem in miscellaneous applications.

3.3   Complexity Analysis

We next analyze the complexity of main procedures in DblpNET. Let N and E denote the number of nodes and the number of edges in the co-author network. Regarding the feature extraction phase, accumulating the number of papers and computing the degree of each node have the same time complexity $O(N)$. To obtain the component size of one node, we need to find the connected component of this node. With a DFS[4]-based algorithm, the time complexity is $O(N + E)$.

---

[4] DFS stands for Depth-First Search.

Notice that since about 80 percent of nodes in the DBLP co-author network compose the largest connected component, they have the same value of the component size without re-finding the connected component of each node. Therefore, the overall time complexity is still $O(N + E)$. As for the closeness centrality, the problem of finding all-pairs shortest paths should be solved[5], and thereby the time complexity is $O(N^3)$. After the completion of the feature extraction phase, we prepare the sorted author list of each feature for the DblpRank algorithm. Since there are N authors in each list, the time complexity of sorting is $O(NlogN)$. Although the worst time complexity is $O(N^3)$, all these procedures can be completed off line.

For the worst case of the DblpRank algorithm, the main execution loop (from line 4 to line 11 of Figure 2) will be repeated N times. Moreover, in line 10 of Figure 2, DblpRank needs to examine whether the rank of an author in each list is equal to or smaller than a specific number. To enhance the efficiency of this step, we utilize a random access array, which is indexed by the identifiers of authors, for each list to record those authors who have been visited. In this way, this step can be done in constant time. Thus, the time complexity of DblpRank is $O(N)$ and the space complexity is also $O(N)$.

4   Experimental Evaluation

To assess the effectiveness of the proposed method, we create a web version of DblpNET [12] and conduct several experiments to explore the performance of efficiency. On the other hand, Figure 4(a) presents the detailed information of the public DBLP dataset [6]. We include all the publications that are published before the end of year 2010, and totally there are about 1.4 million publications. The co-author network constructed by this dataset has close to $900,000$ nodes. Within this network, about 80 percent of authors are in the largest connected component.

We implement DblpNET with C++ programming language, and the programs are executed in Windows XP professional platform with Pentium 4 - 3GHz CPU and 1.5GB RAM. The demonstration of author ranking is shown in Section 4.1. The efficiency issue of DblpNET is discussed in Section 4.2.

4.1   Author Ranking of DblpNET

The demonstration of the top-10 author lists generated by DblpNET is given in this section. Figure 4(b) presents the top-10 author list when the numbers of papers and co-authors are considered, and Figure 4(c) is the list when four features are simultaneously taken into account. The dark columns in Figure 4(b) display the feature information that is not included for ranking. For each feature column, each grid shows the feature value of an author and the rank of this author in terms of this feature. It can be observed that as long as any feature

---

[5] We apply the Floyd-Warshall algorithm.

Snapshot time of DBLP dataset	2011/01/11
File size of dblp xml	775MB
Published years	<= 2010
Number of publications	1,399,341
Number of authors	894,460
Size of the largest connected component	728,632

(a) Detailed information of the dataset for DblpNET

rank	name	numPaper	numCoauthor	closenessCentr	componentSize
1	Wei Wang	521 (rank: 6)	948 (rank: 1)	3.95471 (rank: 117608)	728632 (rank: 1)
2	Wei Zhang	486 (rank: 8)	722 (rank: 3)	4.08538 (rank: 117620)	728632 (rank: 1)
3	Li Zhang	462 (rank: 11)	724 (rank: 2)	4.03448 (rank: 117610)	728632 (rank: 1)
4	Jun Wang	460 (rank: 12)	602 (rank: 6)	4.10723 (rank: 117627)	728632 (rank: 1)
4	Ming Li	469 (rank: 9)	524 (rank: 12)	4.07691 (rank: 117616)	728632 (rank: 1)
6	Yan Zhang	402 (rank: 26)	519 (rank: 13)	4.07612 (rank: 117615)	728632 (rank: 1)
7	Jun Zhang	401 (rank: 28)	562 (rank: 8)	4.10712 (rank: 117626)	728632 (rank: 1)
8	Lei Wang	397 (rank: 30)	664 (rank: 4)	4.07904 (rank: 117619)	728632 (rank: 1)
8	Alberto L. Sangiovanni-Vincentelli	450 (rank: 15)	420 (rank: 30)	4.36798 (rank: 117841)	728632 (rank: 1)
10	Wen Gao	522 (rank: 5)	410 (rank: 31)	4.19493 (rank: 117659)	728632 (rank: 1)

(b) Top-10 author list when the numbers of papers and co-authors are considered

rank	name	numPaper	numCoauthor	closenessCentr	componentSize
1	Wei Wang	521 (rank: 6)	948 (rank: 1)	3.95471 (rank: 117608)	728632 (rank: 1)
2	Li Zhang	462 (rank: 11)	724 (rank: 2)	4.03448 (rank: 117610)	728632 (rank: 1)
3	Wei Li	346 (rank: 45)	658 (rank: 5)	4.05159 (rank: 117611)	728632 (rank: 1)
4	Yan Zhang	402 (rank: 26)	519 (rank: 13)	4.07612 (rank: 117615)	728632 (rank: 1)
5	Ming Li	469 (rank: 9)	524 (rank: 12)	4.07691 (rank: 117616)	728632 (rank: 1)
6	Lei Wang	397 (rank: 30)	664 (rank: 4)	4.07904 (rank: 117619)	728632 (rank: 1)
7	Wei Zhang	486 (rank: 8)	722 (rank: 3)	4.08538 (rank: 117620)	728632 (rank: 1)
8	Li Li	317 (rank: 66)	551 (rank: 10)	4.08925 (rank: 117621)	728632 (rank: 1)
9	Jie Yang	332 (rank: 52)	455 (rank: 23)	4.10337 (rank: 117625)	728632 (rank: 1)
10	Jun Zhang	401 (rank: 28)	562 (rank: 8)	4.10712 (rank: 117626)	728632 (rank: 1)

(c) Top-10 author list when four features are simultaneously considered

Fig. 4. Detailed information of real dataset for DblpNET and top-10 author lists generated by DblpNET when diverse sets of features are considered

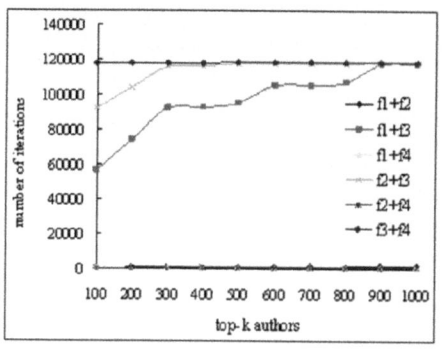 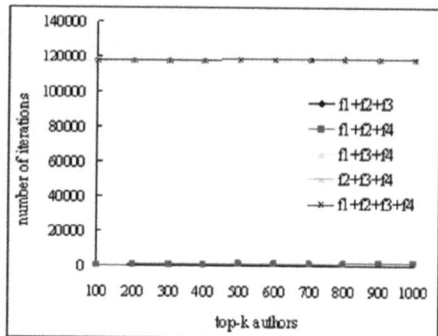

(a) The number of iterations with *k* varied   (b) The number of iterations with *k* varied when
when two features are considered   three and four features are considered

Fig. 5. The efficiency performance of DblpRank. (f1: number of papers; f2: number of co-authors; f3: closeness centrality; f4: component size)

rank of author A is lower than all feature ranks of author B, the overall rank of author A will be lower than that of author B. This is because we simultaneously consider the ranks of all selected features. On the other hand, as shown in Figure 4(b), both author Jun Wang and author Ming Li rank number 4. The reason is that these two authors are added into the top-k list in the same iteration (i.e., 12). The main characteristic of DblpNET is its capability to generate different top-k author lists with diverse sets of features. This trait makes DblpNET more flexible and closer to real situations. More results can be obtained through [12].

It is noted that the names of different authors may be identical. Regarding the issue of name ambiguity, the DBLP dataset uses additional numbers to distinguish different authors with the same name. For instance, when we search the authors with the keyword "Wei Wang" through [12], it can be observed that there are 19 persons named Wei Wang. Nevertheless, it is possible that not all cases have been resolved in the DBLP dataset. Since the issue of name ambiguity is not the focus in this paper, our results are based on the DBLP dataset and we do not particularly deal with this problem.

4.2   Efficiency Issue of DblpRank

Regarding the efficiency performance of DblpRank, we investigate in this section the number of iterations with the number k of top authors varied. Note that the number of iterations is equivalent to the level of rank in each sorted author list that DblpRank needs to examine. Figure 5(a) shows the conditions when two features are considered, and the conditions with three and four features are shown in Figure 5(b). In both figures, we can observe that while closeness centrality (i.e., f3) is included, the number of iterations rises up close to 120, 000. This effect is due to that there are 160, 000 or so nodes not in the largest connected component of the network, and most of these nodes are in relatively much smaller connected components. Thus, the values of closeness centrality are very small and close to 1. However, other features of these nodes are usually not prominent. As a result, when the closeness centrality is considered, DblpRank needs to pass through these authors to find those who predominate over others with every selected feature.

Additionally, for the web version of DblpNET [12], the response time is increased if the feature of component size is selected. This is caused by the fact that the values of component size of more than $720,000$ authors are the same and rank number 1. Namely, such a huge number of authors have to be processed in the first iteration. The time cost mainly originates from the file reading of almost the whole list. With this circumstance, nevertheless, DblpNET can efficiently and rapidly produce the top-k list of significant authors.

5 Discussion and Future Work

In this section, we discuss further extensions and potential future work of this study. First of all, since social networks are usually huge and evolving, the design of update scheme is essential but challenging. To capture the dynamic nature of networks, the update scheme needs to deal with not only the graph structure update of the network, but also the update of features used for the ranking algorithm. We will explore designing specific data structure to facilitate the updating process.

On the other hand, as for the network analysis in academic community, we will focus on broadening the meaning of ranking results. Followings are possible strategies.

- Incorporate other features that cannot be directly derived from the network, such as h-index, citation count, etc.
- Investigate the capability of generating ranking results based on only the data over a period of time.
- Integrate the keyword search into our system. As such, the results of the top-k authors who are experts in certain fields can be produced.
- Consider the significance of publication venue.

The intent of these strategies is to provide a variety of searching capabilities. More research efforts are needed to satisfy these demands.

6 Conclusion

In this paper, we proposed a general node ranking model that aims to produce the top-k list of significant nodes from social networks. We designed an unsupervised learning method to capture the idea that the top significant nodes will vary when different sets of features are considered. To demonstrate the real application of the proposed method, we devised an author ranking system DblpNET that can generate different ranking lists of top-k authors with diverse sets of features. The proposed ranking algorithm DblpRank is able to better satisfy practical requirements and suitable for the node ranking problem in miscellaneous applications.

References

1. Zhang, J., Ackerman, M.S., Adamic, L.: Expertise Networks in Online Communities: Structure and Algorithms. In: Proc. of the 16th International Conference on World Wide Web (WWW), pp. 221–230 (2007)
2. Noll, M.G., Au Yeung, C., Gibbins, N., Meinel, C., Shadbolt, N.: Telling Experts from Spammers: Expertise Ranking in Folksonomies. In: Proc. of the 32nd Annual International ACM SIGIR Conference on Research and Development in Information Retrieval (SIGIR), pp. 612–619 (2009)
3. Balog, K., Azzopardi, L., de Rijke, M.: Formal Models for Expert Finding in Enterprise Corpora. In: Proc. of the 29th Annual International ACM SIGIR Conference on Research and Development in Information Retrieval (SIGIR), pp. 43–50 (2006)
4. Dom, B., Eiron, I., Cozzi, A., Zhang, Y.: Graph-Based Ranking Algorithms for Email Expertise Analysis. In: Proc. of the 8th ACM SIGMOD Workshop on Research Issues in Data Mining and Knowledge Discovery, pp. 42–48 (2003)
5. Zhang, J., Tang, J., Li, J.: Expert Finding in a Social Network. In: Kotagiri, R., Radha Krishna, P., Mohania, M., Nantajeewarawat, E. (eds.) DASFAA 2007. LNCS, vol. 4443, pp. 1066–1069. Springer, Heidelberg (2007)
6. Ley, M.: The DBLP (Digital Bibliography and Library Project) Computer Science Bibliography, http://www.informatik.uni-trier.de/~ley/db/
7. Liu, X., Bollen, J., Nelson, M.L., de Sompel, H.V.: Co-authorship Networks in the Digital Library Research Community. In: Information Processing and Management, pp. 1462–1480 (2005)
8. Nascimento, M.A., Sander, J., Pound, J.: Analysis of SIGMOD's Co-Authorship Graph. ACM SIGMOD Record, 8–10 (2003)
9. Newman, M.E.J.: Coauthorship Networks and Patterns of Scientific Collaboration. Proc. of the National Academy of Sciences, 5200–5205 (2004)
10. Zaïane, O.R., Chen, J., Goebel, R.: Mining Research Communities in Bibliographical Data. In: Zhang, H., Spiliopoulou, M., Mobasher, B., Giles, C.L., McCallum, A., Nasraoui, O., Srivastava, J., Yen, J. (eds.) WebKDD 2007. LNCS, vol. 5439, pp. 59–76. Springer, Heidelberg (2009)
11. Zhou, D., Orshanskiy, S.A., Zha, H., Giles, C.L.: Co-Ranking Authors and Documents in a Heterogeneous Network. In: Proc. of the 7th IEEE International Conference on Data Mining (ICDM), pp. 739–744 (2007)
12. Tseng, C.Y., Chen, M.S.: Significant Nodes Identification in the Co-author Network of DBLP, http://cytseng.no-ip.org/dblpnet.php
13. Newman, M.E.J.: The Structure of Scientific Collaboration Networks. Proc. of the National Academy of Sciences, 404–409 (2001)

Improving Bagging Performance through Multi-algorithm Ensembles

Kuo-Wei Hsu[1,*] and Jaideep Srivastava[2]

[1] Department of Computer Science, National Chengchi University, Taipei 11605,
Taiwan (ROC)

[2] Department of Computer Science and Engineering, University of Minnesota,
Minneapolis, MN 55455, USA

Abstract. Bagging establishes a committee of classifiers first and then
aggregates their outcomes through majority voting. Bagging has
attracted considerable research interest and been applied in various application domains. Its advantages include an increased capability of handling small data sets, less sensitivity to noise or outliers, and a parallel
structure for efficient implementations. However, it has been found to
be less accurate than some other ensemble methods. In this paper, we
propose an approach that improves bagging through the employment of
multiple classification algorithms in ensembles. Our approach preserves
the parallel structure of bagging and improves the accuracy of bagging.
As a result, it unlocks the power and expands the user base of bagging.

1 Introduction

Over the years, ensemble methods have been an active topic of data mining
research. Bagging, *bootstrap aggregating*, is an ensemble method that establishes
a committee of classifiers first and then combines their outcomes through a
majority voting mechanism [4]. It has attracted considerable research interests
and been applied in various application domains, such as business [28,22,27,21],
medicine [10,24], environmental study [35], semantic web [39], engineering [3,2],
linguistics [19,20], and face recognition [37]. Its advantages include better ability
to handle small data sets, less sensitivity to outliers or noisy data samples, and a
parallel structure for efficient implementations. However, bagging is usually less
accurate than some other ensemble methods. Therefore, this paper is primarily
concerned with the classification performance of bagging.

It is commonly admitted that classifiers need to be complementary in order
to improve their performance by aggregation. This is usually referred to as *diversity*, which contributes significantly to the success of ensembles. We leverage
diversity by using different classification algorithms. To begin with, we investigate two special types of diversity in detail and show that considering both
together can provide better diversity. We formulate the diversity caused by the

---

* This paper is a short version of Kuo-Wei Hsu's doctoral dissertation [12], which is
under the supervision of Prof. Jaideep Srivastava.

L. Cao et al. (Eds.): PAKDD 2011 Workshops, LNAI 7104, pp. 471–482, 2012.

difference between training sets and the diversity caused by the difference between classification algorithms on which member classifiers in an ensemble are based. Since there is a lack of research on a theoretical analysis of the use of heterogeneous classification algorithms, it has motivated us to propose a formal definition of heterogeneity, and to systematically investigate the impact of heterogeneity on diversity in ensembles, and further on classification performance such as accuracy of ensembles.

There is a need to theoretically analyze the relationship between diversity and accuracy in order to understand the role and impact of diversity. However, in spite of the importance of such an analysis and the considerable advances made over the last few years in understanding the relationship between diversity and accuracy, we still lack theories to help us understand such a relationship. Rather than directly explore the relationship between diversity and accuracy, we investigate the relationship between diversity and correlation for classification problems to explain the role that diversity plays in ensemble methods. We present a function that describes a relationship between diversity and correlation, which has been related to overall accuracy by others. A direct relationship, of course, remains an open question.

In this paper, we provide a theoretical justification for the use of heterogeneous classification algorithms in ensembles and develop a framework that supports the use of different classification algorithms in bagging. To sum up, the research contributions of this paper are as follows:

– We propose two new types of diversity measures, namely T-Diversity and A-Diversity. The former helps quantify the stability of an algorithm; the latter helps quantify the heterogeneity of two algorithms.
– We theoretically prove that, compared to ensembles of homogeneous classification algorithms, ensembles of heterogeneous classification algorithms have a higher probability to provide better diversity.
– We present a relationship between diversity and correlation among member classifiers, showing that diversity indeed plays an important role in the success of ensemble methods.
– We propose a variant of bagging that preserves the parallel structure but enhances classification performance.

The rest of this paper is organized is as follows: Section 2 provides a theoretical justification for using heterogeneous classification algorithms in ensembles. Section 3 discusses the relationship between diversity and correlation, which has been connected to accuracy. Section 4 proposes the framework of using heterogeneous classification algorithms in bagging. Section 5 discusses experimental results and findings, including comparisons to some other state-of-the-art ensemble methods. Finally, Section 6 concludes this paper and discusses possible directions for future research work.

2   Diversity in Combinations of Heterogeneous Classifiers

In this section, at first, we introduce notations and definitions for our analysis. More details can be found in the paper [13]. In what follows, $\mathbb{E}$ is the expectation operator, $\mathbb{I}$ is the indicator function, and $\hat{y}_i^{(A,S)}$ is the class label obtained from a classifier created using a classification algorithm A and a training set S. In such a notation, the training error of applying A on S is viewed as the number of disagreements that are normalized with respect to the size of the data set S.

Definition 1. *(T-Diversity)* D *is an underlying distribution (which is unknown in practice). Two bootstrap sets obtained from a given data set drawn from* D *are denoted as* S_1 *and* S_2 *(i.e.* S_1 *and* $S_2 \sim D$*). If* S_1 *and* S_2 *are training sets and* $S = S_1 \cap S_2$ *is a probe set used to present common training samples, then T-Diversity of a classification algorithm* A *with respect to* S_1 *and* S_2 *is defined as follows:*

$$
\begin{aligned}
\text{T-Diversity} &= \mathbb{E}_S \left[\mathbb{I} \left(\hat{y}_i^{(A,S_1)} \neq \hat{y}_i^{(A,S_2)} \right) \right] \\
&= \tfrac{1}{|S|} \sum_{i=1}^{|S|} \mathbb{I} \left(\hat{y}_i^{(A,S_1)} \neq \hat{y}_i^{(A,S_2)} \right)
\end{aligned}
\tag{1}
$$

Definition 2. *(Stability with respect to T-Diversity) A classification algorithm* A *is of* (α,β)*-stability with respect to T-Diversity if the following condition holds:*

$$
\begin{aligned}
(\alpha, \beta) - stability &= Pr_D \left[\mathbb{E}_S \left[\mathbb{I} \left(\hat{y}_i^{(A,S_1)} \neq \hat{y}_i^{(A,S_2)} \right) \right] \leq \alpha \right] \geq \beta \\
&= Pr_D \left[\tfrac{1}{n} \sum_{i=1}^{n} \mathbb{I} \left(\hat{y}_i^{(A,S_1)} \neq \hat{y}_i^{(A,S_2)} \right) \leq \alpha \right] \geq \beta
\end{aligned}
\tag{2}
$$

where $s_i \in S = S_1 \cap S_2 \neq \emptyset, 1 \leq i \leq |S| = n,$ *and* $0 \leq \alpha, \beta \leq 1$.

Definition 3. *(A-Diversity)* D *is an underlying distribution (which is unknown in practice) and a bootstrap set obtained from a given data set drawn from* D *is denoted as* S *(i.e.* $S \sim D$*). If* S *is a training set and* S' *is a probe set (i.e.* $S' \subseteq S \sim D$ *and* $S' \neq \emptyset$*) used to investigate classifiers, then A-Diversity of two classification algorithms* A_1 *and* A_2 *with respect to* S *is defined as follows:*

$$
\begin{aligned}
A - Diversity &= \mathbb{E}_{S'} \left[\mathbb{I} \left(\hat{y}_i^{(A_1,S)} \neq \hat{y}_i^{(A_2,S)} \right) \right] \\
&= \tfrac{1}{|S'|} \sum_{i=1}^{|S'|} \mathbb{I} \left(\hat{y}_i^{(A_1,S)} \neq \hat{y}_i^{(A_2,S)} \right).
\end{aligned}
\tag{3}
$$

Definition 4. *(Heterogeneity with respect to A-Diversity) Classification algorithms* A_1 *and* A_2 *are of* (δ,γ)*-heterogeneity if the following condition holds:*

$$(\delta,\gamma) - heterogeneity = Pr_D\left[\mathbb{E}_{S'}\left[\mathbb{I}\left(\hat{y}_i^{(A_1,S)} \neq \hat{y}_i^{(A_2,S)}\right)\right] \geq \delta\right] \geq \gamma$$

$$= Pr_D\left[\frac{1}{|S'|}\sum_{i=1}^{|S'|}\mathbb{I}\left(\hat{y}_i^{(A_1,S)} \neq \hat{y}_i^{(A_2,S)}\right) \geq \delta\right] \geq \gamma \quad (4)$$

where $s_i \in S' \subseteq S \sim D, S' \neq \emptyset, 1 \leq i \leq |S'|$, *and* $0 \leq \delta, \gamma \leq 1$.

Now we consider mini ensembles of two classifiers, where each mini ensemble is of either homogeneous or heterogeneous classification algorithms. Here are three classifiers: The first one is based on a classification algorithm A_1 and trained by a data set S_1; the second one is also based on the classification algorithm A_1 but trained by another data set S_2; the third classifier is based on another classification algorithm A_2 and trained by the data set S_2, the one used in the construction of the second classifier. In view of this, the combination of the first classifier and the second one constructs a homogeneous ensemble, i.e. a pair of homogeneous classifiers, since they both are based on the same algorithm (A_1); the combination of the first classifier and the third one constructs a heterogeneous ensemble, i.e. a pair of heterogeneous classifiers, since the underlying algorithms are different. Theorem 1 is to quantify diversity gain of using heterogeneous classification algorithms in a probabilistic manner. Before we give our theorem, we define the *diversity gain* in Eq. 5.

$$\Delta Div = \mathbb{E}\left[\mathbb{I}\left(\hat{y}_i^{(A_1,S_1)} \neq \hat{y}_i^{(A_2,S_2)}\right)\right] - \mathbb{E}\left[\mathbb{I}\left(\hat{y}_i^{(A_1,S_1)} \neq \hat{y}_i^{(A_1,S_2)}\right)\right] \quad (5)$$

where S_1 and S_2 are two bootstrap subsets obtained from a given data set S.

Theorem 1. *When a given classification algorithm A_1 is of (α_1,β_1)-stability with respect to T-Diversity, another given classification algorithm A_2 is of (α_2,β_2)-stability with respect to T-Diversity, and A_1 and A_2 are of (δ_2,γ_2)-heterogeneity with respect to A-Diversity, then the following holds:*

$$Pr[\Delta Div \geq \delta_2 - 2 \cdot \alpha_1] \geq \gamma_2 \cdot \beta_1 \text{ where } \Delta Div \text{ is defined in Equation 5} \quad (6)$$

The proof of Theorem 1 can be found in the paper [13].

3   Relationship between Diversity and Correlation in Ensembles

Diversity has been studied by many researchers [8,7], but its relationship to accuracy is not clear. One difficulty is that there exists an elegant bias-variance-covariance decomposition framework for regression tasks, but the framework does not directly apply to classification tasks [6]. In this section, we present a relationship between diversity and correlation. More details can be found in the paper [15]. We consider disagreement measure (DIS) representing diversity [17], Q statistic or Q [17,16], and correlation [17] in our analysis. The result is Eq. 7, where N is the number of data samples. Eq. 7 is the most important result of this section, because it gives a non-linear relationship between diversity (represented by disagreement measure, DIS) and Q statistic.

$$Q \leq \frac{(1 - DIS)^2 \cdot N^2 - 4 \cdot DIS \cdot N}{(1 - DIS)^2 \cdot N^2 + 4 \cdot DIS \cdot N} \tag{7}$$

Because we would like to have zero correlation (and we care about the absolute value of $f(x)$ due to $|\rho| \leq |Q|$), we would like to know the interception of $f(x)$ and x-axis. We call the interception the "*critical value of x (x_c)*" or the "*critical point of DIS*", as given in Eq. 8, where x is diversity (represented by disagreement measure, DIS).

$$x_c = \left(1 + \frac{2}{N}\right) - 2 \cdot \sqrt{\frac{1}{N} + \frac{1}{N^2}} \tag{8}$$

We use x_c to divide the graph into two regions. On the one hand, in the left region (between 0 and x_c), the upper bound of Q statistic approaches 0, which implies that classifiers are statistically independent, as the value of disagreement measure representing diversity increases. This means that higher diversity would cause reduce correlation, and this supports the intuition that higher diversity between classifiers is usually associated with a better ensemble. On the other hand, in the right region (between x_c and 1), the upper bound of Q statistic moves away from 0 as the value of disagreement measure increases. When diversity crosses the critical point, increasing diversity would increase correlation while highly correlated classifiers usually correspond to an inferior ensemble.

The derivation of Eq. 7 and that of Eq. 8 can be found in the paper [15].

4   Bagging with Multi-algorithm Ensembles

Section 2 has shown us that heterogeneity could have a positive impact on diversity (i.e. using heterogeneous classification algorithms could improve diversity); Section 3 has shown us that increasing diversity would be related to decreasing correlation between member classifiers in an ensemble, while it has been shown by other researchers that lower correlation between member classifiers in an ensemble is usually associated with better classification performance. Thus, in order to improve classification performance of bagging, the goal is to introduce another source of diversity into bagging by introducing heterogeneity into bagging. Bagging with multi-algorithm ensembles, a new variant of bagging, is presented with this goal in mind.

Bagging works as follows: It first utilizes the bootstrap procedure to generate multiple training sets, and then it creates classifiers by using a classification algorithm and the generated training sets. If it is implemented in a serial fashion, a training set is generated using the bootstrap procedure in each iteration; if it is implemented in parallel, it could simultaneously generate multiple training sets each of which is used to work with a classification algorithm to create a classifier. We call this *classical bagging*.

After realizing that diversity in bagging is completely based on bootstrap data sets, we are motivated to introduce another source of diversity. We accomplish this by transforming heterogeneity into diversity. Algorithm 1 presents the proposed approach, namely bagging with multi-algorithm ensembles. Its inputs include a data set D, a bag of classification algorithms M, and the number of bootstrap training sets (or the number of iterations) T. Its output is an ensemble. For each bootstrap training set (or in each iteration), the proposed approach employs the bootstrap procedure to generate a training set, selects a classification algorithm or multiple ones from the given bag of classification algorithms, and accordingly creates a classifier or multiple ones accordingly. It groups the classifiers created for all bootstrap training sets to form an ensemble (or it adds the created classifiers in all iterations into the output ensemble).

Algorithm 1 (Framework for bagging with multi-algorithm ensembles)

1. Initialize $C^* = \emptyset$
2. For $j = 1$ to T
3. $S_j \leftarrow Bootstrap(D, p = 1)$
4.   Select an algorithm A_i from the given bag of algorithms M
5. $C_i \leftarrow \mathbf{C}(S_i, A_i)$, where $\mathbf{C}$ is a procedure, e.g. API, to create a classifier
6. $C^* \leftarrow C^* \cup C_i$
7. End for

Figure 1 presents the architecture of bagging with multi-algorithm ensembles. In Figure 1, D is a given data set, S_j is a bootstrap training set, and $\sum$ means aggregation. However, there are multiple classification algorithms employed in Figure 1. We denote the bag of classification algorithms as $M = \{A_1, A_2, \ldots, A_i, \ldots, A_m\}$ where $m = |M|$. In Figure 1, for example, A_1 is selected to create classifiers for S_1 and S_{T-1}, while A_m is selected to create classifiers for S_2 and S_T (where $m = |M|$ is the number of algorithms in the bag, and T is the number of bootstrap training sets or the number or iterations). This could be a situation where we randomly select algorithms from the bag. Additionally, this could be a situation where we select all algorithms to create classifiers for each bootstrap training sets and perform pruning to select one of the created classifiers. The first situation suggests *preprocessing*, while the second situation suggests *postprocessing*.

5   Experimental Results and Findings

We use benchmark data sets, and we consider five classification algorithms from "top 10 data mining algorithms" [40], namely C4.5, CART (classification and regression trees), Naive Bayes, kNN (k-nearest neighbor), SVM (support vector machine). We use SMO (sequential minimal optimization[29]) as an implementation of SVM. In addition, we use two types of kernels: Polynomial and RBF (radial basis function) kernels.

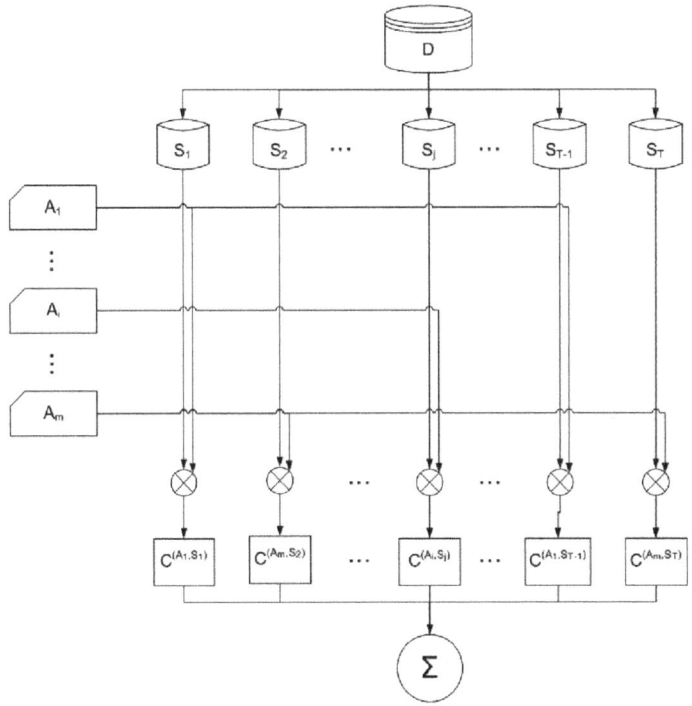

Fig. 1. Bagging with multi-algorithm ensembles

5.1   Impact of Using Heterogeneous Algorithms on Diversity

In this subsection, we discuss some experiments and results related to Section 2. More experimental results can be found in the papers [13,14]. We consider diversity values in various measures for using homogeneous and heterogeneous classification algorithms in mini ensembles. We randomly draw 66% of data samples for training and 33% of data samples for testing the classifiers; for both combinations of homogeneous and heterogeneous classifiers, we use the same pair of training and test sets; we repeat this 10 times.

According to results, we have the following findings: Using heterogeneous classification algorithms in ensembles provide higher diversity values in 9 out of the 10 diversity measures that prefer higher values. The only exception is the diversity measure DFD (distinct failure diversity [1]), but the values given by combinations of heterogeneous classifiers are not much lower than those given by combinations of homogeneous classifiers. Furthermore, using heterogeneous classification algorithms in ensembles provide lower diversity values in 7 out of the 9 diversity measures that prefer lower values. Two exceptions are diversity measures EBPDM (entropy-based pairwise diversity measure [23]) and KP (pairwise kappa [16]). For EBPDM, values given by combinations of heterogeneous classifiers are not much higher than those given by combinations of homogeneous classifiers; for KP, both types of combinations give the same value.

5.2   Simulations for Diversity and Correlation in Ensembles

In this subsection, we discuss some experiments and results related to Section 3. More experimental results can be found in the paper [15]. For a data set, we randomly draw N samples and accordingly train a decision tree (without pruning) for each trial, and we create another classifier to form an ensemble. We generate a disjoint set of N samples and use it as a test set for a data set. In every experiment, we repeat this 100 times, create 100 pairs of classifiers, and use the corresponding test set to evaluate each pair of classifiers.

Figures 2 and 3 illustrate the results for a data set. In each of these figures, the x-axis is the value of disagreement measure (representing diversity) and the y-axis corresponds to values of Q statistic and correlation. Moreover, a diamond (blue) and a square (pink) represent an observed value of Q statistic and an observed value of correlation, respectively; a triangle (yellow) means a theoretical value given by Eq. 7 and it gives an upper bound of the corresponding value of Q statistic.

From these figures we have the following findings: 1) The relationship between disagreement measure and Q statistic is not linear. 2) For some data sets, the theoretical upper bounds of the values of Q statistic are close to the observed values. 3) Exceptional cases where Q is 1 are those that do not follow the assumption in the analysis. 4) As N increases, curves move to the right, and this suggests that we need to increase diversity to obtain low correlation when the number of training samples increases. 5) It is not always the case that we observe critical points in experiments, while we observe that Q statistic and correlation move away from 0 as the diversity increases for those showing critical points.

5.3   Comparison of Bagging with Multi-algorithm Ensembles to Other Ensemble Methods

In this subsection, we discuss some experiments and results related to Section 4. Experimental results of applying bagging with multi-algorithm ensembles to health care data can be found in the paper [11]. We create 10 member classifiers for each of the following settings, except for ADABOOST-M1 in which we create 1,000 member classifiers. We consider pairwise comparison results for F1-measure, mean margin, AUC (area under ROC curve), and training time, respectively. Both mean margin and AUC could be related to the generalization capability of a classifier or a model.

- B-C45+NB-ALT: Bagging with C45 and NB together with the plan ALT, which alternatively selects algorithms.
- B-C45+NB-MAX_F1: Bagging with C45 and NB together with the plan MAX_F1 (or MAX_F1MEASURE), which selects the algorithms that achieve the highest F1-measures in training sets.
- ADABOOST-M1 [33]: AdaBoost (version M1, specific to binary classification) using decision stump as the base algorithm, which is the most common setting, and running for 1,000 iterations.

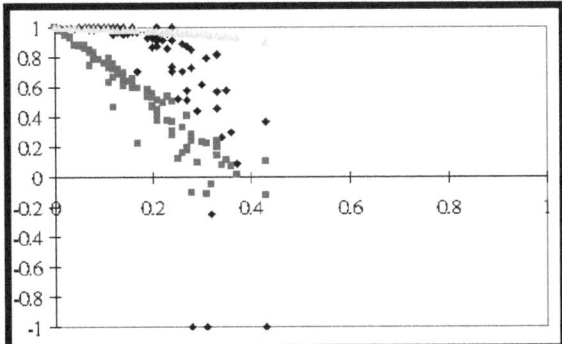

Fig. 2. Simulation results for diversity and correlation on the data set *Letter* with 100 samples

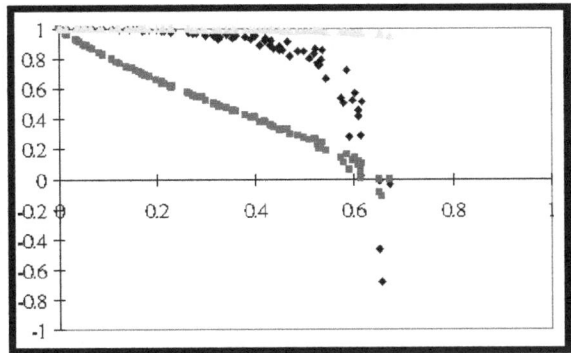

Fig. 3. Simulation results for diversity and correlation on the data set *Letter* with 1000 samples

- RANDOM_FOREST [5]: Random Forest or Random Forests.
- DECORATE [25,26]: DECORATE with C45.
- STACKING [38,34]: StackingC, a two-level hierarchical ensemble design, using a linear regression algorithm in the upper level (i.e. meta level) and using C45 and NB in the lower level with the number of folds set to 10.
- ROTATION_FOREST [32,18]: Rotation Forest with C45.
- RANDOM_SUBSPACE [9]: Random Subspace Method (RSM) using REP-Tree, Reduced Error Pruning Tree [30,31], as the base algorithm.
- DAGGING [36]: A "stacking bagged and dagged model" using SMO with polynomial kernel as the base algorithm.

Compared to B-C45+NB-ALT, STACKING and ROTATION_FOREST are better in terms of F1-measure; B-C45+NB-MAX_F1 is better in terms of mean margin; B-C45+NB-MAX_F1, ADABOOST-M1, and ROTATION_FOREST are better in terms of AUC. AUC is another performance metric that indicated generalization capability of a classifier. B-C45+NB-ALT outperforms B-C45+NB-

MAX_F1 on 5 data sets but is outperformed by B-C45+NB-MAX_F1 on 6 data sets. Similarly, it outperforms ADABOOST-M1 on 7 data sets but is outperformed by ADABOOST-M1 on 8 data sets. B-C45+NB-ALT loses to ROTATION_FOREST on 8 data sets and wins over ROTATION_FOREST on only 1 data set; B-C45+NB-MAX_F1 loses to ROTATION_FOREST on 9 data sets and wins over ROTATION_FOREST on only 2 data sets. Nevertheless, both B-C45+NB-ALT and B-C45+NB-MAX_F1 achieve performance not too far from the best performance achieved by others.

The advantage of B-C45+NB-ALT exists in its simplicity. In terms of training time, B-C45+NB-ALT outperforms all but RANDOM_FOREST and RANDOM_SUBSPACE. These 2 ensemble methods randomly sample subsets of the feature set to generate different training sets, while they run faster because dimensions of these 16 data sets are relatively small. Compared to B-C45+NB-ALT, RANDOM_FOREST and RANDOM_SUBSPACE do not demonstrate better classification on more data sets in F1-measure, mean margin, or AUC. Therefore, we can conclude that our approach provides a better balance between effectiveness and efficiency − it could be as effective as (or even better than) the best ones and, at the same time, more efficient than most others, compared to state-of-the-art ensemble methods.

6   Conclusions and Future Research Directions

In this paper, we investigate two special types of diversities in detail and discover that considering both together could provide better diversity. We formulate the diversity caused by the difference between training sets and the diversity caused by the difference between classification algorithms. We show theoretically that there is a probabilistic relationship between heterogeneity and diversity gain, and we present a non-linear function that describes the relationship between diversity and correlation, whose relationship to accuracy has been studied by others. Moreover, we propose a framework for using heterogeneous algorithms in bagging. Experimental results show that the proposed framework provides comparable performance in terms of the classification performance.

We discuss future work in three directions. The first direction is to study the employment of heterogeneous classification algorithms in other machine learning problems, such as multi-class and multi-label classification problems. The second one is about efficient implementations of our approach on advanced computing platforms. The third direction is to apply our approach to emerging application domains, such as anomaly detection.

References

1. Aksela, M., Laaksonen, J.: Using diversity of errors for selecting members of a committee classifier. Pattern Recognition 39(4), 608–623 (2006)
2. Aljamaan, H.I., Elish, M.O.: An empirical study of bagging and boosting ensembles for identifying faulty classes in object-oriented software. In: CIDM, pp. 187–194. IEEE (2009)

3. Braga, P., Oliveira, A., Ribeiro, G., Meira, S.: Bagging predictors for estimation of software project effort. In: International Joint Conference on Neural Networks, IJCNN 2007, pp. 1595–1600. IEEE (2007)
4. Breiman, L.: Bagging predictors. Machine learning 24(2), 123–140 (1996)
5. Breiman, L.: Random forests. Machine learning 45(1), 5–32 (2001)
6. Brown, G.: Ensemble learning. In: Encyclopedia of Machine Learning. Springer, Heidelberg (2010)
7. Brown, G., Wyatt, J., Harris, R., Yao, X.: Diversity creation methods: a survey and categorisation. Information Fusion 6(1), 5–20 (2005)
8. Ghosh, J.: Multiclassifier Systems: Back to the Future. In: Roli, F., Kittler, J. (eds.) MCS 2002. LNCS, vol. 2364, pp. 1–15. Springer, Heidelberg (2002)
9. Ho, T.: The random subspace method for constructing decision forests. IEEE Transactions on Pattern Analysis and Machine Intelligence 20(8), 832–844 (1998)
10. Hothorn, T., Lausen, B.: Bagging tree classifiers for laser scanning images: a data- and simulation-based strategy. Artificial Intelligence in Medicine 27(1), 65–79 (2003)
11. Hsu, K.W.: Applying bagging with heterogeneous algorithms to health care data (2010)
12. Hsu, K.W.: Improving Bagging Performance through Multi-Algorithm Ensembles. Ph.D. thesis, University of Minnesota (2011)
13. Hsu, K.W., Srivastava, J.: Diversity in Combinations of Heterogeneous Classifiers. In: Theeramunkong, T., Kijsirikul, B., Cercone, N., Ho, T.-B. (eds.) PAKDD 2009. LNCS, vol. 5476, pp. 923–932. Springer, Heidelberg (2009)
14. Hsu, K.-W., Srivastava, J.: An Empirical Study of Applying Ensembles of Heterogeneous Classifiers on Imperfect Data. In: Theeramunkong, T., Nattee, C., Adeodato, P.J.L., Chawla, N., Christen, P., Lenca, P., Poon, J., Williams, G. (eds.) New Frontiers in Applied Data Mining. LNCS, vol. 5669, pp. 28–39. Springer, Heidelberg (2010)
15. Hsu, K.W., Srivastava, J.: Relationship Between Diversity and Correlation in Multi-Classifier Systems. In: Zaki, M.J., Yu, J.X., Ravindran, B., Pudi, V. (eds.) PAKDD 2010. LNCS, vol. 6119, pp. 500–506. Springer, Heidelberg (2010)
16. Kuncheva, L.: Combining pattern classifiers: methods and algorithms. Wiley-Interscience (2004)
17. Kuncheva, L., Whitaker, C.: Measures of diversity in classifier ensembles and their relationship with the ensemble accuracy. Machine Learning 51(2), 181–207 (2003)
18. Kuncheva, L.I., Rodríguez, J.J.: An Experimental Study on Rotation Forest Ensembles. In: Haindl, M., Kittler, J., Roli, F. (eds.) MCS 2007. LNCS, vol. 4472, pp. 459–468. Springer, Heidelberg (2007)
19. Kurogi, S., Nedachi, N., Funatsu, Y.: Reproduction and Recognition of Vowel Signals using Single and Bagging Competitive Associative Nets. In: Ishikawa, M., Doya, K., Miyamoto, H., Yamakawa, T. (eds.) ICONIP 2007, Part II. LNCS, vol. 4985, pp. 40–49. Springer, Heidelberg (2008)
20. Kurogi, S., Sato, S., Ichimaru, K.: Speaker Recognition using Pole Distribution of Speech Signals Obtained by Bagging Can2. In: Leung, C.S., Lee, M., Chan, J.H. (eds.) ICONIP 2009. LNCS, vol. 5863, pp. 622–629. Springer, Heidelberg (2009)
21. Lasota, T., Telec, Z., Trawiński, B., Trawiński, K.: A Multi-Agent System to Assist with Real Estate Appraisals using Bagging Ensembles. In: Nguyen, N.T., Kowalczyk, R., Chen, S.-M. (eds.) ICCCI 2009. LNCS, vol. 5796, pp. 813–824. Springer, Heidelberg (2009)
22. Lemmens, A., Croux, C.: Bagging and boosting classification trees to predict churn. Journal of Marketing Research 43(2), 276–286 (2006)

23. Liu, W., Wu, Z., Pan, G.: An Entropy-Based Diversity Measure for Classifier Combining and its Application to Face Classifier Ensemble Thinning. In: Li, S.Z., Lai, J.-H., Tan, T., Feng, G.-C., Wang, Y. (eds.) SINOBIOMETRICS 2004. LNCS, vol. 3338, pp. 118–124. Springer, Heidelberg (2004)
24. Lu, C., Devos, A., Suykens, J., Arús, C., Van Huffel, S.: Bagging linear sparse bayesian learning models for variable selection in cancer diagnosis. IEEE Transactions on Information Technology in Biomedicine 11(3), 338–347 (2007)
25. Melville, P., Mooney, R.: Constructing diverse classifier ensembles using artificial training examples. In: Proceedings of the IJCAI, pp. 505–510. Citeseer (2003)
26. Melville, P., Mooney, R.: Creating diversity in ensembles using artificial data. Information Fusion 6(1), 99–111 (2005)
27. Perlich, C., Rosset, S., Lawrence, R.D., Zadrozny, B.: High-quantile modeling for customer wallet estimation and other applications. In: Berkhin, P., Caruana, R., Wu, X. (eds.) KDD, pp. 977–985. ACM (2007)
28. Pinheiro, C.A.R., Evsukoff, A., Ebecken, N.F.F.: Revenue recovering with insolvency prevention on a brazilian telecom operator. SIGKDD Explorations 8(1), 65–70 (2006)
29. Platt, J.: Machines using sequential minimal optimization. In: Schoelkopf, B., Burges, C., Smola, A. (eds.) Advances in Kernel Methods - Support Vector Learning. The MIT Press (1998)
30. Quinlan, J.: Learning with continuous classes. In: Proceedings of the 5th Australian Joint Conference on Artificial Intelligence, pp. 343–348. Citeseer (1992)
31. Quinlan, J.: C4. 5: programs for machine learning. Morgan Kaufmann (1993)
32. Rodriguez, J., Kuncheva, L., Alonso, C.: Rotation forest: A new classifier ensemble method. IEEE Transactions on Pattern Analysis and Machine Intelligence 28(10), 1619–1630 (2006)
33. Schapire, R.E.: The strength of weak learnability. Machine Learning 5, 197–227 (1990)
34. Seewald, A.K.: How to make stacking better and faster while also taking care of an unknown weakness. In: Sammut, C., Hoffmann, A.G. (eds.) ICML, pp. 554–561. Morgan Kaufmann (2002)
35. Stepinski, T.F., Ghosh, S., Vilalta, R.: Machine learning for automatic mapping of planetary surfaces. In: AAAI, pp. 1807–1812. AAAI Press (2007)
36. Ting, K., Witten, I.: Stacking bagged and dagged models. In: Proc. 14th International Conference on Machine Learning, pp. 367–375. Morgan Kaufmann (1997)
37. Wang, Y., Wang, Y., Jain, A., Tan, T.: Face Verification Based on Bagging RBF Networks. In: Zhang, D., Jain, A.K. (eds.) ICB 2005. LNCS, vol. 3832, pp. 69–77. Springer, Heidelberg (2005)
38. Wolpert, D.H.: Stacked generalization. Neural Networks 5(2), 241–259 (1992)
39. Wu, F., Weld, D.S.: Autonomously semantifying wikipedia. In: Silva, M.J., Laender, A.H.F., Baeza-Yates, R.A., McGuinness, D.L., Olstad, B., Olsen, Ø.H., Falcão, A.O. (eds.) CIKM, pp. 41–50. ACM (2007)
40. Wu, X., Kumar, V., Ross Quinlan, J., Ghosh, J., Yang, Q., Motoda, H., McLachlan, G., Ng, A., Liu, B., Yu, P., et al.: Top 10 algorithms in data mining. Knowledge and Information Systems 14(1), 1–37 (2008)

Mining Tourist Preferences with Twice-Learning

Chen Zhang and Jie Zhang

Department of Land Resources and Tourism Sciences,
Nanjing University, Nanjing 210093, China
zhangc.nju@gmail.com, jiezhangnju@sina.com

Abstract. Data mining techniques have been recognized as powerful tools for predictive modeling tourist decision-making process. However, two practical yet important problems have not been resolved by the data miners in empirical tourism research. Firstly, comprehensibility-the role of the data mining should not only generate accurate predictions, but also provide insights why certain prediction is made. But most widely used data mining methods that can generalize well are black-box in nature and can provide little information on the tourist decision-making facts. Secondly, the lack of training samples-it is usually rather difficult to collect enough training samples through surveying the tourist on site, especially for surveying the tourist's decision-making facts. Many data mining methods may not achieve satisfactory performance if learned on small data set. In this paper, we show that these two problems can be addressed simultaneously using a twice-learning framework on the travel preference data. The results indicate that by addressing these two problems properly, we can predict tourist preferences accurately as well as extracting meaningful insights which would be useful for tourism marketing.

1 Introduction

There is a rapid growth in the amount of Chinese outbound travelers in recent years. The World Tourism Organization [16] predicted that China would be the fourth largest source market of outbound travel by 2020, with a predicted 100 million travelers per year. Due to its great potential, the prediction of Chinese outbound travel market has drawn worldwide attention in the academics and practitioners.

In the past decades, data mining has been applied to predictive modeling in empirical tourism research. Law and Au [7] applied neural network model to predict Japanese tourist arrivals in Hong Kong. Pai, *et al.* [8] used SVM to predict tourist arrivals in Barbados. Au and Law [1] used rough set to model the relationship of tourism dining. Wong, *et al.* [15] applied Bayesian vector autoregressive model to predict Hong Kong tourism demand. Wong and Yeh [14] adopted structural equation model to analyze tourist hesitation in destination decision making.

However, previous studies neglected two practical yet important problems in empirical tourism research. One problem is that, the predictive models that are essentially useful are expected to not only provide accurate predictions for any unseen example but also explicitly explain why a particular prediction is made, but most widely used data mining methods fail to construct predictive models with strong generalization ability and high comprehensibility at the same time [6] [9] [11]. Neural Network and SVM

L. Cao et al. (Eds.): PAKDD 2011 Workshops, LNAI 7104, pp. 483–493, 2012.

that build predictive models with strong generalization ability are black box in nature, and hence one can hardly gain any insights from the model representation. In contrast, some other methods, such as rough set which can clearly explain the relationship between input factors and target concept, are usually more descriptive than predictive, which means the predictions are usually not as good as those made by the black box methods. Since understanding how and why the predictions are made is crucial for both academics and practitioners , it would be of great importance to apply a method for accurate yet comprehensible modeling the relationship from the input attributes to the target concept.

Another important problem which limits the accuracy of prediction in the empirical tourism research is the difficulty of acquiring sufficient samples, especially when surveying the tourists for their personal information on-site. Usually, the tourists have little time for answering every question of the questionnaire. If the questions are related to the privacy of the tourists, most of them are reluctant to answer them. Thus, only a small number of valid questionnaires can be obtained. However, most data mining methods require a large training set, otherwise, they might not generalize well on unseen data if they learned on a small data set.

To address these tricky problems, in this paper, a general approach named twice-learning [17] to generate comprehensible model is applied first time on tourist preferences data. This paper applies one instantiation of twice-learning called C4.5-Rule PANE [17] to the prediction of Chinese tourist preferences for outbound vs. domestic travel as a test case. The results demonstrate the feasibility of generating accurate and comprehensible prediction simultaneously over small data set, and disclose the relationship between Chinese tourists' personal characteristics (including demographic and psychographic characteristics) and their preferences for outbound vs. domestic travel.

The remainder of this paper is organized as follows: Section 2 describes the C4.5 Rule-PANE in detail; Section 3 presents the mining practice on generate comprehensible model for Chinese tourist preferences; Section 4 concludes this paper.

2   Twice-Learning Framework

2.1   The Algorithm

In many practical data analysis tasks (such as analyzing data in the empirical tourism research), both generalization and comprehensibility of data mining approaches are of great importance, as it denotes the ability not only to accurately predict unseen data, but also to give explanation for decision process, especially when applying data mining to applications where reliability is crucial. However, most of the traditional data mining approaches fail to produce models with strong generalization ability as well as high comprehensibility at the same time.

To tackle this problem, Zhou and Jiang [17] proposed a twice-learning framework, which elaborates on combining of these two abilities in a new way. This learning framework works in two phases, as shown in Fig. 1. In the first phase, a model with strong generalization ability (but usually less comprehensible) is constructed using the original training data, and then, in the second phase, another model with high comprehensibility is constructed from the data preprocessed by the model constructed in the first phase.

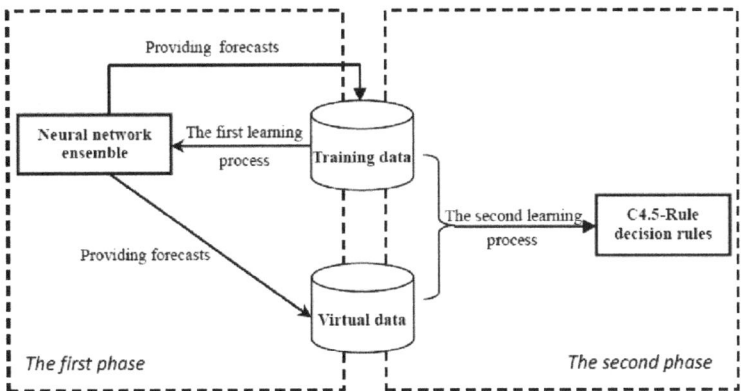

Fig. 1. The twice learning framework

They showed that, by employing such a twice-learning mechanism, one can achieve both strong generalization ability and high comprehensibility at the same time.

One instantiation of this framework, referred to as C4.5-Rule PANE, employs neural network ensemble in the first phase for strong generalization ability and C4.5 Rule in the second phase for good comprehensibility. Neural network ensemble [3], known as one of data mining approaches with the strongest generalization ability, elaborates multiple neural networks to achieve better performance than single neural network, while C4.5 rule [10], known as one of the most widely-used rule learning approaches, generates a set of decision rules that can be essentially understood by the users. Note that C4.5 Rule-PANE has been successfully applied to computer-aided medical diagnosis [17] and gene expression analysis [4]. For the rest of this section, the background information of the algorithms used in different phase is briefly introduced and then the C4.5-Rule PANE algorithm is explained in detail.

Let $S = (x_1, y_1), (x_2, y_2), ..., (x_n, y_n)$ denote the training set. Firstly, neural network ensemble is employed to construct a predictive model h_1 from the training set S. And then, h_1 is used to predict the concept label y_i' for each example $x_i (i = 1, 2, , n)$ in the training set. By replacing the true concept label y_i of x_i with y_i', a new data set S' is obtained. Note that y_i' may not be the same as y_i. Due to the strong generalization ability of h_1, it can usually provide more accurate prediction than other predictive models. By replacing the true label with the predicted label, the labeling noise contained in S would be "smoothed out" by the accurate prediction of produced by h_1, and hence, the S' is less noisy than original training set S, which is beneficial for the learning algorithm with strong comprehensive ability in the second phase.

In the second phase, C4.5 Rule is used to generate a set of rules in the form of "If A AND B THEN C", which can be ultimately understood by human experts. Since C4.5 Rule usually requires a large training set, current training set S' enlarged by generating additional virtual examples. In detail, m random examples $\{x_1', x_2', ..., x_m'\}$ are generated and fed to h_1 for predictions, and then these predictions are used as their true labels. Let y_i'' denote the prediction of x_i', and then a new data set $S'' = \{(x_1', y_1''), (x_2', y_2''), ..., (x_m', y_m'')\}$ is obtained. Finally, the filtered training set S' are combined with the newly generated virtual example set S'' to create a larger training set S^*, and enlarged training set is used for the C4.5 Rule in second phase.

In generally, the whole procedure can be viewed as a model to improve the accuracy of a comprehensible algorithm. According to that, the first phase is aimed to preprocess the original data set using a model with good generalization ability. By using the algorithm, the noise can be eliminated, and the training samples can be enlarged. Thus in the second phase, an excellent model can be established from this enlarged and purified data set. The pseudo code of C4.5 Rule-PANE is shown in Table 1.

Table 1. The pseudo code of C4.5 Rule-PANE algorithm [18]

Algorithm: **C4.5 Rule-PANE**

Input:	training set S
	the number of additional training instance m
Process:	
	$N^* = \text{Bagging}(S)$ /* generate neural network ensemble using Bagging */
	$S' = \Phi, S'' = \Phi$
	for $i = 1$ **to** n **do** /* process the training set by the trained ensemble */
	$\quad y'_i = N^*(x_i : (x_i, y_i) \in S)$
	$\quad S' = S' \cup \{(x_i, y'_i)\}$
	for $j = 1$ **to** m **do** /* generate additional training set using ensemble */
	$\quad x'_j = \text{RandVec}()$ /* generate a random vector */
	$\quad y''_j = N^*(x'_j)$
	$\quad S'' = S'' \cup \{(x'_j, y''_j)\}$
	$S^* = S' \cup S''$ /* construct the training set for rule learning */
	$R = \text{C4.5Rule}(S^*)$
Output:	the rule set R

One question arises here, why can C4.5 Rule-PANE always achieve both excellent comprehensibility and generalization ability simultaneously? Fortunately, Zhou and Jiang [18] analyzed the twice-learning framework and derived an applicability condition which is provided as follows.

Applicability condition [18]:*If the neural network ensemble in the first phase can achieve better performance in prediction than C4.5 Rule if both are trained from the same data set, C4.5 Rule-PANE can be applied to yield a model with strong generalization ability and comprehensibility.*

3   Mining Practice

3.1   Data

The data set obtained from the questionnaire consists of two parts: the input variables including demographic and psychographic variables, and the target concept in terms of

Chinese tourist preferences for outbound versus domestic travel. A five-point Likert-type scale is used, ranging from 1 (very unimportant) to 5 (very important) to evaluate the importance of each item of personal values.

Since it is usually costly to ask someone to fill out the questionnaires, we only managed to collect 400 questionnaires. However, since many respondents were reluctant to disclose some of their personal information, some values of many collected questionnaires are missing. Finally, only 156 complete questionnaires were used. Formally, each questionnaire $i(i = 1, 2, ..., 156)$ is converted into a 18-dim vector of attributes $x_i = (x_i^1, x_i^2, ..., x_i^{18})^T$ and a concept label y_i, where x_i^j indicated the answer to the j-th question in questionnaire i, and $y_i \in \{domestic, outbound\}$. The detail information of the attributes is tabulated in Table 2.

Table 2. The detail information of the attributes (n=156)

	Att_Name	Possible values
1	Gender	Male/ Female
2	Education	High School or below /Junior College/ University/ Postgraduate
3	Age	18-24/25-29/30-39/40-49/50 or above
4	Occupation	Student/ Teacher/ White-collar worker/ Blue-collar worker/
		Self-employed worker/ Manager/Executive/ Retired
5	Life-cycle stage	Single/ Young couple with no children/
		Young couple with infant children/
		Middle-aged couple with dependent children/
		Middle-aged couple with grown up children/Others
6	Personal monthly income(RMB)	2999 or below/3000-5000/5000-10000/10000 or above
7	Bing well-respected	1(very unimportant) 5(very important)
8	Warm relationships with others	1(very unimportant) 5(very important)
9	Fun and enjoyment in life	1(very unimportant) 5(very important)
10	Safety	1(very unimportant) 5(very important)
11	Accomplishment	1(very unimportant) 5(very important)
12	Self-respect	1(very unimportant) 5(very important)
13	A sense of belonging	1(very unimportant) 5(very important)
14	Excitement	1(very unimportant) 5(very important)
15	Self-fulfillment	1(very unimportant) 5(very important)
16	Self-determination	1(very unimportant) 5(very important)
17	Convenience	1(very unimportant) 5(very important)
18	Privacy	1(very unimportant) 5(very important)

3.2   General Mining Process

We applied C4.5-Rule PANE to mine Chinese tourist preferences from the collected data set. The general mining process in this study is shown in Fig. 2.

Firstly, since C4.5-Rule PANE will be effective if the applicability condition is satisfied, before we use it for mining tourist preferences, we have to verify the effectiveness of C4.5 Rule PANE on the collected data. Moreover, in order to show using C4.5-Rule PANE for mining tourist preferences is a good choice, we further compare it with some state-of-art data mining methods that have been widely applied in empirical tourism research for predictive modeling to show whether using C4.5-Rule PANE is beneficial

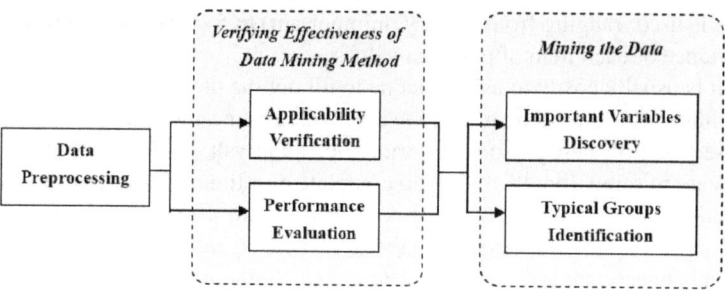

Fig. 2. The twice learning framework

on the collected data using cross-validation. Then, we apply the C4.5-Rule PANE on the collected data and draw some conclusions.

3.3   Results of Applicability Verification and Performance Evaluation

In order to evaluate the applicability according to Applicability Condition mentioned in Section 2, as well as the generalization performance of C4.5 Rule-PANE, four widely-used algorithms, i.e., logistic regression, neural network, neural network ensemble, and C4.5 Rule are compared altogether with C4.5 Rule-PANE before analyzing the current data. Logistic regression is a traditional data mining method which has been widely used for classification in empirical tourism research [1][5], Neural network and Neural network ensemble are data mining methods that has been considered as the powerful classification methods in empirical tourism research. However, they are unable to reveal why a particular prediction is made besides the predictive result itself due to its "black-box" nature. C4.5 Rule is one of the rule learning methods which can gracefully provide comprehensible learning results [10]. Nevertheless, this method may not achieve the generalization ability as high as neural network or neural network ensemble.

  All the methods are implemented using WEKA [13], and the parameters of these methods are set as the default value in WEKA. Ten-fold cross validation is conducted to evaluate the performance of these methods.

  Fig. 3 shows the average error rates for all the compared methods in this study. Pairwise two-tailed t-test at 95% significance level is conducted over the results, and the corresponding p-values between each pair of algorithms are listed in Table 3, where the entries showing significance are boldfaced. Fig. 3 and Table 3 indicate that:

  (1) Neural network ensemble performs much better than C4.5 Rule. This proves that the prerequisite of availability is satisfied.

  (2) C4.5 Rule-PANE is statistically comparable to neural network ensemble but significantly better than Logistic Regression, C4.5 Rule, and Neural Network, which suggests that the non-linear mapping learned by C4.5 Rule-PANE is much closer to the ground-truth mapping than the other compared methods.

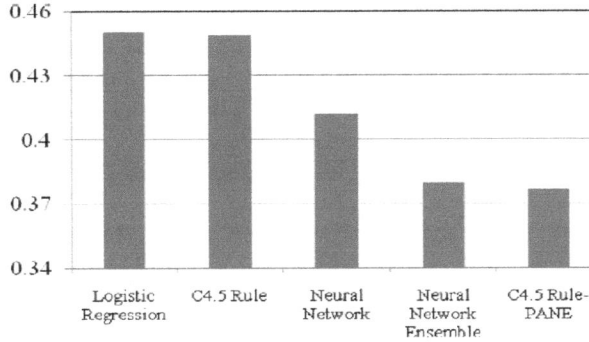

Fig. 3. Error rates of all the compared methods on analyzing the current data set

Table 3. p-values of the pairwise t-test between any two compared methods (significance level 95%)

Methods	Logistic Regression	C4.5 Rule	Neural Network	Neural Network Ensemble	C4.5 Rule-PANE
Logistic Regression	–	–	–	–	–
C4.5 Rule	7.38E-02	–	–	–	–
Neural Network	**1.87E-02**	7.38E-02	–	–	–
Neural Network Ensemble	**1.69E-05**	**2.29E-03**	**1.14E-03**	–	–
C4.5 Rule-PANE	**1.07E-05**	**1.93E-03**	**4.98E-03**	7.98E-01	–

(3) Besides, it is obvious that traditional statistical method (i.e., Logistic Regression) and traditional machine learning approach (e.g., C4.5 Rule) are not as good as the advanced machine learning approaches (e.g., C4.5 Rule-PANE and Neural Network Ensemble).

Above all, C4.5 Rule-PANE is more suitable for modeling tourist preferences, compared with the state-of-art modeling methods that have been widely used in the empirical tourism research. The following discussion on the preferences for outbound and domestic travel is based on the model learned from C4.5 Rule-PANE.

3.4 Discovering Important Variables

Since the applicability and performance of C4.5 Rule-PANE in this study are justified, the whole original data set with demographic and psychographic variables is used as the training data set for C4.5-Rule PANE.

In order to identifying the key factors influencing Chinese tourist preferences, the input variables are ranked according to the gain ratio [10] values produced by C4.5 Rule-PANE. Gain ratio is widely used information-theoretic criterion which measures the ability for discriminating the examples belonging to different concept classes (e.g., preferences for outbound vs. domestic travel) based on the value of this variable. Thus, it can be regarded as an importance measure of each individual variable in predicting the target concept. The higher the gain ratio value is, the more the variable may influence tourist preferences. Besides the C4.5 Rule-PANE, the variables are also ranked by gain ratio produced by C4.5 Rule, the other comprehensible approach, to provide a baseline for comparison. The gain ratio of each individual variable produced by these two methods is shown in Fig. 4 and 5, respectively.

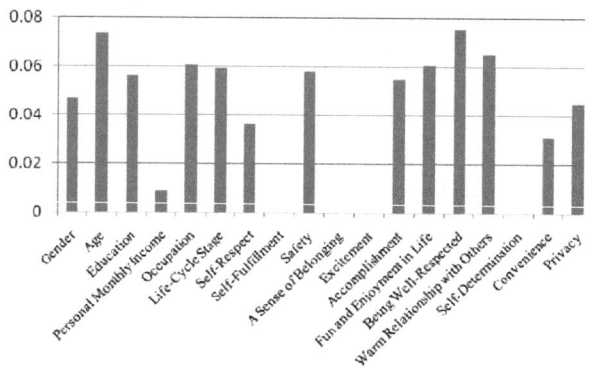

Fig. 4. Gain ratios of the variables obtained from C4.5 Rule-PANE

Fig.4 clearly indicates nine important factors influencing Chinese tourist preferences for outbound and domestic travel, which are (listed in descent order)*being well-respected, age, warm relationships with others, fun and enjoyment in life, occupation, life-cycle stage, safety, education and accomplishment.* Among them, five variables are related to personal values, which indicates the great importance of personal values in influencing tourist preferences, and the finding that the other four variables are demographic variables is consistent with previous studies showing the importance of these demographic variables in influencing tourists' attitudes and choice behavior [3,12,2].

The full ranking list of demographic and psychographic variables based on the gain ratio values in descending order are presented as follows:

(1) Demographic variables:*age, occupation, life-cycle stage, education, gender and income.*

(2) Psychographic variables (personal values): *being well-respected, warm relationships with others, fun and enjoyment in life, safety, accomplishment, privacy, self-respect* and it *convenience.* Notably, the Gain Ratio values of *a sense of belonging, excitement, self-determination*and *self-fulfillment* are zero, which means that these variables appear to be unhelpful in determining target concept (i.e., preferences for outbound vs. domestic travel) individually.

Surprisingly though, by comparing with Fig.5 that represents the ranking of variables provided by C4.5 Rule, different results are found. Without being preprocessed by

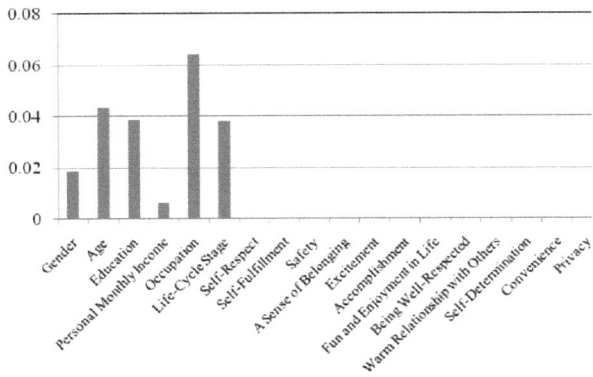

Fig. 5. Gain ratios of the variables obtained from C4.5 Rule

neural network ensemble, the ranking list provided by C4.5 Rule shows that all of the personal values are of little importance in influencing tourist preferences in this study. Since the generalization ability of C4.5 Rule-PANE is much better than C4.5 Rule, the importance ordering of variables disclosed by C4.5 Rule-PANE is much closer to the ground-truth. Therefore, it can be concluded that personal values are also of great importance in influencing Chinese tourist preferences for outbound or domestic travel.

3.5 Obtaining Decision Rules

As indicated, the findings of important variables in previous subsection are based on ability of each variable in predicting tourist preferences (i.e., outbound travel vs. domestic travel) individually. However, since all the variables may interact with each other to influence the final results, only considering the individual generalization ability would not lead to the accurate prediction for a particular tourist. Thus, in this subsection, how these variables interact with each other to produce the final predictions will be further investigated, in order to find what tourist characteristics account for his/her particular preference.

Due to the high compressibility of C4.5 Rule-PANE, the multivariate relationship from the input variables to the target concept is explicitly represented by the learned decision rules. Since C4.5 Rule-PANE can make accurate predictions for tourist preferences, these decision rules reveal the underlying truth of Chinese tourist preferences for outbound and domestic travel. By further exploiting the decision rules produced by C4.5 Rule-PANE, several interesting rules that depict the typical groups of Chinese tourists, who prefer to travel overseas or within China, are obtained. These typical groups are tabulated in Table 4.

Table 4. The induced decision rules

Rule ID	Precondition	Decision Result
$R_1^{(o)}$	**IF** *Life-cycle stage* = "Young couple with no children" **AND** (*Being well-respected* > "Important" **OR** *Warm relationships with others* ≤ "Average")	Outbound Travel
$R_2^{(o)}$	**IF** *Age* = "24 or below" **AND** *Life-cycle stage* = "Single" **AND** *Fun and enjoyment in life* ≥ "Average"	Outbound Travel
$R_3^{(o)}$	**IF** *Gender* = "Female" **AND** *Life-cycle stage* = "Single" / "Middle-aged couple with grown up children"	Outbound Travel
$R_4^{(o)}$	**IF** *Occupation* = "White-collar worker" **AND** *Life-cycle stage* = "Young couple with no children"	Outbound Travel
$R_1^{(d)}$	**IF** *Life-cycle stage* = "Young couple with infant children" **AND** *Fun and enjoyment in life* ≤ "Unimportant"	Domestic Travel
$R_2^{(d)}$	**IF** *Gender* = "Male" **AND** *Life-cycle stage* = "Middle-aged couple with dependent children" **AND** *Safety* ≥ "important"	Domestic Travel
$R_3^{(d)}$	**IF** *Education* = "high school or below" **AND** *Occupation* = "Retired"	Domestic Travel
$R_4^{(d)}$	**IF** *Occupation* = "Blue-collar worker" / "Self-employed worker"	Domestic Travel

4   Conclusions

Data mining techniques have been accepted as powerful tools for data analysis in empirical tourism research recently. However, almost all the previous works failed to build models with strong generalization ability and high comprehensibility at the same time. This paper claims that an important role for data mining techniques to play in tourist decision-making process analysis is to help tourism practitioners to make effective strategies. Moreover, most of the widely used data mining approaches require large data set for good performance, which is usually difficult to achieve in the context of empirical tourism research. To address these entangle problems simultaneous, this paper introduces a general approach called twice-learning to generate predictive model that is accurate in prediction and essentially comprehensible to the data miners from a small number of training data. Mining practice on tourist preferences prediction indicates that the twice-learning framework is effective, and the obtained comprehensible model discloses the relationship between tourists' personal characteristics (including demographic and psychographic characteristics) and their preferences for outbound vs. domestic travel, which would be useful for tourism marketing.

The property of twice-learning framework makes it rather appealing to both academics and practitioners for intelligent data analysis in the tourism context. Applying twice-learning methods such as C4.5-Rule PANE for predictive modeling in other empirical tourism research problems is the interesting future work.

References

1. Au, N., Law, R.: Categorical classification of tourism dining. Annals of Tourism Research 29, 819–833 (2002)
2. Gibson, H., Yiannakis, A.: Tourist roles: Needs and the lifecourse. Annals of Tourism Research 29, 358–383 (2002)
3. Hansen, L.K., Salamon, P.: Neural network ensembles. IEEE Trans. Pattern Analysis and Machine Intelligence 12, 993–1001 (1990)
4. Jiang, Y., Li, M., Zhou, Z.-H.: Generation of Comprehensible Hypotheses from Gene Expression Data. In: Li, J., Yang, Q., Tan, A.-H. (eds.) BioDM 2006. LNCS (LNBI), vol. 3916, pp. 116–123. Springer, Heidelberg (2006)
5. Kim, H., Gu, Z.: Financial features of dividend-paying firms in the hospitality industry: A logistic regression analysis. International Journal of Hospitality Management 28, 359–366 (2009)
6. Kon, S.C., Turner, W.L.: Neural network forecasting of tourism demand. Tourism Economics 11, 301–328 (2005)
7. Law, R., Au, N.: A neural network model to forecast Japanese demand for travel to Hong Kong. Tourism Management 20, 89–97 (1999)
8. Pai, P.F., Hong, W.C., Chang, P.T., Chen, C.T.: The application of support vector machines to forecast tourist arrivals in Barbados: An empirical study. International Journal of Management 23, 375–385 (2006)
9. Palmer, A., Jose Montano, J.J., Sese, A.: Designing an artificial neural network for forecasting tourism time-series. Tourism Management 27, 781–790 (2006)
10. Quinlan, J.R.: C4.5: Programs for machine learning. Morgan Kaufmann, San Mateo (1993)
11. Song, H., Li, G.: Tourism demand modeling and forecasting A review of recent research. Tourism Management 29, 203–220 (2008)
12. Cao, L.: In-depth Behavior Understanding and Use: the Behavior Informatics Approach. Information Science 180(17), 3067–3085 (2010)
13. Witten, I., Frank, E.: Data Mining: Practical Machine Learning Tools and Techniques with Java Implementations. Morgan Kaufmann, San Francisco (2000)
14. Wong, J.-Y., Yeh, C.: Tourism hesitation in destination decision making. Annals of Tourism Research 36, 6–23 (2009)
15. Wong, K.K.F., Song, H., Chon, K.S.: Bayesian models for tourism demand forecasting. Tourism Management 27, 773–780 (2006)
16. World Tourism Organization. Chinese outbound tourism, Madrid (2003)
17. Zhou, Z.-H., Jiang, Y.: Medical diagnosis with C4.5 rule preceded by artificial neural network ensemble. IEEE Transactions on Information Technology in Biomedicine 7, 37–42 (2003)
18. Zhou, Z.-H., Jiang, Y.: NeC4.5: neural ensemble based C4.5. IEEE Transactions on Knowledge and Data Engineering 16, 770–773 (2004)

Towards Cost-Sensitive Learning
for Real-World Applications*

Xu-Ying Liu[1,2] and Zhi-Hua Zhou[2]

[1] School of Computer Science and Engineering, Southeast University, China
[2] National Key Laboratory for Novel Software Technology, Nanjing University, China
`liuxy@seu.edu.cn, zhouzh@nju.edu.cn`

Abstract. Many research work in cost-sensitive learning focused on binary class problems and assumed that the costs are precise. But real-world applications often have multiple classes and the costs cannot be obtained precisely. It is important to address these issues for cost-sensitive learning to be more useful for real-world applications. This paper gives a short introduction to cost-sensitive learning and then summaries some of our previous work related to the above two issues: (1) The analysis of why traditional Rescaling method fails to solve multi-class problems and our method Rescale$_{new}$. (2) The problem of learning with cost intervals and our CISVM method. (3) The problem of learning with cost distributions and our CODIS method.

1 Introduction to Cost-Sensitive Learning

1.1 Unequal Costs

Machine learning and data mining methods often aim at minimizing error rate. This implies that the costs of different misclassification errors are all equal. But in many real-world applications misclassification costs are often different. For a guard system, it is very dangerous to let in a stranger by mistake while a false alarm can be endured. In medical diagnosis, the cost of misdiagnose a patient having a life-threatening disease being healthy is much larger than the cost of misdiagnose a healthy person as a patient.

Here are two real applications of unequal costs. The first one is the network intrusion detection problem of KDD Cup 1999 [11]. The goal is to detect four types of network intrusions from normal connections: DOS (denial-of-service), R2L (unauthorized access from a remote machine), U2R (unauthorized access to local superuser (root) privileges) and probing (surveillance and other probing). The costs of different types of classification errors are different. For example, the consequence of giving access to a R2L connection is much more serious than to a probe one.

---

* The content of this paper is mainly from the Ph.D dissertation of the first author. This research was supported by Startup Foundation of Southeast University (4009001126) and Open Foundation of National Key Laboratory for Novel Software Technology of China (KFKT2011B01).

L. Cao et al. (Eds.): PAKDD 2011 Workshops, LNAI 7104, pp. 494–505, 2012.

In this application, the costs are only dependent on classes. That is to say, the costs of misclassifying the examples of one class to another class are all the same. This kind of cost is called class-dependent cost, and it can be represented by a matrix which is called cost matrix.

The second one is the donation problem of KDD Cup 1998 [11]. The task is to send promotions by mail to potential donors to achieve maximum benefit. The donation amounts are different for different person. Some people will donate more than $100, some people will donate about $5, while some others will not donate at all resulting a loss of $0.68 of sending a mail. Therefore, the cost of missing a major donor is quite larger than missing a common one, and sending to those who will not donate will lose money.

In this application, there are two classes, i.e., donor and non-donor. The cost of sending a mail to a non-donor is $mail\_cost$, and the benefit of sending a mail to a donor is $donate\_amount - mail\_cost$. So the costs dependent on examples. This type of cost is called example-dependent cost. It is different from class-dependent cost and cannot be represented by a cost matrix.

1.2 Formulation

Cost-sensitive learning tries to minimize total cost instead of error rate to handle unequal costs.

Suppose X is the d-dimensional input space and Y is the output space with $y \in \{1, 2, ..., c\}$. When costs are class-dependent, a training set $S = \{(x_i, y_i)\}_{i=1}^{n}$ are i.i.d. drawn from distribution D over $X \times Y$. Assume the cost of misclassifying an i-th class example to j-th class is $cost_{ij}$. The learning goal is to learn a hypothesis $h : X \to Y$ to minimize the expected cost:

$$\arg \min_{h} E_{(x,y) \sim D}[cost_{yh(x)}]. \tag{1}$$

When costs are example-dependent, let $cost(x, y, y_1)$ denote the cost of predicting example x belong to class y to class y_1, and $\bar{C}_{x,y}$ denote the cost vector of $[cost(x, y, 1), \ldots, cost(x, y, c)]$. Then, a training set $T = \{(x_i, y_i, \bar{C}_{x_i, y_i})\}_{i=1}^{n}$ are i.i.d. drawn from distribution $\mathcal{D}$ over $X \times Y \times R^{+c}$. The learning goal is to learn a hypothesis $h : X \to Y$ to minimize the expected cost:

$$\arg \min_{h} E_{(x,y,\bar{C}_{x,y}) \sim \mathcal{D}}[cost(x, y, h(x))]. \tag{2}$$

1.3 Evaluation

It is reasonable to use total cost for evaluation when costs are fixed for a specific application. To evaluate a cost-sensitive learning method, it should be tested with different costs since learning methods are designed for general usage. Cost curve [6] is a popular evaluation method for cost-sensitive learning. The x-axis is the probability-cost function for positive examples, which is defined as:

$$PC(+) = \frac{p(+)cost_{+,-}}{p(+)cost_{+,-} + p(-)cost_{-,+}}, \tag{3}$$

where, $p(+)/p(-)$ is the prior distribution of positive/negative class. The y-axis is the total cost normalized by the cost of every example being misclassified:

$$Norm(E[Cost]) = \frac{fnr * p(+) * cost_{+,-} + fpr * p(-) * cost_{-,+}}{p(+)cost_{+,-} + p(-)cost_{-,+}}, \quad (4)$$

where, fpr is the false positive rate and fnr is the false negative rate. The lower a cost curve, the better the performance.

ROC curve [9] can also be used to evaluate cost-sensitive learning methods. It plots (fpr, tpr) (tpr is true positive rate) pairs. ROC curve is a dual representation of cost curve. But it cannot directly show how a method performs with a specific cost. While Cost curve is designed particular for cost-sensitive learning, and it has many good properties that ROC curve does not have.

1.4   Learning Methods

Cost-sensitive learning has attracted much attention from machine learning and data mining communities and has become an important research field [3,31,37,32,2]. The inclusion of costs into learning has been regarded not only as one of the most relevant topics of future machine learning research [22], but also one of the most challenging problems in data mining research [34].

Many cost-sensitive learning methods are developed, which can be roughly categorized into two types: (1) general methods which can adapt any standard classification methods minimizing error rate to handle unequal costs, including threshold-moving, sampling, and instance-weighting; (2) embedded methods which embed cost sensitivity specifically for a particular learning method.

Threshold-moving is a very popular cost-sensitive learning method and it is guaranteed by Bayes risk theory [25]. It lowers the decision threshold of posterior probability $p(y|x)$ for the class with higher cost so that expensive examples are easier to be predicted right. Suppose there is a positive class and a negative class, an example x should be predicted as positive when $p(+|x) \geq 0.5$ according to Bayes theory. When positive class has higher cost, the decision for x to be positive is made when $p(+|x) \geq p_0$, where p_0 is smaller than 0.5. This is because Bayes risk theory predicts an example to the class with the minimum expected loss. MetaCost [4], one of the pioneering work, is a threshold-moving method. It uses bagging on decision trees to predict posterior probability $p(y|x)$, then relabels training examples to the class with the minimum expected risk. After that, the relabeled data is used to train a classifier minimizing error rate. Many work devoted to improve the quality of probability estimation since threshold-moving relies greatly on it, such as [35,20].

Sampling methods gain cost-sensitivity by altering the class distributions $p(y)$. This type of methods increases or decreases examples for the class with higher cost or lower cost, respectively. Then a classifier trained to minimize error rate of the new data is sensitive to unequal costs of the original problem. Sampling methods are guaranteed by Elkan theorem [7]. Over-sampling can increase examples and under-sampling can decrease examples. Random over-sampling has the risk of over-fitting since it replicates examples and the new data is not i.i.d.

sampled. Costing is proposed [36] to overcome this problem by using rejection sampling which samples an example from the data and then keep it with a probability proportional to its misclassification cost. The size of the sampled data by rejection sampling is usually small. Costing uses bagging to ensemble classifiers trained by multiple sampled data to ease this problem. Roulette sampling [23] is proposed to solve this problem further. It can generate sampled data sets with different sizes.

Instance-weighting methods assign weights to examples proportional to their misclassification costs. The examples with higher costs have larger weights so that they are easier to be predicted correctly. C4.5CS [28] is very popular in cost-sensitive learning and it is one of instance-weighting methods. It exploits C4.5's ability of handling missing values [21] to utilize weighted examples. That is, the weights are used to calculate the probability of node t belong to the j-th class $p(j|t)$. There are also other instance-weighting methods, such as cost-sensitive Naive Bayes [36] and cost-sensitive support vector machines [36,1].

Though threshold-moving, sampling and instance-weighting make classifiers sensitive to unequal costs in different ways, they are closely connected to each other via Bayes risk theory. These general cost-sensitive methods are called Rescaling methods [41], since they all rescale the influence of different examples in the learning process in proportion to their misclassification costs. This will be introduced with more details in Section 3.1.

There are many embedded methods which design cost sensitivity in a particular way. Many efforts were devoted to make AdaBoost [10] cost-sensitive, such as CSB0, CSB1, CSB2 [27], AdaC1, AdaC2, AdaC3 [24], AdaCost [8], Asymmetric-AdaBoost [30] and Asymmetric Boosting [18]. And there are many cost-sensitive decision trees [5,29], cost-sensitive neural networks [14,40], and cost-sensitive Naive Bayes [12,13].

2   Towards Real-World Applications

Though cost-sensitive learning has gained some achievements, there are often strong assumptions for these methods to be applied successfully in real-world applications.

Before the year of 2004, most of the cost-sensitive learning methods were designed for binary-class problems. While many real-world applications have multiple classes, such as the network intrusion detection problem of KDD Cup 1999 (see Section 1.1). Simple extension of the methods designed for binary-class cases failed to reduce total cost for these multi-class problems.

In 2004, Lee et al. [15] proposed a multi-class cost-sensitive SVM and Abe et al. proposed an iterative method GBSE for multi-class cost-sensitive learning. In 2006, our previous work [39] analyzed why Recaling method fails to solve multi-class problems and proposed $Rescale_{new}$[1], which will be introduced in Section 3. These are some early work on multi-class cost-sensitive learning. Recent advances on this problem include a logistic regression method mcKLR proposed by Zhang

---

[1] A longer version is [41].

and Zhou [38], a Boosting method L_p-CSB proposed by Lozano and Abe [17], a threshold-moving method proposed by O'Brien et al. [19], and a reduction method proposed by Xia et al. [33]. Though our method Rescale$_{new}$ is one of the early methods, it still achieves good results compared to many others [33].

On the other hand, the cost information is provided by domain knowledge and is assumed to be precise. The classifiers will then be well tuned to reduce the total cost w.r.t. this particular cost value. However, in many real-world situations, although the user knows that one type of mistake is more severe than another type, it may be difficult to specify a precise cost value. The aspects that can lead to imprecise costs include but not limited to:

- Inherent impreciseness. Some information is naturally stochastic, e.g., one may donate $1 or $5 randomly.

  Unknown information. Sometimes we can't know everything about a system. For example, customers' SSN and salary information can't be obtained because of privacy.
- Variations in modeling. In the process of modeling risk, the approach may sample data, transform input space, or using different parameters. Due to these variations, the model may not provide precise assessment for costs.
- Expert opinions. Experts may have different opinions. And sometimes, experts can just give fair estimates of real risks.
- Dynamic environments. Environments always change. The risk assessed today may change tomorrow, e.g., due to the appearance of a new competitor.

Our previous work [16] quantifies imprecise costs with cost intervals and cost distributions and then proposed methods for both cases, which will be introduced in Section 4. Current cost-sensitive methods can be applied only when precise costs are given. To the best of our knowledge, there is no methods learning with cost intervals or cost distributions. A related work is [26], which considers that costs change over time. But it assumes true cost is known at time of classification. In our assumption, true cost is always unknown. ROC curve can evaluate classifiers under imprecise class distributions or misclassification costs. The classifiers with larger AUC (area under ROC curve) are regarded as better ones. This essentially assumes that nothing whatsoever is known about the relative severity of costs, which is a very rare situation in real-world problems. In our problem settings, cost interval or cost distribution is known.

3   Extending Rescaling to Multi-class Problems

Our previous work [39] analyzed why traditional Rescaling methods are not effective for multi-class problems and then proposed Rescale$_{new}$ method.

3.1   Analysis

Suppose the costs are class-dependent. Recall the notations defined in Section 1.2. We further assume that correct predictions cost zero, i.e., $cost_{ii} = 0$

for $i = 1, \ldots, c$. Let n_i denote the size of the i-th class. To simplify the discussion, all classes have the same size, that is, $n_i = n/c, (i = 1, \ldots, c)$.

Bayes risk theory predicts x to the class with the minimum expected cost. When there are 2 classes, the optimal decision of x to be the 1st class when the following holds:

$$(1 - p) \times cost_{21} \le p \times cost_{12}, \tag{5}$$

where $p = p(class = 1|x)$. The left and right term is the expected cost of predicting the 1st class and 2nd class, respectively (recall that correct predictions cost zero). When the inequality of Eq. 5 becomes equality, predicting either class is optimal. And the p value making the equality holds is called decision threshold, which is denoted by p^*:

$$p^* = \frac{cost_{21}}{cost_{21} + cost_{12}}. \tag{6}$$

When $p \ge p^*$, the optimal decision for x is the 1st class. Threshold-moving directly applies Bayes risk theory. It firstly estimates posterior probability $p(y|x)$, then makes decisions according to Eq. 5. In threshold-moving, the reverse of the decision threshold for a class reflects how important this class is, which satisfies:

$$\frac{1/p^*}{1/(1 - p^*)} = \frac{cost_{12}}{cost_{21}}. \tag{7}$$

Sampling methods are guaranteed by Elkan theorem, which can be easily derived from Bayes risk theory. Due to page limit, please refer to [7] for details.

Theorem 1. *Elkan Theorem [7]: To make a target probability threshold p^* correspond to a given probability threshold p_0, the number of the 2nd class examples in the training set should be multiplied by $\frac{p^*}{1-p^*} \frac{1-p_0}{p_0}$.*

When the classifier has no bias to any class (i.e., minimizing error rate), the threshold p_0 is 0.5. Then according to Elkan Theorem and Bayes risk theory, when the number of the 2nd class examples is multiplied by $cost_{21}/cost_{12}$, p^* is the optimal solution of Bayes risk theory. This means when the 2nd class has higher cost, its size should be increased. Thus, a cost-sensitive problem can be reduced to a standard classification problem by altering class distributions accordingly. Suppose n'_i is the altered class size, then we have:

$$\frac{n'_1/n_1}{n'_2/n_2} = \frac{1}{\frac{cost_{21}}{cost_{12}}} = \frac{cost_{12}}{cost_{21}}. \tag{8}$$

That implies the change of the size of a class reflects its importance.

Instance-weighting uses examples' weights to reflect their importance. Assuming w_i is the weight for the examples in the i-th class, then,

$$\frac{w_1}{w_2} = \frac{cost_{12}}{cost_{21}}. \tag{9}$$

Therefore, threshold-moving, sampling and instance-weighting can be represented in a unified framework, which is Rescaling method. It rescales the 1st class against the 2nd class according to:

$$\tau_{opt}(1,2) = \frac{cost_{12}}{cost_{21}}. \tag{10}$$

Where, $\tau_{opt}(1,2)$ is called the rescaling ratio.

When there are c classes with $c > 2$, the rescaling ratio should satisfy:

$$\tau_{opt}(i,j) = \frac{cost_{ij}}{cost_{ji}}. \tag{11}$$

The traditional Rescaling method uses $cost_i$ to reflect the importance of a class:

$$cost_i = \sum_{j=1}^{c} cost_{ij}, \tag{12}$$

$$\tau_{old}(i,j) = \frac{cost_i}{cost_j}. \tag{13}$$

When $c = 2$, $\tau_{old}(i,j) = \tau_{opt}(i,j)$. When $c > 2$, $\tau_{old}(i,j)$ is usually not equal to $\tau_{opt}(i,j)$. This explains why tradition Rescaling method fails to solve multi-class problems.

3.2   Rescale$_{new}$

Suppose each class is assigned a weight w_i by instance-weighting. In order to appropriately rescale all the classes simultaneously, the weights are expected to satisfy:

$$\frac{w_i}{w_j} = \tau_{opt}(i,j) = \frac{cost_{ij}}{cost_{ji}}. \tag{14}$$

This implies the following $\binom{2}{c}$ constraints should hold in the meanwhile:

$$
\begin{array}{lll}
\frac{w_1}{w_2} = \frac{cost_{12}}{cost_{21}}, & \frac{w_1}{w_3} = \frac{cost_{13}}{cost_{31}}, \cdots, & \frac{w_1}{w_c} = \frac{cost_{1c}}{cost_{c1}} \\
& \frac{w_2}{w_3} = \frac{cost_{23}}{cost_{32}}, \cdots, & \frac{w_2}{w_c} = \frac{cost_{2c}}{cost_{c2}} \\
\cdots & \cdots \cdots & \\
& & \frac{w_{c-1}}{w_c} = \frac{cost_{c-1,c}}{cost_{c,c-1}}
\end{array}
\tag{15}
$$

When these constraints hold simultaneously, Rescaling$_{new}$ directly applies the rescaling method, with the weights being the values that satisfy these constraints. When they can not hold simultaneously, Rescaling$_{new}$ splits the multiclass problem into a series of binary-class problems via pairwise coupling, then uses Rescaling to solve them one by one. The pseudo code can be found in [41].

4   Handling Imprecise Costs

Our previous work [16] studies two forms of imprecise costs: cost intervals and cost distributions.

Cost intervals is the cost information represented by intervals. There are several ways to obtain cost intervals, including but not limited to:

1. Natural cost intervals. In some applications, costs naturally have upper and lower bounds, e.g., stock investigating.
2. Expert opinions. It is much easier for domain expert to provide a cost interval than "precise" costs.
3. Transforming from confidence intervals. When there's no clear upper and lower bound of cost, we can use a confidence interval of cost instead. For example, the 95% confidence interval indicates cost will appear in the interval with a probability of 0.95.

In additional to cost intervals, sometimes we can known more information about costs, such as cost distributions. In some applications, experts can provide the costs distributions according to their experience. For example, the normal and uniform distributions are very popular and can be easily recognized from experience. For complex distributions, we can build models to assess costs. Then the values provided by different models can be regarded as samples from the underlying cost distribution.

4.1   Learning with Cost Intervals

We consider class-dependent costs and binary classification problems with $y \in \{= 1, -1\}$. Assume correct prediction cost 0, and let c_+ and c_- denote the cost of misclassifying a positive and negative example, respectively. Assume positive class has higher cost, i.e., $c_+ \geq c_-$. Since the optimal decisions are unchanged when a cost matrix is multiplied by a positive constant [7], we can simplify the costs by fixing the cost of negative class so that we only need to consider the cost of positive class, i.e., $c_- = 1$, $c_+ = c$ ($c \geq 1$). Let $C_\mu = 0.5(C_{min} + C_{max})$ denote the mean cost. The empirical risk w.r.t. a cost value c of a classifier h is:

$$\tilde{R}(h, c) = \sum_{i=1}^{n} l(c, h(x), y), \tag{16}$$

$$l(c, h(x), y) = cI(h(x) \neq y \wedge y = +) + I(h(x) \neq y \wedge y = -) \tag{17}$$

where, $l(c, h(x), y)$ is the real loss of x, $I(a) = 1$ if $a = true$ and 0 otherwise.

When the true cost is unknown and a cost interval is available, we assume that the unique true cost C^* is a random value in the interval $[C_{min}, C_{max}]$. The goal is to learn a classifier H^* minimizing the *true risk* $\tilde{R}^* = \tilde{R}(h, C^*)$. Unfortunately, since the true cost is unknown, $\tilde{R}^*$ can't be obtained to guide the learning process. To overcome this difficulty, some risk $\tilde{R}_s$ can be used instead to learn a classifier, which is in fact determined by some cost C_s, i.e., $\tilde{R}_s = \tilde{R}(h, C_s)$. Since in general $\tilde{R}_s$ and C_s are different from the true risk $\tilde{R}^*$ and the true cost C^*, respectively, we call them "surrogate risk" and "surrogate cost". By minimizing surrogate risk $\tilde{R}_s$, the optimal classifier h_s^* is expected to minimize the true risk $\tilde{R}^*$. But this is infeasible since $\tilde{R}^*$ is unknown. However, since the true cost can be any value in the cost interval, it is expected that any possible risk of h_s^* should be small enough. Obviously, not all surrogate risk will be good enough for this purpose, so an appropriate surrogate cost C_s must be carefully chosen. Thus, in order to learn a classifier making any possible risks small enough, we can formulate the problem of learning with cost intervals as Eq. 18, by considering C_s as a variable for learning.

$$\min_{h,C_s} \tilde{R}(h, C_s) \tag{18}$$

$$\text{s.t. } p(\tilde{R}(h, c) < \epsilon) > 1 - \delta, \ \forall c \in [C_{min}, C_{max}]$$
$$C_{min} \leq C_s \leq C_{max}.$$

Since there are infinite constrains in Eq. 18, it is intractable to get optimal solutions. To overcome this difficulty, CISVM tries to solve a relaxation with a small number of informative constraints. The first one is the worst case risk which is the upper bound of the risks w.r.t. any c in $[C_{min}, C_{max}]$. its optimal solution can make all the constraints in Eq. 18 hold. So, the worst case risk is appropriate to be used as surrogate risk $\tilde{R}(h, C_s)$ to guide the learning process. However, the worst case risk could be far away from the true risk. So its optimal solution could not make the true risk small enough sometimes. CISVM overcomes this difficulty by minimizing a second risk, the "mean" risk (the risk w.r.t. the mean cost C_μ) in the meanwhile to avoid overfitting to surrogate risk. This is because when cost distribution is unknown, the mean risk has the smallest maximal distortion of the true risk, so it is the best choice to reflect how good a classifier performs on the entire interval.

Assuming that the prediction function is $f = w^T x + b$, CISVM utilizes a surrogate loss in the following form:

$$L(C_p, f(x), y) = I_{y=+}[C_p - yf(x)]_+ + I_{y=-}[1 - yf(x)]_+, \tag{19}$$

where $[a]_+ = max(a, 0)$ and $I_a = I(a)$. It means that, the loss for a negative and positive example is $L^- = [1 - yf(x)]_+$ and $L^+ = [C_p - yf(x)]_+$, respectively. $L(C_{max}, f(x), y)$ is the worse case risk and $L(C_\mu, f(x), y)$ is the mean risk. This form of loss has theoretical guarantee to have smaller risk distortion than SVM's hinge loss.

CISVM involves two parts: (1) minimizing the regularized worst case risk by learning a variation of SVM:

$$\min_{w,b,\xi \geq 0} \|w\|^2/2 + \lambda \sum_{i=1}^{n} \xi_i \tag{20}$$

$$\text{s.t. } y_i(w^T\phi(x_i) + b) \geq C_{max} - \xi_i, \ \forall i : y_i = +1$$
$$y_i(w^T\phi(x_i) + b) \geq 1 - \xi_i, \ \forall i : y_i = -1$$

where $\phi(x)$ is a feature map induced by a kernel function. (2) minimizing the mean risk by parameter selection on a validation set. The pseudo code can be found in [16].

An intuitive way to handle cost intervals is that, taking some value in a cost interval as the true cost, such as the minimal value, mean value, or maximal value, and then applying standard cost-sensitive learning methods. We showed theoretically that, they are not the best solutions. Experiments showed that CISVM is significantly superior to all of them.

4.2 Learning with Cost Distributions

Assume that cost c is independently drawn from distribution v with domain C, which is independent of X. Then the goal is to find a classifier h minimizing the expected risk over v:

$$R_{CD}(h, v) = E_{c \sim v}[R(h, c)] \tag{21}$$
$$= E_{c \sim v} E_{D(X,Y)}[l(c, h(x), y)].$$

Note that, this is different from learning with example-dependent costs (see Section 1.2) because costs are independent of examples in our settings.

An intuitive way to handle cost distributions is taking the expected cost $E[c]$ as true cost then exploiting standard cost-sensitive learning methods to minimize $R(h, E[c]) = E_{D(X,Y)}[l(E[c], h(x), y)]$. However, minimizing the risk is not equivalent to minimizing $R_{CD}(h, v)$ generally. So the intuitive way is not the optimal solution.

CODIS handles cost distributions by reducing the problem to a special case of example-dependent cost-sensitive learning problem, which has a theoretical guarantee (see Theorem 3 in [16]). Firstly, a cost sample c_i is drawn from v (or provided by a risk model) for each example (x_i, y_i) in training set S to form a new example set $\hat{S} = \{(x_i, y_i, c_i)\}_{i=1}^n$. Secondly, a standard example-dependent cost-sensitive method is called to learn a classifier minimizing the risk of $\hat{S}$. Furthermore, to reduce the variance caused by sampling from v, cost are sampled multiple times from v for a single example (x_i, y_i) since v and D are independent. Thus, the first two steps are repeated several times and all the classifiers form an ensemble. The pseudo code can be found in [16].

5 Conclusion

Many research work in cost-sensitive learning focused on binary class problems and assumed that the costs are precise. But these assumptions cannot hold in many real-world applications. This paper summaries some of our previous work towards relaxing these assumptions: (1) The analysis of the failure of traditional Rescaling method in multi-class problems and our method Rescale$_{new}$. (2) propose two methods to learn from cost intervals and cost distributions, respectively. Due to page limit, we only introduce the analysis and the proposed methods in this paper. Please refer to [39,41,16] for the detailed experimental results.

References

1. Brefeld, U., Geibel, P., Wysotzki, F.: Support Vector Machines with Example Dependent Costs. In: Lavrač, N., Gamberger, D., Todorovski, L., Blockeel, H. (eds.) ECML 2003. LNCS (LNAI), vol. 2837, pp. 23–34. Springer, Heidelberg (2003)
2. Chawla, N., Japkowicz, N., Zhou, Z.-H. (eds.): Proceedings on PAKDD 2009 Workshop on Data Mining When Classes are Imbalanced and Errors Have Costs (2009)

3. Dietterich, T., Margineantu, D., Provost, F., Turney, P. (eds.): Proceedings of the ICML 2000 Workshop on Cost-Sensitive Learning (2000)
4. Domingos, P.: MetaCost: A general method for making classifiers cost-sensitive. In: Proceedings of the 5th ACM SIGKDD International Conference on Knowledge Discovery and Data mining, San Diego, California, pp. 155–164 (1999)
5. Drummond, C., Holte, R.C.: Exploiting the cost of (in)sensitivity of decision tree splitting criteria. In: Proceedings of the 17th International Conference on Machine Learning, pp. 239–246. Morgan Kaufmann, San Francisco (2000)
6. Drummond, C., Holte, R.C.: Cost curves: An improved method for visualizing classifier performance. Machine Learning 65, 95–130 (2006)
7. Elkan, C.: The foundations of cost-sensitive learning. In: Proceedings of the 17th International Joint Conference on Artificial Intelligence, Seattle, Washington, pp. 973–978 (2001)
8. Fan, W., Stolfo, S.J., Zhang, J., Chan, P.K.: AdaCost: Misclassification cost-sensitive boosting. In: Proceedings of the 16th International Conference on Machine Learning, Bled, Slovenia, pp. 97–105 (1999)
9. Fawcett, T.: ROC graphs: Notes and practical considerations for researchers. Tech. rep., HP Laboratories, Palo Alto, CA (2004)
10. Freund, Y., Schapire, R.E.: A decision-theoretic generalization of on-line learning and an application to boosting. Journal of Computer and System Sciences 55(1), 119–139 (1997)
11. Hettich, S., Bay, S.D.: The UCI KDD archive. University of California, Department of Information and Computer Science, Irvine, CA (1999), http://kdd.ics.uci.edu
12. Kolcz, A.: Local sparsity control for Naive Bayes with extreme misclassification costs. In: Proceedings of the 11th ACM SIGKDD International Conference on Knowledge Discovery and Data Mining, Chicago, Illinois, pp. 128–137 (2005)
13. Kolcz, A., Chowdhury, A.: Improved Naive Bayes for Extremely Skewed Misclassification. In: Jorge, A.M., Torgo, L., Brazdil, P.B., Camacho, R., Gama, J. (eds.) PKDD 2005. LNCS (LNAI), vol. 3721, pp. 561–568. Springer, Heidelberg (2005)
14. Kukar, M., Kononenko, I.: Cost-sensitive learning with neural networks. In: Proceedings of the 13th European Conference on Artificial Intelligence, pp. 445–449 (1998)
15. Lee, Y., Lin, Y., Wahba, G.: Multicategory support vector machines, theory, and application to the classification of microarray data and satellite radiance data. Journal of American Statistical Association 99(465), 67–81 (2004)
16. Liu, X.-Y., Zhou, Z.-H.: Learning with cost intervals. In: Proceedings of the 16th ACM SIGKDD Conference on Knowledge Discovery and Data Mining, Washington, DC, pp. 403–412 (2010)
17. Lozano, A.C., Abe, N.: Multi-class cost-sensitive boosting with p-norm loss functions. In: Proceedings of the 14th ACM SIGKDD International Conference on Knowledge Discovery and Data Mining, Las Vegas, Nevada, pp. 506–514 (2008)
18. Masnadi-Shirazi, H., Vasconcelos, N.: Asymmetric boosting. In: Proceedings of the 24th International Conference, Corvalis, Oregon, pp. 609–61 (2007)
19. O'Brien, D.B., Gupta, M.R., Gray, R.M.: Cost-sensitive multi-class classification from probability estimates. In: Proceedings of the 25th International Conference on Machine learning, pp. 712–719 (2008)
20. Provost, F., Domingos, P.M.: Tree induction for probability-based ranking. Machine Learning 52(3), 199–215 (2003)
21. Quinlan, J.R.: C4. 5: Programs for machine learning. Morgan Kaufmann (2003)

22. Saitta, L., Lavrac, N.: Machine learning - a technological roadmap. Tech. rep. University of Amsterdam, The Netherland (2000)
23. Sheng, V.S., Ling, C.X.: Roulette Sampling for Cost-Sensitive Learning. In: Kok, J.N., Koronacki, J., Lopez de Mantaras, R., Matwin, S., Mladenič, D., Skowron, A. (eds.) ECML 2007. LNCS (LNAI), vol. 4701, pp. 724–731. Springer, Heidelberg (2007)
24. Sun, Y., Wong, A.K.C., Wang, Y.: Parameter Inference of Cost-Sensitive Boosting Algorithms. In: Perner, P., Imiya, A. (eds.) MLDM 2005. LNCS (LNAI), vol. 3587, pp. 21–30. Springer, Heidelberg (2005)
25. Theodoridis, S., Koutroumbas, K.: Pattern Recognition, 3rd edn. Elsevier (2006)
26. Ting, K.M., Zheng, Z.: Boosting Trees for Cost-Sensitive Classifications. In: Nédellec, C., Rouveirol, C. (eds.) ECML 1998. LNCS, vol. 1398, pp. 190–195. Springer, Heidelberg (1998)
27. Ting, K.M.: A comparative study of cost-sensitive boosting algorithms. In: Proceedings of the 17th International Conference on Machine Learning, Standord, CA, pp. 983–990 (2000)
28. Ting, K.M.: An instance-weighting method to induce cost-sensitive trees. IEEE Transactions on Knowledge and Data Engineering 14(3), 659–665 (2002)
29. Turney, P.D.: Cost -sensitive classification: empirical evaluation of a hybrid genetic sensitive classification. Journal of Artificial Intelligence Research 2, 369–409 (1995)
30. Viola, P., Jones, M.: Fast and robust classification using asymmetric AdaBoost and a detector cascade. In: Advances in Neural Information Processing Systems, vol. 14, pp. 1311–1318 (2002)
31. Weiss, G.M., Saar-Tsechansky, M., Zadrozny, B. (eds.): Proceedings of the 1st International Workshop on Utility-Based Data Mining. ACM Press, Chicago (2005)
32. Weiss, G.M., Saar-Tsechansky, M., Zadrozny, B.: Special issue on utility-based data mining. Data Mining and Knowledge Discovery 17(2) (2008)
33. Xia, F., Yang, Y., Zhou, L., Li, F., Cai, M., Zeng, D.D.: A closed-form reduction of multi-class cost-sensitive learning to weighted multi-class learning. Pattern Recognition 42(7), 1572–1581 (2009)
34. Yang, Q., Wu, X.: 10 challenging problems in data mining research. International Journal of Information Technology and Decision Making 5(4), 597–604 (2006)
35. Zadrozny, B., Elkan, C.: Learning and making decisions when costs and probabilities are both unknown. In: Proceedings of the 7th ACM SIGKDD International Conference on Knowledge Discovery and Data Mining, San Francisco, CA, pp. 204–213 (2001)
36. Zadrozny, B., Langford, J., Abe, N.: Cost-sensitive learning by cost-proportionate example weighting. In: Proceedings of the 3rd IEEE International Conference on Data Mining, Melbourne, Florida, pp. 435–442 (2003)
37. Zadrozny, B., Weiss, G.M., Saar-Tsechansky, M. (eds.): Proceedings of the Second International Workshop on Utility-Based Data Mining. ACM Press, Philadelphia (2006)
38. Zhang, Y., Zhou, Z.-H.: Cost-sensitive face recognition. IEEE Transactions on Pattern Analysis and Machine Intelligence 32(10), 1758–1769 (2010)
39. Zhou, Z.-H., Liu, X.-Y.: On multi-class cost-sensitive learning. In: Proceedings of the 21st National Conference on Artificial Intelligence, pp. 567–572 (2006)
40. Zhou, Z.H., Liu, X.Y.: Training cost-sensitive neural networks with methods addressing the class imbalance problem. IEEE Transactions on Knowledge and Data Engineering 18(1), 63–77 (2006)
41. Zhou, Z.-H., Liu, X.-Y.: On multi-class cost-sensitive learning. Computational Intelligence 26(3), 232–257 (2010)

Author Index

Adams, Brett   53
Akehurst, Joshua   15
Alam, Shafiq   316
Albayrak, Sahin   100
Amphawan, Komate   124
Anh, Duong Tuan   148

Chen, Ming-Syan   459
Chen, Tao   353
Christen, Peter   171
Ciou, Cin-Siang   65
Clausen, Jan   100
Cuxac, Pascal   209

Dang, Yanzhong   243
Denny,   171
Detterer, Dion   361
Dobbie, Gillian   316

Estivill-Castro, Vladimir   197

Fan, Jun   234

Garriga, Joan   279
Gu, Wei   420
Guo, Chonghui   136

Hadzic, Fedja   221
He, Xianmang   111
Hirasawa, Kotaro   243
Hsiao, Hui-Fang   77
Hsu, Kuo-Wei   471
Huang, Tiejun   234
Huang, Yin-Fu   65

Ishikawa, Masahiro   160
Ivanescu, Anca Maria   185

Janviriyasopak, Uraiwan   431
Jiang, Sheng-yi   339
Jin, Cheqing   304
Jönsson, Arne   40

Kay, Judy   15
Kim, Hyoungnyoun   3

Kiran, R. Uday   254
Koh, Yun Sing   327
Koprinska, Irena   15
Kunegis, Jérôme   100
Kwan, Paul   361

Lamirel, Jean-Charles   209
Lenca, Philippe   124
Li, Hailin   136
Li, Lixiang   40
Li, Yujia   111
Liu, Wenhuang   293
Liu, Xu-Ying   494
Luo, Zhe   384

Mabu, Shingo   243
Mak, Peng Un   408
Mak, Pui In   408
Mall, Raghvendra   209

Nguyen, Thin   53
Nhon, Vo Le Quy   148

Painuall, Sumran   431
Park, Ji-Hyung   3
Pears, Russel   327
Pechsiri, Chaveevan   431
Phung, Dinh   53
Pizzato, Luiz   15
Poon, Simon   384

Reddy, P. Krishna   254
Rej, Tomasz   15
Riddle, Patricia   316

Safi, Ghada   209
Saravanan, M.   397
Seidl, Thomas   185
Shalini, S.   397
Shanthi, S.   397
Shi, Baile   111
Shie, Bai-En   77
Shimada, Kaoru   243
Spiegel, Stephan   100

Srivastava, Jaideep   471
Surana, Akshat   254
Surarerks, Athasit   124

Tsai, Flora S.   28
Tseng, Chi-Yao   459
Tseng, Vincent S.   77

Vai, Mang I   408
Venkatesh, Svetha   53

Wan, Feng   408
Wan, Li   267
Wan, Miao   40
Wang, Baijie   420
Wang, Boyu   408
Wang, Cong   40
Wang, Lian-xi   339
Wang, Qing   111
Wang, Wei   111
Wang, Xin   420
Wang, Yi   304
Wang, Yi-guo   372
Wang, Yiwen   89
Wichterich, Marc   185

Williams, Graham J.   171
Wu, Jianjun   267

Xiao, Yanghua   111
Xu, Zeren   267

Yacef, Kalina   15
Yang, Guangfei   243
Yang, Libin   136
Yang, Yixian   40
Yao, Min   89
Yu, Dong-lin   372
Yu, Philip S.   77
Yuan, Bo   293

Zhang, Chen   483
Zhang, Jie   483
Zhang, Lei   372
Zhang, Nevin L.   353
Zhang, Qi-ming   372
Zhang, Runshun   384
Zhang, Runsun   353
Zhang, Yu   447
Zhou, Aoying   304
Zhou, Minqi   304
Zhou, Zhi-Hua   494